Complete BUILDING EQUIPMENT MAINTENANCE Desk Book

SECOND EDITION

Edited by
SHELDON J. FUCHS, P.E.

PRENTICE HALL
Englewood Cliffs, New Jersey 07632

Prentice-Hall International (UK) Limited, *London*
Prentice-Hall of Australia Pty. Limited, *Sydney*
Prentice-Hall Canada, Inc., *Toronto*
Prentice-Hall Hispanoamericana, S.A., *Mexico*
Prentice-Hall of India Private Limited, *New Delhi*
Prentice-Hall of Japan, Inc., *Tokyo*
Simon & Schuster Asia Pte. Ltd., *Singapore*
Editora Prentice-Hall do Brasil, Ltda., *Rio de Janeiro*

10 9 8 7 6 5 4 3

Disclaimer

The techniques and methods presented in this book are the result of the authors' experiences in working with certain materials and tools. The information contained in this book is of broad general usefulness. The reader should not use the information in this book without first having a thorough understanding of his or her particular equipment, environment, and organization, and care should be taken to use the proper materials and tools as advised by the authors. The information contained in this book is as up-to-date as it can be and has been carefully checked by the authors. Nevertheless, it is difficult to ensure that all the information given is entirely accurate for all circumstances. The Publisher disclaims liability for loss or damage incurred as a result of the use and application of any of the contents of this book.

Library of Congress Cataloging-in-Publication Data

Complete building equipment maintenance desk book / edited by Sheldon
J. Fuchs. — 2nd ed.
p. cm.
Includes index.
ISBN 0-13-157462-0
1. Buildings—Mechanical equipment—Maintenance and repair.
2. Plant maintenance. I. Fuchs, Sheldon J.
TH6013.C66 1992
696—dc20 92-6729
CIP

ISBN 0-13-157462-0

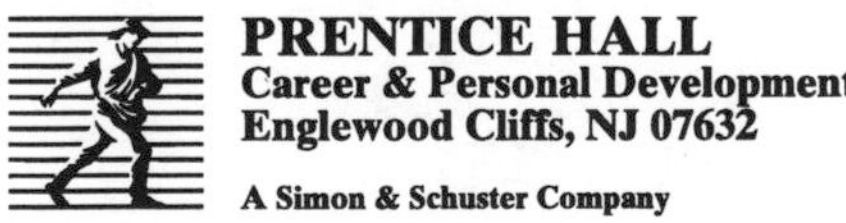

Printed in the United States of America

DEDICATION

This book is dedicated to the memory of my mother, father, and wife, Myrna, whose love, devotion, inspiration and dedication made this undertaking possible.

And also

To the eight people in my life who make waking up in the morning have meaning and dreams still possible—Linda, Laura, Ron and Dave and my second team—Julie, Allison, Michael and Jacqueline.

And also

To Sylvia, a very special person who has a special part in my life and makes it worthwhile.

Sheldon J. Fuchs, P.E., is an associate with CFW Associates in Merrick, N.Y. His background includes Plant and Facilities Manager for the Baldwin School District, Superintendent of Buildings and Grounds, Director of the Facility Maintenance Institute and adjunct professor at Hofstra University, and plant maintenance engineer at Kollsman Instrument Corporation.

Mr. Fuchs holds a bachelor of civil engineering and a master's degree in business administration from the City College of New York. He has continued his graduate study at New York University and Hofstra University and is a licensed professional engineer in New York State and a certified plant engineer.

His articles have appeared in many trade journals and he has been a frequent speaker at seminars and meetings of various professional engineering societies. He is a contributing author to the *Plant Engineers Manual and Guide* and the *Encyclopedia of Professional Management.* As past president of the New York and Metropolitan Manhattan chapters of the American Institute of Plant Engineers (AIPE), he has been active in various professional organizations. Mr. Fuchs is founder of the Hofstra University Annual Plant Maintenance Seminar and Exhibit and is currently coordinating a similar program for AIPE Chapter #4 in New York. Mr. Fuchs has been awarded the Certificate of Merit from Nassau County (State of New York).

The Editor

Contributing Authors

Alfred T. Baker, P.E. is the president of A.T. Baker Electrical Consulting, Inc. He holds a B.E.E. degree from Manhattan College and is a professional engineer in the states of New York, Connecticut and New Jersey. His company is heavily involved in engineering and design associated with facility power and lighting systems and cogeneration projects.

Robert M. Barr holds degrees from the University of North Carolina and Yale and is a member of Phi Beta Kappa. Mr. Barr has been involved with the HVAC industry for 35 years and with compressor remanufacturing and parts supply for 20 years. He is currently vice-president of United Rebuilders, Inc. of Mt. Vernon, New York, and has conducted many seminars on the avoidance of compressor failure for many professional groups.

Jerold I. Berkley has a B.S. in chemistry from Long Island University. He was an analytical chemist for Fisher Scientific, an industrial filtration marketing manager for Multi-Metal, Inc. and a general sales manager for Stewart Hall Chemical Corp. He is currently the president and partner of North Eastern Chemicals, a water treatment and metal finishing chemical company in Hackensack, N.J.

William B. Birkel is a graduate of Farmingdale College. He worked for the Grumman Corporation where he was a tool designer, as a sales engineer for the Carrier Corporation, and for Kelly Trane as a sales engineer before starting the company of Birkel Downes Air Conditioning, Inc. where he is the current president. Birkel Downes of Hicksville, N.Y. provides service for all types of air conditioners and water towers.

Robert Burger has a degree from Cornell and is president of Burger Associates, Inc. of Dallas, a company specializing in the rebuilding and upgrading of cooling towers. Mr. Burger has published over 50 articles that have appeared in national magazines as well as a comprehensive textbook on cooling towers entitled *Cooling Tower Technology*. He has lectured throughout the country and overseas, and is a member of the Cooling Tower Institute, NAPE, NAC, ASHRAE, AEE, ASME, and other engineering societies. He has a contract to produce four two-day seminars a year for the Association of Energy Engineers and was named "Engineer of the Year" in 1989 by the National Association of Power Engineers for his educational efforts.

Raymond Combs has a B.B.A. from Hofstra University. He is a member of the Long Island chapter of ASHRAE, and past chapter president, a member of NAPE, NAOHSM, and CSI. He is active in training programs for private companies, and in the field of combustion and boiler room safety. He is an adjunct professor and member of the Advisory Committee at SUNY Farmingdale. He joined Volkmann Corporation as a principal of that company, later forming the firm of Boiler Room Specialties, Inc., of Patchogue, N.Y., who are suppliers of equipment and technical services in the field of combustion, boilers and related equipment, and is currently president and CEO.

Salvatore T. Cordaro is a graduate of New York State Maritime College and holds a degree in chemical engineering from Brooklyn Polytechnic Institute. He is president of Cord Associates, a management consulting firm. Mr. Cordaro is an experienced manager, having worked 30 years in the process industries. As a consultant, he has served many major corporations designing and implementing successful maintenance policies and programs. He is in great demand nationally as a seminar lecturer presenting regular maintenance seminars for North Carolina State University and Fairleigh Dickinson University.

William Dumper holds a B.E.E. and an M.E.E. from Polytechnic Institute of Brooklyn with a major in rotating machine design. He is a registered professional engineer in New York. His 38 years of practical experience includes the design of large motors and generators for Westinghouse Electric Corp. At Consolidated Edison Co. of New York he specialized in motor applications and engineering assistance to maintenance groups in addition to responsibilities for plant design. He was Chief Electrical Engineer for Gibbs & Hill, Inc., responsible for design of power plants and industrial facilities. He has served on numerous engineering society committees including IEEE, ANSI, and IEC. At present he is President of Wit-Craft Electric Service of Plainview, N.Y. which is a motor and pump repair facility.

Ed Feldman has a bachelor of industrial engineering from Georgia Tech and is a registered professional engineer. He is the author of 10 books, a contributing author to 5 others, and has written over 250 articles. After experience in industry as a plant engineer, Mr. Feldman founded and operated an international facilities consulting and training firm for 30 years. Mr. Feldman now operates as an independent consultant and trainer for Ed Feldman, P.E. and Associates in Atlanta, GA.

Clyde H. Gordon, Sr. has a B.S. in architectural engineering from the University of Texas and an M.B.A. from the University of New Orleans. He is the principal and founder of Gordon & Associates in Concord, CA., a facilities management consulting firm that designs and installs systems for various aspects of facilities such as P.M., budget forecasting, etc. He is a charter member of the Assoc. of Energy Engineers and has been accepted for Who's Who in the West in 1989-90.

Michael Heit has a B.E.E. from the City College of New York. He was a sales manager with the Trane Company and is currently the vice-president of Ashcor New York, Inc.

Michael J. Kerr, former fire chief, is the founder and president of Fire Command Company of Long Beach, New York, which is a fire protection consulting firm which provides a wide range of fire protection services and equipment. Mr. Kerr has a Certificate of Fire Technology from S.U.N.Y. at Farmingdale, has attended Brooklyn Polytechnic University and has a certificate from Hofstra University in occupational safety and health.

William Krause was president and CEO of Tribology, Inc. He purchased Tech-Lub Corp. and merged the two into Tribology/Tech-Lub Inc. of Islip, N.Y. He is now president and CEO of this company which specializes in gear oils, speciality greases, aerosols for chain lubes, and dry solid lubes. Mr. Krause has a B.S. from C.W. Post and is a member of STLE, SAE and NLGI.

Evans J. Lizardos, P.E., is a principal of Lizardos Engineering Associates, P.C., a consulting engineering firm that specializes in heating, ventilating, air conditioning, electrical, plumbing, fire protection, and energy conservation for commercial, institutional, municipal, and industrial projects. Mr. Lizardos holds a B.M.E. from Polytechnic University (formerly Brooklyn Polytechnic Institute), is a fellow of ASHRAE, and is a member of ISA. Mr. Lizardos has been heavily involved in the conceptual planning and implementation of automatic temperature controls.

Robert Mayer is vice-president of engineering of Tribology/Tech-Lube, Inc. and has developed many products, some of which have received U.S. patents. He is a graduate of Princeton University and holds an M.E. from Stevens Institute and Brooklyn Polytechnic Institute. He is a member of STLE formerly ASLE, SAE and NLGI. For two years he was chairman of the Lubrication Equipment and Practices committee of ASLE, as well as a member of the executive committee of the Petroleum and Chemicals Council.

Calvin D. MacCracken founded CALMAC Manufacturing Corporation in 1947 and remains as president and CEO. A holder of 84 U.S. patents, he has degrees from Princeton in science (magna cum laude) and The Massachusetts Institute of Technology in engineering. His company is the leader in off-peak air conditioning by ice storage. He is a Fellow in the American Society of Heating, Refrigerating and Air Conditioning Engineers (ASHRAE) and has been elected to both the New Jersey Inventors and Business Hall of Fame. He is the author of *A Handbook for Inventors* (Macmillan).

Douglas J. Pavone, P.E., is a principal of Lizardos Engineering Associates, P.C., a consulting engineering firm that specializes in heating, ventilating, air conditioning, electrical, plumbing, fire protection, and energy conservation for commercial, institutional, municipal, and industrial projects. Mr. Pavone is an engineering graduate of Syracuse and Cornell Universities. He has more than 20 years of experience in HVAC and building mechanical systems design, including the design of several building automation systems.

William Scales holds a bachelor of electrical power engineering degree from the Polytechnic Institute of New York, and is a licensed professional engineer in many states. He is the president of Scales Air Compressor Corporation, which sells, services, and rebuilds air compressors throughout the United States. He is a member of the ANSI Oil Panel, which has written specifications for compressor lubricants for the International Standards Organization. In addition, he has lectured on air compressors before many world organizations, and as president of Air Compressor Systems Consultants, has done consulting work for many companies. He is a member of NSPE, AIPE and two engineering honor societies.

Alvin Soffler is founder and president of Dynaire Corp. of Mineola, N.Y. which specializes in air conditioning, heating, ventilating, humidification, commercial and industrial building automation and related services offering design, installation, and full maintenance services.

Mr. Soffler has a B.M.E. from Cooper Union, an M.M.E. from Stevens Institute and is a New York state licensed professional engineer. He is a member of ASHRAE, NSPE, AIPE and ACCA.

Ronald J. Texel has a degree from New York Institute of Technology with over 20 years of experience in the power systems engineering field. Mr. Texel has broad experience in utility power engineering and as a facilities plant engineer. He is a member of the IEEE, an associate member of the International Association of Electrical Inspectors (IAEI) and past president of the American Institute of Plant Engineers (AIPE) chapter 4.

Paul D. Tomlingson is a veteran of 22 years of worldwide maintenance management consulting, the author of 3 textbooks and 75 trade journal articles on maintenance management. Tomlingson is a 1953 graduate of West Point, and in addition to a B.S. in engineering, holds an M.A. in government and an M.B.A., both from the University of New Hampshire. He is listed in *Who's Who in the West.*

If You Have Maintenance Responsibility This Book Is for You

The 21 contributors to *The Complete Building Equipment Maintenance Desk Book* are outstanding examples of specialists who are well versed not only in theory but also in the how-to aspect of plant engineering and maintenance. They have pooled their talents to produce a practical information guide that will be used as a constant source of information by those who are involved in all aspects of maintenance.

This manual presents 16 chapters packed with ideas, checklists, guides, maintenance procedures, and concepts that will enable you to improve your operation and get the maximum from every dollar spent.

Provided are proven ideas and techniques that can double, triple, or quadruple profits—resulting from implementing a moderate, cost-effective equipment maintenance program. Every idea, every method has been fired in the furnace of real world application. This cornucopia of practical answers offers the best thinking of a cadre of experts in the field, people who have been faced with the same problems you confront and found workable, manageable solutions. Collectively, the cost-saving, equipment-saving, manpower-saving examples from which they tap a rick bedrock of experience, have boosted the bottom line of actual companies by hundreds of millions of dollars.

The first chapter covers a period that is overlooked by most technical references—the period spanning preparation for construction until the takeover after construction. Specific recommendations are given regarding working with the architect, how you can be of most benefit during construction, and how you can develop maintenance manuals using techniques such as videotaping. The chapter contains a section on blueprint retention and preservation, with specific means of accomplishing this. It accents how you can assume a pro-active rather than a reactive role. Your input at this time can have a beneficial impact on the maintenance and operations throughout the entire facility life. The savings that can be achieved can be reaped for many years thereafter.

In Chapter 2, Paul Tomlingson outlines the essentials of getting a preventive maintenance program underway and making it work. He gives emphasis to the 'detection orientation' of PM—inspection and testing—to ensure that equipment problems are uncovered before failure. In turn, this early discovery minimizes repairs and ensures there is adequate lead-time to plan work. Lubrication, cleaning, adjusting, and minor component replacement (belts, filters, and so on) are placed in perspective as well as the needed participation of equipment users in conducting PM-related checks and reporting problems quickly.

Sal Cordaro discusses the basics of predictive maintenance and contrasts it to preventive maintenance in Chapter 3. The two main objectives of the chapter are to help you understand the difference between the two reliability improving methods and to show how you can use predictive maintenance surveillance systems in a practical program. Some of the more important predictive tools are discussed to show how fundamental predictive maintenance concepts are actually used. Finally, Cordaro employs the vibration analysis method to show in detail how to develop a surveillance program. After reading this chapter you will have a clear picture on how to use this state-of-the-art method.

Chapter 4 covers the maintenance, theory, and application of cooling towers. Robert Burger, an expert in this field, presents various checklists for the different sections of a cooling tower. In addition, recommended maintenance procedures at the beginning and at the end of the cooling seasons are included. Many suggestions for tower modifications and their advantages are presented. In these days of high energy costs and shortages, managers are becoming more and more aware of the energy savings potential of colder water cooling their equipment. For example, enthalpy charts indicate that for every 1°F put over compressors, 2¼% of the input electric energy can be saved.

In Chapter 5, Bill Scales tells you what steps to follow to select, install, operate, and maintain an effective compressed air system at the lowest overall cost. Mr. Scales based his comprehensive checklists for air compressors and accessories on practical in-the-field experience as a mechanic, engineer, and consultant. Chapter 5 is a must for all those concerned with the maintenance of rotary screw, reciprocating and centrifugal air compressors, dryers and filters, or any components found in compressed air systems, whether you are an expert or not.

Chapter 6, entitled Effective Motor and Automatic Control Maintenance by Evans J. Lizardos and Douglas J. Pavone, reveals the fundamental electrical diagram (also known as a ladder diagram) as it is used to provide easy understanding of the logic of equipment control systems. The chapter leads you from the very simple control arrangement through complex programmable controller applications into building automation systems. We explain in a step-by-step manner the construction of elementary diagrams that frequently reveal solutions to the majority of the control problems encountered. To further clarify the maintenance procedures for motor and automatic controls, numerous illustrations are included.

In Chapter 7, six well-known experts, Bill Birkel, Bill Downes, Robert M. Barr, Al Soffler, Cal MacCracken and Michael Heit, cover a most complex topic. The maintenance and proper operation of air conditioning systems is divided into six sections:

- Absorption
- Centrifugal
- Reciprocating
- Heat Pumps
- Off Peak Air Conditioning
- Modular Air Conditioning

Each of these component parts is treated in clear, precise terminology giving step-by-step procedures for operating, trouble-shooting, and maintaining the various types of equipment. These procedures are guaranteed to lessen the amount of breakdowns and to stretch the maintenance dollar. This chapter is vital for any individual involved with any aspect of air conditioning. This guide will be used not only as an occasional reference but as a day-to-day guide to efficient air conditioning maintenance procedures. To ensure easy practical help, many charts, schedules, and timetables for various overhaul procedures are included. A reciprocating Air Conditioning guide shows you how to start up and operate the system; how to set up superheat; precautions to take concerning moisture; how to avoid electrical problems; spotting voltage imbalance; the importance of crankcase heaters; the three ultimate compressor failure modes and what causes them; how to clean a system after burnout; CFCs and the atmosphere; and how to reduce compressor failures by ⅔ by following the monthly methodical checkout. In the heat pump section, Al Soffler points out the versatility of staging the maintaining temperature control that is standard in heat pump systems.

Chapter 8 deals with the various types of boilers by construction, the important considerations of proper maintenance, and improving boiler operating efficiency. An acknowledgement of the importance of boiler safety and testing is presented with a suggested operating log.

Alfred Baker and Ron Texel present an in-depth discussion of the importance of the maintenance of electrical facilities in Chapter 9. Checklists are provided for the acceptance testing and preventive maintenance that should be performed on the facility's transformers, switchgear, cables, protective relays, emergency generators, UPS equipment, batteries, and other electrical systems. They recommend the development of a strong electrical maintenance staff along with the use of outside testing companies for the more complex systems. The chapter truly demonstrates that "electrical maintenance expenditures are monies well spent."

Electric motors are found everywhere in our society and procedures for their care and maintenance are covered in Chapter 10. Bill Dumper introduces this subject with a brief, non-technical description of the many types of motors found in practice. The presentation is directed towards your need to recognize and understand the specific care required by each type. Problem areas such as improper electric supply, incorrect enclosures, hostile environments, ambient temperatures, lubrication, alignment, and so on are thoroughly discussed. Extensive preventative maintenance and trouble-shooting checklists are included. Other topics covered are winding testing and maintenance, bearing and commutator maintenance, repair versus replacement considerations, manufacturers warranties, and recommended test and service equipment.

Chapter 11 is an aid to individuals who are responsible for preventive maintenance, including lubrication of all types of industrial equipment, heating, and ventilation found in industrial plants and commercial and residential buildings. The necessary information is provided to set up a successful lubrication program as part of plant maintenance in a modern facility where the maintenance manager does not have a large engineering staff to develop lubrication requirements.

In addition, information is also provided regarding storage of lubricants in the plant as well as their distribution for use, and a section on central lubrication systems for the application of lubricants in the equipment itself.

The filter is the first line of defense in protecting air side heat exchange surfaces. In Chapter 12, Clyde Gordon describes the most common types of air filters in HVAC. He gives instructions for maintaining and maximizing useful life and economy for each type. Case histories for upgrading obsolete systems illustrate the economics available using modern technology. The very important role of filtration maintenance in indoor air quality is discussed. Finally, Gordon explains what to do when filtration has not been maintained, including remedial cleaning for coils, plenum chambers, ducts, and diffusers.

Chapter 13 covers the different types of belts and bearings, their care, installation, and maintenance. It also describes conditions to avoid and provides troubleshooting guides for both belts and bearings.

Chapter 14 covers the movement of fluids and gases and is divided into two parts. The first part covers the movement of fluids and semi-fluids by pumps and auxiliary piping, and the second part covers the movement of gases, particularly air, by fans, and the necessary auxiliary equipment to deliver the air to its desired destination. This second section is known as ventilating systems. This chapter considers different kinds of pumps and fans, their installation, the auxiliary piping systems connected with both their installation, and troubleshooting guides for maintenance of each in operation.

Before you can purchase and supervise a good water treatment program you must become an informed buyer and administrator. Chapter 15 offers advice about what information you must have and where to get it. The latest water treatment chemicals, equipment, and administrative record keeping techniques are summarized to help decide what kind of a water treatment program you need.

The area of fire protection and safety is one of the most important aspects facing a property manager or engineer today. We live in a litigious society today and our emergency safety equipment must be installed and maintained by properly qualified personnel. In Chapter 16, Michael Kerr discusses fire suppression systems, sprinkler and standpipe systems, fire equipment, employer training, housekeeping and fire alarm systems.

Sheldon J. Fuchs

Acknowledgments

This author is proud to be associated with individuals and organizations that have dedicated themselves to serve their profession.

Over the years, the American Institute of Plant Engineers (AIPE) has attracted individuals who are not only professionally capable, but sensitive to the needs of fellow engineers. Several of the contributors to this manual are active participants in this society and have demonstrated their capabilities in numerous ways. Another organization that promotes high ideals and constantly updates current practices is the American Society of Heating, Refrigerating and Air Conditioning Engineers, Inc. (ASHRAE). As specialists, they have set standards and codes for the industry. Several contributors and reviewers of this book are active participants in this worthwhile society.

The members of AIPE New York Chapter No. 4 are as dedicated and professional as one can find. Their contributions to this endeavor have been many and varied. Their contribution and dedication to the profession is a pleasure to behold.

The contributing authors, who themselves are experts in their fields, express their gratitude to all those who have generously assisted in the preparation of this book and have made a significant contribution to the text. These include, among others:

Gregg Homeyer, Raytheon Company

Glen L. Ponczak, Johnson Controls, Inc.

Harold Schneider, P.E., Office of Management & Budget, City of N.Y.

Gus Sortino, SKF USA Inc.

Jerry Donovan, Gates Rubber Company

Michael Cunningham, Waukesha Pump Company

Gordon D. Duffy, Publishing Director, Engineered Systems

Lori Nardello and Stephanie Zimmerman, Tribology/Tech-Lube Corp.

Bernie McAllister, Farr Company

Jim Fox, Air Exchange Inc. 1185 San Mateo Ave, San Bruno, Ca. 94066

Dr. Roy Takekawa, Director, Environmental Health and Safety, University of Hawaii

Robert S Kirk, Interboro Sprinkler Corp.

Samuel K Harpwick, Ocean Coatings
Thomas Fritz, Armstrong World Industries
James Goerl
Stan Slonski, National Assn. of Fire Equipment
Lee Divito, Kidde-Fenwal
John Manown, Micro Film Specialties
Kohler Generators, Kohler, Wisc.
Alber Engineering, Boca Raton, Fla.
Peak Power Inc., Copiague, N.Y.
Mark Waller, Computer Facilities Consultant, Glendale, Cal.
Emanuel Calligeros, CFW Associates
John Willenbrock, CFW Associates
Vincent Monaco, Avis Rent-a-Car.

Table of Contents

1

Making the Transition from Building Construction to Operation and Maintenance*

Sheldon J. Fuchs, P.E.
Edwin B. Feldman, P.E.

This chapter describes how you can take a pro-active—rather than reactive—role in maintenance operations. Positive activities at the beginning can have a considerable beneficial impact on maintenance and operations throughout the entire life of the facility. The savings that you can achieve at the beginning will be reaped annually thereafter. You will learn not only to become involved in construction or renovation design, but also to play an important role during construction. You will learn, further, how to control, preserve, and organize the overwhelming instructions, manuals, guides, shop drawings, contracts, and as-built drawings that are turned over to you. Years after construction, when the architect, the engineer, and the many contractors who worked on the project are no longer available, will the maintenance manager have meaningful documents of the building and equipment? The documents that are referred to are not the various drawings and specifications, but rather, detailed information on how the equipment is intended to operate, its interrelationship with the building, and miscellaneous information that is not in written form.

---

* This chapter includes excerpts from *Building Design for Maintainability*, by Edwin B. Feldman, available from Ed Feldman and Associates, P.O. Box 52729, Atlanta GA 30355. Reprinted with permission of the author.

There are many specific ways you can ensure a meaningful grasp of the entire building. This chapter deals with the importance of your participation before, during, and after the construction of both new structures and alterations.

INFLUENCING BUILDING DESIGN TO IMPROVE MAINTAINABILITY

One of the greatest opportunities available to you is in the concept "design for maintainability." Prevention is always preferable to cure, and here is one of the few places where this is possible. The amount of money lost each year in building maintenance is astronomical. For most buildings, over a period of two or three decades, maintenance costs will equal the original cost of construction. We must not only design for easy maintenance, but also for lack of maintenance. As wages and benefits increase, there is the tendency for management to make budget cuts in maintenance; good design will help us survive this.

The greatest possible return on the maintenance dollar is realized when the maintenance program is developed during the planning stages of building. Maintenance personnel should work with the architect, engineer, and decorator in selecting materials and surfaces for easy maintenance.

Ease of maintainability means the condition of an item or a surface which permits its repair, adjustment, or cleaning with reasonable effort and cost. Reasonable effort and cost also means, by inference, that the maintenance must not require unusual worker skills or expensive equipment which is rarely used (although specialized equipment used on a regular basis can be very economical), that it must not involve a procedure which will not permit the reuse of the item in a short time, and that it must not change the item's original appearance nor require overly frequent attention.

HOW TO SELL MANAGEMENT ON MAINTAINABILITY

The objective in new construction is a coordinated effort of all involved parties, to develop an optimum building.

Bear in mind and emphasize to management the following benefits from building design for maintainability, *in addition to cost savings:*

- Fire prevention
- Safety
- Health
- Energy management
- Internal relations
- Public relations
- Reduced down time
- Waste recycling opportunities.

It is difficult for management to ignore economic implications. The sale to management of this concept is best made when a complete economic analysis is made, including these factors:

- Initial cost
- Anticipated labor savings
- Anticipated wage rates
- Anticipated cost of fringe benefits and support costs
- Comparative life expectancies
- Cost of money (interest costs)
- Salvage value
- Effect on insurance, if any
- Effect on taxes, if any
- Effect on depreciation, if any
- Secondary benefits

8 WAYS TO ENSURE BUILDING EQUIPMENT MAINTAINABILITY

Keep a case history file on design for maintainability, obtaining information from tours of your own facility, other facilities, publications, professional organizations, and the like. Slides or color photos are useful to convince management of the problems and opportunities. The file should be built on the basis of category, such as grounds, building exterior, maintenance, custodial, and so on.

In the larger organization, a design committee might be established for this purpose, including the assistant plant engineer, maintenance manager, custodial manager, and grounds manager.

For the organization that constructs many buildings, a guideline for architects and interior designers should be developed, to indicate that certain things would simply not be accepted in building design.

Ensure Accessibility to Building Equipment for Ease of Maintenance

Because of an architect's oversights, economics, or a contractor's lack of concern for later consequences, the lack of accessibility can cause serious maintenance problems and affect the life of the structure. Pay special attention to the following items, which are often overlooked:

- Provide metal access doors in acoustical tile ceilings with a minimum of $2' \times 2'$ dimension.
- Provide access doors at valves, regulators, and steam traps.

- Provide access to mechanical equipment rooms, fan and pump rooms, and electric rooms from a corridor or public space—not through an office, lounge, classroom, or janitor's closet.
- Ensure that all HVAC and other operating equipment is easily accessible for maintenance functions.
- Do not hide valves in ceilings or seal traps in walls. Make sure panels do not have to be removed to lubricate bearings.
- Install air, gas, and water lines for easy accessibility.
- Use appropriate ladders to connect all roof levels to afford safe accessibility.
- Locate filters and light bulbs so that they can be conveniently changed.
- Consider a plumbing chase—36″ wide—between two large restrooms.
- Be sure that exterior items—such as plantings, light standards, signs—are accessible.

Isolate Systems to Prevent Total System Failure

Make necessary provisions for adequate valves and bypasses in HVAC, plumbing, and electrical systems. Failures will occur; however, you should not have to shut down an entire system. The installation of appropriate shut-off valves, unions and bypasses is not expensive during construction, but must be considered to isolate reasonable areas in case of a failure. All major machinery and equipment should have valves at their approach.

Ensure Adequate Clearances

In many facilities, the necessary clearances are completely inadequate. One facility used a 5′ aisle for all food deliveries for the feeding of 7,000 people daily. The heavy food-hand trucks continually damaged the walls, doorknobs, and other items in the corridor. The aisle should have been 8′ to 10′ wide. In another facility, removing the boiler tubes required the removal of a wall section. Mechanical doors and aisles should be sized adequately to ensure that all equipment can be removed when necessary. Ramp angles are extremely critical. An oversized bookmobile that was purchased for a large library system had to go down a ramp and pass under an overhead door. Due to the angle and the length and height of the unit, it was unable to enter the garage. In an other case, a hospital had to return several food carts to the manufacturer, since they were two inches longer than the elevator cab!

Standardize Building Components to Minimize Maintenance Costs

Standardization is imperative if operating costs are to be minimized. The advantages of having all partitions, lockers, bookshelves, HVAC equipment, controls, lighting fixtures, hardware, and other building components standardized is obvious to all maintenance managers, but not to all architects. For instance, one school building of 175,000 square feet was designed to use 42 different types of light bulbs and fluorescent tubes. This information can be conveyed only by you.

Facilitate Replacement and Service of Building Equipment

One of the most frustrating problems faced by maintenance personnel is having a recently installed piece of equipment fail and then finding that parts are unavailable because the model has been discontinued. Check out newly installed equipment to make sure that parts will be available and maintenance expenses are not unreasonable. A possible solution to any questionable piece of equipment is to require an extended maintenance contract as part of the contractor's construction obligation.

Provide for Adequate Storage and Work Space

Too often, because of oversight or other priorities, important storage space is not included. Maintenance shop areas, custodial areas, general office equipment, and miscellaneous storage areas are undersized or completely neglected. This results in significant losses because of insufficient inventories, poor organization, work inefficiencies, and numerous other losses that cannot be categorized. For example, in equipment rooms, the installation of various overhead ceiling anchors for hoists and room for A-frames to expedite the moving of heavy equipment could pay for itself the first time it is used. These areas should be carefully planned and every effort made to include them in the facility.

Special attention should be given to spaces for a possible work control center, computerized operations center, and zone (or satellite) shops. Also provide space for battery chargers, washing tanks, and spray booths where necessary.

If you give the architect the direction to include these space requirements along with equipment requirements, he or she may be able to offer the owner solutions that meet all needs.

Pay Attention to Special Equipment Requirements

The following lists some recommendations for special equipment requirements:

- Identify all components, piping, and controls by color, arrows, and tags with a key for ease of location.
- Include necessary feeders for water treatment.
- Set clean-out plugs in piping with approved lubricant.
- Provide air chambers on hot and cold water piping at each toilet fixture.
- Use recessed entrance mats.
- Provide hose bibs in all toilets, machinery spaces, and 150′ spacing on exterior walls.

- Use plugged tees in lieu of some elbows.
- Make sure windows are accessible for washing and have approved anchors. Consider using rotating windows.
- Design the custodial equipment room—one room per each 20,000 square feet of floor space at each level—with lockable storage areas; floor-type service sink with hot and cold water not less than 2′–0″ above the floor and a basin curb with a minimum of 6″ above the floor; a floor drain; a 3′–0″ door opening outward; and recessed light fixtures at least 8′ above the floor. Suggested size: 6′ × 8′

Check for Omissions in Construction Plans

Blunders or omissions are not unusual. Electrical receptacles are often forgotten, as well as appropriate pitch in roofs, sufficient numbers of roof drains, vibration dampening, and, in general, adequate provision for future expansion. These represent only some of the items to check when you review preliminary plans and specifications. Develop your own specific list according to experience. However, remember that the cost to correct items after a contract has been awarded or after construction has been completed could be considerable, unless picked up during preliminary review.

SUPERVISING THE CONSTRUCTION STAGE

Too often, owners look to the architect for complete representation during all phases of construction. In some cases, the owner may request or allow you to check in occasionally on the construction. Yet few owners feel the necessity for a qualified employee to be assigned full-time to the construction site. Then, when problems arise, they wonder why. One important factor is the absence of the maintenance manager's formal input during the building phase. More and more, the need for the regular owner's employees to be assigned to construction along with the architect's staff is gaining acceptance. Ideally, as the owner's representative, you should have an in-depth knowledge of mechanical and electrical building systems, a knowledge of basic construction, an ability to maintain files, and a good retention for details. You have a vested interest in the contractor's performance, and the future maintenance of the facility.

There should be a clear understanding as to what your responsibilities are. First and foremost, you must not interfere with the architect's clerk-of-the-works. Nor should you be the general liaison between the contractor and the architect. You should not involve yourself with scheduling, payments, making of tests, or other routine construction matters, but you must communicate any information you may have to the architect and not to the contractor.

There are many duties that only you can perform. You can be an involved party during the entire construction process, making sure future maintenance requirements are always considered. Often last-minute changes are made that may solve a construction problem, but will create a future maintenance nightmare.

Watch for errors of construction. For example:

- Porous concrete
- Concrete floor troweling
- High floor drains
- High roof drains
- Overly stiff or overly liquid floor tile mastic
- Mortar in piping.

You could also monitor the following activities during construction:

- Control of construction soils
- Use of wet rather than dry grinding
- Removal of accumulations of soil and waste
- Use of "duck boards" on wet day
- Protection of adjacent occupied areas with matting or curtains
- Pre-occupancy cleaning by the construction contractor
- Regular changing of air conditioning filters
- Protection of trees, and so on, that are to be preserved
- Use of adequate signage and supervision to control traffic flow
- Mapping of all buried utilities and drains

Careful monitoring of these construction activities will facilitate maintenance of the building and the equipment.

After the concrete is poured, the walls and ceilings installed, and the masonry and roofing completed, you should know the location of all piping such as gas, water, oil, electric, and vacuum. You should also know where valves, traps, and anything else that may be permanently concealed in walls and ceilings are located. You should take your own set of job photographs and maintain your own set of as-built drawings.

It is also important to make measurements, such as for the locations of valves behind drop ceilings, to save a lot of time later on.

During construction, you should constantly be thinking in terms of maintenance. You can begin to measure items such as filters, collect information on equipment maintenance, and begin to find out what mechanical manuals are available. You can check to see what spare parts are readily available, and what parts will require a long wait. In this way appropriate inventories of spare parts, including minimum and maximum levels, can be established.

Make sure the needs of each department, as outlined in the planning stages, are being met. Any changes in the amount of storage space or in the layout should be brought to the attention of the prospective department head and cleared.

Collect, file, catalog and protect drawings, specifications, and important documents. Become involved in any governmental agency, fire department regulations and various manufacturers' relationships that will affect future maintenance and operations. Work through the architect's representative and have your suggestions, deletions, and substitutions cleared through the architect. Under no conditions should you issue orders, approve change orders or shop drawings, or perform any of the regular functions of a clerk-of-the-works.

At the conclusion of the construction, you should be completely familiar with all of the equipment and its operation, receive all manuals, drawings, and items required of the contractor, and be prepared to take over the operation of the plant. All guarantees, warranties, and guaranty bonds should be received, and you should not only have a clear concept of the maintenance program, but have determined the number of maintenance personnel required and their duties.

Devise a master schedule for occupying the facility, making sure that all furniture, phone installations, decorations, specialized equipment, and so on, are delivered and installed. During this transitional period, you will be the liaison between the new occupants and the contractors.

DEVELOPING AN EQUIPMENT INVENTORY

A basic requirement in any maintenance program, whether formal or informal, is to have a complete inventory of all maintained equipment. The maintenance program is covered in Chapters 2 and 3; however, the sooner the inventory is completed, the sooner the program can be set up.

The ideal time for the inventory to be taken is during the construction phase while the equipment is being installed. The systems used for marking equipment with an identification number vary from company to company. Some companies burn the number into the unit; others punch the numbers in. Other techniques include the use of metal plates, gummed plastic labels, tape, and stenciling the identifying numbers. If identification numbers are wired on, the wires should be stainless steel.

Manufacturers' manuals and equipment nameplates are excellent sources of information for the collection of inventory data and the maintenance requirements of the equipment. The equipment manuals indicate recommended frequency of various kinds of maintenance. However, there are many factors that may affect the recommended frequency. Typical examples are: changing of filters in a dirty environment versus the frequency of filter changes in a clean environment, and the protective painting frequency in a corrosive atmosphere versus the frequency required in a noncorrosive atmosphere. Therefore, it is essential that you use the manufacturer's recommendations as a guide, while keeping in mind your specific problems and conditions.

Another major source of information is the equipment nameplates. This information should be copied as soon as possible before the nameplates become inaccessible, worn, or difficult to read. The best and most logical time for the collection of this information is when the inventory is being taken during construction. A good procedure is to photograph all nameplates when new. The nameplates contain a wealth of information such as: apparatus name; manufacturer's name and city; serial number; model number; various rating information, such as nominal output and maximum ambient conditions at which output can be achieved. Some nameplates include information on installation, operation, maintenance, and replacement; others point out various cautions or warnings. Nameplates on generators often include much significant data, such as: KW, KVA, power factors, rated armature amperes and volts, rated field currents, excitation volts, maximum temperature rises, and ambient temperatures for the armature and the field. Diesel engine-driven generator nameplates will also often include the firing order.

Nameplates can become worn, obscured, or for many reasons unreadable through the years. Therefore, when setting up the equipment files, the maintenance procedure, and the stockroom, record information from the nameplates on a permanent record file for the specific piece of equipment. If nameplate data refers to a manual, check to see if the manual is available. Many manufacturers can identify the equipment from the model and serial number on the nameplate and, in turn, can provide replacement parts lists, instructions, and rating information. If not already furnished, request the information while setting up the program. The information can be tabulated and collected easily when the equipment is installed. During this time, the manufacturers' representatives are available and any desired information can be requested.

CONTROLLING THE SPARE PARTS INVENTORY

An excellent opportunity is available during construction to determine what the stockroom inventory should be. During installation, check the following items:

- availability of replacement items
- various sources of supply
- the time required to obtain the items
- standard packaging practices
- overhaul requirements
- shelf life
- space limitations
- the replacement items' values.

With this information available, you can incorporate the replacement items for the new or remodeled structure into the general maintenance storeroom or whatever policy is used for the entire facility.

Keep in mind that the spare parts inventory is an insurance against prolonged shutdowns; overstocking and understocking must be avoided. Overstocking results in high carrying charges and possible losses due to obsolescence. Understocking can result in excessive downtime in case of equipment failure. It is not normally necessary to stock items for one machine when other units are in the same plant. Many manufacturers recommend minimum parts inventories for groups of machines.

As the owner's representative on the construction site, you can be instrumental in collecting the various data and assisting in setting up the stockroom for the facility or incorporating its requirements into the overall stores operation. You can also check to find what spare parts the contractor must turn over to the owner and make sure all are received in good condition. You may recommend that the owner purchase additional items that may be difficult to acquire after construction has been completed.

DEVELOPING MAINTENANCE MANUALS AND VIDEO SYSTEMS

During the construction period, it is wise to plan the maintenance of the facility and, after the acceptance and operation, compare the plan with the actual results. A maintenance team can be well motivated, properly trained, and inspired by working in a new facility. Without the necessary direction, many problems can develop that could have been avoided.

Communicating information about various equipment systems can be difficult and, especially where a lot of information is involved, often confusing. Additionally, relying solely on verbal communication is not wise, considering the high percentage of personnel turnover in the maintenance field. For this reason the more progressive maintenance managers use devices to record (for posterity) information relative to the maintenance of a facility. The maintenance manual is that instrument.

During construction, the owner's representative on the job should work with an experienced writer to compose the maintenance manual. The manual should not merely copy material that is readily available from any supplier. It should be practical, not theoretical, and should indicate in simple terms *what* is to be done, *when* it is to be done, and *how* it is to be done. The information relating to operating equipment must be readily available when needed. Basic drawings and graphs should be included. The manual must refer to the overall system as well as to the individual components. Schematic flow and/or control diagrams of each system should be indicated. The operation section may be kept separate from the maintenance section and can include a general description of the system, starting and stopping information, specific operating information, special emergency data, and various troubleshooting information.

It is a tremendous asset when all maintenance personnel assigned to a new or rebuilt facility can refer to a maintenance manual. Revise the initial maintenance manual constantly to ensure the incorporation of any updating or changes that occur. Actively solicit comments from the maintenance personnel and incorporate them into the revised edition. The maintenance personnel will feel part of the revision process, and that in itself can be a substantial motivational benefit.

Much of the information conveyed by the architect, engineer, and manufacturer on the system design and intended operation can be videotaped. Specific review of the mechanical and electrical systems and material that is unique or possibly difficult to comprehend can be explained verbally and graphically on videotape. Typical questions can be asked and answered to clarify the demonstration.

There are many advantages of video documentation. Some of them are:

- Photos can only cover a certain amount of area without losing detail. Videotape can, for example, trace complete electrical harnesses, single out and trace 220 lines or trace piping for water, gas, and sewage with continuous visual images from one room to the next. Tracing wire or piping in this manner gives another visual dimension that still photographs simply cannot provide.
- Information as to what a particular photo relates to may get lost or misinterpreted, whereas videotape allows for an audio track explaining what you are viewing, while you view it.
- Video is much easier to follow than photos, as well as being more realistic. Videotape actually *feeds* you information audibly as well as visually.
- Video is inexpensive. There is no need to bring a professional photographer on site. Videos can be taken at any time during the construction progress by construction inspectors or any of the owners' representatives. The video can be discarded and reused if it is felt that what was taped is not of value.
- Printers are now on the market that will allow you to freeze any video picture or group of pictures and print them on paper in less than two minutes.

Proponents of videotaping claim the videotape can replace the maintenance manual. This is questionable, but together the maintenance manual and the videotape programs describing the maintenance and operation of the facility are an excellent combination that can maximize the best of both systems. The combination of the videotape lectures, a meaningful maintenance manual, and an effective supervisor who has witnessed the entire construction progress provides the ideal foundation to cost effective maintenance and construction.

RETAINING AND STORING PLANS, SPECIFICATIONS, AND MISCELLANEOUS MATERIALS

When construction is completed, volumes of information are turned over to the owner. Some of the information is of no continuing importance as a future reference and can be discarded. However, particularly on large construction projects, hundreds of drawings, manufacturers' literature, construction specifications, catalogs, spare parts data, maintenance and operation instructions, operational test data, reports, guarantees, warranties, and other information are very valuable for future reference. This material should be incorporated into a master facility file and retained. The speed and ease with which information can be found when needed indicates the efficiency of the filing system.

Drawings in particular present many difficulties in efficient filing. Drawings may range in size from 8½″ × 11″ to 42″ × 60″. Various techniques can be used to store the original and copies. Many types or combinations of cabinets and racks are designed for this purpose. Wall racks of different sizes housing the prints either parallel or perpendicular to the wall are available. Various types of stands consisting of both stationary and portable models can be purchased. There are also many various models of cabinets that hold the drawings in an enclosed environment. Most of the cabinets can be locked if so desired. Plans can be stored either in a hanging position or in rolls. A cabinet of drawers in which drawings are stored horizontally is also quite common. Cabinets can be connected to drawing boards; some are designed for executive offices, and others are appropriate for drafting rooms. Cabinets can be permanent or made out of corrugated fiberboard.

There are many disadvantages in using actual drawings:

- Any natural disaster such as a flood or fire can completely destroy all documentation if the files are in one location.
- The filing and cataloging of the plans for particularly large companies may require a full-time clerk. Without this control, many drawings are borrowed and never returned.
- The continual use of drawings, particularly if no one person is responsible, results in general wear and tear of the plans. Plans will deteriorate over the years and become unreadable.
- A large area is required to store the actual plans and they usually wind up in basement areas where the dampness accelerates the deterioration.
- Good storage cabinets are expensive, and must be fireproof.

In most cases, drawings are covered by insurance. Yet insurance companies will only cover the cost of replacing the bond, vellum or mylar that the drawings were created on, *not* the hours it took a draftsman to create the drawings. In the event of a catastrophic loss of drawings from fire or water damage, cost of re-drawing those lost documents would be astronomical in most cases.

But if drawings were microfilmed and original drawings were lost or destroyed, cost and time of replacements would be minimal. You would have your drawings back quickly and at the time you need them most. Fire damage could be assessed and reconstruction of damaged areas can begin right away.

To overcome the cost and problems of storage and use of actual drawings, many companies are utilizing alternatives. In the last few years, the science of micrographics has been greatly improved. Micrographics is making the storage, retrieval, and dissemination of large amounts of data possible with increasing speed, efficiency, and economy. It has tremendous flexibility and can be used by both small and large companies. There are many possible variations, and the selection of the appropriate combination depends on your needs and requirements.

There are many forms in which microfilm is made, stored, and used. They can be 16mm roll film, 16mm cartridge, aperture cards, 35mm film, microfiche, microfilm jackets, and micropositives. All systems have a file security, which is the ability to reproduce an inexpensive duplicate set and store it in another area so that a natural disaster does not destroy the files. Another advantage of the system is that the storage area required for the files is insignificant. A small fireproof cabinet can house thousands of copies of drawings, since microfilm reduces information 96 percent and still retains all details. Retrieval can be manual or automated. An important advantage is that once microfilmed, drawings can be reproduced at will. Therefore, assuming that the films are kept protected properly, there is no fear of loss or damage to the drawings due to misuse or age, which is so destructive to original plans.

The master microform is stored in a secure central area and duplicates are used as the active file for routine day to day use. The originals should never be used in the field.

Upon the completion of a project, manufacturers' literature, descriptive material, operational information, wiring drawings, and maintenance instructions are submitted to the owner. Several copies are often submitted. It is essential that all material be filed for ready access. One copy should be maintained as a source document and never allowed to leave the office in case the other copies are lost or misplaced. The copies can be kept in files or put in loose-leaf books, although mechanics, foremen, and maintenance contractors, when using this information, generally do not preserve the copies. Like the drawings, the copies become unreadable, misplaced, or unavailable through the years. One procedure that has been particularly successful is the organizing of the material and the binding of the entire document. In this way, no one section or page could be used without taking the entire volume. The volume can be logged out to the individual like a library book. Since it is bound, no pages or drawings are lost. This procedure is inexpensive and should be considered. If the material is bound, inserts cannot be added as in a loose-leaf binder; however, a second volume can be made if new material is to be added. In one building alone, four volumes represented our library of manufacturers' information, data, maintenance recommendations, instructional sheets, parts information, bulletins, and installation information. In the rear of the volume, we inserted a library card, so that when someone borrowed the volume, the card was retained until the volume was returned.

PREPARING FOR COMPLETION OF THE CONSTRUCTION PROJECT

At or near the completion of the construction project, a series of meetings are scheduled with the owner, the architect, the engineer, and selected suppliers or manufacturers. In addition to reviewing punch list items, basic scheduling of occupancy, and other routine matters, you should ensure that various design criteria are explained, along with the many operational and maintenance recommendations, comments and suggestions regarding the equipment. This series of meetings often incorporates a training program for the use of the equipment. The information that is communicated is extremely important, and in many cases it is irreplaceable and is not in any written form.

Because of the high turnover rate, it is possible that the original maintenance team will no longer be employed by the firm, and a substantial amount of information that was communicated at the above-mentioned meetings is lost forever. A solution to this dilemma is to carefully structure the project completion meetings or any training sessions and to videotape these meetings.

By structuring the meetings, questions can be prepared in advance and answered, and general comments and recommendations can be made during the course of the meetings. The benefits of videotaping can continue for the entire life of the structure. This valuable information will be available for individuals who may need this material years later. The costs are insignificant in light of the advances in videocassette equipment.

An alternative to videocassette recording, which has built-in problems but is far less expensive, is the use of the super-eight sound movie systems now on the market. A second system to consider is the use of an instant super-eight movie system coupled with the various tape recording devices.

OBTAINING MAINTENANCE GUARANTEES AND WARRANTIES

The final segment of the transition phase is the period after the construction has been completed. Here, a final responsibility exists for the contractor: his responsibility regarding guarantees and warranties.

Make sure that all guarantees and warranties required by contract are submitted. These documents should be stored in a safe place. If any premature failure of items that are covered occurs, it should be made known to the responsible party. Within the first year, maintenance supervisors should be instructed to report all improperly working equipment. They should also report all structural defects, even if minor, to make sure that they are not indicative of major defects. It is a good practice to check items before their guarantee and warranty periods run out. You can keep track of guarantee or warranty expiration dates on a computer, or for smaller operations, on a supervisor's calendar. This procedure is particularly important for items that are rarely or never used. It sometimes becomes obvious that such items have operational problems after the guarantee or warranty has expired.

All well-written specifications require a contractor to remedy any defects caused by faulty materials or workmanship, and to pay for any damages to other work resulting therefrom, which shall appear within a period of time from the date of final payment.

An owner can require a guarantee bond for longer periods on any phase of the job. In this case, the contractor must secure a surety (a third party who will ensure that the work is completed if the contractor defaults). Guarantee bonds can be required on any section of the work. For example, guarantee bonds on roofs used to be common practice. In the past few years, however, the practice of bonding roofs has been discontinued by many owners.

A warranty is frequently required with the furnishing of machinery items such as motors and pumps and in connection with the use of certain processed materials where designs are furnished by the vendor and the products are manufactured by him to meet the performance specifications. By means of a warranty the contractor certifies that the material or equipment will perform as required. The contractor is liable to a suit for damages on the grounds of a breach of warranty, if what is covered fails to perform. Surety bonds may be required to guarantee the compliance of a warranty.

Warranties can be full or limited. A product can carry more than one written warranty. For example, it can have a full warranty on part of the product and a limited warranty on the rest. Be sure to fulfill the provision of all warranties, such as regular roofing inspections and reports, to avoid voiding the warranty.

Implied warranties are rights created by state law, not by the company. All states have them.

The protection that an owner has after completion of construction is proportional to your follow-up. Assuming reasonable contractual protection, if you and your staff are diligent and all protection afforded by the contract including guarantees and warranties is pursued, it is likely that the facility under construction and its equipment will operate with minimum troubles.

2

Preventive Maintenance: Key to an Effective Maintenance Program

Paul D. Tomlingson

Preventive maintenance (PM) is any action that can be taken to prolong the life of equipment and prevent premature failures. Typically, PM includes equipment inspection, lubrication, adjustment, cleaning, nondestructive testing (predictive maintenance), and periodic maintenance—usually component replacements. Rebuilds and overhauls are generally not classified as PM. Rather, these are actions needed when equipment has deteriorated beyond the need to conduct PM services. Rebuilding and overhauling restore equipment to an effective operating condition. Only then is PM effective.

Preventive maintenance aims at preserving the useful life of equipment and avoiding premature failures. In addition to the routine aspects of cleaning, adjusting, lubricating, and testing, the *detection orientation* of regular equipment inspection and testing yields a vital means of improving the overall maintenance program. Early detection of impending problems prevents a critical level of deterioration. Thus, you have the opportunity to utilize this lead time to deliberately plan repair work and carry it out with more effective use of its resources. As a result, preventive maintenance inspections can become a primary means of increasing planned work, enhancing labor effectiveness, and sharply reducing emergency repairs. These conditions, in turn, reduce cost. This chapter describes the means available to achieve these objectives and the specific means of organizing and carrying out preventive maintenance functions.

PM TERMINOLOGY

The following are terms often used by maintenance managers to explain the development of a PM program:

- *Route*—The path followed by the PM worker that will ensure all equipment is reached, yet avoid needless backtracking for complete coverage.
- *Frequency or Service Interval*—The period of time between PM services. It can be expressed as a frequency: weekly, monthly, quarterly, and so on.
- *Fixed Interval / Variable Frequency*—A fixed interval service would be carried out at the completion of a specific time period: a week, a month, or a quarter. A variable frequency service would be carried out at the end of a specific time; for example, 1,000 hours or 3,000 miles.
- *Equivalent Scheduling Day*—Many times it is easier to schedule PM services when operating hours or miles are equated to days or calendar periods (frequencies). For example, 1,000 miles could be a 30-day (monthly) service.
- *Service Inspection Time*—The amount of time (hours) allocated to perform the PM service.
- *Checklist*—The list of actions to be performed as part of the PM service.
- *Detection-Orientation*—Refers to making inspection the most prominent part of the PM program.
- *Visual Inspection*—The observation of equipment to ascertain its deficiencies.
- *Dynamic or Static Inspection*—Inspections carried out while equipment is, respectively, in motion or at rest.
- *Drift or Chance Statistics*—The basic means for determining when a unit should be serviced. Drift means that the unit gradually deteriorates, and at the end of a specific period must be serviced or inspected because it is in the danger zone of failing. Chance means that there is no particular deterioration pattern; therefore, the equipment is looked at frequently enough to catch any problems occurring.
- *Workload*—The amount of labor necessary to carry out the whole PM program or any of its parts, such as inspection or lubrication.
- *Nondestructive Testing*—Testing techniques to help uncover problems using technical resources such as vibration-analysis, sonic testing, and dye testing (for cracks). Nondestructive testing is sometimes called predictive maintenance.
- *Exception Reporting*—PM inspection whose results list only the problems.

BENEFITS OF A SUCCESSFUL PM PROGRAM

A successful preventive maintenance program can pay considerable dividends to both the maintenance department and the facility fortunate enough to have an effective PM program. To cite a few results:

- *Fewer Failures*—A timely PM inspection program usually uncovers problems before they become serious enough to cause equipment failure. As a result, routine adjustments and minor repairs take the place of failures. Breakdowns can be reduced by 50 percent.
- *More Planned Work*—The timeliness of preventive maintenance inspections usually uncovers those major jobs that require planning. Sufficient lead time allows planning to be done.
- *Fewer Emergencies*—An effective PM program has every maintenance employee on the alert for those things that cause problems. As a result, fewer problems that would cause an emergency situation escape detection. Emergency work can be reduced by half.
- *Reduced Overtime*—One of the largest contributing factors to overtime is the need to perform emergency work. A reduction in emergency work usually produces a corresponding decrease in overtime.
- *Extended Equipment Life*—PM invariably rewards its users by extending equipment life by as much as 20 percent and increasing dependability.
- *Better Use of Maintenance Personnel*—A job done under emergency conditions is 15 percent more costly in labor than a similar, well-planned job. A maintenance department with a good PM program commits more staff to planned work. The result is more effective, productive use of this manpower.
- *Improved Equipment Operation*—Well-cared-for equipment is its own reward—it runs better.
- *Less Downtime*—Because the PM program can reduce emergency work, it follows that downtime can also be reduced. In the manufacturing environment, it is essential that an investment in scheduled downtime for PM is better than unscheduled downtime for emergency work. Once demonstrated, the PM idea becomes an accepted "better way." Then, reduced downtime can result in more production output.
- *Reduced Maintenance Cost*—Personnel tend to work more effectively with less lost motion in a well-planned job environment. This pattern soon emerges into a cost reduction trend, usually reaching about 15 percent.

PM costs will equal about 8 percent of the hours used by a typical maintenance department and about the same percent of total maintenance costs. The overall cost of a PM program is equal to about 4 percent of total equipment value. Generally, the cost of installing a good PM program will equal about 10 percent of the total savings realized by the program.

It should be clearly understood that effective PM programs are the result of a well-organized, carefully executed effort. In short, they are hard to come by. If these benefits are to be realized, they must be earned.

MAXIMIZING PREVENTIVE MAINTENANCE INSPECTIONS

The most essential activity of the preventive maintenance program is equipment inspection. It is this portion of the PM program that generates advance information on the status of equipment. This information provides the lead time that permits your maintenance department the opportunity to plan necessary repairs.

It follows then that the preventive maintenance program should be *detection-oriented*. That is, its principal aim should be to uncover problems before they reach the crisis stage of equipment failure or breakdown. The sooner the problems are found, the greater the opportunity for planning: gathering materials, coordinating the shutdown, estimating, and allocating the staff.

The effect of conducting timely PM services is illustrated in Figure 2-1. As the deterioration of equipment increases, PM inspections, when performed on a timely

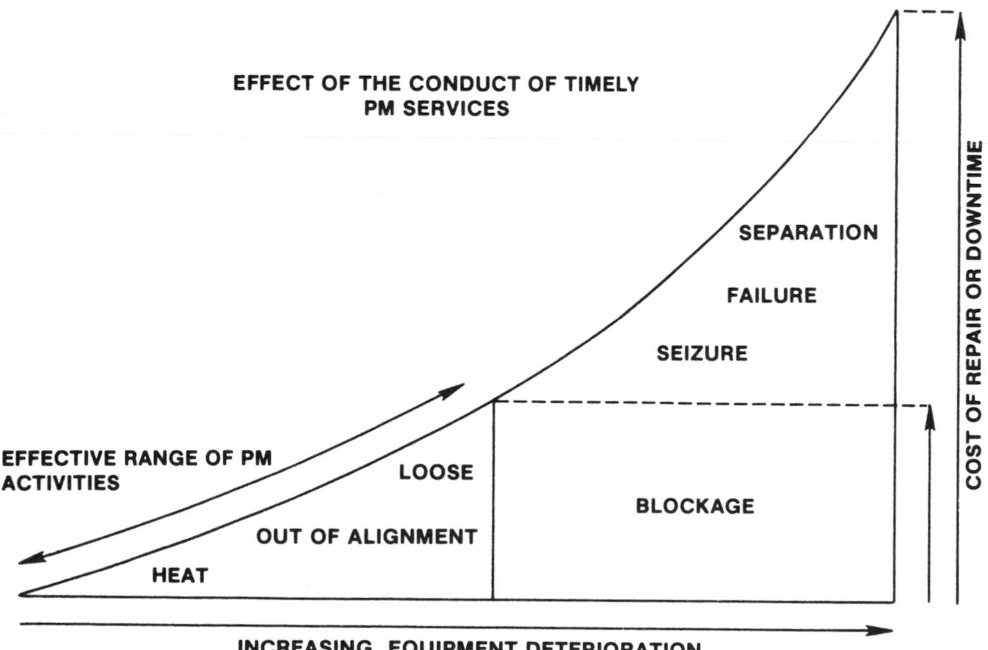

Figure 2-1. Effect of the Conduct of Timely PM Services

basis, can uncover less serious problems before they cause damage. As a result, downtime and costs are reduced.

As many PM inspection programs get underway, there is a great tendency to forget the detection-orientation. The most common action is to attempt to repair each deficiency as it is found. This tendency must be resisted because, in the long run, it undermines the PM inspection program. (See Figure 2-2.) If there were some 20 units to be inspected on a specific route, the number of deficiencies recorded would be far greater on the earlier units when repairs are made as the inspection progresses. Units at the end of the route may never get inspected because the inspector has used all allocated inspection time on repairs.

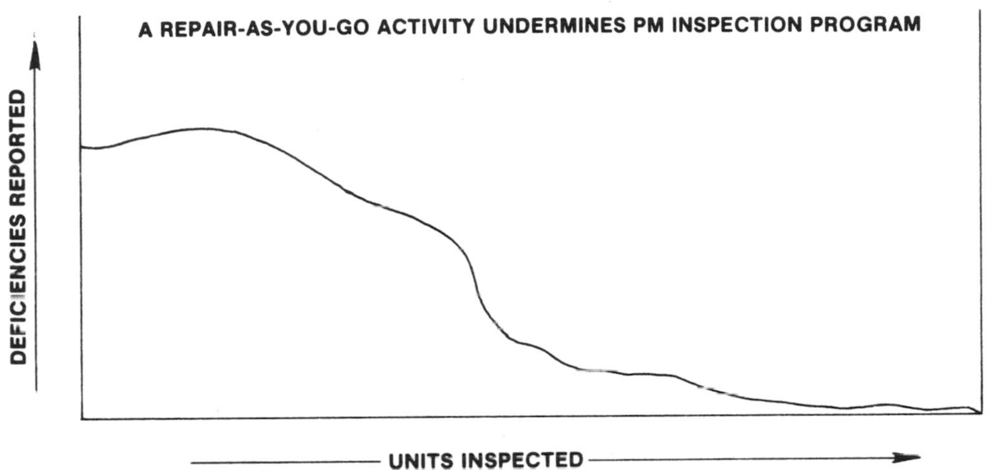

Figure 2-2. A Repair-as-You-Go Activity Undermines PM Inspection Program

USING PM INSPECTIONS TO HELP PLAN AND SCHEDULE MAINTENANCE

When PM inspections are carried out properly—on a timely basis—the lead time developed before repairs must be made provides the basis for better planning and scheduling. As a result, the amount of effort used on scheduled work gradually increases.

As the amount of scheduled work increases, so does the productive use of staff effort. Essentially, as the scheduled work displaces the unscheduled and emergency work, more effective use is made of the maintenance staff. Thus, with a fixed work force, more jobs are able to be performed until a point is reached when the available labor exceeds the work that must be done. Thus, labor savings can be generated and used elsewhere. See Figure 2-3.

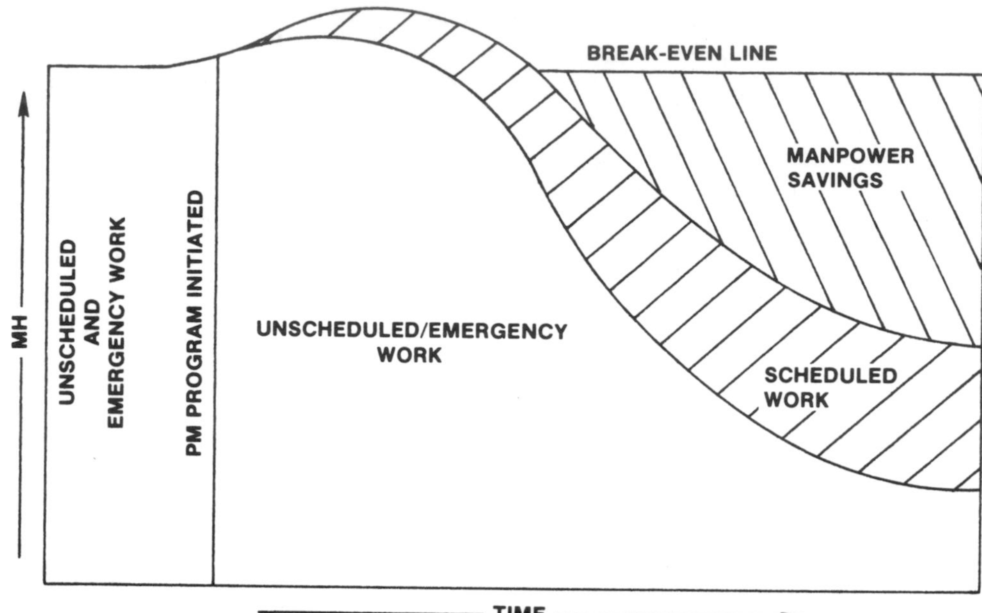

Figure 2-3. Labor Savings Resulting from a Good PM Program

MAKING PM A PART OF THE MAINTENANCE MANAGEMENT PLAN

Just as maintenance is a key element to the production strategy, so is PM a key element to the maintenance program. In fact, to ensure that this is the case, you should seek strong plant management support for PM. Instead of a lukewarm endorsement such as "I'm for it," a strong endorsement is desired. A typical position might be a policy statement: "PM should take precedence over every aspect of maintenance except bona fide emergency work."

Preventive maintenance is one of several actions carried out by a maintenance department. Others include handling complaints, performing emergency or safety work, or performing periodic actions: overhauls, component replacements and rebuilds, and supporting project work.

Since these elements are fed into a maintenance department, preventive maintenance should be kept in perspective. However, because preventive maintenance inspection results usually require some planning support for major jobs, planning support is needed.

Figure 2-4 illustrates a typical Maintenance Management Network. PM figures prominently within the total scope of a maintenance management network.

Within the maintenance management system is the preventive maintenance inspection control network. This network, showing the sequence of events of PM inspections, links together the actions of supervisors, planners, and personnel. See

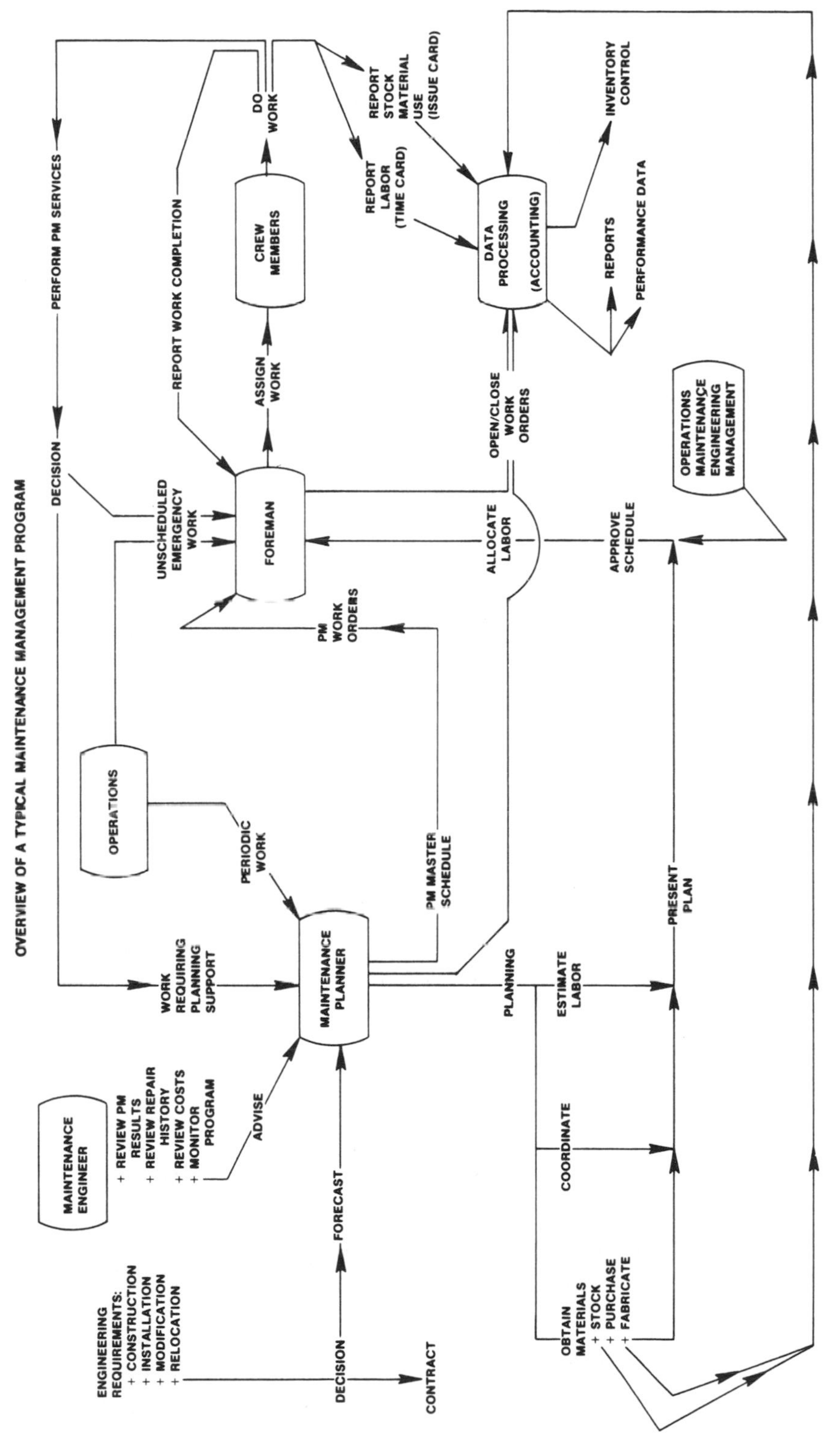

Figure 2-4. Overview of a Typical Maintenance Management Program

Figure 2-5, which traces PM from the master schedule through each event leading to completion of the work. Inspection control is illustrated.

It is absolutely necessary to ensure that the PM program fits correctly into the overall plan of the maintenance management system.

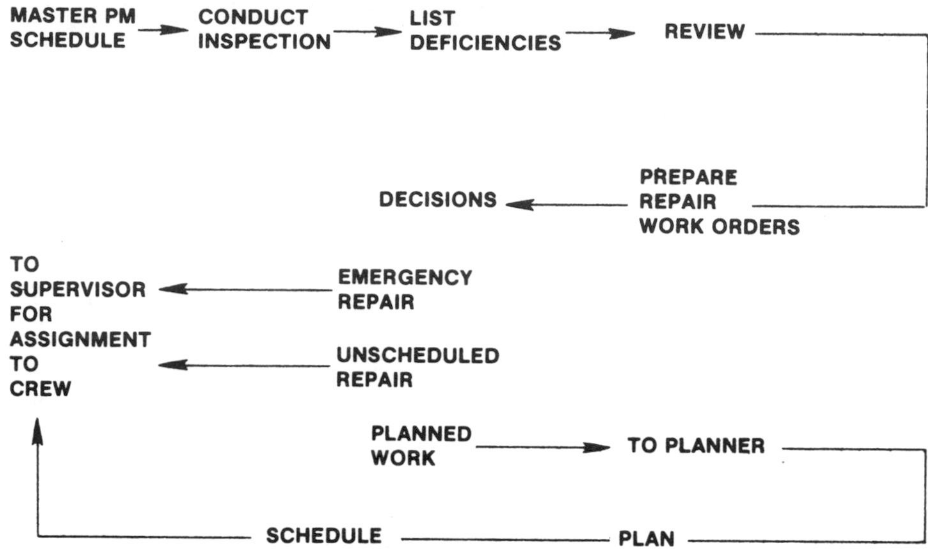

Figure 2-5. Preventive Maintenance Control Network

STARTING UP A PM PROGRAM

There are two broad aspects in the start-up of a PM inspection program: organizing and operating.

Organizational steps for starting up a PM inspection program would include:

- *Listing All Equipment to Be Inspected*—The equipment to be inspected should be listed so that a basic plan can be prepared.
- *Developing Routes for Fixed Equipment*—Establish the route to be followed to bring the inspector to all units of equipment that must be inspected. Avoid backtracking.
- *Preparing a Program for Mobile Equipment*—Determine the basic plan for inclusion of mobile equipment in the program.
- *Establishing Standard Times for Inspections*—Determine how much time will be allocated to complete each service.

- *Establishing Intervals*—Spell out how long it will be between services.
- *Determining Labor Requirements*—The annual labor requirements (workload) are equivalent to the time (hours) for each service, times the number of repetitions of that service per year.

Operating steps for starting up a PM inspection program would include:

- *Preparing and Issuing PM Schedules*—The PM Master Schedule prescribes the week-to-week plan for the conduct of PM services. When a service is scheduled, the serviceperson uses a previously prepared checklist to conduct the service.
- *Conducting Service*—This is the actual performance of prescribed services.
- *Reporting Results*—As PM services are performed, results are reported. These should be translated into work orders to correct any deficiencies uncovered.
- *Monitoring Repairs from PM Program*—The number of repairs generated from the PM program is an indication of the thoroughness. As a result, more opportunity for scheduled work is created. Scheduled work should go up, emergencies down.
- *Monitoring Actual Versus Planned Times*—Labor reported as a result of performing PM services should be able to be compared with the planned times. This helps to regulate the PM workload.
- *Adjusting Service Intervals*—A lack of deficiencies when inspections occur at a certain interval is adequate license to extend inspection intervals—and the reverse is true. As this happens, review all inspection intervals and adjust them.
- *Checking Methods of Inspection*—Determine how inspections are being done. Is there, for example, too much backtracking? Have these times checked. Industrial Engineering can help.
- *Adding or Deleting Items on Checklist*—Few checklists survive without change. Just as a good plan must bend to changing conditions, so must PM checklists be altered to ensure they are accurate and complete.
- *Balancing Labor Needs*—Constantly adjust the labor needs to satisfy PM requirements. Keep in mind that much of the PM start-up is a trial-and-error effort. Changes in labor needs merely acknowledge this.

See Figure 2-6 for a schematic illustration of these steps.

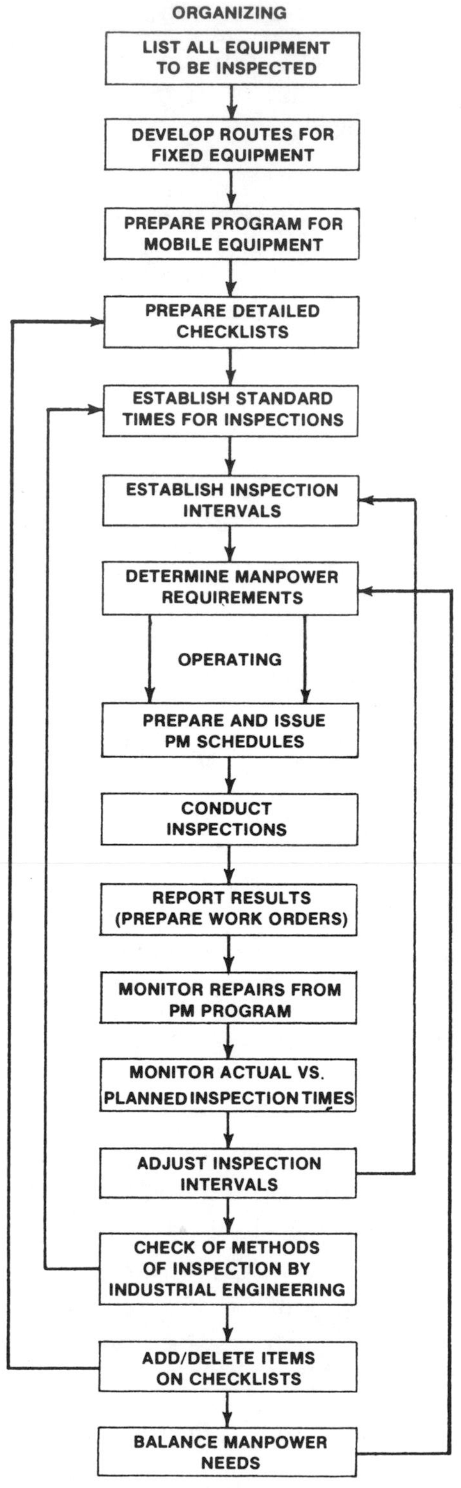

Figure 2-6. Steps in Starting a PM Program

The Details of Start-Up

Workloading is an important first step in starting up the PM program. Essentially, it is the determination of the labor necessary to carry out the program. See Figure 2-7, which illustrates how to estimate staff resources, by craft, for each preventive maintenance service. When multiplied by the number of repetitions annually, it yields the total annual labor per service for each craft. Summarizing these produces the total labor required for preventive maintenance. Week-to-week staff leveling is required to establish the average weekly staff to support the program.

MASTER PREVENTIVE MAINTENANCE SCHEDULE

LUBRICATION	MONTHS																						
DESCRIPTION OF WORK	WEEKS																						
LUBRICATION - DEPT 269	1	2	3	4	5	6	7	8	9	10	11	12	13	14	15	16	17	18	19	20	21	22	23
ROUTE 1 WEEKLY	2	2	2	2	2	2	2	2	2	2	2	2	2	2	2	2	2	2	2	2	2	2	2
ROUTE 2 WEEKLY	4	4	4	4	4	4	4	4	4	4	4	4	4	4	4	4	4	4	4	4	4	4	4
ROUTE 3 MONTHLY				2				2				2				2				2			
ROUTE 4 QUARTERLY											6												
ROUTE 5 QUARTERLY						3												3					
ROUTE 6 SEMI-ANNUALLY														4									
ROUTE 7 ANNUALLY																							
SUBTOTAL MANPOWER	6	6	6	8	6	9	6	8	6	6	12	8	6	10	6	8	6	9	6	8	6		
LUBRICATION - DEPT 270																							
ROUTE 1 WEEKLY	3	3	3	3	3	3	3	3	3	3	3	3	3	3	3	3	3	3	3	3	3	3	
ROUTE 2 MONTHLY			4				4				4				4				4				
ROUTE 3 QUARTERLY					6												6						
ROUTE 4 QUARTERLY												6											
SUBTOTAL MANPOWER	3	3	7	3	9	3	7	3	3	3	7	9	3	3	7	3	9	3	7	3	3	3	
LUBRICATION - DEPT 22																							
ROUTE 1 MONTHLY		6				6				6				6				6					
ROUTE 2 QUARTERLY				6												6							
ROUTE 3 QUARTERLY	4												4										
SUBTOTAL MANPOWER	4	6	0	6	0	6	0	0	0	6	0	0	4	6	0	6	0	6	0	0			
TOTAL MANPOWER BY WEEKS	13	15	13	17	15	18	13	11	9	15	19	17	13	19	13	17	15	18	13				

Figure 2-7. Determining Labor Requirements for Preventive Maintenance

Controlling PM Inspection Assignments

A master preventive maintenance schedule prescribes the timing for the conduct of services. See Figure 2-8, which shows how PM services for fixed equipment are scheduled in increments of weeks. The master schedule covers an entire year.

PM SERVICE DUE				WEEKS												
SWO	FQ	CR	LH	01	02	03	04	05	06	07	08	09	10	11	12	. . .
934911	2	MW	02	*		*		*		*		*		*		
934912	3	EL	03		*			*			*			*		
934921	2	ME	04		*		*		*		*		*		*	
934922	4	EL	03	*				*				*				
934923	2	MW	02		*		*		*		*		*		*	

Legend: SWO – Standing Work Order Number
FQ – Frequency of service in weeks
CR – Craft performing the service:
MW – Millwright
El – Electrician
Me – Mechanic
LH – Labor hours required for each repetition
* – Indicates the week the service is due

Figure 2-8. Master PM Schedule (Extract)

Creating PM Inspection Checklists and Instructions

Maintenance is sometimes criticized for promising improvements as a result of its PM program but not always delivering. Their progress is sometimes slowed by a perceived notion that the program cannot start until all the checklists are prepared.

On the contrary, visual inspections, without the checklists—conducted by well-trained personnel, permit the investigative action of PM inspections to get underway before the PM checklists are written. Not only does this help to get the program started, but it identifies inspectors who can later help prepare the checklists.

One organization asked its PM inspectors to tape-record observations as the first visual inspections were made. The results provided the basis for preparing written checklists.

Because preventive maintenance is a routine, repetitive task, the same checklist is used each time the service is carried out. It is not necessary to reprint the checklist each time the service is performed.

There are instances of equipment inspection when only the condition of the equipment is important. Reporting temperatures and pressures is not required. Therefore, exception reporting, in which only the problems are reported, is simpler and quicker. This technique requires the inspector's supervisor to review each deficiency and develop a definitive action for each. Figure 2-9 shows a deficiency list in which the inspector has reported the problems found. Action is required on each item listed.

SWO NO 22 10 910 Tradesman: J. Doak Date 2/2/92

Machine Centre Name: Quad - Saw

Equip. No.	List Deficiencies Found During Inspection	UR
C22	RPL SPROCKET ON L. TURN ROLL	M
C23	WLD WEAR STRIP ON SHARP chain	W
C32	RPL C. RIG BELTS	E

Time to Complete PM Tour 2.5 Hours

I have been informed of Equipment Deficiencies

Supervisor R. Johnson

Urgency of Repair (UR)
L - more than one month
M - Within one month
W - Within one week
E - Emergency

Figure 2-9. Sample Equipment Deficiency List

Involving Equipment Users in the PM Program

Involve the users of the equipment when getting started. They can perform many helpful activities, including:

- Determining the scope of PM work—That is, how much PM work can be done and how non-maintenance personnel can help sustain it by making equipment available. One of the prevalent causes of the failure of PM programs is the absence of operator participation at the outset. Few PM programs ever succeed as a unilateral maintenance effort. A joint effort with equipment operators is essential to identify PM services required, their frequency, and the potential assistance of operating personnel.
- Determining the frequency of PM work—How often will services be performed? Can users make units available? If not, what alternate scheduling arrangements can be made? Equipment operators can help to determine this.
- Making equipment available for service—When visual inspections are made, there is little problem in making equipment available. However, some PM service requires that the equipment be shut down to perform services. Equipment users who understand this need can make such services much easier to schedule.
- Allocating downtime for PM—Operating time can be attained only by equipment that operates. A particular level of production, for example, must be accompanied by a specific period of utilized time of equipment. Therefore, equipment users must be as aware of scheduled PM downtime as they are of scheduled running time.
- Providing supplemental support for PM—Operators know their equipment well. They are aware of its many quirks and are prepared to tell PM inspectors of many suspected problems. Thus, they are an excellent source of information on equipment condition. They can also check lube or oil levels, test hydraulic controls, and verify safety devices. Many times, they can carry out daily lube services when it is difficult for regular lubricators to reach the equipment on a timely basis. Operators are an excellent source of supplemental PM support and should be used.
- Reviewing equipment deficiency lists—When serious problems are found during a PM inspection, the equipment operator and his/her supervisor should be informed so they can scheduled the equipment down to accommodate repairs.
- Facilitating scheduling and repairs—Once major deficiencies are uncovered, repairs must be made. These repairs also need downtime—and they must be coordinated with equipment users.

Assigning PM Inspectors

Your PM inspectors are often more than mere inspectors. They are goodwill ambassadors for the PM program. What they say and do and how they do it can be critical to the PM program.

Typically, they should drop in at the start of the inspection and let the operating personnel know why they are there, and what they will be doing. Often this pre-inspection visit is worth the time. Operating supervisors may have some real concerns about their equipment and are anxious to point them out. An understanding PM inspector—by merely listening to these problems—can help make the PM program a success. The inspector can gain cooperation and credibility.

During the inspection, the PM inspector should talk to operators and learn of their concerns, to be made aware of problems. As the PM inspection is completed, the inspector should make a special effort to reassure operating personnel that the problems they pointed out were looked into, solved, or that necessary repair actions are underway.

The Maintenance Supervisor

The maintenance supervisor is the key to the whole PM effort. He or she gets it started. The inspectors usually report to the supervisor. Deficiencies come back to the supervisor for decisions. The supervisor advises the operating supervisors on repair decisions. The supervisor also briefs the maintenance planner on major jobs so that planning details can be taken care of. The PM program's success often hinges on the maintenance supervisor.

In smaller maintenance organizations, the supervisor performs a dual role. He or she must evaluate reported deficiencies and then plan for and schedule follow-up repairs. Under these circumstances the maintenance supervisor's role is even more important in PM.

The Maintenance Planner

In maintenance organizations with planners, major jobs developed from PM inspections should flow through the maintenance planner. To ensure that repairs are made promptly, the planner must work closely with the supervisor to accomplish the following tasks:

- Clarify the job requirements
- Agree on a job scope and plan
- Gather materials
- Estimate the labor required for the job

- Complete fabrication
- Coordinate with equipment operators for scheduled shutdown.

When these actions are completed, the job is ready for scheduling.

Putting the PM Inspection Effort in Perspective

Preventive maintenance inspections are pivotal to the whole PM program. They uncover problems in time to avoid failure and wasteful emergency repairs. If inspections can be made on a regular basis, there is more lead-time between when problems are found and when the work must be done. This yields a greater opportunity for planning work. In turn, the increase in planned work means that maintenance labor is used more productively. Jobs use less labor and are completed in less elapsed downtime.

SCHEDULING PREVENTIVE MAINTENANCE

Computers can be used to schedule PM services at different frequencies. To facilitate scheduling, each PM service should have a unique Standing Work Order (SWO). This permits the development of a separate file relating that number to the type of service, craft performing it, frequency of the service, hours needed per service, and so on.

Typically, for a specific week, all PM services due that week would be announced as in Figure 2-8. If the PM service can be fitted in anytime during the week, the maintenance supervisor may coordinate them informally and assign them to the maintenance staff. However, if the PM service requires equipment shutdown, it should be included in the weekly maintenance-operations scheduling meeting for proper coordination. In either case, as deficiencies are identified, the supervisor makes decisions to obtain planner support or alternately assign work to crews. Labor use reporting, in turn, generates information on PM compliance and labor used on PM. If the inspection yields important deficiencies, operations should be advised.

PERFORMING PREDICTIVE MAINTENANCE

Predictive maintenance is part of the overall maintenance program. The use of the phrase "predictive maintenance" has tended to displace the use of the phrase "non-destructive testing." Many organizations now call their program "PPM"—for Preventive, Predictive Maintenance. The particular phrase used depends on the mix of services performed.

Predictive maintenance provides techniques whereby impending failures can be predicted. They pinpoint specific components that are performing abnormally and are likely to fail. Likewise, these techniques are able to isolate those components that are performing normally and do not have to be replaced. Among them:

- Vibration Analysis—vibration is measured at various critical points in rotating equipment. Variations noted from a previously recorded norm or baseline help to isolate the specific problem or component in need of replacement.
- Dye Testing—the use of dyes to help spot cracks not otherwise visible.
- Sonic Testing—the use of sonic equipment to measure thickness on certain critical wear points. For example, a pumping station handling abrasive slurry tends to rupture when fast-moving slurry wears piping at sharp turning points. Sonic testing would reveal excessive wear and alert staff to the need for replacing the piping at that point.
- Acoustical Analysis—sound measurements taken repetitively at fixed locations in order to detect abnormal noise level variations.
- Infrared Analysis—location of unusual "hot spots" using infrared techniques.
- Other Techniques—moisture absorption, flow or temperature measurements, or chemical analysis.

Preventive and predictive maintenance techniques equip the maintenance department with a wide range of diagnostic tools. Generally, preventive maintenance techniques are broader and less sophisticated than non-destructive testing techniques, but both have value.

Predictive maintenance is discussed in detail in Chapter 3.

SUMMING UP

The preventive maintenance program can prolong equipment life and prevent premature equipment failure.

Other less tangible benefits accrue as well. When done effectively, equipment inspection, testing, and cleaning can uncover deficiencies in the earlier stages of equipment deterioration. Thus, the time available for making repairs before failure is greater. In many cases, these deficiencies relate to major jobs that require planning before they can be performed. Thus, the greater lead-time available before work must be done can be used to plan the work. In turn, this creates opportunity to perform more work on a planned basis. Herein lies one of the principal advantages of a "detection oriented" PM program.

It follows that as more work is planned, better use is made of maintenance labor. Generally, savings of 12 to 15 percent in labor are possible in performing similar planned jobs versus unplanned jobs. In turn, this aspect of labor savings applies to the whole maintenance program.

For those who have installed successful PM programs, the rewards are well known. This chapter is intended for those who have yet to do it.

Perhaps the best advice for those who are committing to PM is the need for planning the program before getting underway. The most important step is to secure the necessary commitment from management, equipment users, and maintenance staff. Developing and installing the program and then making it work is never easy. However, it is easier with this commitment, which can make the PM program succeed.

3

Putting Predictive Maintenance to Work

S.T. Cordaro

Predictive maintenance is often confused with *preventive maintenance*. The confusion probably results from the common objectives of these maintenance activities. This chapter focuses on predictive maintenance as a tool for improving the reliability of building equipment. It is, however, difficult to describe predictive maintenance without contrasting it to other reliability-improvement methods including preventive maintenance. In these discussions, some reference will be made to these other techniques.

The purpose of both predictive maintenance and preventive maintenance is to extend the useful life of the equipment and to detect defects before they lead to catastrophic failure. To optimize the use of these methods, you should have a clear understanding of the differences in techniques and the conditions under which each is appropriately applied.

This chapter has two main objectives: to help maintenance managers understand the difference in the two methods and to show how you can use predictive maintenance surveillance systems in a practical maintenance reliability program.

UNDERSTANDING THE DIFFERENCE BETWEEN PREDICTIVE MAINTENANCE AND PREVENTIVE MAINTENANCE

The differences between preventive and predictive maintenance methods may be simply stated as the differences between a time-based system (preventive maintenance) and a performance-based system (predictive maintenance).

Preventive Maintenance—A Time-Based System

Preventive maintenance is a time-based surveillance method in which periodic inspections are performed on equipment to determine the progress of wear in its components and sub-systems. When wear has advanced to a degree which warrants correction, maintenance is performed on the equipment to rectify the worn condition. The corrective maintenance work can be performed at the time of the inspection or shortly following the inspection. The decision of when to perform the corrective repair related to PM inspections depends on the length of shutdown required for the repair. You must consider the impact of the shutdown on the operation caused by the repair versus how immediate is the need for repair. If it is judged that the worn component will probably continue to operate until a future repair can be scheduled, it may be better to postpone major repairs until they can be planned and scheduled. In effect, preventive maintenance applications change the shape of the wear-out portion of the failure rate curve. Figure 3-1 illustrates this concept.

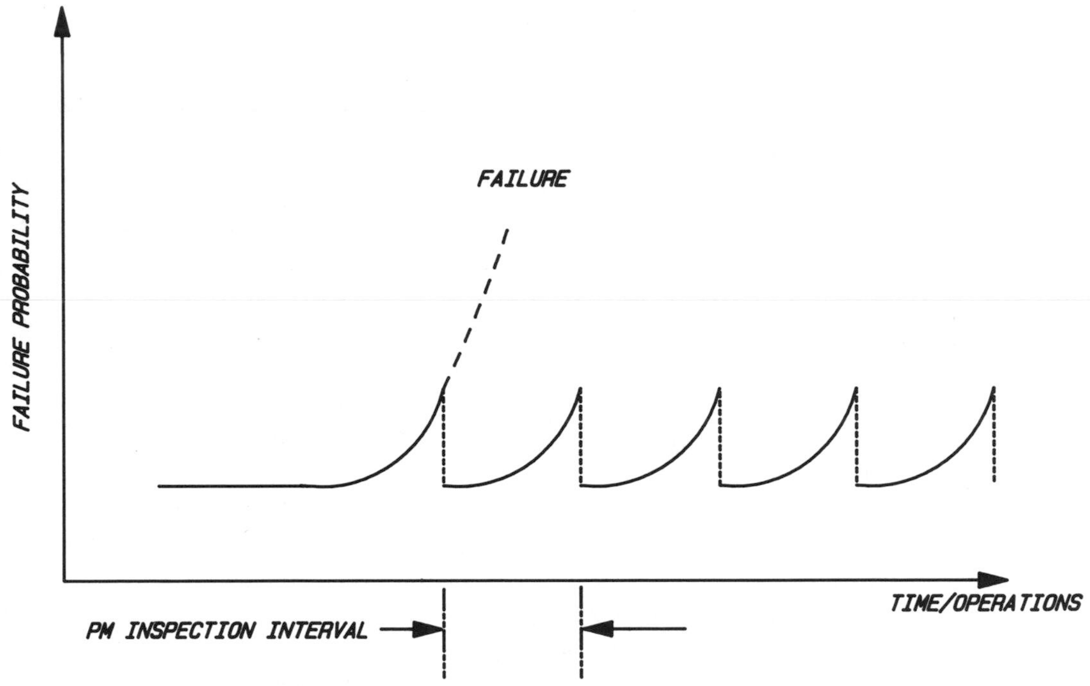

Figure 3-1. How Preventive and Predictive Maintenance Influence the Wear-Out Portion of the Failure Rate Curve

Predictive maintenance is designed to accomplish exactly the same objective. Figure 3-1 illustrates that, as the defect in the component or sub-assembly develops over time (or use), the probability of failure increases. Periodically, a preventive maintenance inspection is made. If the inspection reveals serious wear, some maintenance operation is performed to restore the component or sub-assembly to a good state of repair. This reduces the probability of failure, thus changing the shape of the wear-out portion of the failure rate curve. By utilizing the preventive maintenance system, you will increase the probability that the equipment will perform as expected without failure until the next inspection is due.

To determine the interval between inspections, consider the history of maintenance for the equipment in a particular facility. The decision of time between preventive maintenance inspections will ultimately be guided by a number of resources. These include manufacturer's recommendations, feed-back information from breakdowns, and the subjective knowledge of the maintenance workers and supervisors who daily maintain the asset.

A central characteristic of preventive maintenance is that in most major preventive maintenance applications, the asset must be shut down for inspection. For example, a heat exchanger must be shut down and isolated when a non-destructive eddy current inspection is made on its tubes. The inspection process would require a discreet amount of down time for the unit, which is typical for preventive maintenance.

The loss of operational time when significant preventive maintenance inspections are made is one of the reasons PM programs are often less than successful. This is especially true in applications where there are few redundant units and equipment must operate at 100 percent of capacity. In some situations the loss from shut-down is considered too high a penalty and preventive maintenance inspections are resisted.

The need for down time during significant preventive maintenance inspections however, should not be construed to mean that preventive maintenance is not a valid and effective reliability tool. Preventive maintenance, when properly applied, unquestionably increases overall equipment availability. You must be aware, however, that in exchange for reducing breakdowns, there is a price that must be paid in addition to the maintenance-related costs. This cost is expressed in terms of lost productive time for the asset during the inspection. It is the Mean Time for Preventive Maintenance (MTPM). The total price paid for preventive maintenance equals MTPM plus maintenance labor and materials cost involved in the PM inspection process. This may be described as the total price for PM (TPPM).

Predictive Maintenance—A Condition-Based System

On the other hand, predictive maintenance has been called a condition-based system. The method measures some output from the equipment that is related to the degeneration of the component or sub-system. An example might be metal fatigue on the race of a rolling element bearing. The vibration amplitude produced by the rolling element as it passes over the degenerating surface is an indicator of the degree of severity of wear. As deterioration progresses, the amplitude of vibration increases. At some critical value, the vibration analyst concludes that corrective action should be taken if catastrophic failure is to be avoided.

The method usually permits discrete measurements which may be trended, compared to some pre-defined limit, or tracked using statistical control charting.

When an anomaly is observed, you are warned in sufficient time to analyze the nature of the problem and take corrective action to avoid failure. Thus predictive maintenance accomplishes the same central objective as preventive maintenance. It changes the shape of the wear-out portion of the failure rate curve. (See Figure 3-1.) Confidence in the continued "on specification" operation of the asset is increased. By early detection of wear, you can plan for and take corrective action to retard the rate of wear, or to prevent or minimize the impact of failure. The corrective maintenance work restores the component or sub-assembly to a good state of repair. Thus the equipment will operate with a greater probability of trouble-free performance.

The enhanced ability to trend and plot numbers collected from predictive measurements also gives the method greater sensitivity than traditional preventive maintenance methods. The technique yields earlier warning of severe wear and thus provides greater lead time for you to react. Corrective actions may be scheduled so that they have minimum impact on operations.

A principle advantage of predictive maintenance is the capability it offers the user to perform inspections while the equipment is operating. In particular, in order to reflect routine operating conditions, the technique requires that measurements be taken when the equipment is normally loaded in its production environment. Since the machine does not need to be removed from the production cycle, there is no shutdown penalty. The ability to conduct machine inspections while equipment is running is especially important in continuous operations such as in utilities, chemical, and petro-chemical manufacture.

Another advantage of predictive maintenance is that the cost of surveillance labor is much less than the cost of preventive maintenance activities. Although the technical knowledge required for predictive maintenance inspections is usually higher than those for preventive maintenance, the inspection time required per machine is much less. With predictive maintenance, the machines does not have to be disassembled for inspection. For example, with vibration analysis, 50 to 60 machines may be inspected in a single day using modern computer data collectors.

When comparing cost advantages of predictive maintenance over preventive maintenance, consider production downtime costs, maintenance labor costs, maintenance materials costs, and the cost of holding spare parts in inventory. Table 3-1 illustrates cost advantages of the three different maintenance approaches—i.e., breakdown, preventive maintenance, and predictive maintenance.

HOW TO DETERMINE WHEN TO USE PREDICTIVE MAINTENANCE

If predictive maintenance methods are superior to preventive maintenance, why use preventive maintenance at all? The answer is simple. The character of your facility will determine which methods are most effective. In actual practice, some combination of preventive and predictive maintenance is required to assure maximum reliability. The degree of application of each will vary with the type of equipment and the percent of the time these machines are operating. Pumps, fans, gear reducers, other rotating machines, and machines with large inventories of hydraulic

	FREQUENCY OF APPLICATION	MAINTENANCE COST	PRODUCTION COST	INVENTORY CARRYING COST
BREAKDOWN MAINTENANCE	NO SET FREQUENCY - AT ANY TIME	HIGHEST - DUE TO DESTRUCTIVE FAILURE, PREMIUM PAY, OT, CALL IN	HIGHEST - LONGER EQUIPMENT SHUTDOWN	HIGHEST - REQUIRES LARGEST INSURANCE INVENTORY
PREVENTIVE MAINTENANCE	BASED ON WEAR RATE - PRIMARILY COMPILED FROM HISTORY	LOW - WORK IS PLANNED AND BREAKDOWN AVOIDED	LOW - MINIMUM SHUTDOWN COSTS RESULTING FROM PM INSPECTION	LOW - BETTER CONTROL, SCHEDULED REPAIR PERMITS FEWER PARTS IN INVENTORY
PREDICTIVE MAINTENANCE	DETERMINED BY MEASUREMENT OF ASSET'S CONDITION	LOWEST - CATOSTROPHIC FAILURE AVOIDED PLANNED REPAIR IMPROVED PRODUCTIVITY OF PERSONNEL	LOWEST - INSPECTION PERFORMED WHILE ASSET IS OPERATING REDUCED REPAIR TIME	LOWEST - LONG LEAD TIMES PERMITS FORWARD PLANNING OF MATERIALS REQUIREMENTS HIGH INVENTORY NOT REQUIRED

Table 3-1. Comparison of Cost for Different Maintenance Approaches

and lubricating oils lend themselves to predictive maintenance surveillance methods. On the other hand, machines such as those which might be involved in high-speed packaging may be better inspected using traditional preventive maintenance methods. Machines which have critical timing adjustments, which tend to loosen and require precision adjustments, or have many cams and linkages which must be reset over time, lend themselves to preventive maintenance activities.

The strategy for selecting the appropriate preventive or predictive approach involves the following decision process:

- Consider the variety of problems (defects) that develop in your equipment.
- Use the predictive method if a predictive tool is adequate for detecting the variety of maintenance problems you normally experience. One or a combination of several predictive maintenance methods may be required.
- Use preventive maintenance if it is apparent that predictive maintenance tools do not adequately apply. Inspection tasks must be developed that reveal the defects not adequately covered by predictive maintenance.
- After you have decided the combination of inspection methods, then determine the frequency at which the particular inspection tasks must be applied.

Some equipment will be satisfactorily monitored using only predictive maintenance. Other equipment will require preventive maintenance. Ultimately some combination of methods will provide the required coverage for your facility to assure reliable performance. In most facilities, it is wise to apply a combination of methods to ensure that equipment defects do not go undetected.

THREE OF THE BEST PREDICTIVE MAINTENANCE TOOLS AND HOW TO USE THEM

Vibration Analysis

As was discussed above, this predictive method employs a signal generated by a machine defect as the source which indicates that a defect is under development. In other applications the predictive indicator may simply be some measurement of machine performance. The change in the indicator reflects a change in the magnitude of the defect or increase in degeneration. Interpretation of the signal permits diagnosis of mechanical health. Obvious from the name given to the method, vibration analysis uses vibration as the signal that measures the relative deterioration of machine components.

Using vibration to detect machine problems is not a new concept. The "old timer" maintenance craftsman applied a crude form of vibration analysis long before the advent of the electronic instrument. If you have been involved with maintenance for many years, you probably have seen senior maintenance people holding the blade of a screw driver against a machine surface while placing his ear to the handle of the tool. He was listening to the transmitted vibration. He understood the basic concept that mechanical objects vibrate in response to the pulsating force of a machine defect. He knew that as the defect grew in magnitude, the vibration would increase in amplitude.

Today, electronic instrumentation is available that goes far beyond the human limitations with which our old time craftsmen had to contend when trying to interpret vibration signals. These new instruments can detect with accuracy and repeatability, extremely low amplitude vibration signals. They cannot only assign a numerical dimension to the amplitude of vibration but can isolate the frequency at which the vibration is occurring. When measurements of both amplitude and frequency are available, diagnostic methods can be used to not only determine the magnitude of a problem but its probable cause.

When you use electronic instruments in organized and methodical programs of vibration analysis you are able to:

- detect machine problems long before the onslaught of failure
- isolate conditions causing accelerated wear
- make conclusions concerning the nature of defects causing machine problems

- execute advance planning and scheduling of corrective repair so that catastrophic failure may be avoided
- execute repair at a time which has minimum impact on production activities.

Vibration analysis is probably the most comprehensive and universally applicable of the various predictive maintenance methods. Its applications have been commercially developed to a sophistication and maturity that permits its successful use almost anywhere rotating machinery exists. Vibration analysis can be used in any industrial or commercial environment. If you have rotating assets which can cause financial loss if maintenance failure occurs, vibration surveillance can reduce or prevent that loss. Vibration can reveal problems in machines involving mechanical unbalance, electrical unbalance, misalignment, looseness and other degenerative problems. It can uncover and track the development of defects in machine components such as those occurring in rolling element and journal bearings, gears, belts, drives and other power transmission components. Any system deterioration which shows up as an increase in vibration amplitude can be disclosed with the use of vibration analysis.

The simplest way to illustrate how vibration works is to use a spring-mass system to represent a machine. (See Figure 3-2.) This is a good analogy because machines have mass and since they are made of metal, they have spring-like properties. When a

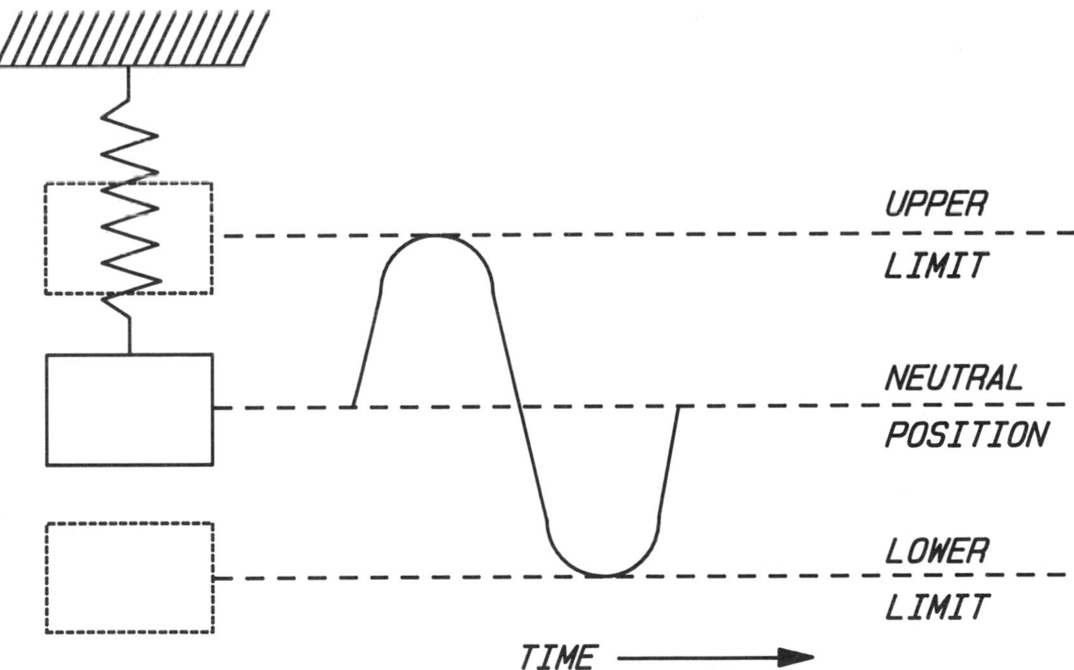

Figure 3-2. Spring/Mass Model of a Machine in Vibration

machine develops a defect, this defect acts as a force on the machine. Using the analogy illustrated in Figure 3-2, the force acts on the weight to move it in a direction that would extend the spring. The spring in turn tries to resume its original length and pulls the mass in an opposite direction. The force of the defect persists and displaces the mass once again. Thus the driving force of the defect, opposed by the spring-like action of the metal of the machine, causes a movement back and forth across some neutral point which describes the location of the mass at rest. This oscillating motion, the displacement of a point on the machine, describes vibration. If this motion is graphically represented over time, a sinusoidal plot results which is characteristic of vibration. (See Figure 3-3.)

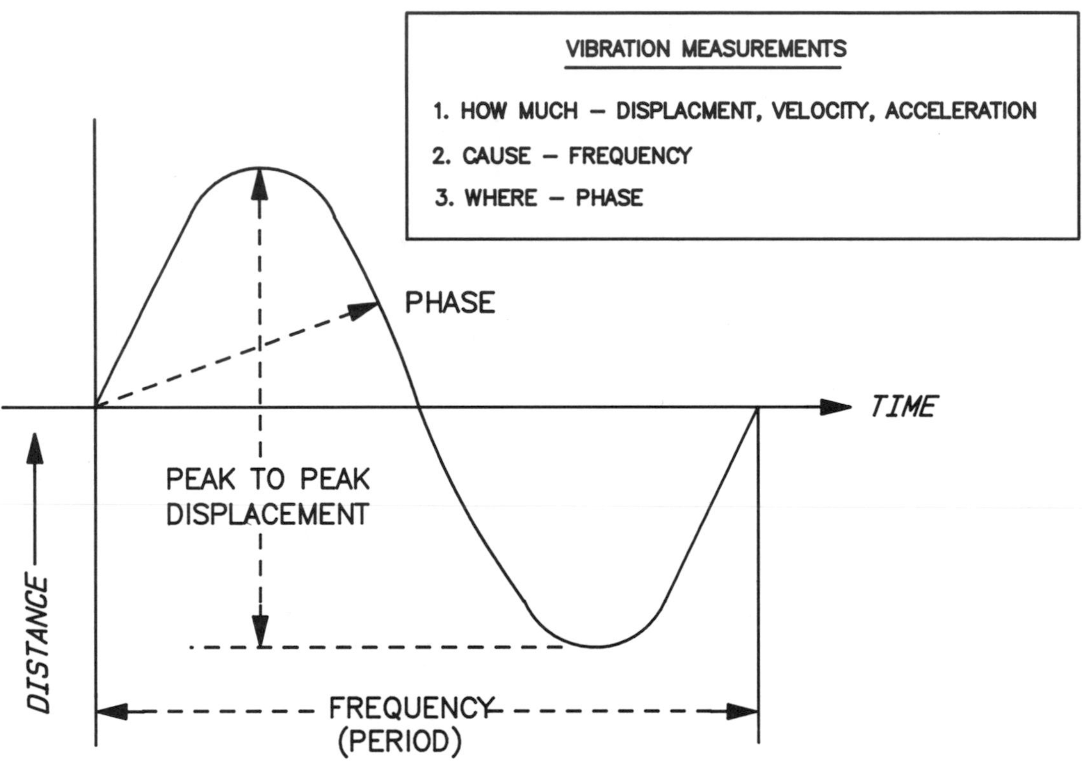

Figure 3-3. Graphical Illustration of Common Vibration Measurements

Measuring Vibration Amplitude

Using Figure 3-3, you can now define the important vibration measurements that are used in standard surveillance programs. When you want to know how much vibration exists in the machine, (amplitude of vibration) the measurement that is taken is called displacement. Displacement describes the linear motion of a point on the machine that moves back and forth across its neutral. A common practice in the United States is to measure the *peak to peak* vibration. Since displacement is a linear measurement, a linear dimension is assigned. Miles, yards, or feet could be used for displacement. However, it is obvious these dimensions are ridiculously large for machine vibration. A more practical dimension used when English units are assigned to vibration displacement is mils (1 mil = 0.001 inch).

Two other measurements are used to describe the amplitude of vibration. These are velocity and acceleration. Both are mathematically related to displacement in that velocity is the change in displacement over time (▲D/▲t), and acceleration is the change in velocity over time (▲V/▲t). Velocity is expressed as inches per second (IPS) and acceleration in "g's." The "g" refers to the acceleration of gravity at the earth's surface. Vibration producing a force twice that of gravity would be 2 g's, and a force of ½ of that of gravity would be 0.5 g's.

Under conditions of dynamic stress, displacement may be a better measurement than velocity. When parts are subject to failure due to brittleness, even though frequency may be small (thus velocity not very high), large vibrations due to displacement can be significant. Therefore displacement measurements become more important at low frequencies. Displacement measurements are generally useful at frequencies below 1,000 CPM.

Acceleration is closely related to the vibratory forces applied to the machine. At very high frequencies, these forces can be very high even though displacement and velocity are not. At high frequencies failure can occur due to the excessive force applied to the machine component. Acceleration measurements are recommended for frequencies above 60,000 CPM.

Measuring Vibration Frequency

To determine the nature of the defect causing the excessive vibration, isolate the frequency at which the major vibration amplitude is occurring. As Figure 3-3 illustrates, frequency is merely the number of cycles of vibration which occur in a given unit of time. Cycles per minute (CPM) and cycles per second (CPS) are commonly used as the unit of measurement for vibration frequency. Hertz (Hz), which is cycles per second, is also frequently used.

How Vibration Is Used to Diagnose Machine Problems

Various machine problems manifest themselves by generating certain frequencies that are related to the rotational speed of the machine and its physical construction. For example, unbalance will cause a machine to vibrate at a high amplitude at a frequency that is equal to the rotational speed of the unit. If a machine rotates at 3,600 RPM, and it is out of balance, an unbalance frequency will occur at 3,600 CPM. Thus by noting the amplitude of the vibration, the frequency, the rotational speed of the machine, the type of construction, and the direction at which the vibration pickup is placed on the machine, you can judge the nature of the problem causing high vibration. A collection of diagnostic information exists which permits the analysis of problems. The analysis of problems in a machine using vibration is analogous to a doctor when examining a patient for a disease. The doctor looks for symptoms, and then compares these symptoms with a history of information collected by the profession. A similar approach is taken by the vibration analyst, although with vibration, the complexity is obviously far less.

Measuring Vibration Phase

Another important measurement in vibration is phase. Phase tells us where in the cycle of vibration the vibration measurement is occurring with respect to some fixed reference or some other vibration.

The units of measurement for phase is degree or sometimes radian. Once again referring to Figure 3-3, phase is indicated by the arrow originating at the start of the sine wave. If the start is the reference or 0°, the arrow is pointing at an angle of approximately 140°. One cycle of vibration would traverse 360°. Phase measurements are important for two reasons:

- They permit the comparison of the relative motion of two or more parts of a machine. For example, if a machine's vibration is out of phase with its foundation, looseness of the foundation bolts or similar problem is suggested.
- They are used to balance rotating equipment. Phase shifts with a change in the center of gravity. By noting the direction and amount of phase shift, when using a trial weight for balancing, the machine may be balanced.

Diagnostic guidelines are shown in Table 3-2. Problem analysis is obviously more complex than suggested by the simple relations shown in the table.

Shock Pulse

Shock pulse is a method of surveillance that is specific to rolling element bearings. Since rolling element bearings (sometimes referred to as anti-friction bearings) are so common in machines, the method has many applications. A secondary but important feature of the shock pulse method is that it permits maintenance workers to judge the adequacy of the lubrication program applied to this type of bearing.

The shock pulse instrument measures a transient high frequency wave produced when the moving elements in a bearing strike a defect and release mechanical energy. The device to measure this transient signal, which occurs over an extremely short time (several microseconds), is a piezo-electric accelerometer. The accelerometer measures the initial compression (shock) wave due to the impact before detectable deformation takes place which generates wave impulses normally measured with vibration instruments. The transducer is mechanically and electronically tuned to a frequency of 32 KHz and thus differentiates this type of signal from that of normal vibration.

The shock pulse instrument electronically filters out other signals and compares the peak amplitude of the "shock" signal with empirical data. The empirical data is based on "shock" measurements of similar bearings, which are properly installed, properly lubricated and in good condition. The difference between the signal produced by the good bearing referred to as dB_i and the signal from the measured bearing identified as dB_{sv}, indicates the severity of damage to the measured bearing. The

MACHINE PROBLEM	VIBRATION AMPLITUDE	VIBRATION FREQUENCY	PHASE AS SHOWN BY STROBE
UNBALANCE	V OR H READING MODERATE	1 X RPM	STEADY, SINGLE MARK
MISALIGNMENT	V OR H READING MOD., AXIAL 0.5 TO 1 OF V OR H	1 AND 2 AND SOMETIMES 3 X RPM	OFTEN 2 MARKS
DEFECTIVE ROLLING ELEMENT BEARING	LOW, STEADY	6 TO 15 X RPM UNSTEADY	MANY MARKS
BENT SHAFT	V, H AND AXIAL MODERATE	1 OR 2 X RPM	1 OR 2 MARKS
DEFECTIVE GEARS	LOW	EQUAL TO NUMBER OF TEETH X ENGAGEMENTS PER MINUTE	
MECHANICAL LOOSENESS	MODERATE TO HIGH	1 OR 2 X RPM	SINGLE OR DOUBLE USUALLY UNSTEADY
ELECTRICAL	LOW TO MODERATE	POWER LINE FREQUENCY X 1 OR 2 (3600 OR 7200 CPM)	
AERODYNAMIC OR HYDRODYNAMIC FORCES	LOW TO MODERATE	1 X NUMBER OF FAN BLADES, VANES OR PISTON STROKES	

Table 3-2. Diagnostic Guidelines for Vibration Analysis

measurement which states the condition of the bearing is identified as dB_n. Thus the equation which provides bearing condition is empirically based as follows:

$$\text{BEARING CONDITION} = dB_n = dB_{sv} - dB_i$$

Based on statistical data, a maximum value below $20dB_n$ is interpreted as a low probability of bearing damage, and a maximum value of $35dB_n$ is considered as a high probability of bearing damage. (See Figure 3-4.)

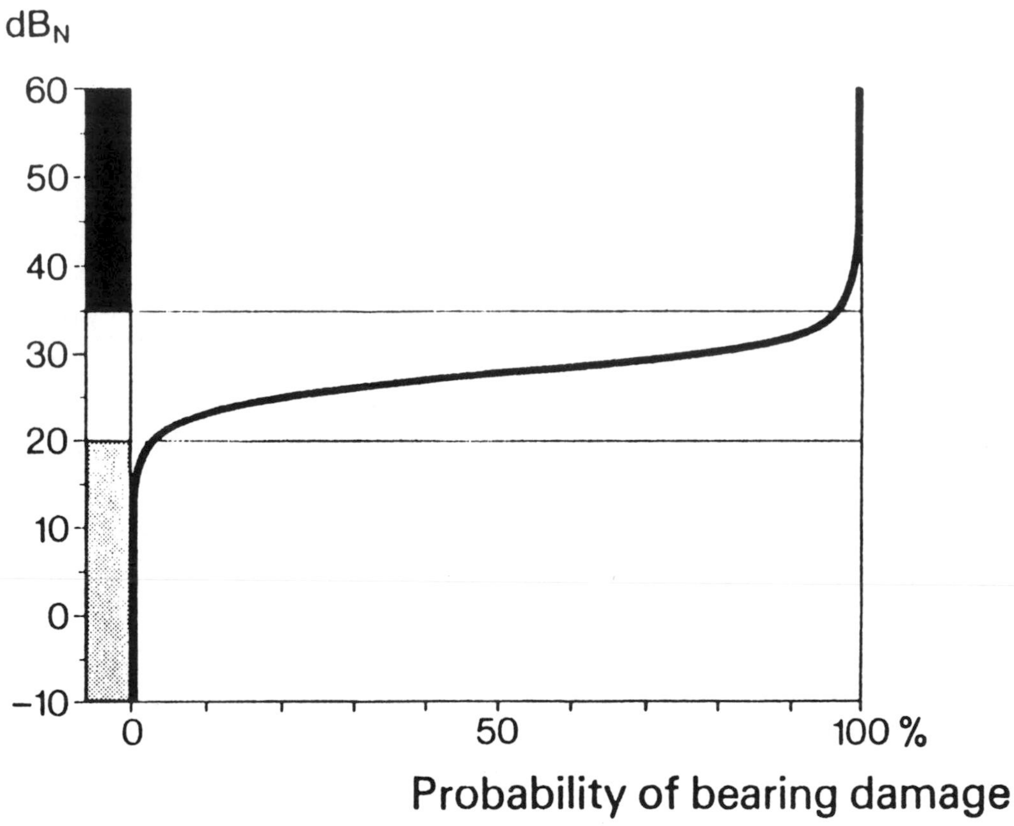

Figure 3-4. Shock Pulse Reading Versus Probability of Bearing Damage—*Courtesy of SPM Instrument, Marlborough, CT*

Several models of the shock pulse instrument are available. The more recent models of SPM instruments measure the quality of lubrication in undamaged bearings by analyzing the shock pattern put forth by the bearing. When a measured bearing is adequately lubricated and has a good lubrication film thickness, the moving surfaces of the bearing are separated sufficiently to reduce the incidence of impact. If the lubrication film thickness is inadequate, the surfaces are more likely to strike each other and thus the output of shock impact energy increases. The built-in microcomputer in these instruments analyzes the shock pattern and reports two factors pertaining to the quality of lubrication. These are "Condition Number" which relates to the oil supply to the bearing, and "Lubrication Number" which is a statement on the thickness of the oil film that carries the load.

Spectrometric Oil Analysis

Oil in machines carries the products of deterioration resulting from wear and mechanical failure. By analyzing the oil resident in a machine, or the debris the oil carries, you can predict the state of health of that machine. The critical measurement reflecting the condition of machine wear is the amount of microscopic metal wear particles that are suspended in the oil system of the machine.

A number of laboratories throughout the country perform these analysis for client locations at a nominal fee ranging from perhaps $10.00 to $35.00 per sample depending on the number and types of analysis performed on that sample. The sampling price usually includes a sample bottle and a mailing package that permits both ease of collection and handling. The analysis results are returned to the client by mail with comments and recommendations. If an extremely severe wear process is detected which requires immediate action, the lab will telephone the client.

The spectrometric oil analysis process, sometimes referred to as SOAP, is a laboratory technique which uses various instruments to analyze a used oil sample from a machine. Several instruments are used in the process to determine not only the status of the wear process but also the condition of the oil (i.e., whether it needs to be changed or purified).

A spectrometer is the instrument that is used to show when a significant wear mode is underway. Some varieties of these instruments can isolate up to perhaps 80 different types of metals in the sample at levels as low as one part per million. The spectrometric result is compared to a base line level of metal found to be typically suspended in the oil under normal operating conditions. When the wear is meaningful, the sample will show high levels (in parts per million) of wear metals compared to the base line oil sample.

An important feature of the spectroscopic method is that it not only determines the amount of metal in the sample but also the type. Using a knowledge of the machine construction, you can match a particular metal type found to be present in the sample to the machine component undergoing excessive wear. Thus the analysis not only permits the discovery of severe wear but analysis of the possible location of this process in a machine. With this information you can take timely action to prevent further deterioration, either by oil purification, oil replacement, or some other means appropriate to the problem.

Two illustrations may be provided to show how spectrometric oil analysis might work.

Example 1—A machine containing a hydraulic system shows high concentration of iron particles in the sampled hydraulic fluid. This high concentration suggests that a severe wear mode is underway, probably in the hydraulic pump because it is made of iron. The total particle count and size in the hydraulic fluid is also noted. Large amounts of large particles (15μ or larger) are indicated. (See sample date 9/22/90 in Figure 3-5.) The advice provided by the SOAP lab is:

WEAR CHECK®

C6

OIL SAMPLE ANALYSIS REPORT

Spectro/Metrics, Inc.
35 Executive Park Drive, N.E.
Atlanta, Georgia 30329
(404) 321-7909

HYDRAULIC SYSTEM

MACHINEMANUF.-> CINCINNATI 500
PRESSURE------> 3000 PSI

PUMP-MANUFACT.> VICKERS
FILTER-TYPE--->

UNIT-NUMBER---> 11
OPERATINGTEMP.> 110 F

SAMPLE DATA	WEAR METALS AND ADDITIVES (in parts per million)																	CONTAMINANTS				PHYSICAL TESTS		
	ALUMINUM	CHROMIUM	COPPER	IRON	LEAD	TIN	NICKEL	SILVER	SILICON	SODIUM	POTASSIUM	BORON	MAGNESIUM	MOLYBDENUM	CALCIUM	PHOSPHORUS	ZINC	WATER %	PART WGT.			VIS. 40C	TAN	ISO CODE
Sample Date: 10/27/90 Sample No.: 044787 Lab No.: 200076 Total Mis/Hrs: 680 Mis/Hrs on Oil: 0 Mis/Hrs on Flt: 623 HOURS	1.0 N	1.0 N	1.0 N	6.3 N	2.0 N	1.2 N	0 N	0 N	1.0 N	2.1 N	0 N	0 N	0 -	0 -	0 -	251 -	512 -	<.03 N	12.1 (MG)			68 N	N/A	17 /13
	PARTICLE COUNTS: >5MICRONS=1300 >15MICRONS=68 >25MICRONS=10 >50MICRONS=1 >100MICRONS=0																							
	IRON WEAR RATES NORMAL. AFTER ADDITIONAL FILTERING, GENERALLY ACCEPTABLE AMOUNT OF PARTICLES PRESENT. SYSTEM AND FLUID CLEANLINESS SATISFACTORY. RESAMPLE NEXT SERVICE INTERVAL TO FURTHER MONITOR.																							
Sample Date: 09/22/90 Sample No.: 042364 Lab No.: 200075 Total Mis/Hrs: 480 Mis/Hrs on Oil: 0 Mis/Hrs on Flt: 403 HOURS	1.0 N	1.0 N	7.0 N	71 A	7.0 N	7.2 N	0 N	0 N	3.1 N	2.0 N	0 N	0 N	0 -	0 -	0 -	236 -	475 -	<.03 N	24.1 (MG)			67 N	N/A	21 /18
	PARTICLE COUNTS: >5MICRONS=10223 >15MICRONS=2490 >25MICRONS=300 >50MICRONS=71 >100MICRONS=8																							
	IRON LEVEL ABNORMAL. VERY HIGH AMOUNT OF PARTICLES (5->100 MICRONS) ARE CREATING WEAR. ADVISE FILTER SERVICE AND USE OF OFF-LINE PURIFICATION EQUIPMENT. RESAMPLE NEXT SERVICE INTERVAL TO FURTHER MONITOR.																							
Sample Date: Sample No.: Lab No.: Total Mis/Hrs: Mis/Hrs on Oil: Mis/Hrs on Flt:																								
Sample Date: Sample No.: Lab No.: Total Mis/Hrs: Mis/Hrs on Oil: Mis/Hrs on Flt:																								

ACME UTILITIES
ATTN:MAINTENANCE MANAGER
STREET ADDRESS
CITY, STATE 99999

COMPUTER-CODE--> ANYANY 11HY

MICROMETERRATING-->
SYSTEM CAPACITY---> 26 GALLONS
OIL MAKE AND TYPE-> CHEVRON AW 68
HISTORY/REMARKS--->

A service of Spectro/Metrics, Inc., a member of the Spectrometric Oil Analysis Laboratory Association

LEGEND:
N = Normal
A = Abnormal
S = Severe
> = greater than
< = less than
Neg = Negative
Nil = Negligible
N/A = Not Applicable

-->VC<--DATAENTRY RGT

EXPLANATION OF YOUR REPORT

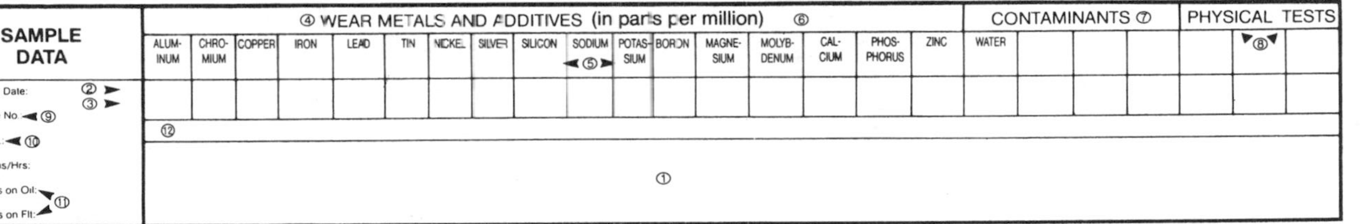

SAMPLE DATA	④ WEAR METALS AND ADDITIVES (in parts per million) ⑥																	CONTAMINANTS ⑦				PHYSICAL TESTS	
	ALUMINUM	CHROMIUM	COPPER	IRON	LEAD	TIN	NICKEL	SILVER	SILICON	SODIUM ◄⑤►	POTASSIUM	BORON	MAGNESIUM	MOLYBDENUM	CALCIUM	PHOSPHORUS	ZINC	WATER				▼⑧▼	
Sample Date: ②► ③►																							
Sample No. ◄⑨	⑫																						
Lab No. ◄⑩																							
Total Mis/Hrs:	①																						
Mis/Hrs on Oil: ⑪																							
Mis/Hrs on Flt:																							

1. A valuable part of your report is the Diagnostic statement which tells you, in plain language, if you have a problem, the level of severity, the probable cause, and maintenance action recommended to correct the problem.
2. The values under Wear Metals and Additives are in parts per million (1 ppm = .0001%) and are used to track the wear of component parts and measure the amounts present. The numbers become relative only when compared to Sample Data information such as Total Hours on Unit, Hours on Oil, Hours Since Last Filter Change and Overhaul History. The wear metals can then be interpreted as a rate of wear.
3. Under each ppm value, there is "N" for NORMAL, "A" for ABNORMAL, or "S" for SEVERE. Usually "A" indicates unit should be monitored closely for possible increased wear rates. "S" means immediate action is necessary.
4. WEAR METALS are the key metals worn from the friction surfaces of your system. As your unit runs, bacteria-size particles are produced. The oil holds them in suspension. Energizing a sample of oil in a spectrometer produces a measurement in ppm for each of the Metals, Additives, and Contaminants simultaneously. Examples are:
   Aluminum: Pistons, blocks, pumps
   Chromium: Piston rings, roller/taper bearings, coolant additives
   Copper: Bearings, oil cooler, clutches, thrust washers, thrust plates
   Iron: Crankshaft, valve train, cylinders, gears, bearings
   Lead: Bearings, additives in some fuels, oil additives
   Tin: Bearings, pistons
   Nickel: Exhaust Valves
   Silver: Bearings, wrist-pin bushing
5. ADDITIVES/CONTAMINANTS:
   - SILICON may indicate the presence of dirt, (a contaminant) which is the most common cause of system wear. It may also be a sealant or an indicator of the presence of a silicone based additive in the lube oil and/or coolant.
   - SODIUM can be found as an additive in engine coolant but is also frequently used as a lube oil additive.
   - POTASSIUM and BORON, also common coolant additives.
6. ADDITIVES such as magnesium, molybdenum, calcium, phosphorus and zinc are ingredients that can be blended into the oil to improve its life and performance.
7. CONTAMINANTS:
   - WATER in light concentrations can be condensation due to a cold running system. High concentrations may be from the cooling system or outside environmental contamination.
   - GLYCOL is in the formulation of most commercial antifreezes. Its presence in the oil indicates a serious coolant leak. Under certain conditions, glycol may no longer be present in the oil due to evaporation and chemical change.
   - SOOT is an indication of the buildup of dispersed carbon that has occurred due to the excessive blow-by in the engine. The value given is a relative measure reported in absorbance units using an infrared spectrometer.
   - FUEL in the oil indicates faulty combustion, rich air/fuel mixture or poor injection.
8. PHYSICAL TESTS:
   - VISCOSITY is a measure of the oil's resistance to flow indicating if the oil is too thick or too thin. Oil thickens with use due to oxidation and contamination, but thins with fuel dilution.
   - TOTAL ACID NUMBER (TAN) is a measure of oxidation.
   - TOTAL BASE NUMBER (TBN) is a measure of the oil's alkalinity or "acid resistance" provided by certain additives in the oil.
   - PENTANE/TOLUENE INSOLUBLES AND RESINS are insoluble materials including dirt, soot, wear metals, oxidation resins, fuel carbon, and carbonized materials from oil and/or additive degradation.
   - OXIDATION is a measurement of the reaction of oxygen with oil under high system operating temperatures. The value given is reported in absorbance units, using an infrared spectrometer.
   - NITROGEN FIXATION is a measured value mainly for natural gas engine and compressor applications. A rapid increase in this value usually indicates an imbalance of fuel-air ratio or improper inlet/outlet crankcase oil temperatures. The value given is reported in absorbance units, using an infrared spectrometer.
9. SAMPLE IDENTIFICATION CODE NUMBER. This number is printed on the information form that accompanied the sample and appears on the tear off stub for your record keeping.
10. A number we assign the sample bottle for internal processing at our facility.
11. Miles or hours that the oil and oil filter (if applicable) have been in service.
12. PARTICLE COUNT:
    Number of particles per 1 ml of sample.
    Number and size of particles > 5, > 15, > 25, > 50, and > 100 microns.

Your computer code at the bottom of the report is a combination of company name or surname, address, city and unit/vehicle number. It is essential for communication and to link histories of successive samples. Always use the same name and address. If your address changes, please inform us.

For specific system information, contact our Technical Service Department at 404-321-7909.

VISCOSITY EQUIVALENTS

KINEMATIC VISCOSITIES cSt/40C cSt/100 C | ISO VG | AGMA GRADES | SAE GRADES CRANKCASE OILS | SAE GRADES GEAR OILS | SAYBOLT VISCOSITIES SUS/210 F SUS/100 F

GENERIC (4/90)

Figure 3-5. A Sample Spectrometric Data Sheet—*Courtesy of Spectro/Metrics Inc., Atlanta, GA*

that a severe wear mode is in progress, and that the client should change filters, purify the hydraulic oil off-line and re-sample after next service interval.

The facility follows the lab's recommendations. Results from the sample collected after off-line purification (sample date 10/27/90) show that iron wear concentration is now at a normal level and particle count and size is acceptable. The lab recommends that the oil be sampled at the next service interval to continue monitoring.

A data sheet illustrating how the results of the above oil analysis might look is shown in Figure 3-5.

Example 2—An oil sample from an air compressor shows a high level of silicon that is not part of any additive in the oil formulation. The SOAP lab concludes that the silicon is dirt (silicon dioxide) originating from the atmosphere around the compressor. They suggest that it is entering the system through a leaking air filter. The oil sample also contains a high amount of iron, showing a severe wear mode is under way in the compressor. If something is not done soon, premature failure may result.

The facility shuts down the compressor and inspects the air filter. It is discovered that a gasket on the filter housing has failed. The gasket is replaced along with a new filter element and the oil in the machine is changed. A subsequent oil sample analysis by the SOAP lab reveals that wear metal particle level in the oil is normal and the problem has been corrected.

It is common for the oil analysis laboratories to maintain a large data base of their findings in computer files. Thus wear metal levels in machine systems operating with normal wear may be statistically evaluated. These same laboratories will also build a history on machines unique to your facility. After a number of samples that establish a baseline they may make judgments concerning normal versus abnormal wear conditions for distinctive equipment you may have.

Standard Oil Analysis

In addition to the spectrometric analysis the oil laboratories also check the oil using common oil analysis techniques. For example, the oil is usually checked for viscosity. If the viscosity of the oil has changed 5% from new oil of the same type, it is probably time for an oil change due to contamination. Typical of these types of analysis are: total acid number, percent moisture, particle count (for hydraulic systems), total solids, percent silicon (representing dirt from the atmosphere in the form of silicon dioxide or perhaps just from an additive). With prior agreement, special tests are performed at additional cost.

The relatively low costs of SOAP makes it a very valuable and commonly used reliability method. It is practical for machine systems that have a reasonable inventory of oil and are provided with "reuse" methods of oil application such as circulating, bath, splash, flood, and ring oiling designs.

In addition to providing advance warning when severe wear occurs, SOAP methods can give important assistance in machine lubrication programs.

Conventional oil change frequencies are arrived at according to the traditional preventive maintenance approach. A best estimate (or guess) is made of the time or calendar period over which the oil in the system will degenerate and thus require changing. SOAP enables you to change oil only when the actual condition of the oil requires! This approach can save you a great deal of money since you do not have to change oil unless analysis shows it is degenerated beyond safe use. By prolonging the useful life of the oil you obviously reduce your oil change labor and oil replacement costs. You also save on oil disposal costs since there will be less oil to discard over time.

Other Predictive Maintenance Techniques

There are a large number of predictive maintenance methods available to you. For example a more recent oil analysis method called Ferrographic Oil Analysis, uses a magnetic technique to isolate ferromagnetic particles suspended in the oil sample. This method has greater sensitivity than spectrometric analysis because it considers not only wear metal quantity but particle size in predicting the advance of severe wear. More recently, oil laboratories are offering a combination of spectrometric and ferrographic tests in order to maximize the potential of each method.

Infrared and ultrasonic devices may be employed both as non-destructive testing tools and in predictive applications. As time goes on and our technology advances, the opportunities for the use of new monitoring processes increases.

The methods discussed above however are only a few of the more commonly applied predictive maintenance systems. You do not have to restrict yourself to a commercially available conventional technique. Since the predictive system is based on the measurement of some signal or output which is indicative of the condition of the machine you are monitoring, many possibilities exist for original applications. Within your organization there may be potential for developing monitoring methods which are distinctive and one of a kind. The particular applications that are best for your facility can vary significantly and will depend on your own inventiveness and initiative. To be effective you merely have to apply the predictive maintenance concept in a manner which best fits your facility and equipment characteristics.

For example, one facility developed an in-house method using a common oscilloscope. The shape of patterns displayed on the oscilloscope provided a signal when a system in a sophisticated computer directed manufacturing operation had to be recalibrated. The operation involved a precision machine which fastened gold wire to form terminations on micro chips. The computer directed the mechanism in a "X", "Y", and "Z" axis so that the termination could be bonded to a tiny point on the computer chip. This bonding action had to occur at the precise moment the chip was positioned under the head of the bonder by a conveyor. Over time the electronic technician responsible for maintenance of this area noticed that a significant number of product defects were manufactured before the machine (one of several) was put out of service for recalibration. Concluding that this was an expensive way to discover the problem, he devised a method to detect significant drift before a large number of defects were produced. He retrofitted the control computer with a multipin connection so that at regular intervals he could observe the output signal from the control circuit. He eventually learned the normal and abnormal patterns of the output signal which was displayed on the CRT of an oscilloscope. The abnormal pattern revealed when recalibration was necessary. This unique idea thus saved many thousands of dollars by reducing off-specification product. It permitted swifter reaction by the maintenance department so that they could help reduce the incidence of quality problems. The system the technician devised was a model example of predictive maintenance since he was forewarned of an approaching problem while the bonder was operating and producing product. He was able to correct the defect with minimum impact on the operation.

HOW TO IMPLEMENT A PREDICTIVE MAINTENANCE PROGRAM

Since vibration analysis is the most universally applicable predictive maintenance technique, we will focus on this method when discussing how to implement a predictive maintenance program. If you have rotating equipment in your facility, you have an application for vibration analysis.

The procedure to follow in the start of any predictive surveillance program is probably best described by the flow chart shown in Figure 3-6. As the illustration suggests, the first step in program start-up is to select those items of your facility's equipment which will be included in the monitoring program. More will be said below about equipment selection.

Once having selected equipment, the next step is to choose the appropriate surveillance method. When selecting the surveillance method you must first consider the kinds of defects and problems that develop which must be avoided. Having defined these defects and problems, you are then in a position to decide which of the predictive maintenance methods will best do the job for those problems in that particular asset under consideration. Perhaps you will require more than one predictive method for complete surveillance assurance. For example, it may be desirable to conduct both vibration analysis and oil analysis inspections on a major, high, horsepower gear reducer in your facility.

This would provide you with high confidence that you are providing full protection to your equipment. You would not only detect early signs of mechanical degeneration in the system through the use of vibration analysis, but perhaps have warnings of severe wear before damage occurs through the use of oil analysis.

With some items of equipment, it may also be necessary to conduct some traditional preventive maintenance task inspections in conjunction with predictive applications. As shown in the flowchart, preventive maintenance should be used only if predictive applications can't do the entire job. This is a logical approach because pre-

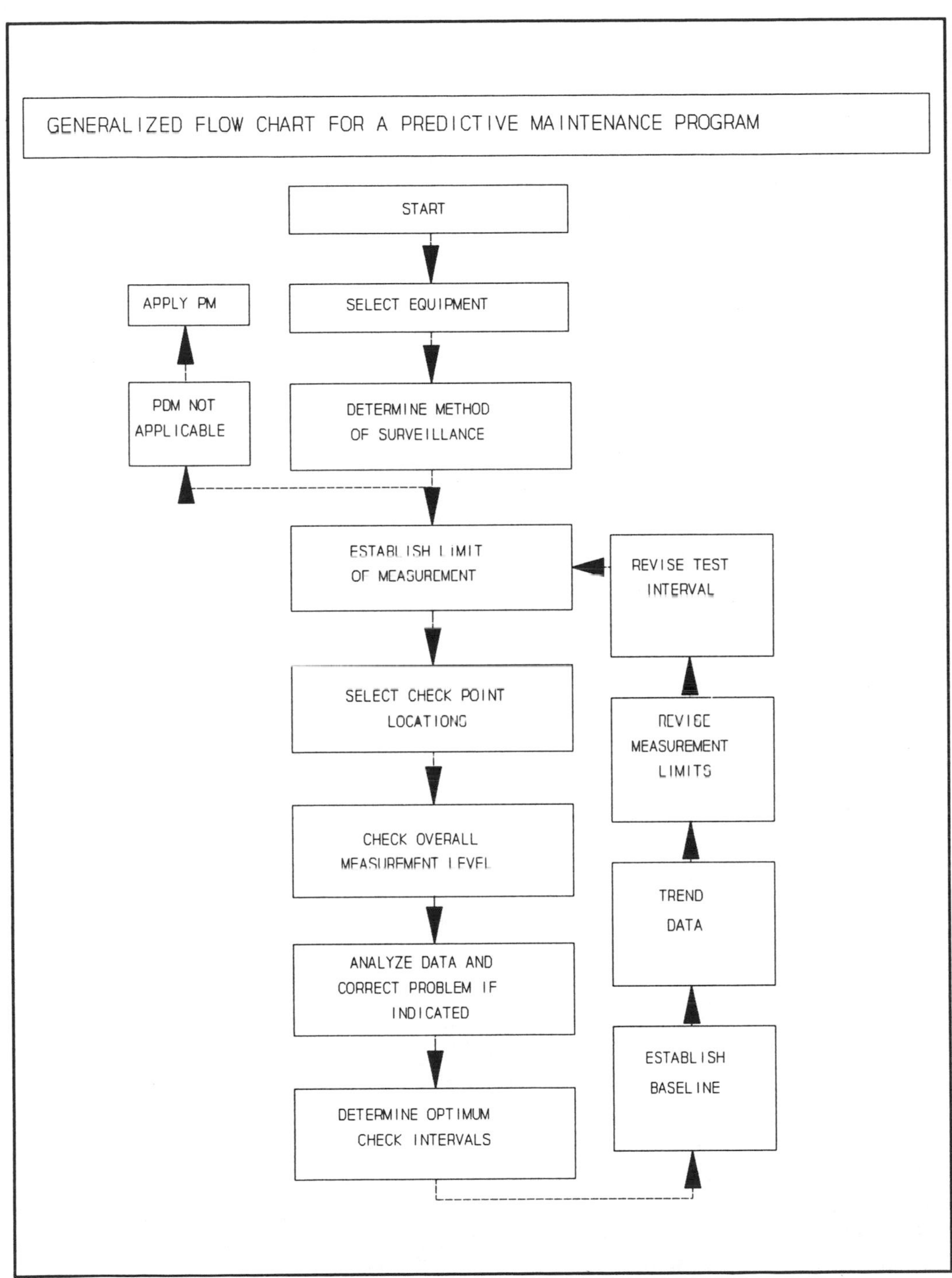

Figure 3-6. Flow Chart for Implementing a Predictive Maintenance Program

ventive maintenance activities are more costly than predictive maintenance. It is possible, however, that because of some unique wear modes experienced by a unit, supplemental preventive maintenance inspections will also be required.

Selecting the Proper Instruments

Before proceeding further with a discussion of how to establish a vibration program, you should be aware of the general classes of vibration instruments that are available to you. When you buy an instrument you must consider the cost of the instrument versus the return it will provide. You should also understand how the features of the instrument can be appropriately applied in your facility. It can be a needless waste of resources to purchase an instrument which is beyond the sophistication needs of your facility. I have seen many situations in which expensive instrument purchases were made without personnel to operate them. As a result, the instruments remained idle and program interest and efforts faded. The potential for considerable savings in operating costs was lost.

A large variety of vibration instruments are sold in the market today. This diversity of kind may be generalized into four classes of instruments. It must be remembered, however, that by simplifying to four classes I am overlooking the very great variations that exist within these classifications. The four classes of instruments are:

- Meters
- Analyzers
- Monitors
- Automated Data Collectors

Meters

A meter is a vibration instrument that reads the vibration signal across a wide range of vibration frequencies. The meter has no capability of discriminating vibration amplitude by frequency.

The primary purpose of the meter is to screen machines which may have mechanical problems from those that are operating satisfactorily. An individual using the instrument would follow a pre-defined collection route and collect and record vibration data from the machines included in this route. If the vibration signal collected shows some irregularity, analysis of that apparent problem must be made by another type of vibration instrument called an "analyzer." Analyzers will be discussed later.

Meters are generally very portable and are battery operated. They indicate what the total vibration signal is by either an analog or digital display or perhaps both. The full scale range of the display and the units of measurement for the vibration can usually be adjusted with controls on the body of the instrument. These controls revise the electronic circuits in the unit to integrate the input signal and calculate the requested units of measurement. With the appropriate instrument the turning of a dial will permit the vibration reading to be displayed as displacement, velocity, or acceleration.

The meter is a basic tool in vibration analysis and has application in surveillance programs in which data is collected manually. Because of their relatively inexpensive price, perhaps in the $1,000 to $3,000 range, they have application in smaller facilities. Meters are being rapidly displaced as a functional instrument by the considerably more sophisticated automated data collectors, discussed below. The automated data collector eliminates the need for manual recording of data and provides other powerful advantages over the meter.

Analyzers

Vibration analyzers, as the name implies, are instruments which permit analysis of the nature of the vibration signal and thus the fundamental mechanical problem causing it. When you know at what frequency of vibration the major amplitude of vibration is occurring, you have fundamental data which you can use to analyze the defect.

The principal difference between the meter and the analyzer is that the meter can discriminate between vibration frequencies. Analyzers permit either the display or plot of frequency versus amplitude graphs. They also can display amplitude over time, referred to as "time domain" plots. The distinction however can be much greater since there is a very large variety of features available in analyzers. Modern technology has produced analyzers of imposing sophistication. For example, analyzers may have the following features:

- a backlit supertwist LCD display with a viewing area of 225 × 100 mm,
- in addition to vibration inputs, one for an infrared temperature probe,
- frequency analysis using FFT calculations,
- field balancing capabilities with a strobe or photocell,
- internal memory for balancing calculations,
- phase data display,
- an internal printer for strip charts, and/or
- on-line help.

This type of analyzer permits you to take a sophisticated instrument into the field and make a thorough analysis of problems which may have been detected by a simple monitor. The monitor alerted you to an anomaly; the analyzer permits you to collect diagnostic data so that you can decide what corrective action must be taken, and may also permit field balancing of rotating equipment. This can be a very important feature since a large percentage of vibration problems are due to unbalance.

Monitor

The monitor is really an application in which the vibration instrumentation is dedicated to one or several items of equipment. It is in reality a system of instruments. In this arrangement the vibration pickups are fixed to the machines they are monitoring. Via cable, their outputs are directed to a panel containing equipment which can scan the multiple signals from the various pickups. The monitoring instrumentation conditions the signals received and sends them to other devices which store, display and alarm. Thus the monitoring system provides full time surveillance on a single or group of machines. If an alarm level is reached an operator who would be in attendance of the equipment, would be immediately alerted to a problem. Since monitoring installations are expensive, the critical nature of failure in the units protected must justify the cost.

Automated Data Collectors

Automated data collectors are battery operated vibration instruments which have companion microcomputers housed in the instrument package. The unique feature of the automated data collector is that the instrument can store and carry into the field vibration collection information. When a vibration signal is read in the field, the instrument accepts and stores that input, eliminating the need for manual recording of the data. The automated data collector (ADC) does more than this however. The ADC is part of a system that consists of a host computer, software, and peripheral devices such as printers and secondary storage media. When a technician is ready to collect vibration data in the field, he connects the ADC to the host computer and transfers key information to the ADC. He calls up the appropriate software program and downloads to the ADC information on the route he is scheduled to follow. For example, the information that will be transferred to the ADC will be similar to the following:

- the identification of all machines that are included in the collection route,
- the identification of the points on the machine from which a vibration reading is to be taken,
- the unit of vibration measurement to be read (mils, IPS, g's, and so on). With some instruments manufactured, several types of readings are collected simultaneously,
- the vibration limit or limits for that check point,
- the previous reading or readings for that check point, and
- other data to be collected such as process parameters, observations of machine problems.

After the technician finds the first check point, he or she places the transducer on the machine, notes if the vibration signal is reasonable, and then presses an enter key to collect the vibration signal. If the current reading exceeds the pre-established vibration limit, or is significantly larger or smaller than the previously collected reading by a predefined amount, a FFT is taken. The FFT is a mathematical process which permits the collection of vibration amplitude over a range of vibration frequencies. This information is some times referred to as a spectral plot. The spectral information is stored in the instrument along with individual data readings.

When the technician has completed his route, he takes the ADC back to the host computer and transfers his collected data to the computer's data files via a direct link through a computer input port. With the appropriate software he can store, report, and analyze the information.

The sophistication of analysis provided by the ADC software makes the ADC a very powerful instrument. Assisted by powerful analysis software, the analyst is more likely to make accurate diagnoses of machine problems, and/or detect problems that might other wise go undetected until failure. This returns real savings in terms of fewer catastrophic failures and increased equipment operating time.

Another significant advantage with this instrument is the speed with which responses may be made to machine problems. Since data and machine history is stored in computer files, information retrieval is much faster than when the same details are on paper. The information which can be retrieved is also formatted in a manner which makes isolation of facts faster. For example, a common report available in such software is the "exception report." Of the hundreds of data points for which vibration signals were collected in a particular route, the exception report will only print or display those data points for which an anomaly is detected. In this manner the technician does not have to scan large tables of numbers and can concentrate immediately on the problem areas.

Beyond this there are cost savings produced by the speed and accuracy of data collection. Field collection is much faster than with the use of a meter since manual recording of the data is made unnecessary. The number of data points that can be collected in a day with the ADC is perhaps 2 to 3 times the amount collectable with manual systems.

Although automated data collectors are more expensive than meters they often can be justified in terms of the above stated advantages. Even smaller organizations should consider the purchase of an ADC, considering the advantages. Care must be taken however. These units are in the approximate cost range of $12,000 to $20,000 depending on the associated software and training provided. This does not include the host computer. If you are going to spend that amount of money, you want to be sure that someone in your organization will put the instrument to work and that it will not remain idle on the storeroom shelf. If you use it properly, it will return your investment many times over.

EIGHT STEPS FOR IMPLEMENTING A VIBRATION ANALYSIS PROGRAM

Step 1. Select Equipment to Be Monitored

Referring to Figure 3-6, the first step in any predictive maintenance program is to select the assets to be included in the program. Although a payout can probably be justified for even moderate size horsepower equipment, a limit must be placed in an initial program. When embarking on a new surveillance method like vibration analysis, take care that you are not overwhelmed with too much data. You should start with a relatively small and manageable group of machines. Limit yourself to critical equipment. When you have resolved the problems and are comfortable with the successes of the prototype program, then expand your efforts to a second phase effort and beyond.

How to Select Candidate Equipment

Critical equipment to be selected for the first phase predictive maintenance program can be ranked for selection using four evaluating considerations. These are:

- The impact that failure of a candidate unit has on other equipment. If the machine is one of a train of equipment in a continuous operation and its failure means it will shut down an entire system or operation it must be ranked high for selection in the program. For example, a utility such as an instrument air compressor can affect an entire facility if it fails because all pneumatic controls in the facility cease to operate.
- The amount of machine capacity used—consider redundancy or installed spares. If a machine is one of a kind, and is functioning at full capacity 24 hours per day, 7 days per week, 52 weeks per year, it is a good candidate for selection. Under such conditions, lost production due to mechanical failure cannot be made up. Thus machines used at full capacity should be ranked higher than those which are not loaded in that manner. A one-of-a-kind machine should be ranked higher than machines which have installed redundancy.
- The profitability of that equipment or machine. A candidate unit would rank high for selection if a serious profit loss occurs when it fails. Perhaps equal to profit loss would be the impact failure would have on delaying fulfillment of the facility's mission.
- Other factors such as safety, environmental, government mandated requirements, repair history, and so on, should also be considered in this ranking process.

Some factors cannot be evaluated on the basis of cost alone. If a unit's failure increases unsafe exposure or hazards the environment, it should receive a high ranking for selection in the monitoring program.

Another consideration when ranking equipment is its frequency of failure. Obviously if a unit frequently breaks down it is contributing to high maintenance costs. Reduction of breakdown through predictive maintenance control will provide a program return.

Another factor to consider is whether the equipment has a high potential for demonstrating success. If you do not have all the resources you would like to have for vibration analysis activities, and have yet to fully convince management on the viability of the method, you need evidence to prove the effort is practical. You might select a unit because it has a high potential for demonstrating program success. Those machines with high downtime may also be good candidates in this latter case. In this respect you should try to select equipment for which you have good cost and repair documentation. This data would be essential as base line reference when you attempt to show cost reduction improvements.

Step 2. Establish Vibration Limits

A second step in setting up your predictive maintenance program would be the selection of reasonable vibration limits which define the boundary between normal operation and the need to take corrective action.

Every predictive maintenance program has three basic steps. These are:

- Detection—Under normal surveillance, you collect vibration readings periodically from the equipment. When a unit vibration signal exhibits some anomaly, you have detected a possible machine problem. This constitutes the detection phase.
- Analysis—When a problem is noted, it is then necessary to ascertain the underlying cause of the abnormal vibration signal. This may require the collection of additional information. Quite often if not usually, the vibration analyst takes his vibration analyzer to the machine in the field and collects data at the source in order to isolate the cause or causes of the problem. In many cases there may be a combination or "stack up" of problems which create the excessive vibration signals. The high vibration signal may be a machine defect in development or may simply be due to some aberration such as an error in taking the reading. If the cause is a mechanical problem, then the appropriate corrective action must be planned. The process from problem discovery to corrective plan can be considered the analysis portion of the predictive application.
- Correction—Finally, and the point of the whole effort assuming there is a problem, is correction of the condition producing the machine problem. It is thus necessary to perform some repair function on the unit if the program is to fulfill its role of avoiding catastrophic failure or reducing or retarding the impact of failure. The repair may be as simple as an in-place balancing of the rotating element of the machine, or as complex as a complete dismantling of

the machine for component replacement. This step constitutes the corrective phase of the predictive maintenance process.

The establishment of a vibration limit for each check point on a machine is therefore a critical program element. A vibration reading above this limit is the signal for the start of the three-phase process.

Determining How Much Vibration Is Too Much Vibration

"How much vibration is too much vibration" is a question often asked by beginners in a vibration analysis program. Unfortunately the answer to this question is not simple. Several observations must be made when attempting to answer this question.

First, keep in mind the objective of the program. Remember that the program's primary purpose is to provide sufficient advance warning of a machine problem so that you can schedule and execute corrective action before catastrophic failure. The vibration limit selected therefore does not have to be precisely the point prior to failure. It must only provide sufficient warning of the development of a real problem so that you can appropriately analyze the nature of the problem and plan and execute proper corrective action with minimum impact on the operation.

Second, a conventional rule of thumb instructs that when a vibration signal increases to twice the value of the original vibration signal, it is a clear warning of degeneration. Some action must be taken because the significant change in signal gives evidence to the potential of ultimate machine failure. The machine should receive attention.

Beyond these observations you may use a number of different resources for setting up the vibration limits. Prediction of impending failure is based on the comparison of current vibration readings with vibration information of the machine in consideration, or similar machines, when in a condition of satisfactory operation. Comparisons are made several ways:

- by comparison to standards developed by engineering standards organizations, manufacturer's associations, or governmental bodies (e.g., API, American Gear Manufacturer's Association, NEMA, ANSI)
- by comparison to previously measured baseline—This technique, sometimes referred to as "trending," can be applied graphically. (See Figure 3-7.) At measurement 26, there is a determination that the unit requires repair, the repair is carried out between measurements 28 and 29, bringing the measured vibration to original satisfactory levels,
- deviation from a statistical control chart. Using statistical quality control charting methods, the baseline vibration information can be plotted over time, using $\pm 3\sigma$ limits; deviation beyond that limit alerts the technician that machine degeneration is probable.

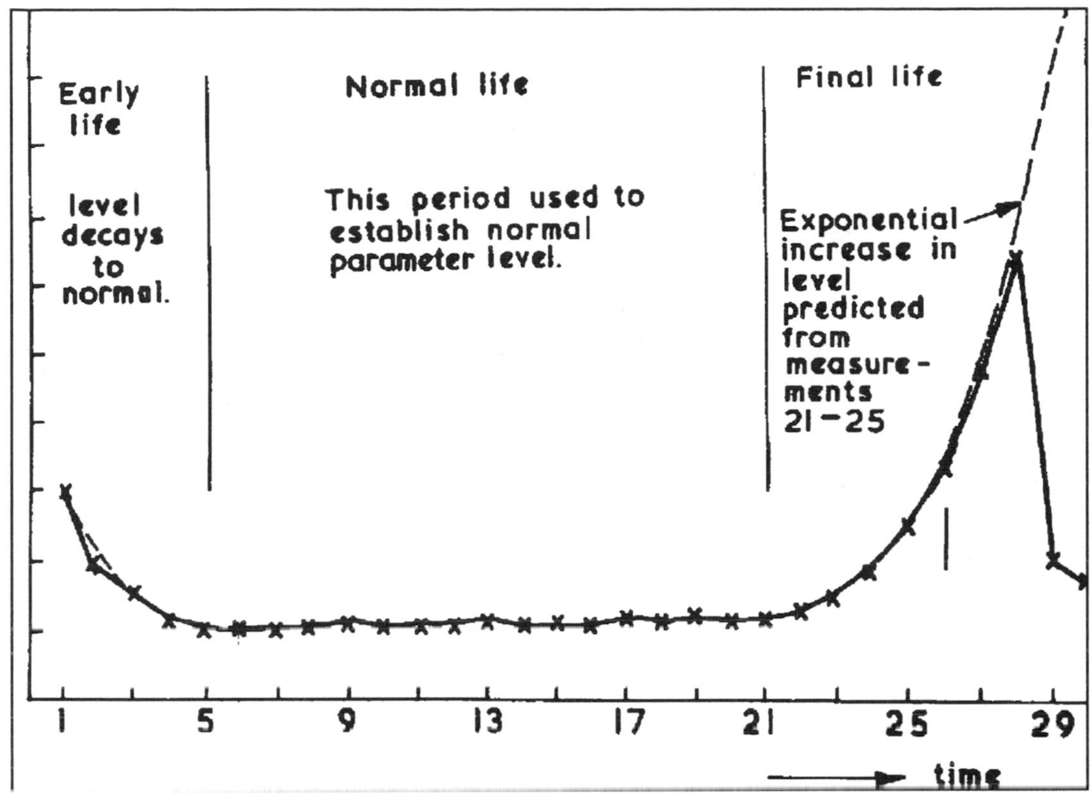

Figure 3-7. Trending of Predictive Maintenance Data—*Courtesy Butterworth Inc., Publishers, Boston, Management of Industrial Maintenance, A. Kelly & M. J. Harris, 1978, p. 159*

When automated data collectors are used, the selection of vibration limits may be more complex. This is because many of these instruments can collect several different types of readings from a single check. For example, the DC+ data collector (a product of Vitec Inc., Cleveland, OH) takes three readings simultaneously. One of these readings is what is termed "bearing units." This is an acceleration reading for frequencies above 18,000 CPM. Selection of vibration limits for these three readings is obviously somewhat more complex than for a simple single velocity reading.

Step 3. Select Check Points

For the same reason that it is necessary to draw representative oil samples from a lubrication system, it is important to properly locate vibration check points. The resulting data (analysis) should truly reflect the condition of the unit. The checkpoint is the spot on the machine that the vibration transducer is placed in order to collect a vibration reading. Since the dynamic forces of the machine are transmitted through the bearings, bearing caps are a logical point. If this is not possible you should get as close to the bearing caps as possible. For example, the bell ends of a motor where stiffeners are located. The coupling bearings of a driver/driven machine set are more important than the outboard bearings. On the coupling side, the driven bearing is the most important of the two.

To properly compare current readings with past data, you should always take the vibration readings from the same location on the machine. Vibration transmissions are absorbed and will vary as you move over the surface of the machine frame. Therefore, having selected a checkpoint, you should permanently mark the spot so that the transducer can be placed in the same location for each reading.

You should initially take readings in all radial directions. That direction which provides the highest reading should be the checkpoint spot that is selected. It is the plane of most sensitivity for a given load and RPM.

Take five readings per machine set including four radial, one for each bearing location and one axial, to check thrust.

Step 4. Check Overall Vibration Level and Correct Existing Deficiencies in Startup Machines

At this point, a basic requirement would be to collect baseline information. This step would first involve collecting overall vibration amplitudes from each of the checkpoints established for each machine. If the overall amplitude exceeds the vibration limit originally selected, it will be necessary to analyze why. If mechanical problems are indicated, you must correct them.

Experience tells us that in an initial program of this type you may find as many as 30 percent of your machines will have mechanical problems. Conditions such as unbalance and looseness are common. If corrected at this stage you will have already started to receive a return from your effort! At the very beginning of the vibration program, the life of your equipment will be extended by reducing the destructive forces of vibration acting upon them.

Step 5. Determine Optimum Check Intervals

If your staff resources permit, try taking one vibration reading at each checkpoint each week. This frequency will help in detecting a degenerating machine before damage progresses too far. If your staff resources do not permit weekly checks, you should space the readings on some logical determination. Include in your analysis:

- how fast deterioration in a machine can occur
- what the consequences would be of missing a problem and permitting a machine to go to failure
- history of failure.

Weigh considerations such as machine speed, service factor and operating environment when evaluating the speed at which a machine may degenerate. The outside limit for surveillance checks should be about one month intervals. Intervals beyond this are not considered good practice.

Step 6. Collect Baseline Information

Once all machines have been brought into proper operating condition, take baseline spectral plots at each check point. A spectral plot is a graph showing frequency in the *x*-axis versus amplitude in the *y*-axis. These plots are usually generated over a wide span of frequencies. The range covered would include output typical for the problems developed at the machine's operating speeds. These baseline "signatures" serve several important purposes.

If you are monitoring your equipment with the use of a meter, the spectral plot will help you determine if a simple unfiltered reading will indicate the problems in the machine under surveillance. The spectral plot will show if vibrations from nearby sources are being transmitted, collected, and disguising the true signals from the machine being monitored. If this is the case, simple meter monitoring will not be adequate.

The signature also provides a baseline which can be used for comparison with future plots. The growth in amplitude at a particular frequency can be compared to the baseline frequency so that a better judgment can be made concerning developing problems.

With automated data collectors, baseline signatures and subsequent spectral plots are stored in electronic files. Comparisons are thus made more quickly and computer displays aid in clarity and interpretation.

Step 7. Design A Data Collection Sheet

The data sheet shown in Figure 3-8, illustrates the format for a good manual collection system. The major parts of the form, starting from the upper left and moving clockwise, are as follows:

- Schematic Section—a box for showing a simplified sketch of the machine—checkpoint locations are identified as "A" through "E",

- Equipment Data Section—a box containing key information including pump name and number, initial vibration readings for each of the checkpoints and vibration limits established for these check points,

- Vibration Trend Chart Sections—a vibration trend chart with the vertical axis showing time,

- Data Section—the date and results of period vibration checks.

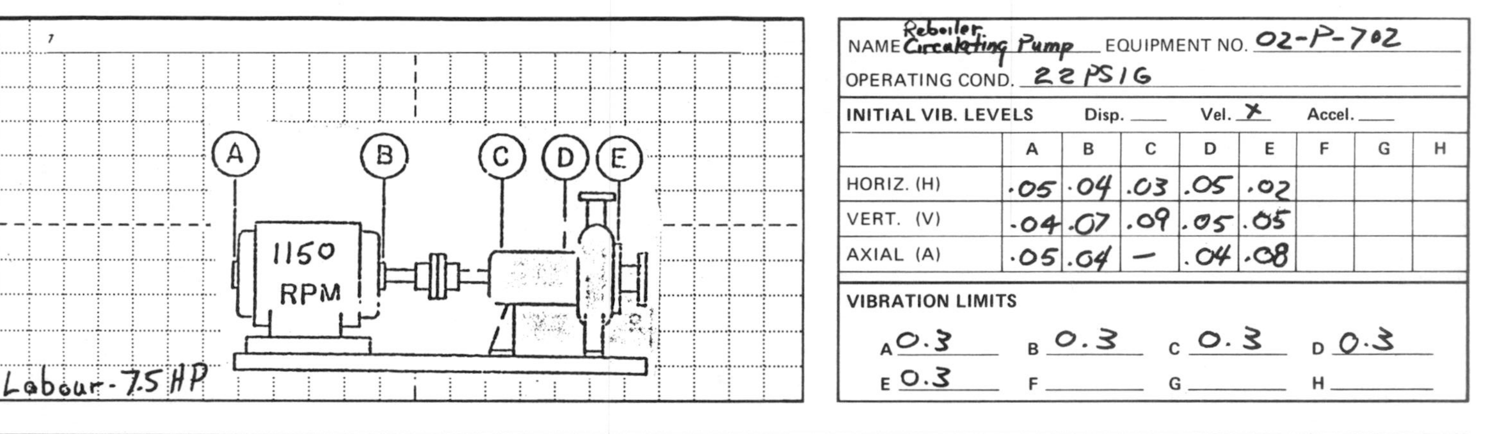

NAME Reboiler Circulating Pump EQUIPMENT NO. 02-P-702

OPERATING COND. 22 PSIG

INITIAL VIB. LEVELS	Disp. ___	Vel. X	Accel. ___				
	A	B	C	D	E	F	G	H
HORIZ. (H)	.05	.04	.03	.05	.02			
VERT. (V)	.04	.07	.09	.05	.05			
AXIAL (A)	.05	.04	—	.04	.08			

VIBRATION LIMITS

A 0.3 B 0.3 C 0.3 D 0.3
E 0.3 F ___ G ___ H ___

CHECKED		PERIODIC VIBRATION CHECKS Disp. ___ Vel. X Accel. ___							
DATE	BY	A(H)	B(H)	C(H)	D(H)	E(H)	F( )	G( )	H( )
1/3	S	0.05	.04	.03	.02	.02			
1/10	S	.05	.05	.04	.02	.02			
1/17	S	.05	.05	.04	.02	.02			
1/24	S	.05	.10	.09	.03	.02			
1/31	S	.06	.16	.13	.03	.02			
2/7	S	.06	.19	.17	.04	.02			
2/14	S	.07	.22	.19	.05	.03			
2/21	S	.07	.29	.25	.05	.03			
2/28	S	.07	.30	.29	.06	.04			
3/7	S	.07	.32	.31	.07	.05			
3/14	S	.05	.05	.04	.02	.03			
3/21	S	.05	.05	.04	.02	.03			

VIBRATION TREND CHART

A(H) B(H) C(H) D(H) E(H) F( ) G( ) H( )

1 2 3 4 (scale for each column)

Figure 3-8. Sample Vibration Data Sheet for Manual Data Collection

If you are going to use an ADC for data collection, the system will provide the data collection format for you. Design of a data collection sheet will be unnecessary. The data will be stored in the host computer data files and data reports generated will be determined by the software blueprint. ADC systems usually include trend reports both as screen displays and as printed charts.

Step 8. Implement Training for Vibration Analysis

Since a vibration analysis program requires specialized knowledge, you must obviously provide training for your personnel who will be assigned to run the program. There are two levels of training required in most facility vibration analysis programs. These are defined by the tasks of data collection and data analysis. In many facilities, the person who collects data may be the same person who does the analysis. In this discussion, however, I will treat these tasks as separate responsibilities involving different people.

Training for Data Collection

The person collecting the data should have the following minimum abilities:

- be familiar with facility layout and machine location,
- be able to identify unusual conditions in machines,
- know the objectives of the predictive maintenance program and how it works,
- know how to use the data collection instrument, how to place the transducer for collecting representative readings, and how to take the readings,
- know how to react to indicated problems and what to do with the data after it is recorded.

The person doing the data collection does not have to be highly technically qualified. This person however should have enough knowledge of the machines from which data is to be collected to be aware of symptoms of malfunction. That is, he or she should understand what excessive heat at a bearing cap might indicate or understand that a leaking oil seal is a problem that should be reported. This person should have enough understanding of machine operation to be aware of strange sounds abnormal to the operation.

The person collecting data should have a minimum of training in vibration fundamentals so that he or she is aware of basic vibration measurements and what they mean. He or she should certainly have received training in the use of the instruments with which data is collected. If the vibration data in a particular measurement is not consistent with what would be expected, he or she should be able to discern this and seek out the causes.

The type of training in specific vibration subjects required for this person can probably be covered in a two-day course typically provided by the instrument vendor. With several weeks of on-the-job training, this person will soon be fully qualified to perform the tasks of data collection.

Good candidates for this responsibility are maintenance mechanics who have worked in the operation and who have a strong desire, if not enthusiasm, in becoming involved in predictive maintenance activities.

The rank and file maintenance worker is often rotated in the task of data collection in organizations in which small team problem solving and worker involvement is the norm. In these and in more traditional maintenance cultures, all maintenance personnel should be trained in predictive maintenance techniques. The knowledge of how predictive maintenance systems work will bring with it a strong desire for success of these programs. Extensive training of all maintenance personnel will help make the predictive maintenance program successful. Full worker participation is a sign of maintenance excellence.

On the other hand, be careful that the wide distribution of data collection responsibilities does not lead to carelessness and disinterest. If there is no single "champion" of the system, it may be cast aside because of the pressures of other activities. Therefore even with worker involvement environments, make sure someone is charged with ongoing responsibility for the success of the predictive maintenance efforts. This person may be the one selected to analyze the data.

Training for Data Analysis

The person who is assigned the responsibility for analysis of vibration data should as a minimum:

- have a good understanding of rotating machinery,
- be familiar with the types of troubles these machines experience and how these troubles may be corrected,
- be trained in methods for analyzing vibration signals and how to interpret these signals and the mechanical problems they suggest (vibration analysis),
- be trained in in-place dynamic balancing (the most common cause of vibration),
- be computer literate (with today's technology, the analyst must be able to freely interface with computers and be well-versed in their application and nomenclature),
- have enthusiasm for the job, have analytical ability, be thorough, methodical and patient, and above all have a high degree of personal integrity in the decision process.

A good candidate for a job of this type is a graduate from some technical military training such as aircraft mechanic or electronic technician, or a graduate from a two-year technical college. It would be desirable for this individual to have several years of hands-on experience in the repair of machinery, and specifically, in the type of machinery in your facility. As a minimum, this person should have a reasonable technical background.

Many organizations assign degree engineers to this type of activity, especially those larger ones that can support the staff. When an engineer is involved he or she generally is given broad responsibility for establishing and overseeing the operation of reliability programs. However, a degree is not a necessary requirement for a good vibration analysis program. In some situations it is a greater advantage to have a non-degree technical person in this position. This is especially true if this person has broad hands-on field experience and understands the machines under his or her area of responsibility.

Training for this level of proficiency involves more advanced vibration subjects. Principal in this area is the need for an in-depth knowledge of analysis. The person doing analysis should have a clear understanding of all the varied types of signals different machine problems project. Since many vibration outputs overlap, rarely are the details of the data collected simple and straightforward. The analyst needs to separate many confusing and sometimes contradictory information. It often takes persistent effort and step-wise elimination of misleading signals. Detailed understanding should therefore be provided not only in methods of diagnosis, but also in the electronic characteristics of the instrument systems used. The analyst should have a thorough understanding of the capabilities and limitations of the measurement and analysis instruments used, and should know what details are captured, and what details are lost. Critical in this respect is an understanding of the features of vibration transducers and the impact the various methods of holding the transducer to the check point will have on the signals collected. The analyst should also be well-versed in the significance of the spectral plots collected. Since the task of dynamic balancing often falls on the analyst, he or she should also be trained in these methods.

Many (if not most) manufacturers and vendors of vibration instruments offer training to their customers. In most cases, they charge for this training only if it goes beyond simple instructional information concerning the use of the instrument purchased. Fundamental classroom and hands-on training involving analysis and balancing would require probably a minimum of five 8-hour days of instruction. This training, often provided in separate sessions of basic and advanced short courses, is a starting point of training for the analyst.

Beyond this formal instruction it is important that the analyst embark on a program of self instruction. He or she should read and learn from the many magazine articles available and the many texts published on the subject. There are also a number of short course (seminar) training programs advertised to the trade. For example, North Carolina State University has conducted such courses as part of their continuing education curriculum. Another non-profit organization offering such training is the Vibration Institute located in Clarendon Hills, Illinois. This organization, dedicated to the advancement of vibration studies, can also provide publications, library searches, bibliographies, and surveys on particular topics. For example,

I at one time was able to obtain a reprint of a paper published by the director of the Institute. This paper contained a survey on national and international sources of vibration standards.

A five-part self study course can also be purchased from a company specializing in such training, such as Applied Learning, in Scranton, PA. There are many consulting organizations also willing to provide their service in assisting you to get started. If your organization lacks the experience or training in maintenance engineering applications, there is plenty of help around.

ADDITIONAL SOURCES OF INFORMATION ON PREDICTIVE MAINTENANCE

A full discussion of predictive maintenance applications would require several volumes of text. This chapter can only be considered an introduction to the subject. The information presented hopefully is sufficient to enable those interested in the subject to understand the technique and to justify further action in its pursuit. To fully grasp the subject, you are advised to learn more about the various methods, including the more common oil analysis methods, infrared detection techniques, and ultrasonic applications. Resources for learning more about the subject include texts and magazine articles in the technical literature; seminars and other short courses are available, including the following:

The Vibration Institute, 101 55th Street, Clarendon Hills, IL 60514, Tel. (312) 654-2254.

Cord Associates, 2423 Allen Street, Allentown, PA 18104, Tel. (215) 433-5020

North Carolina State University Office of Continuing Education, Raleigh, NC 27695 Tel. (919) 515-2261

4

Maintaining Cooling Towers

Robert Burger

To maintain a cooling tower, or any mechanical equipment, you must be aware of the technological theory. You must clearly understand the various elements of the machinery, plus their functions and relationships to each other, so that the device can be utilized in a professional manner at maximum levels of efficiency. Therefore, this chapter on cooling towers consists of three sections:

- Inspection and maintenance of cooling towers
- Cooling tower design considerations
- Case studies of upgraded cooling towers.

Since the purpose of the book is *maintenance,* this portion appears first, followed by theory and application for those who desire to obtain better understanding and utilization of their equipment.

Cooling towers are manufactured in a great variety of sizes and shapes ranging from small, five-ton, factory-assembled packages to field-erected units measuring up to, and sometimes exceeding, 150 feet long by 50 feet wide by 50 feet high. The service of cooling towers can vary from air conditioning applications, where the water is put on the tower at 95° F for mechanical refrigeration, to 102°F to 105.5° F. The wet bulb, depending upon the area of the country, is between 65° to 80° F for absorption equipment, returning in most cases from the tower at 85° F. Chemical process and power generation temperatures run the gamut of 95° F to 150° F on the tower discharging at 83° to 90° F off the tower, where the wet bulb again varies depending upon the geographic location.

Acceptable limits of operation are determined by the requirements of the installation. It is obvious that when the mechanical equipment breaks down, it is past its acceptable limits. When the water is returned 2 to 5 degrees—or more—warmer than manufacturer requirements for the compressors and condensers, corrective actions must be taken to avoid excessive power use and to keep the equipment on line. Again, strict guidelines cannot be published since each individual case is determined at the site by the operating requirements.

HISTORICAL DEVELOPMENTS

The modern cooling tower developed in an orderly progression. In the "dark ages" of plentiful resources and the absence of ecological conservation, cooling water was used merely "once through," discharged, and forgotten. Where topographical considerations were analyzed and available, large ponds or lakes (artificial or natural), or canals were utilized to hold, cool, and recirculate the water.

To facilitate cooling and reduce the amount of real estate used, spray systems were installed to aerate the water in the ponds and canals. This obtained faster cooling; not by relying on top layer evaporation and sensible heat exchange, but by offering more water surface to the atmosphere in the form of spray for evaporative cooling.

In order to obtain better control, it was soon determined that by rotating the nozzles 180 degrees and spraying down in the box instead of up, lower temperatures could be obtained. Soon, instead of being dependent upon prevailing winds, fans or air movers were designed to master and assist nature's unreliable aerodynamics.

As more was understood of the mechanics and hydrodynamic process of water cooling, fill packing was added to slow the vertical fall of water and to provide a longer air/water interface contact for greater levels of cooling and heat transfer.

Today, all of these techniques are used, however, some are used in modified forms.

WASTE HEAT: THE COMMON DENOMINATOR

The New Orleans Superdome air conditioning/refrigeration system is rated at 10,000 tons, circulating 30,000 gallons per minute (GPM) of water, while a small-town shoe store cools its premises with a 10-ton central air conditioning system circulating 30 GPM. There is one common denominator.

The fabric of our society would collapse without electric generating plants, whether they be one megawatt (MW) installations, for a small apartment house complex burning fossil fuel, or 1,000 MW nuclear steam-electric facilities for a large public utility. Again, there is one common denominator.

Chemical plants throughout the country yield an unimaginable number and variety of products that our civilization utilizes. In all of our activities, day and night, we rely upon them, and in these chemical operations, there is one common denominator.

This common denominator cannot be ignored: It is *waste heat.*

In the above examples, waste heat is absorbed by water having a temperature colder than the process. This hot water must now either be discharged into a body of water or cooled and recycled. The common alternative for all systems is to lower the water temperature in a device called a cooling tower. Here, the waste heat is rejected into the atmosphere, and the colder water is then recirculated throughout the system, thereby conserving our water resources.

The problem of waste heat is enormous, since industry uses approximately 500 billion gallons of water a day, which, if not recirculated, would seriously deplete the available water supply. The unwanted heat from the water must be restored to the environment without creating ecological havoc or adding unnecessary expenses to the inflationary spiral.

The development of cooling towers solved the problem of possible ecological damage of returning hot water to the environment, and enabled industry to recycle the greater proportion of the water it required at reasonable capital investment costs.

Operating procedures, costs, and upgrading potential for greater efficiencies must be investigated constantly to provide economic returns for the installation. To optimize the upgrading of the tower, you must first understand the internal elements and their functions. A comprehensive maintenance program must also be instituted to obtain maximum equipment utilization.

COOLING TOWER MAINTENANCE AND INSPECTION

Cooling towers are mass heat transfer devices that also consume energy in the course of their performance. Properly functioning cooling towers make money by producing colder water for refrigeration, cooling, or process water. Colder water returning from the cooling tower, under similar conditions of load and temperature, can add up to many dollars saved in energy consumption and/or product results. In order to keep a tower functioning efficiently, all of its components must be put in optimum working order so that excessive energy charges and performance slippages are eliminated.

Refrigeration equipment can utilize over 450 horsepower for motors to drive compressors; or steam is utilized to power the turbines that consume energy and money to operate. A typical 1,000-ton refrigeration system (circulating 3,000 GPM of water) can use over $425,000 a year in electrical and/or steam energy.

Chemical plants manufacturing a saleable condensate can turn out over $3,500 a day in additional materials, which will more than pay for an extensive upgrading, rebuilding, or maintenance program of large industrial cooling towers.

Power generating plants use their electricity to generate power, and the degree of consumption is called the energy penalty. Again, colder water uses less energy and reduces the penalty, thereby producing more saleable electricity for the customer.

Obviously, it is necessary to carefully inspect your tower to eliminate any deficiencies that prevent optimum cold water temperatures returning off the tower to the equipment, to make note of these problems, and to budget money and time to effect repairs.

There are two ways to obtain better performance from a cooling tower:

- Inspection
- Maintenance

Over the years, the equipment loses some of its efficiency because of wear and tear, misalignment of parts, and general deterioration that occurs when inspection and maintenance are not effectively provided.

The oldest and best tools available to every maintenance department are its personnel's *eyes*. Proper inspection and adequate attention paid to discrepancies will more than pay for themselves in a smoother, more efficient operation.

When the cooling season draws to a close, operators of air conditioning and refrigeration systems plan shutdowns and schedule normal maintenance of the compressors, turbines, shells, and tubes. The cooling tower, an important element of the entire system (sometimes because of its location) is often ignored until "five minutes" before turn-on. Industrial towers, due to the nature of the operation, are never turned off except for scheduled turnarounds that are usually spaced many months apart.

Inspecting Cooling Tower Components

One of the quickest methods of improving the operation of a cooling tower to produce colder water—which translates into added product and energy savings—is good, common sense inspection. The checklist in Figure 4-1 is designed to assist you in optimizing the performance of your tower.

Considerable sums of money can be expended to upgrade the performance of the tower by changing to more efficient fill or higher horsepower motors and larger fan diameters. Before this expenditure is contemplated, good maintenance should be considered as an inexpensive alternate.

Many owners and large industrial users rely on periodic inspections and reports by a professional field engineer, who spends at least one to two days at the site and submits a detailed evaluation of discrepancies noticed and suggested repair programs, together with the possible capability of thermal upgrading. This section discusses the procedures that parallel the methodical manner in which field inspections are made. While there are a great variety of cooling tower types, the elements of all units are the same, and an individual checklist can be worked out covering all items as a preplanning procedure for developing a comprehensive inspection program. Figure 4-2 is typical of large industrial installations, but can easily be modified for

PROBLEM	*CAUSE*	*RESOLUTION*
Cold Water is too hot	Antiquated wet decking heat transfer surfaces.	Investigate installation of state-of-the art high efficiency cellular film fill. Consider "V" bar fill replacement for large cross flow type.
	X-flow overspilling basin curb	Balance all hot water distribution basins. Caulk basin curbs at junctures.
	Too much flow per cell	Reduce flow in each cell to Design by activating flow control valves.
	Insufficient cooling air	Pitch blades up to maximum allowable plate amperage. Check fan tip clearance and hub cover security. Add motor horsepower and/or larger fan.
"Dirty Water"	Scale, sludge, and/or microbiological buildup.	URGENT! Contact Water Treatment Consultant.
Excessive Drift	Improper elimination	Investigate installing cellular drift eliminators. Retrofit with higher velocity resistant units. Check installation for gaps in eliminator wall.
Fans Vibration	Unbalanced	Tighten all bolts with torque wrench to bolt mfgr specifications. Check shaft alignment. Call I.R.D. specialist to balance fans and shafts.
Motor—hot and/or noisy	Wrong voltage Bearings overgreased Bearings worn Poor ventilation Wrong grease	Check voltage and current vs plate requirements. Purge excesive grease. Replace bad bearings, check lubrication. Check ventilation openings, clean motor. See manufacturer's instructions.

Figure 4-1. Cooling Tower Problem-Solver

PROBLEM	*CAUSE*	*RESOLUTION*
Gear Box Noise	Bad bearings Sludged oil	Check oil for metal filings, replace bearings, if necessary. Flush, replace with manufacturer's recommended lubricant.
Wood Deterioration	Old age Chemical treatment	Replace with modern plastic high heat transfer fill. Renew chemical treatment with biocide interior spray.

Figure 4-1. Cooling Tower Problem-Solver (*continued*)

COOLING TOWER INSPECTION CHECKLIST

Robert Burger Associates, Inc.

RBA Job No.

Owner ______ Date Inspected ______

Plant ______ Inspected By ______

Location ______ Tower Manufacturer ______

Owner Designation ______ Installed ______ 19 ______

Water Treatment Used ______ Model No. ______

Design Conditions ______ GPM ______ HW ______ CW ______ WB

Condition: 1-Good; 2-Repair; 3-Replace	1	2	3
EXTERIOR STRUCTURE:			
1. Endwall Casing & Access Doors			
2. Louvers (Type)			
3. Drain Boards			
4. Stairway			
5. Fan Deck			
6. Fan Deck Supports			
7. Handrails			
8. Ladders & Walkways			
9. Distribution System			
Headers (Type)			
Distribution Basin			
Water Level			
Flow Control Valve (Size)			
Nozzles (Size)			
Water Distribution			
10. Spray System & Spray Nozzles			
11. Fan Cylinders (Type)			
INTERIOR STRUCTURE:			
12. Fill (Type)			
13. Columns			
14. Girts			
15. Diagonals			
16. Partitions & Doors			
17. Eliminators (Type)			
18. Walkway			
19. Cold Water Basin (Type)			
Water Depth			
20. Mech. Equip. Support (Type)			

Condition: 1-Good; 2-Repair; 3-Replace	1	2	3
MECHANICAL EQUIPMENT			
21. Drive Shafts (Type)			
22. Speed Reducer			
Series ______ Ratio			
Oil Level			
Oil Seals			
Vent			
Back Lash			
Pinion Shaft Play			
Fan Shaft End Play			
Last Oil Change (Date)			
Oil Used			
23. Fans			
Dia. ______ Type			
Hub			
Blades			
Hub Cover			
Tip Clearance			
No Vibration ______ Vibration			
Additional Components (If installed on tower)			
Fan Guards			
Oil Gauge & Drain Lines			
Vibration Limit Switches			
Other:			

24. Motor: Mfr. ______
Name Plate ______ HP ______ RPM
Phase ______ Cycle ______ Volts ______
Amperes ______ Frame ______

REPLACEMENT PARTS REQUIRED:

QUANTITY	DESCRIPTION	ORDER FROM	DATE REQ'D

MAINTENANCE WORK REQUIRED: DESCRIPTION	REQ'D COMPLETION

(Use back of this sheet for additional requirements or notes.)

Figure 4-2. Sample Cooling Tower Inspection Checklist

any specific installation, whether it be a 50,000 gallon per minute crossflow or 50-ton (150 GPM) blow-through counterflow.

There are three major divisions of a cooling tower: exterior structure, mechanical equipment, and interior structure. We will investigate each division as a unit, marking down discrepancies as they appear.

Inspecting the Exterior Structure of the Cooling Tower

This inspection is the visual walk-around examining the following components.

- *Louvers*
  - Determine the number of missing louvers.
  - Measure accurately the existing louver size, length, and width, and measure the thickness very carefully, since a 1/16″ difference can prevent the new louver blades from fitting in the existing support slots.
  - Check out-of-position louvers for repositioning.
  - If the louvers are wood, look for splits or breakage and check for wood rot and decay.
  - Examine asbestos, cement, or fiberglass reinforced polyester louvers for breaks, cracks, and corner damage, which can prevent proper alignment.
  - Examine wood louver supports for rot and decay; if made of galvanized steel, observe the condition of the protective coating and make notes of broken and rotted supports.
  - See that attaching hardware is adequate and properly supports the louver posts.
  - Examine louver supports for ice damage.
- *Casings*
  - On corrugated or flat cement asbestos board, transite, or fiberglass reinforced polyester casing, look for cracks, leaks, condition of attaching hardware, and other physical damage.
  - Check corner rolls for integrity and security of attachment.
  - On wood casing towers, look for loose boards, evidence of wood decay, leaks, and wood rot. Wood for casing of towers is a rather old technique and usually consists of a double layer. This should be checked from the interior to determine if wood rot dictates entire replacement with modern fiberglass sheathing.
- *Access Doors*
  - Check for missing doors, broken or loose hinges or latches, general conditions of doors in operation, and any evidence of physical damage.
  - Measure doors accurately if replacement is necessary.

- *Distribution System*
  - Observe flange connectors at top of riser for security and gasketing and make notes of effectiveness of corrosion control coating on supply lines and risers.
  - On counterflow towers, check where header enters casing to determine if it is properly caulked and no apparent leaks exist.
  - On crossflow towers, look for deterioration in distribution basins, splash guards and associated piping.
  - With water trough configuration, check boards for warpage or splitting on basin sides and leakage caused by gaps in trough.
- *Drain Boards*
  - Look for evidence of water being diverted outside the drain to tower exterior.
  - Check for displaced or damaged drain boards.
  - If drain boards are metal, observe for corrosion and rot.
  - Look for missing boards and condition of fasteners on all types of materials.
- *Stairways*
  - Note deterioration or loose handrails and kneerails.
  - Check stringers and stair treads for missing components, breakage, and cracks.
  - Examine structure for looseness, evidence of rot, and condition of fastening devices.
  - Check steel stairways for evidence of corrosion or acid attack.
  - Examine all fasteners for security (in most installations, it is prudent to shake the stairway before using).
  - Pay special attention to condition and security of upper braces of cantilevered platform supports.
- *Ladders and Walkways*
  - Before climbing any ladder, check for security of fasteners by vigorously shaking ladder rails, then make sure that all the rungs are in position.
  - Note missing or deteriorated rungs.
  - Check treads on walkways for security, rot, or corrosion.
  - Check stringers and rails for corrosion.
  - To comply with O.S.H.A. safety requirements, make sure any ladder over 20 feet has a cage or safety device to protect the climber. Further, a cage above 20 feet must have welded platforms at 20 feet intervals or be protected by a safety device. Check these appliances for security and evidences of corrosion, rot or weakness. An old surviving "ladder climber trick" is to grasp ladder rails rather than rungs, since rungs could be rotted and weak.

- *Fan Decks*
  - Examine fan decks for decay, missing or broken members, and gaps between boards that can cause short-circuiting of air circulation.
  - Check fan decks for security so that air does not bypass through leaks. Weakened fan decks are also dangerous for personnel walking on them.
  - On steel decks, check for corrosion, holes, and condition of corrosion control coating.
  - Examine fan deck supports for decay, loose or broken members, and condition of fastening devices.
- *Fan Cylinders*
  - Make sure that fan cylinders are securely anchored to the fan deck supports.
  - Observe for looseness of fan cylinder and condition of fastening devices.
  - Look for damage, missing structure parts, and deterioration of components.
  - Observe wood rot, or where applicable, corrosion of steel sides, rings, or support strap banding. Destructive corrosion can move metal into the path of rotating fan blades.
  - Inspect for proper tip clearance between fan blade and interior of the cylinder walls. O.S.H.A. safety requirements dictate a protective fan guard on cylinders less than five feet high. If a guard is installed, check condition of components for corrosion, sagging, and missing parts.

Inspecting the Interior Structure of the Cooling Tower

- *Distribution System*
  - Inspect metal piping headers for decay, rust, or acid attack.
  - Check flange connections for tightness and condition of fasteners and gaskets.
  - See that all branch arms are secure; note any corrosion on flange connectors.
  - Check for missing nozzles and, if practical, observe spray pattern (Figure 4-3) to see if nozzles are in good operating order and are not clogged.
  - For crossflow water distribution, make sure that nozzles are cleaned and operating, and, if nozzles are not used, check to ascertain that all distribution holes are clear and are not clogged.
  - Note condition of redistribution section beneath hot water basin for necessary repairs or replacements.
  - If piping is fabricated from wood, check condition of retaining bands, look for warpage or split boards, and check for excessive leakage.

Figure 4-3. Check spray pattern to determine nozzle performance

- Observe water trough distribution for condition of wood and fasteners, for excessive gaps between sides and bottom of trough, and for alignment of splash cups.

- *Mechanical Equipment Supports*
  - Inspect the mechanical equipment supporting steel for evidence of corrosion, and condition of the fasteners, straps, or hangers.
  - Look at wood structural members in contact with the steel supports to ensure soundness or evidence of weakness that could result in loss of structural strength.
  - If springs or rubber vibration absorption pads have been installed, check conditions of springs, mounting and adjusting bolts, corrosion of ferrous members, and condition of rubber pads.
  - Check for iron rot where steel and wood are in contact. Iron rot is a possible cause of deterioration.
- *Fire Detection*
  - Test the valves and the operating condition of wet or dry sprinkler system.
  - Check for evidence of corrosion on pipes and connectors.
  - See to it that activating devices are in apparent good order.
  - If a thermocouple wiring system is used as an alarm indicator, check for broken wiring.

- *Eliminators*
  - Check drift eliminators for broken, displaced or deteriorated blades, and for gaps or misalignment that will permit excessive drift.
  - Check supports for breakage or deterioration.
  - Look for clogging caused by algae, debris, or slime.
  - Examine split or expanded crossflow fiberboard or plywood vertical support members for possible replacement.
- *Fill*
  - Inspect tower fill for any breakage, deterioration, and misplaced or missing splash bars. Worn fill splash bars create gaps and very inefficient cooling.
  - Look for damage to splash bars supports and fill supports.
  - If sections of fill have collapsed, look for damage from icing.
  - Obtain design conditions of water and operating temperatures so that thermal calculations can be made later to determine degree of upgrading capability available by changing fill to a more efficient design. Fireproof asbestos fill should be considered as an alternate to expensive automatic deluge system, since piping and controls of a fire alarm system need continual checking and maintenance while, on the other hand, asbestos fill, by the nature of its fireproof construction, does its fire protection job without further attention.
- *Interior Structural Supports*
  - Test columns, girts, and diagonal wood members for soundness by striking with a hammer. A high-pitched, sharp sound indicates good wood, while a low-keyed, dull sound indicates softness of wood.
  - Wood rot will soften wood, so, if an area indicates possible rot, probe it with a screwdriver to further determine condition of the wood.
  - Check carefully around metal fasteners for iron rot on wood in contact with metal.
  - Check steel internals for corrosion and deterioration caused by rusting.
  - Check bolt condition and tightness on both wood and steel interiors.
- *Partitions*
  - Examine bolts and nuts in partitions for looseness and corrosion.
  - Inspect for deteriorated or loose partition boards.
  - Note if partitions go from basin to roof deck dividing individual cells so as to prevent windmilling of idle fans or short-circuiting of air between cells.
  - See that wind walls parallel to air intake louvers are in position, and that boards or transite members are securely fastened.
  - Check supports, either wood or steel, for condition of rot or corrosion and condition of the protective coating.

- *Doors*
  - Make notation of missing doors, condition of hinges and fasteners, and look for corrosion, rot, or deterioration of door material.
  - If cell partition does not have doors at basin level, indicate it on the report.
- *Cold Water Basins*
  - For wood basins, check for deterioration, warped or split basin sides, open joints, and soundness of wood. During winter operations, leaks in basins are quite discernible: they freeze, so check for ice.
  - Inspect steel basins for corrosion, holes, and condition of protective coating.
  - Inspect concrete basins for cracks, breaking joints, and acid attack.
  - Observe all sumps for accumulation of debris, condition of screens, anti-turbulator plates, and freely operating drain valves. If the basin is one continuous unit, partitions between cells, together with necessary sumps and piping, should be recommended. This adds flexibility for utilization of equipment by operating only the number of cells required for the heat rejection load, thereby conserving power.

Inspecting Mechanical Equipment

- *Fans*
  - Check fan hubs for corrosion.
  - Inspect hub covers for corrosion and condition of attaching hardware.
  - Inspect blade clamping arrangement for tightness and corrosion.
  - Look at fan blades carefully for corrosion and erosion.
  - Check fan blade pitch with bubble protractor for uniformity; check blades for build-up of solids that would change blade moment weight or air foil characteristic, thereby causing excessive vibration.
- *Gear Boxes*
  - Check gear reducers for proper oil level. Oil levels can be determined by external dip stick or looking at the sight glass on the gearbox.
  - Inspect oil for moisture and sludge.
  - Rotate input shaft by hand back and forth against gear tooth contact to feel for backlash.
  - Check input pinion shaft bearing for wear, by attempting to move the shaft radially.
  - Look at fan shaft bearings for excessive end play by applying force up and down on a fan blade tip and note the movement of output shaft. A running clearance is built into some output shafts, which should not be confused with excessive end play.

- *Power Transmission*
  - Inspect drive shaft keys and set screws; check assembly hardware for tightness.
  - Examine drive shaft couplings for corrosion, wear, or missing elements.
  - Look at exterior of drive shaft visually for corrosion, and check interior by lightly tapping draft shaft for sound along its length for dead spots that could indicate internal corrosion.
  - Carefully observe flexible connectors of both ends of shaft. A shaft guard is imperative to prevent fan wrecks. The weakest link in the power transmission train is the coupling connectors. Invariably when a coupling breaks, the rotating shaft swings upward and wrecks the fan blades, possibly doing damage to the gear box and motor too. This can be prevented by the installation of simple, inexpensive shaft guards to absorb the upward thrust of the rotating broken shaft. (See Figure 4-4.)
  - Observe belt-driven units for alignment, tension, and condition of wear of belts.

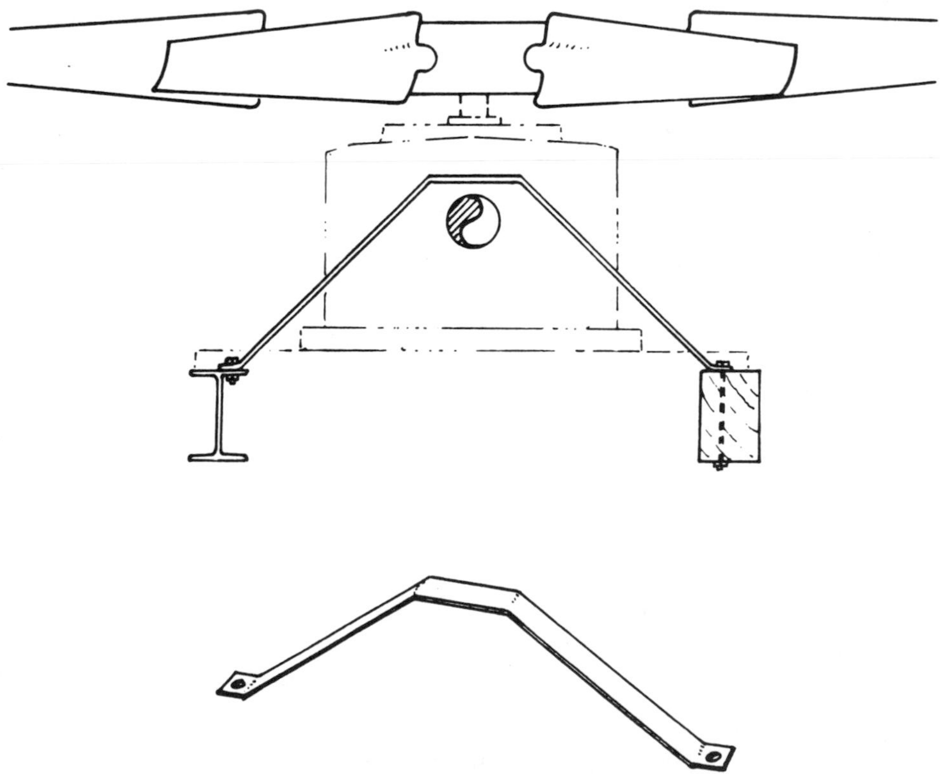

Figure 4-4. Installing a simple fan shaft guard.

- *Motors*
  - Look for evidence of motor heating, and observe proper lubrication of the bearings. Excessive grease or oil applied to ball bearing motors can damage the unit by forcing grease into windings and causing deterioration of insulation.
  - Look for deposits of dirt or dust at motor air intakes and check TEFC motors for conditions of air passages and fans.
  - The drain moisture plug is supplied with motor; see if it is operational.
  - Measure operating loads with amp probe and volt meter as compared with nameplate data; recommend pitching of fan blades up or down to compensate.
  - Examine mounting bolts and attachments for security and corrosion.

All of the above items might seem excessively complicated and time-consuming. However, if a thorough investigation and description is to be made, it must cover all elements of the cooling tower. To facilitate the actual inspection, a checklist can be compiled, which will greatly speed up the job and ensure that all aspects of the inspection were accomplished. This will also serve as a reference to writing and submitting your report.

Extending mechanical equipment life by proper maintenance, timely repairs, and intelligent utilization of colder water generated by a cooling tower rebuilt with modern engineering principles, are essential.

Using a Preventive Maintenance Service Checklist

A routine preventive maintenance service checklist should be made for all of the equipment you are responsible for. A three-card system could be put together for the cooling tower as follows.

Prior to the start-up of the cooling tower, perform the following:

- Clean out tower basin.
- Caulk cracks and openings in the T&G joints with mastic.
- Tighten all bolts of columns, beams, connectors, and equipment.
- Clean, prime, and coat all areas of steel corrosion.
- Clean and lubricate electric motor bearings in accordance with the manufacturers' recommendations.
- Inspect and adjust gearbox oil level.
- Check shafts and couplings for security or tension and alignment of V-belts.
- Check starters, relays, coils, clean contacts.
- Operate fan driver and check bearings for overheating.
- Consult your chemical treatment firm for start-up.
- Fill system to basin level and check float and make-up equipment.

During the cooling season, check tower under operation as follows:

- Observe fan and motor for unusual noise or vibration.
- Examine V-belts and/or shaft and couplings.
- Inspect basin water level, float, and bleed-rate.
- See that water pattern is uniform and that spray nozzles or metering orifices are not clogged.
- Check water treatment for proper operation.
- Clean intake strainer; check tower basin for accumulation of debris.
- Shut down equipment periodically to observe gearbox lubrication, fan bearing condition, or shaft coupling connectors for security.
- Observe steel casing and steel parts for rust and corrosion.
- Perform any urgent repairs that cannot be scheduled when tower is shut down at the end of the season.

At end of cooling season, perform the following:

- Clean and flush out tower leaving drains open.
- Drain only parts of tower subject to freezing.
- Clean drift eliminators, fill decks, louvers and spray heads of observable debris.
- Schedule repairs necessary before placing tower into operation.
- When tower is completely dried, clean, prime, and coat areas requiring it, using recommendations from a coating specialist.
- Fill gearbox with oil to prevent condensation.
- To protect the owner's investment in this tower, go back to the complete inspection checklist.

Cold Water Makes Profits—Your Operational Responsibility

Because of insufficient time and limited personnel, most maintenance managers pay a minimum of attention to the cooling tower, taking its function—to cool hot water—for granted.

Many managers are not even aware of, or do not consider, the role of the cooling tower in the company's profit or loss picture. It is your function to employ the equipment in an efficient a manner as possible to produce money and profits for management.

The major area of profit to analyze is the utilization of the cold water. In the refrigeration/air conditioning system, the colder the water, the less energy is required to power the equipment. In chemical product manufacturing plants, the colder the water, the more efficient the condensation and the greater the volume of saleable condensate product is yielded for the plant at lower cost. In power plants, colder water reduces the electrical generation penalty (the cost of electricity used in the process of producing saleable or usable power).

The second area of cost to consider is the electrical and/or steam dollar expenditure which is purchased to power the systems. The cooling tower is a high energy utilization machine as well as a mass heat transfer device.

After the cooling tower has been erected, proper operation and maintenance is required to yield the design conditions of cooling. By not maintaining towers adequately, they can lose appreciable performance as soon as they are turned on.

This chapter on cooling towers is aimed at one basic end result—COLD WATER.

COOLING TOWER DESIGN

The criteria of cooling tower performance are outlined in the design conditions (see Figure 4-5), specified when the cooling tower is purchased and/or rebuilt to mean the cooling of a specific quantity of circulating water entering the tower at a particular temperature and leaving at a definite value. The difference between these two temperatures is called *The Range* or *Delta T.* This cooling of the specific amount

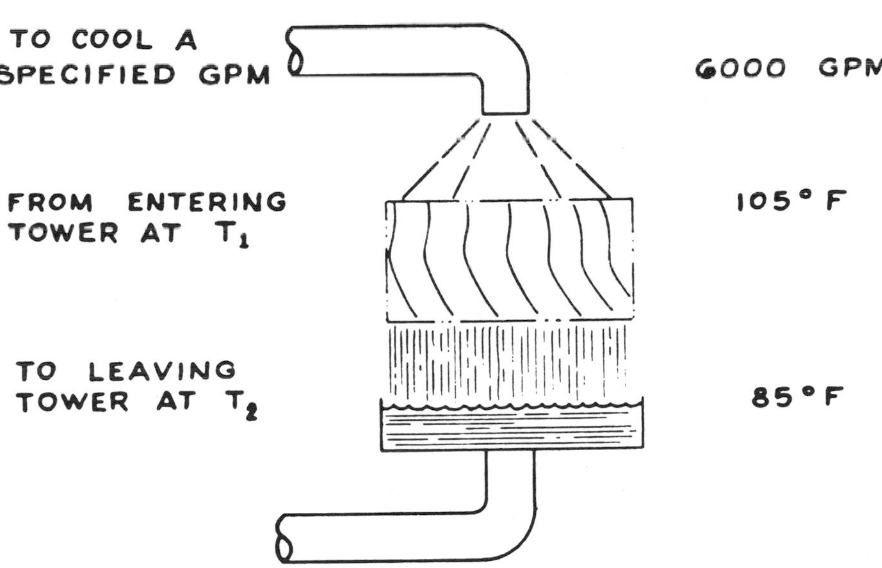

Figure 4-5. Cooling Tower Design Conditions
Figure 3-15.

of water is also to be performed at a wet bulb temperature that is noted in the specifications. The difference between the cold discharged water and wet bulb is called *the approach to the wet bulb.*

A 1°F, or 0.6°C, colder water returned to the compressors and condensers in air conditioning refrigeration equipment calculates, by use of Enthalpy Charts, a 3¼ percent savings in electrical energy input to these machines. Therefore, a little more than 3°F (1.7°C) colder water off the tower can save 10 percent of electrical energy and resulting charges thereof at any given time. A 2,000-ton (1,815 metric ton) refrigeration system circulating 6,000 GPM (22,700 liters/minute) could use $425,000 in electrical power a year. A 10 percent savings of $42,500 obtained by sending colder water to the machinery will be a significant factor towards rapid payback for the upgrading of the tower.

Cooling Tower Fundamentals for Maintenance

The basic function of cooling tower operation is that of evaporative cooling and exchange of sensible heat. The air and water mixture releases latent heat vaporization. Water exposed to the atmosphere evaporates, and as the water changes to vapor, heat is consumed at approximately 1,000 BTUs per pound of water evaporated. The heat is taken from the water that remains by lowering its temperature.

However, there is a penalty involved—the loss of water that goes up to the cooling tower and is discharged into the atmosphere as hot moist water vapor (Figure 4-6). Under normal operating conditions this amounts to approximately 1.2 percent for each 10°F (5.5°C) of cooling range. Sensible heat that changes temperature is also responsible for part of the cooling tower's operation. When water is warmer than the air, there is a tendency for the air to cool the water. The air gets hotter as it gains the sensible heat of the water and the water is cooled as its sensible heat is transferred to the air.

Cooling due to the evaporative effect of the release of latent heat of vaporization amounts to approximately 75 percent, while 25 percent of the heat exchange in the cooling tower is sensible heat transfer. The cooling tower, like any other device, does not escape the unchangeable law of the indestructability of matter. A cooling tower is merely a machine that takes a mass of heat from one area and moves it to another area. In technical terms, it is referred to by thermal engineers as "the heat rejection solution" or "correction of the heat penalty generation of compression equipment." The cooling tower is a machine that moves heat from point "A" to point "B" and ultimately discharges this heat into the atmosphere.

Understanding Counterflow Versus Crossflow Cooling Tower Design

All cooling towers, except the small packaged specialty models, are custom-made. The manufacturer varies the shape, size, configuration, and input to meet the

Figure 4-6. Cooling Tower Plumes (Fog) Are the Byproduct of Evaporative Cooling Resulting in Loss of Water Volume

particular set of thermal parameters required by the customer. Since they are available in two basic operational designs—counterflow and crossflow—it might be helpful at this time to list a few terms and their definitions to aid in investigating these differences. The letters in the circles on the schematic diagrams in Figures 4-7 and 4-8 refer to the elements that perform the same function and their location in the structure. The giant natural draft or hyperbolic cooling towers operate on the same principles as a small 50-ton tin-box package; only the size is vastly different.

How to Upgrade the Performance of Your Cooling Tower

To upgrade the performance of an existing cooling tower, the three major areas to investigate for retro-fit are:

- Wet decking fill
- Water distribution system
- Drift eliminators

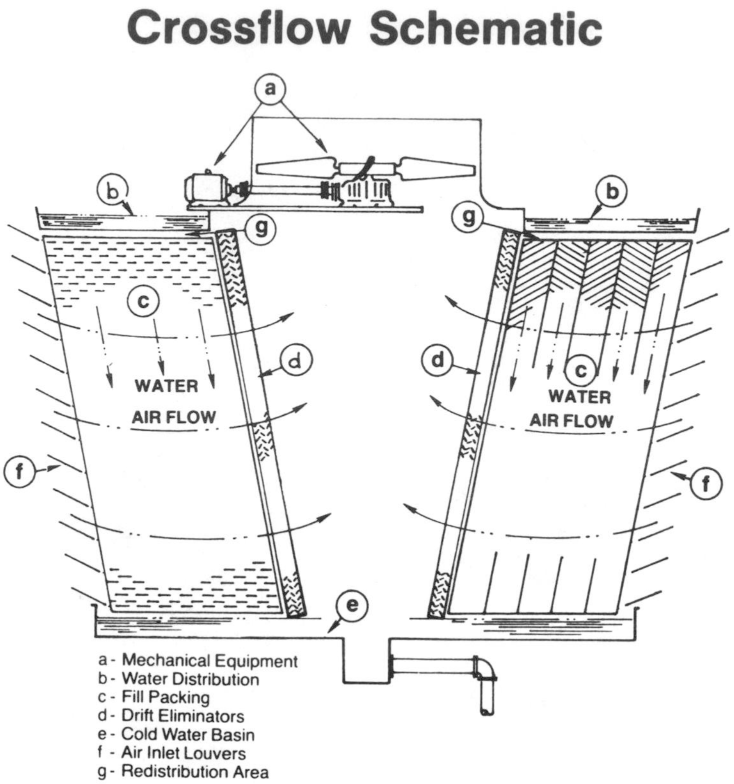

Crossflow Schematic

a - Mechanical Equipment
b - Water Distribution
c - Fill Packing
d - Drift Eliminators
e - Cold Water Basin
f - Air Inlet Louvers
g - Redistribution Area

Figure 4-7. Crossflow Schematic Cooling Tower Indicating the Various Elements

Changing the Wet Decking Fill

Generally, the most significant improvements can be made simply by changing the wet decking fill. This, however, is not done capriciously. The heat transfer must be investigated from a thermal engineering point of view, in conjunction with the fill characteristics determined by the manufacturer's performance curves. These are developed very painstakingly by trial and error experimentation, which is expressed as KaV/L* or heat transfer characteristics.

The basic function of a cooling tower is to cool water by intimately mixing it with air. This cooling is accomplished by a combination of sensible heat transfer between the air and the water and the evaporation of a small portion of the water. This type of transfer is represented by the Merkel equation:

$$\frac{{}^{*}KaV}{L} = \int_{T_2}^{T_1} \frac{dT}{hw-ha}$$

Where:

a area of transfer surface per unit of tower volume; expressed in square feet per cubic feet

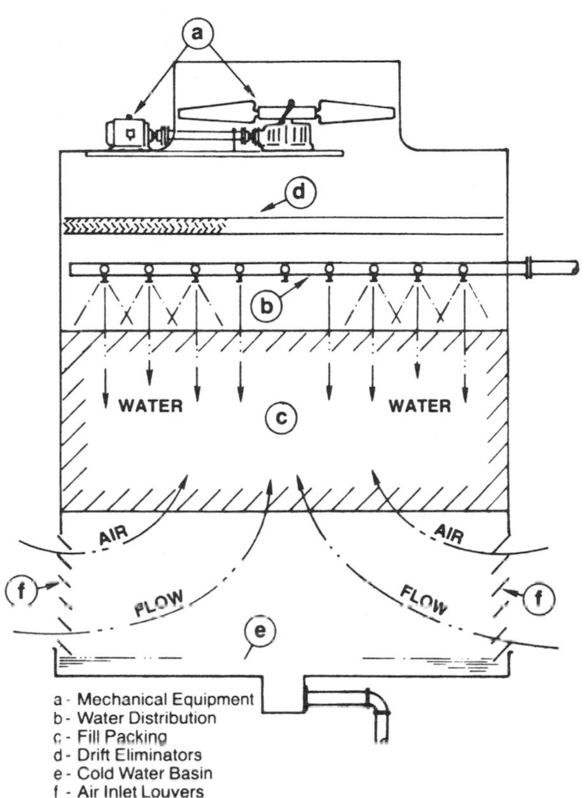

Figure 4-8. Counterflow Schematic Cooling Tower Indicating the Schematic Elements

dT	(delta T) difference between hot and cold temperature expressed in °F
h_a	enthalpy of air-water vapor mixture at wet bulb temperature; expressed in BTU per pound of dry air
h_w	enthalpy of air-water vapor mixture at bulk water temperature; expressed in BTU per pound of dry air
K	overall enthalpy transfer coefficient; expressed in pound per hour per square feet per pound of water per pound of dry air
$\dot{L}$	mass water flow; expressed in pound per hour per square feet of plan area
T_1	hot water temperature; expressed in °F
T_2	cold water temperature; expressed in °F
V	effective cooling tower volume; expressed in cubic feet per square feet plan area

It is quite obvious that wood fill is extremely inefficient compared to cellular fill; cellular fill has a lower resistance to air flow, which further enhances the heat transfer by more efficient utilization of the existing air. In wood slat splash bars (illustrated in Figure 4-9), droplets of water bounce from one layer of wood to the other and the rising air cools the outside of each sphere of the water droplets. Cellular fill takes the same droplets of water and spreads it out in a very thin molecular film where the air can now affect the entire surface of the film (Figure 4-10). Considerably more surface is then available to the flowing air for vaporization and sensible heat exchange to take place. The film pack contains more surface area than splash bars, and since the design of cellular fill permits air to flow through it with less resistance, it is extremely efficient compared to the old-fashioned splash bar mixture system for the same parameters of cooling.

Crossflow cooling towers also have a rebuilding capability to improve the performance, by changing wood splash bars to more efficient fill. Old-fashioned stacked fill is rather inefficient compared to the new grid type splash bar packing consisting of corrosion proof fiberglass polyester grids for higher capacity installation. A less expensive approach is to use metal grid hangers that are coated for corrosion protection. The installation of this type of fill is quite laborious, but is relatively inexpensive to install for the percentage of improvement that can be obtained compared to stacked fill.

The greatest improvement of performance in crossflow fill is obtained by changing the wood splash bars to cellular fill, (Figure 4-11), installing redistribution

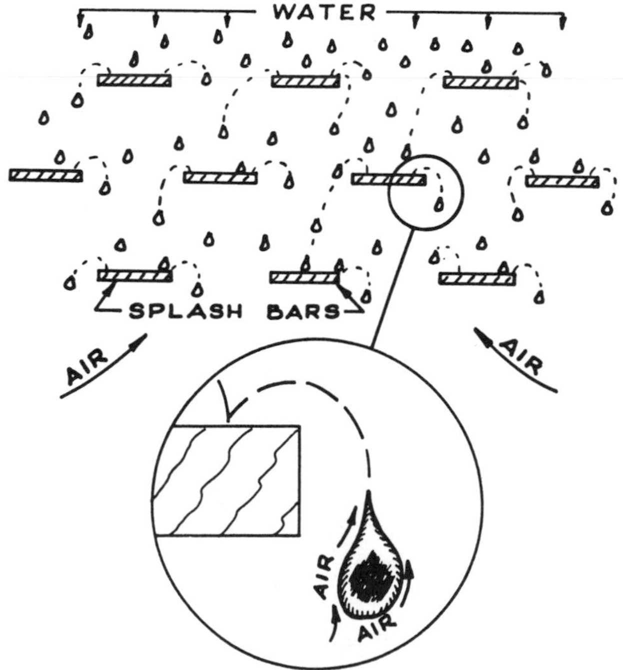

Figure 4-9. Schematic Operation of Splash Bar Cooling

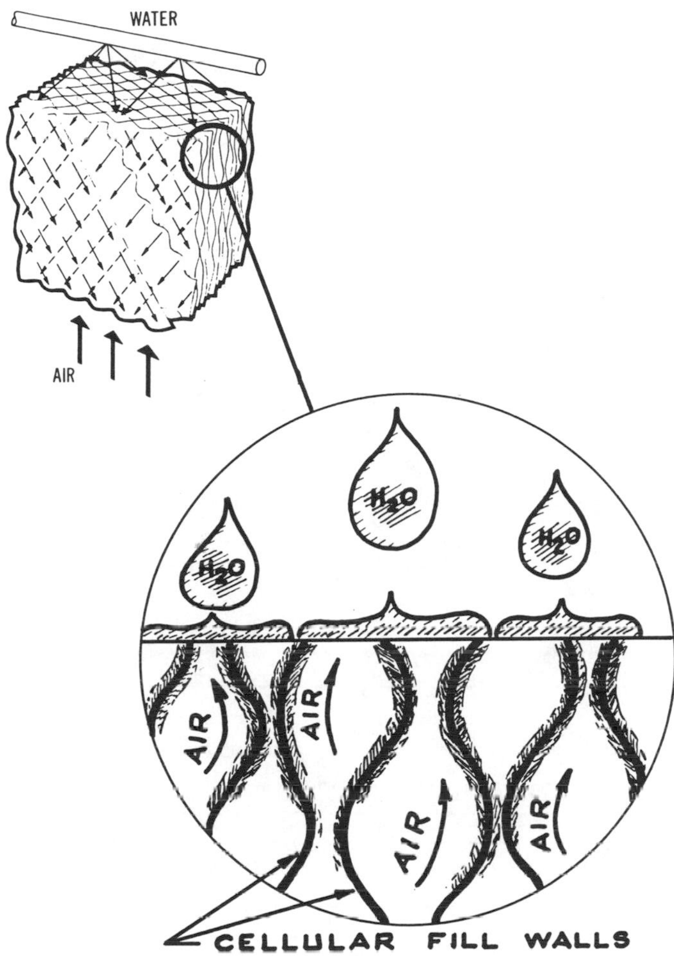

Figure 4-10. Schematic Illustration of Cellular Film Fill Cooling

sections of predetermined levels, and adjusting the flow rate throughout the hot water distribution basin to provide proper water loading through the fill.

Here again, wood splash bars are replaced by the efficient cellular fill sections, thereby reducing air flow resistance through the tower. This provides a more effective use of the input energy, and results in colder water from the tower. Whether it be crossflow or counterflow, *uniformity is the key to success:* uniformity of water distribution, uniformity of fill configuration, and uniformity of air velocity—even though in a crossflow tower, the air pressure differential will vary with the height.

Changing the Water Distribution System

For maximum performance, the water distribution system must provide a uniform pattern over the fill for optimum air-water interface. The older type trough has a very uneven splash distribution based upon columns of water falling vertically and hitting cups that, when accurately placed, distribute the water throughout the

Figure 4-11. Crossflow Tower Converted to Cellular Fill

fill. By replacing this water trough with ceramic spray nozzles and PVC piping, tremendous upgrading in capacity is available. Over the years this delicate balance is destroyed as the tower deteriorates. The end result is a vertical column of water, in many places, dropping up to four feet through the fill before it is broken up, losing entire areas of efficiency. When a nozzle is clogged, it leaves a dry spot on the fill; the air, being lazy, follows the path of least resistance and rushes up this dry spot, wasting energy and cooling potential.

If a given cooling tower is performing well, its pipes are not corroding, the nozzles are not clogged up, and the maintenance costs are negligible, then there is no requirement to spend money on the installation. On the other hand—if better performance through colder water is urgent, the maintenance department spends too many labor-hours cleaning nonfunctioning nozzles, and management is considering spending capital investment money for new facilities, now is the time to consider upgrading the existing plant by changing the water distribution to a highly efficient method—utilizing cheaper tax deductible maintenance dollars.

Experience and many installations have proven the high efficiency of water distribution of the specially designed ceramic nozzle, together with noncorroding polyvinylchloride (PVC) piping. The labor savings in not having to clean the 12 ceramic 1″ orifice units periodically will more than offset the installation cost against the old-fashioned ⅛″ openings on the 385 nozzles.

Also the large diameter 3″ PVC piping will not clog with rust readily. If by chance it does, it is simple to unscrew the end cap for blowdown. The nozzle snaps out of a bayonet slot holder by hand pressure.

The above portion has been devoted to small package units; however, the same techniques apply to larger field-erected installations.

The necessary hydraulic calculations will determine the pipe sizes and orifice openings. Figure 4-12 shows a 14″ diameter pipe assembly with 6″ diameter arms reduced to 4″ diameter, with the nozzles on 42″ centers, 42″ between arms. The water distribution pattern on Figure 4-13 indicates a slight overlap of the cone in the previous installation. This depth is calculated and located at the optimum point by the manufacturer's curves and charts.

If existing steel header pipes and arms are corroded to the point of replacement, the labor cost of installing PVC piping over galvanized steel is less. A man with a rope can hoist the plastic pipe and have an assistant help lock it in place. The weight of the steel will require riggers and a large team of workers to do the same thing.

Water volumes from 15 GPM to 250 GPM per nozzle are available in the two to seven psi-pressure range, well within the operating parameters of today's cooling tower requirements.

Figure 4-12. 14″ Diameter PVC Main Header with 6″ Diameter Branch Arms Having Nozzles on 42″ Center Distances

Figure 4-13. Spray Pattern from Ceramic Nozzle Provides Complete Wetting Coverage of Fill

Upgrading Checklist

There are many methods of upgrading cooling towers for higher performance. A common procedure to save dollars is a "Do It Yourself Program." This may or may not be practical. A strongly recommended procedure would be to hire the services of a cooling tower consultant who has access to many different types of fill materials and procedures. A sales rep from a cooling tower firm might be inclined to advise the cooling tower owner only to use materials that are proprietary with the rep's company. The knowledgeable consultant, however, can recommend procedures developed by independent suppliers and upgrading innovations that are not necessarily provided by many manufacturers.

If you employ the services of a consultant, use the cooling tower fact sheet in Figure 4-14 to make your job easier and to save your company some money.

CASE STUDIES IN UPGRADING COOLING TOWERS

Using the previous information about the various elements, let us investigate examples of how various cooling towers can be upgraded to permit the operation to produce additional profits after a payback of the retro-fit costs.

Cooling Tower Fact Sheet

Supply as much data as is available

Requested by: ________________ Date: ________________

Name, Title: ________________

Company: ________________ Phone: ________________

Address: ________________ Fax: ________________

City: ________ State ______ Zip ______

Cooling Tower Manufacturer ________________ Model No. ________ Number of Cells ________

Counter Flow ☐ Cross Flow ☐

Mechanical Equipment

Fan Diameter (Feet) ________

No. Fan Blades/Fan ________

Fan RPM ________

CFM/Fan ________

Motor H.P. ________

Motor RPM ________

Gear Ratio ________

Riser Pipe Dia ________

Original Design Conditions

GPM Each Per Cell ________

GPM Total System ________

Temperatures of Water Degrees F:

On Tower ________

Off Tower ________

Amb WBT ________

New Conditions Desired

GPM Each Per Cell ________

GPM Total System ________

Temperature of Water Degreess F:

On Tower ________

Off Tower ________

Amb WBT ________

DIMENSIONAL INFORMATION

Indicate Air Travel Distance for Crossflow Towers ______ Feet

Fill Height

TYPICAL CELL

Length ______ Width ______ Height ______

Indicate Cells and Fan (s) Position on Both Sketches

Figure 4-14. Thermal Analysis Checklist

Case Study #1: Upgrading the Cooling Tower for an Industrial Plant Refrigeration System

At a food processing plant on the East Coast, the total water on the tower is 3,750 GPM (14,200 liters per minute) for two cells. Design conditions were to cool the entering water from 95°F (35°C) leaving at 85°F (29.5°C) at an ambient wet bulb temperature of 78°F (25.5°C). One cell at a time was converted, to maintain continuous plant operation.

The first indication that engineering calculations and predictions would be exceeded occurred when the total amount of gallons was put on the newly converted cell, while the second cell was rebuilt. With 50 percent of the tower in operation and the same electrical input to the 40 HP motor, the one cell cooled the water to design with a 10°F (5.5°C) range and 7°F (4°C) approach. When both cells were operated with maximum water 3,750 GPM (14,000 liters), temperatures were returned almost 4°F (2.3°C) colder.

Records kept by the chief engineer indicated there was a direct fuel oil savings of approximately $375 a day, since in generating their own steam for the turbines, the 6,000 gallons-a-day (22,700 liters) oil requirement was cut down to 5,700 gallons (21,500 liters)—a savings of 300 gallons (1,200 liters) a day.

Further, the turbines were running at 350 GPM less and head pressures were four to five pounds per square inch (0.28 to 0.35 kg. per square cm) lower than previous records indicated. Those small increments added up to approximately $70,000 a year savings. With escalation of all costs prevalent, this figure should also increase.

The ultimate savings, however, is that there is a 50 percent plant expansion contemplated. Calculations indicate that, with the addition of 12″ (30 centimeters) of cellular fill and piping changes to larger diameters for the water distribution system, the existing cooling tower with two 40 horsepower motors and 16′ (4.9 meters) diameter fans would be more than adequate to take care of the new requirements. This should result in an economy of approximately $87,500, which is the cost between new OEM equipment and additional rebuilding work.

Case Study #2: Upgrading the Cooling Tower for an Office Building

This upgrade involves a counterflow tower in a large government office building in Washington, D.C., where original design conditions were marginally met. G.S.A. required a 25 percent increase in performance from 12,000 to 15,000 GPM (45,420 to 56,770 liters) per minute due to higher heat loads. The successful bid was $125,000 less than a proposal to install new original equipment manufacturers towers. Since the cooling tower was built using the vertical building columns as the main tower supports, cost of demolition and installation, including reconstructing the building around the new equipment, dictated that the prudent way would be to rebuild the existing tower.

One of the keys to improvement is a ceramic nozzle costing $15.50. This replaces a dozen or more bronze units $10 to $12 each, plus the cost reduction for using PVC piping at $18 per foot installed for 12″ diameter (30 centimeters diameter), versus $27.00 installed for galvanized steel. The weight differential is 90 percent plus in saving labor during installation.

The elements of rebuilding this tower were as follows:

- Cellular Fill
- Drift Eliminators
- PVC Piping
- Ceramic Nozzles
- 30 inches (76 centimeters) of cellular fill replaced 144 inches (365 centimeters) of wood splash bars.
- Efficient fireproof drift eliminators that can be walked on, provided mist protection and also acted as a fireproof barrier.
- A tremendous aid in upgrading this structure was the replacement of old style gravity trough water distribution to PVC low pressure piping.
- In conjunction with the PVC piping, ceramic nonclogging, noncorroding large orifice nozzles were installed.

Figure 4-15 illustrates the upgraded and rebuilt tower.

This type of rebuilding is an open-ended funnel. Additional capacity may be obtained by pumping more water over the fill, increasing the depth of fill, or increasing the airflow; any combination of these events will add up to more work for less investment. In this case, the savings in dollars were generated because the four 60 horsepower motors, after conversion, produced an additional 3,000 GPM of cooling (11,330 liters per minute). If the four-cell tower was not rebuilt and an additional cell installed, then 300 horsepower would have been required to produce the same

Figure 4-15. Elements of Upgraded Rebuilt Tower:

amount of work that the presently rebuilt installation is doing using only 240 horsepower. Multiplying the current Washington, D.C. power rates by the tower utilization results in a KWH reduction of 1,575 and a $16,200 savings per year.

Case Study #3: Upgrading the Cooling Tower at The Interchurch Center

The design conditions of this four-cell, double flow crossflow tower were originally to cool 5,400 GPM from entering the tower at 95°F leaving at 85°F at a 78°F wet bulb temperature.

According to the manager, design conditions were never met and the deficiency of the cooling tower created high heads requiring excess electricity to power the system.

Since the Interchurch Center is not a profit-oriented organization, funds were limited. Thermal engineering calculations indicated that if two of four cells were rebuilt with highly efficient cellular fill, the results would bring the tower up to design.

All four cells were stripped of wood and the interior metallic surfaces were sandblasted to remove corrosion and the old coating, primed with synthetic Parlon®, and coated with two applications of Corro-tect®, liquid chlorinated rubber corrosion-protective coating.

The wood splash bars were carefully separated so that sufficient quantities were selected to rebuild two cells. The remaining two cells were then installed with cellular fill.

Operating records indicate that the condenser water is now two degrees colder than last year, which is more than adequate to provide proper operating conditions. The head pressures are running five to six psi lower, which reduces electrical consumption. One of the major savings is that in their two-refrigeration machine operation, one machine is now utilized 25 percent less than previous seasons. Future planning calls for changing the remaining two wood cells. A further improvement will produce extra colder water, which will more than pay back in energy conservation, and includes less projected maintenance wear and tear on the equipment.

Case Study #4: Upgrading the Cooling Tower at Sage Intercontinental Defense Missile Command

This cooling tower was 18 years old and required extensive rebuilding as an alternate to demolition or replacement. Government specifications called for rebuilding the existing drift eliminators, taking out and putting back new wood wet decking fill, installing gravity splash nozzles attached to the wood water distribution trough orifices, miscellaneous carpentry work, and repairs.

As an alternate bid, the U.S. Air Force engineers, after an educational session, re-advertised the bid, allowing an alternate for modern retro-fit. Drift eliminators fabricated from cellular materials were installed, water troughs were taken out, and PVC piping with ceramic square spray nozzles were engineered and installed together with wet decking cellular fill.

This new system saved the Air Force $29,500 over the second bidder, and upon testing the installation with full load, the cooling tower performed 32 percent over the previous year's operation of cooling 5,000 GPM from entering the tower at 95°F, leaving at 85°F at 75°F ambient wet bulb temperature.

One of the main upgrading elements was changing the old water trough distribution system to PVC piping and ceramic nozzles.

CONCLUSION

A blanket statement cannot and should not be made that all cooling towers can be rebuilt to save large quantities of money and power or to return water colder by a significant amount. Each installation must be treated on an individual basis with thermal, hydraulic, and aerodynamic calculations studies of existing conditions to see where areas of improvement can be largely anticipated. Since significant sums of money can be spent on larger cooling towers, it is strongly recommended that a testing requirement be made part of the contract. The Cooling Tower Institute (CTI) has formulated ATC-105, which contains rigid testing procedures for determining whether or not a cooling tower will produce what the customer has been promised and is expected to pay for.

Acceptance Test Procedure Code is recognized throughout the industry as the standard measurement of a tower's efficiency. The CTI, in the interests of obtaining a strictly impartial test, contracts the testing to be done by the Midwest Research Institute (MRI), whose only concern is to conduct the procedure according to the code, collect field data, using authorized and calibrated instruments, and obtain a computerized readout of the percent of the cooling tower's performance.

All interested parties—the contractor, the owner, and the manufacturer—are invited to observe to see that impartiality is conducted; and regardless of who pays the fee, the owner receives a report as to the operation of his cooling tower.

A cursory inspection by maintenance managers or consulting engineers of a deficient and seemingly old cooling tower should not automatically bring forth the snap judgment that this structure should be removed and a new unit put in. Examination of the tower's structural strength and its components could indicate that, with the applications of sound modern engineering principles and proper maintenance, the cooling tower could be given an extended useful life of service, producing colder water for the added profit of the facility and saving input energy.

5

Maintaining Air Compressors and the Compressed Air System

William Scales, P.E.

Many problems with air compressors and the compressed air system can be avoided with proper selection of the equipment. This chapter deals with selection, applications, and evaluations, and provides details on what and how to do maintenance to prolong the life of air compressors and accessories.

Ask most non-technical people about applications of air compressors and perhaps one or two can think of an answer. Most people are unaware of the diversity of compressed air—that it affects their everyday lives from the aeration of the drinking water in a reservoir, to the making of artificial snow, to the operation of air tools, spray equipment, and packaging machinery in manufacturing plants throughout the world.

There are two basic types of air compressors in use today. The positive displacement type compresses air by admitting successive volumes of air into a closed space and then decreasing the volume. Reciprocating, rotary screw, sliding vane, two lobe rotary, and liquid piston are all examples of a positive displacement compressor. Dynamic compressors refer to the centrifugal or axial flow compressors that apply mechanical force to the air to increase its velocity, which is then converted to pressure. The majority of industrial plants have positive displacement compressors; therefore, the major portion of this chapter is devoted to this type, although the same principles apply to maintenance for all types. The sections dealing with the compressed air system and accessories are applicable to any plant regardless of the type of compressor used.

UNDERSTANDING THE TERMINOLOGY—BASIC DEFINITIONS

Before you can select the compressor which is best suited for an application, there are certain basic terms that should be understood.

The pressure in a system is defined in psia (pounds per square inch absolute) and psig (pounds per square inch gauge). The absolute pressure is the sum of the gauge pressure and the atmospheric pressure. The basic metric unit used is kilograms per square centimeter; more recently the kilopascal has been accepted as the unit of measure of air pressure.

Volume is expressed in cubic feet per minute (cfm) in the English system, and in cubic meters per hour or per minute in the metric system. However, because there are modifications to the normal term *volume,* you must clearly understand whether reference is being made to actual cfm (acfm), actual cfm at inlet conditions, standard cubic feet per minute (scfm), free air (cfm), etc.

Brake horsepower (BHP) is the measured horsepower delivered to or required at the compressor shaft. One of the most commonly used measurements of a compressor's efficiency is the brake horsepower required to produce 100 cubic feet of air at a given pressure (BHP/100). The lower the figure, the more efficient the compressor. This figure enables you to compute energy costs. At 100 psig, the BHP/100 can vary between 18.5 and 27 depending on the type of compressor selected. In many cases, the annual cost of electric energy exceeds the initial price of the air compressor. Therefore, careful selection of the compressor coupled with accurate application information can save a company thousands of dollars per year in energy costs.

Become familiar with the terminology used in the industry to be able to compare the performance of different compressors. (See Figure 5-1.)

TYPES OF COMPRESSORS

Compressors fall into two broad categories: positive displacement and dynamic. Positive displacement compressors admit successive volumes of air into an enclosed space and then decrease the volume, thereby increasing the pressure. Dynamic compressors increase pressure by translating kinetic energy or velocity into pressure. The reciprocating, rotary sliding vane, and rotary screw are examples of positive displacement machines, and the centrifugal and axial flow are dynamic compressors. Because of structural differences which can vary widely from unit to unit, each type of compressor is prone to different types of problems and requires a different approach to maintenance to keep it in working order. A description of each compressor type and illustrations appear in Figures 5-2 and 5-3.

This chapter first describes how to maintain positive displacement compressors—both reciprocating and rotary—which come in many types, and then how to maintain and troubleshoot dynamic (centrifugal) compressors.

ABSOLUTE PRESSURE (psia) is the total pressure measured from absolute zero, i.e., from an absolute vacuum. It equals the sum of the gauge pressure and the atmospheric pressure corresponding to the barometer. At sea level atmospheric pressure is 14.7 psia and 100 pounds gauge pressure would be 114.7 psia.

ACTUAL CAPACITY (acfm) is the actual volume rate of flow of air compressed and delivered at the discharged point but referred to conditions of total pressure, temperature, and humidity at the inlet.

ADIABATIC COMPRESSION is achieved by preventing the transfer of heat to or from the air during compression.

AFTERCOOLERS remove heat from the air after compression is completed and are used to effect moisture removal.

AIR RECEIVERS or "tanks" provide a volume of stored air. They are used as pulsation dampeners on reciprocating compressors and are generally sized in cubic feet for a minimum of one-seventh the cfm of the compressor for constant speed operation and one-third for automatic start-stop control.

BLOWERS generally refer to the broad category of rotating compressors which operate below 30 psig.

BOOSTER COMPRESSORS receive air that has already been compressed and deliver it at a higher pressure.

BRAKE HORSEPOWER (BHP) is the measured horse power delivered to the compressor shaft.

CAPACITY is often used interchangeably with actual capacity or inlet capacity and is stated in cfm.

CENTRIFUGAL COMPRESSORS are dynamic type compressors where velocity is translated to pressure, and discharge is through a collector ring at the outside rim of the housing. It is often built as a multi-stage unit.

CLEARANCE is the volume inside a compression space that contains air trapped at the end of the compression cycle—it is between the piston and the head and under the valves in a reciprocating compressor.

COMPRESSION EFFICIENCY (adiabatic) is the ratio of the theoretical horsepower to the horsepower imparted to the air actually delivered by the compressor. The power imparted to the air is brake horsepower minus mechanical losses.

COMPRESSION RATIO is the ratio of the absolute discharge pressure to the absolute inlet pressure.

DISCHARGE PRESSURE is the absolute total pressure at the compressor's discharge flange; it is generally stated in gauge pressure (psig) but should include reference to the barometic pressure.

DISCHARGE TEMPERATURE is the total temperature at the discharge flange of the compressor.

Figure 5-1. Compressor Terminology

DISPLACEMENT of a compressor is the volume expressed in cubic feet per minute (cfm) displaced per unit of time. The term is generally used for reciprocating compressors and is derived by taking the product of the net area of the compressor piston times the length of stroke and the number of compression strokes per minute. The displacement of a multi-stage compressor is the displacement of the low pressure cylinder only.

DOUBLE-ACTING COMPRESSORS compress air on both strokes of the piston, therefore twice per revolution.

DRYERS are machines or devices used to remove water vapor from the compressed air. The most common types are deliquescent, refrigerated and regenerative.

DYNAMIC COMPRESSORS refer to the centrifugal or axial flow compressors that translate velocity into pressure, i.e., kinetic energy to pressure energy.

FREE AIR is defined as air at atmospheric conditions at any specific location. Because the altitude, barometer and temperature may vary at different localities and at different times, it follows that this term does not mean air under identical or standard conditions.

INLET PRESSURE is the absolute total pressure existing at the intake flange of a compressor.

INLET TEMPERATURE is the total temperature at the intake flange of a compressor.

INTERCOOLING is the removal of heat from the air between stages. Ideal intercooling exists when the temperature of the air leaving the intercooler equals the temperature of the air at the intake of the first stage.

ISOTHERMAL COMPRESSION is achieved by maintaining the air at constant temperatures during compression.

LIQUID PISTON COMPRESSORS displace the air in an elliptical casing with water or other liquid, which acts as the "piston" to compress the air.

LOAD FACTOR is the ratio of the average compressor load during a given period of time to the maximum rated load of the compressor.

MECHANICAL EFFICIENCY is the ratio of the horsepower imparted to the air to the brake horsepower expressed in percent.

MOISTURE SEPARATORS are small vessels that are usually used with intercoolers and aftercoolers to separate the condensed moisture from the air through mechanical or cyclonic action.

MOISTURE TRAPS automatically discharge moisture from separators or receivers. There are many types available such as float, inverted bucket, solenoid valve with timer, etc.

Figure 5-1. *(continued)*

MULTI-STAGE COMPRESSORS achieve final discharge pressure in more than one stage.

OVERALL EFFICIENCY is the ratio of the theoretical horsepower to the brake horsepower. It is equal to the product of compression efficiency times mechanical efficiency. Electric motor and transmission losses should also be considered in computing cfm out versus kilowatts in.

PACKAGED COMPRESSORS are complete units consisting of all components, i.e., compressor, motor, aftercooler, receiver and controls mounted on one base for "single point" electrical and air and two point water connections. Installation is simplified with this type of purchase.

PORTABLE COMPRESSORS are usually mounted on skids or wheels to be moved from one location to another. They may be electric, gas or diesel driven.

POSITIVE DISPLACEMENT compressors compress air by increasing pressure by admitting successive volumes of air into a closed space and then decreasing the volume. Reciprocating, rotary screw, sliding vane, two lobe rotary, and liquid piston are all examples of a positive displacement compressor.

RECIPROCATING COMPRESSORS achieve compression by a piston moving in an enclosed cylinder and decreasing the volume.

ROTARY SCREW COMPRESSORS utilize two close clearance helical lobe rotors turning in synchronous mesh and compress the air by forcing it into a decreasing inter lobe cavity until it reaches the discharge part.

SINGLE-ACTING COMPRESSORS compress air on only one side of the piston. For example, in vertical configuration the downward stroke is suction, and the upward stroke is compression.

SINGLE-STAGE COMPRESSORS accomplish final discharge pressure in one step.

STANDARD AIR (scfm) is air at defined conditions. Two standards are presently in use. The first is a temperature of 68 degrees F, a pressure of 14.7 psia and a relative humidity of 36 percent. The second is 60 degrees F, 14.7 psia and dry.

VACUUM PUMPS compress air from sub-atmospheric pressures to atmospheric.

VANE TYPE COMPRESSORS have a rotor and vanes or blades that slide radially in an offset housing forming sealed sectors. Volume is decreased by the convergence of the cylinder walls with the vanes and the rotor body.

VOLUMETRIC EFFICIENCY is the ratio of the capacity of the compressor to the displacement expressed in percent. It is usually between 60 and 80 percent and is always reduced as pressure increases in a given compressor.

Figure 5-1. (*continued*)

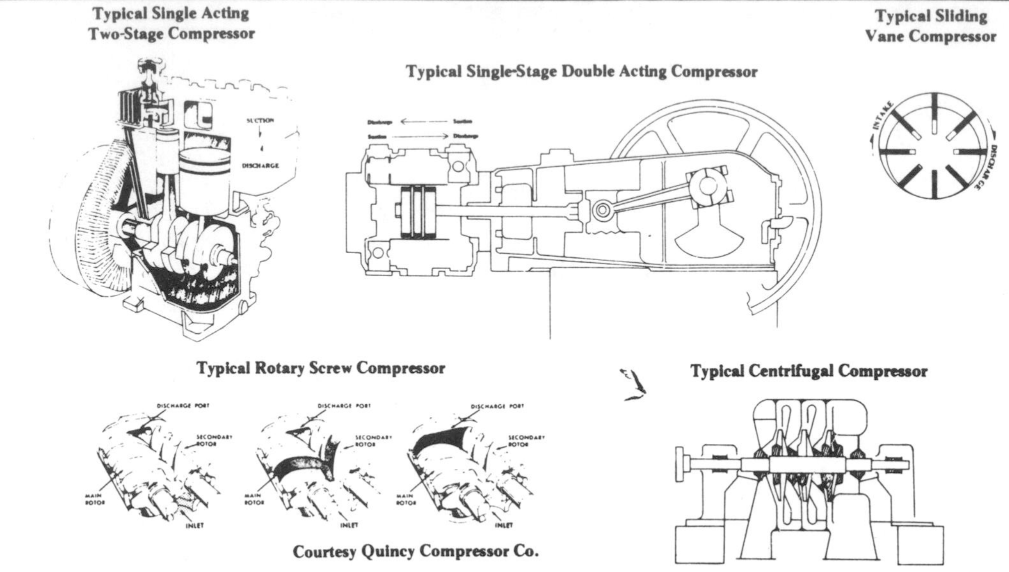

Reciprocating Type (a, b)

Compression achieved by a piston moving in an enclosed cylinder and developing pressure is a reciprocating compressor. Cylinder arrangement may be vertical, horizontal or angular, and may be either air-cooled, water-cooled, lubricated or non-lubricated. Compression may take place in either one end (single-acting) or both ends (double-acting).

Single-stage compressors consist of one or more cylinders taking air in from the atmosphere and compressing it to the final discharge pressure required. Normally single-stage units are used for pressures up to 100 psig for continuous service and can sometimes be used for intermittent service at higher pressures.

Two-stage compressors have a minimum of two pistons; the first compresses air from atmosphere to an intermediate pressure (25-40 psig). The air is then cooled in an intercooler before entering the second stage where it is compressed from the inter-stage pressure to the final discharge pressure (100-200 psig). Due to the dissipation of heat between stages (because of the intercooler), above 100 psig, two-stage units allow for more efficient and cooler operating compressors, which increases compressor life.

Rotary Sliding Vane (c)

A rotary sliding vane compressor is a machine with a rotor and metallic or non-metallic vanes which slide radially in an offset housing forming sealed sectors. As the rotor turns, the vanes are held to the cylinder walls by centrifugal force. As a sector rotates toward the discharge port, its volume is decreased by the convergence of the cylinder walls with the vanes and rotor body. Units can be single or two-stage, water or air-cooled.

Rotary Screw (d)

Two close clearance helical-lobe rotors turn in synchronous mesh. Normal is four lobe main male rotor with a six groove mating female rotor (gate). Air is drawn into the suction port into a space between the rotors. As the rotors revolve, air is forced into decreasing inter-lobe cavity until it reaches the discharge port. In lubricated units the male rotor drives the female and oil is injected into the cylinder serving as a lubricant, coolant and as an oil seal to reduce back slippage. On non-lubricated types, timing gears are used to drive the rotors and multi-staging is necessary for pressures above 50 psig.

Centrifugal Compressor (e)

A compressor in which air is introduced at the center of a rotating impeller which then accelerates the air radially. This velocity is translated to pressure, and discharge is through a collector ring at the outside rim of the housing. It is often built as a multi-stage unit.

Axial Flow Compressor

A compressor in which the air is introduced at one end of the unit and accelerated along the axis of the compressor. This velocity is translated to pressure, and discharge is from the end opposite the intake. It is often built as a multi-stage unit.

Figure 5-2.

a. Typical Single-Acting Two-Stage Compressor
b. Typical Single-Stage Double-Acting Compressor
c. Typical Sliding Vane Compressor
d. Typical Rotary Screw Compressor
e. Typical Centrifugal Compressor

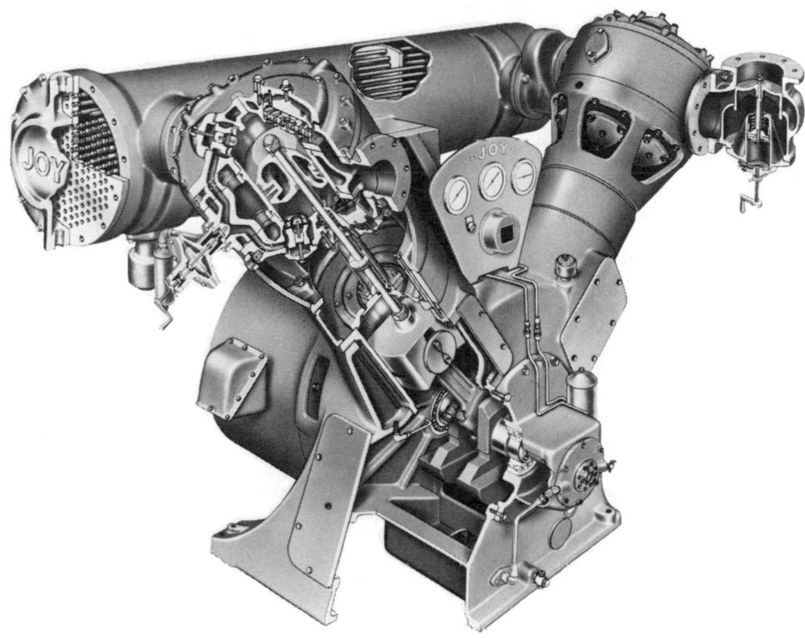

Courtesy of Joy Manufacturing Company Division of Cooper Industries

Figure 5-3. Two-Stage, Double-Acting, Water-Cooled Compressor. Available in Lubricated or in Teflon Ring Construction Through 400HP. Different Configuration Available in Larger Sizes.

RECIPROCATING COMPRESSORS

In considering the purchase of a reciprocating compressor, you should evaluate single-acting, double-acting, single-stage, two-stage, air-cooled, and water-cooled units. Each plant is different, so you must weigh initial cost, installation, maintenance, and operating expenses to arrive at a fair comparison. In addition, if water is required for a water-cooled unit, or fresh air ducts are needed to keep the ambient temperature for an air-cooled unit below 100°F, then these costs must also be calculated. At 100 psig, multi-stage units generally result in a power savings of up to 15 percent versus single-stage units. In addition, multi-staging with cooling of the air between stages reduces the final discharge temperature in the cylinders, which reduces the problems with carbon deposits and lubrication breakdown. Generally, the higher the discharge temperature, the higher the maintenance and operating costs of a compressor, and the greater the possibility of discharge line explosions in lubricated compressors.

Types of Controls

The types of controls that are available to regulate the delivery of the reciprocating compressor fall into basic categories: automatic start-stop, constant speed control, and dual control.

Automatic Start-Stop

In this form of control, a pressure switch automatically controls the starting and stopping of the compressor. This type of control is generally used on smaller compressors (tank mounted type) or on larger compressors when the demand for air is quite low (less than 50 percent). It is important when using start-stop control that the air receiver be sized to limit the number of starts per hour to a figure that is acceptable to the motor manufacturer. Frequent starts can reduce motor life and create control problems and rapid contact wear in motor starters.

Constant Speed Control

In this mode, the compressor operates continuously at its normal operating speed and the compressor is "unloaded" (compressor stops pumping but continues running) at a set point. When pressure drops to a preadjusted point, the unloading mechanism allows the compressor to compress air again. The primary method of unloading is inlet valve regulation. This system uses a device (air-operated pilot valve or a pressure switch with a three-way electric solenoid valve), which allows air to feed back from the receiver to a plate actuated by a diaphragm or unloading piston on top of the inlet valve. This plate holds the intake valves open whenever there is no demand for air (Figure 5-4). With the inlet (also called intake or suction) valves open, the air will sweep in and out of the cylinder and very little horsepower is required to operate the compressor. Generally, the combination of the friction horsepower and horsepower required in the air to force it through the inlet valve is less than 20 percent of the full load horsepower.

On smaller compressors, loading and unloading is normally accomplished in two steps, i.e., either fully loaded or full unloaded. On larger water-cooled compressors, three-step control (full load, half load, no load) and five-step control (100 percent, 75 percent, 50 percent, 25 percent, no load) is available.

Dual Control

Dual control combines the automatic start-stop and constant speed control so that the operator can select the mode required based on the plant demands. For example, constant speed operation may be required during the normal workday and start-stop may be used at night and on weekends when demand for air may be lower.

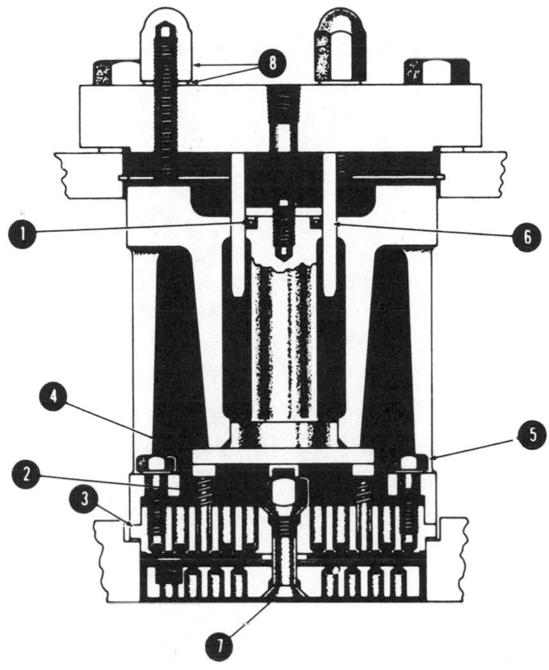

Courtesy of Gardner-Denver Division of Cooper Industries

Figure 5-4. Suction Unloading Valve Assembly:

1. The plunger seal is used to seal the unloading air.
2. Finger pin springs return the unloader mechanism when unloading air is released.
3. Clamp rests on full circumference of valve seat to eliminate warpage.
4. Plunger acts on finger pins to open valve disc fully and evenly.
5. Clamp is secured to the valve assembly by hex head cap screws to aid in installation and removal.
6. Separate valve cover is guided in clamp to provide proper alignment of assembly.
7. Complete lower valve assembly is held together by one flat head screw and hex nut permitting easy assembly and disassembly. Numerous suction and discharge valve parts are interchangeable.
8. Acorn nuts and gaskets prevent leakage around locking screws.

Automatic Dual Control

Automatic dual control is a further refinement. This system has constant speed control as the basic mode of operation. If the compressor unloads and remains unloaded for a preset period of time, the compressor stops and is placed in the standby position. If pressure continues to drop, the compressor will start automatically and begin the cycle again.

Dual and automatic dual control are very inexpensive compared to the purchase price of a larger compressor, and these options should be requested as part of the quotation when specifying or purchasing new compressors.

Safety Devices

The number of devices used in a compressor system depends upon the maintenance manager and the compressor vendor. However, the most commonly used are described in the following sections.

High Air Temperature Switch

This is installed in the air compressor discharge line as close to the discharge flange as practical. At 100 psig, the switch is set approximately 400-425°F on a water-cooled single-stage compressor and 350°F on a two-stage compressor. This switch is very important and will normally shut down the compressor because of inadequate cooling or bad valves. On air-cooled machines, the settings may be somewhat higher.

Low Oil Level and Pressure Switches

The low oil level switch monitors oil level on a "splash" lubricated unit and will not allow the compressor to start unless adequate oil is in the crankcase.

The low oil pressure switch monitors oil pressure on a pressure lubricated compressor. On an oil pressure switch, the contact is normally open and closed on pressure rise. Since this contact is in series with the coil of the magnetic starter, a normally closed contact, which is timed to open after 10 or 15 seconds, (enough time for oil pressure to develop), is shunted across the oil pressure switch contact to allow the compressor to start. If oil pressure has not developed, the compressor will shut down and should be locked out through a proper relay.

Safety Valves

Safety valves should be used on all receivers and before any isolation valves that may be installed between the compressor head and the receiver.

Safety valves should be of the ASME type and should not be set above the working pressure of the receiver it is protecting. The safety valve should also be used between the compressor and any isolation valve between the compressor head and the receiver. This should be set at the maximum safe operating pressure of the compressor or below the maximum safe operating pressure of the isolation valve. A safety valve on an interstage cooler will not protect the system against the danger of an unprotected isolation valve. The interstage valve merely protects the low pressure cylinder from damage that may be caused when a valve malfunction occurs on the high pressure cylinder. Safety valves are also often installed before aftercoolers to protect against heavy carbon buildup restricting the flow of air through the cooler. Carbon restriction could result in excessive head pressure and possible damage to the compressor or fire and explosions in the discharge line.

Maintaining Your Compressor's Valves

The valves are the heart of the compressor and must be maintained properly for the compressor to operate properly. Valves in a compressor operating at 900 rpm for only eight hours per day and 220 days per year open and close 54,000 times per hour or 95,000,000 times per year. Whether valves are of concentric ring design or rectangular leaf or channel type, they should be inspected yearly as a minimum and rebuilt as experience dictates. Keeping records of valve repairs and what has worn is of primary importance in determining causes of failure. Replacement parts should be kept in stock to reduce downtime that may result because of a valve failure. The parts should be greased if necessary and wrapped in a protective coating before storing in a dry area.

Follow the manufacturer's instructions carefully when rebuilding valves and use only genuine replacement parts. Well-designed valves should not require frequent repairs. However, poor repair work or foreign matter will increase the frequency of repairs. Dirty intake filters, excessive lubrication, or the wrong lubricant and moisture in the air all contribute to valve failures.

In water-cooled compressors, cylinder jacket water temperatures should be maintained 15-20°F above the incoming air temperature to insure that condensation of the air does not cause water (condensate) in the valves and cylinder areas. The water temperature can be maintained automatically with a thermostatic valve or monitored by plant personnel and adjusted manually as required. (Because of varying conditions, manual adjustment may not be practical.) The selection of the proper intake filter and lubricant will result in increased valve life.

Valve repairs require careful attention. Intake and discharge valve assemblies and parts may look alike but be different. It is important that these valves and parts be assembled *exactly* as they were removed.

Other important maintenance tips:

- Be sure the original valve lift is maintained.
- Make sure all valve parts are clean—this is absolutely essential.
- Inspect valves and springs carefully and replace if any wear is evident.
- Replace valve seat gaskets and cover gaskets. In some cases these gaskets can be reused, but good practice dictates new gaskets.
- Be certain inlet valves are placed in the inlet cavity and discharge valves in the discharge cavity. Severe problems can develop if the valves are inadvertently interchanged.
- Valves can be checked for leakage after rebuilding by pouring a commercial solvent into the valve disc and observing if any liquid leaks past the valve disc(s). Be sure all solvent is removed before the valves are installed.

- An often used method to test discharge valves is to install them in the cylinder or head without the suction valves and apply back pressure to allow the system to act on the discharge valves. If the valves leak, air will be in the cylinder and you will feel which valves leak.
- Above all, read the instructions.

A Guide to Clearances and Tolerances

Valves

Lift is the primary dimension that is considered.

- Concentric ring valves generally require .80 to .100 inches. However, one manufacturer states .060 to .070 inches for his valves.
- Channel valves, feather valves and any other rectangular-shaped valves should be checked with the manufacturer's recommendations.

Rings

Side clearance—.002 to .004 inches.

End gap clearance—.002 to .004 inches times piston diameter.

Piston to Cylinder

Aluminum—.003 inches times the bore diameter plus .010 inches.

Cast iron—.001 inches times the bore diameter plus .005 inches.

Piston End Clearance

Allow for expansion of the piston rod. From the piston face to the frame end should be one-third of the total, and opposite the piston face to the outer head should be two-thirds of the total clearance. This approximation should be done when the machine is cold. It can be done with a piece of lead or solder compressed by the piston against the head when the machine is rotated one revolution by hand. The lead or solder is inserted through a valve cavity.

- Packing side clearance:
Cast iron—.005 to .010 inches.
Teflon—.009 to .015 inches.
- Cross head clearance:
.00075 to .001 per inch of diameter of the crosshead guide.
- Reboring cylinders:
Reboring is practical in many cases where cylinders are out of round. The normal maximum reboring limits are .060 to .125 inches. Oversize pistons should be used in cases where boring is done.

In many cases where cylinders are worn less than .010 to .020 inches, reboring is not necessary and oversize rings are all that is required.

Trouble-Shooting Reciprocating Compressors

Trouble-shooting can be accomplished effectively only if the operator has records of what normal operations should be. For example:

- Record normal full load current at a specific voltage. These should be measured several times during the first week of operation of the compressor.
- If the compressor has its own receiver, allow the compressor to fill this receiver from zero to the cutout point. Record the cutout pressure and the time required to fill the receiver.
  The above can be used as a guide to check compressor efficiency in the future.
- Do not overlook an operator's instinct or feeling that something doesn't sound right. This knowledge of how a machine should sound has often prevented major damage.

Maintaining Reciprocating Compressors

The best advice I can give is to read the instruction book. No manufacturer wants problems with his equipment and he knows his machinery better than anyone else. Follow his advice.

Basic maintenance and trouble-shooting instructions for both single-acting and double-acting reciprocating compressors follow. (See Figures 5-5 through 5-7). However, there is no substitute for the owner's instruction manual; at all times the owner's manual should be followed whenever there is any conflict with the guide in this book.

Lubricating Reciprocating Compressors

A good lubricant should minimize wear, reduce friction and minimize deposits. Provide the oil company with the specifications given by the compressor manufacturer, and have the compressor company agree that this oil is acceptable for the application.

The proper oil is probably neither the most expensive nor the least costly. Oil selection should never be based on price; it should be based on performance. If experience has resulted in low maintenance and carbon deposits, continue using the oil.

A single-acting lubricated compressor uses the same oil for the crankcase and the cylinder. While the function of the rings is to control and reduce oil carryover, some oil will pass. The oil in contact with the hot valve areas may leave carbon. A low carbon residue rating for the oil is recommended, although this doesn't necessarily guarantee low carbon deposits. If the compressor begins to pass oil, it is important that the problem be corrected rapidly. The danger of fires and explosions is present

RECOMMENDED INSPECTION, MAINTENANCE AND SERVICE PROCEDURES FOR SINGLE ACTING AIR COMPRESSORS

DAILY

1. Check oil level. Replenish, if necessary, with proper oil and correct viscosity for surrounding conditions. Check oil pressure if unit is pressure lubricated.

2. Drain receiver of accumulated moisture. If automatic drain is used, manually drain tank weekly to check operation of automatic drain. Recommended that discharge of automatic drain be visible to check for operation.

The above can be done at bi-weekly intervals if operating conditions indicate this is adequate. Special attention should be given to changing seasons requiring more or less frequent attention.

MONTHLY

1. Check distribution system for leaks.
2. Manually operate safety valves to be certain they are functioning.
3. Clean intercooler fins and cylinder with a jet of air.
4. Replace or clean intake filter.
5. Inspect oil for contamination and change if necessary.
6. Check belts for correct tension.
7. Check operation of controls.
8. Check efficiency of compressor (pump up time check).

EVERY THREE MONTHS

1. Change oil.
2. Tighten all bolts.

YEARLY

1. Inspect valve assemblies. (May be necessary more frequently under heavy duty conditions)
2. Inspect motor starter, controls and wiring.
3. Check electric motor and lubricate if necessary.

TROUBLE SHOOTING

Problem	Causes
Fails to attain pressure	1, 2, 3, 4, 13, 14, 16
Knocking	2, 7, 11, 13, 15
Oil Pumping	2, 6, 9, 11, 13, 16
Compressor overheating	1, 2,
Motor draws excessive current	2, 5, 8, 10
Compressor doesn't unload when stopped	2, 3, 12
Abnormally high pressure	3,12,14
Compressor won't attain full operating speed	2, 3, 4, 10, 12
Excessive starting and stopping	1, 2, 14

1. Air leaks in distribution system.
2. Broken, leaking or carbonized valves.
3. Faulty unloader valve.
4. Loose belts.
5. Tight belts.
6. Oil level high.
7. Oil level low.
8. Oil viscosity high.
9. Oil viscosity low.
10. Unbalanced or low voltage, loose connections or bad contacts.
11. Worn bearings or connecting rods.
12. Worn unloader parts.
13. Piston, rings or cylinder worn.
14. Faulty pressure switch.
15. Loose flywheel, motor pulley or other loose parts.
16. Dirty intake filter.

The above is a general chart applicable for many compressors.

Figure 5-5. Recommended Inspection, Maintenance and Service Procedures for Single-Acting Air Compressors

Checklist for Maintenance and Service of Double-Acting Compressors

Daily

☐ If pressure lubricated, check oil pressure.
☐ Check oil level in compressor crankcase.
☐ On lubricated units, fill the lubricator at least once per shift.
☐ Observe cylinder lubricator sight feeds to be sure all feeds are operating at the proper feed to all cylinders and packing.
☐ Check pressure controls for proper operation.
☐ Drain moisture trap to pressure controls.
☐ Drain water from bottom of receiver.
☐ Check discharge temperature of cylinder cooling water.
☐ If equipped with clearance pocket, drain moisture.
☐ Check and investigate any unusual operation or noise.

Monthly

☐ Drain oil filter and clean oil filter element.
☐ Check and adjust (if necessary) all pressure and temperature shutdown devices.
☐ Check and clean (if necessary) intake filter element.
☐ Check all automatic moisture traps for proper operation.
☐ Check operation of safety valves.
☐ Check all pressure gauges for accuracy.

Semi-annually

☐ Remove cylinder valves and inspect for wear, broken valve springs, broken valve discs, and damaged valve seats. If necessary, clean and repair.
☐ Drain crankcase oil, clean interior of crankcase, and refill with new oil.
☐ Inspect cylinder bore to be sure cylinder is receiving proper lubrication.
☐ Check for loose foundation bolts and tighten if necessary.
☐ Drain oil and clean force feed lubricator oil reservoir.
☐ If compressor piston assembly is of the nonlubricated type, remove piston and check piston rings, piston, piston rod and compressor cylinder bore for wear.
☐ Remove piston rod packing and piston rod oil scraper ring. Check for wear and clean (if necessary). *Important:* Don't intermix packing ring or oil scraper ring segment.

Annually

☐ Remove, clean and inspect intercooler tube, bundles, aftercooler tube bundles and clean interior of cooler shells.
☐ Drain water and clean cylinder and cylinder water jackets.
☐ Check and clean (if necessary) compressor motor.
☐ Check and inspect compressor drive belts for wear and adjustment.
☐ Check for wear and inspect main bearing, crank bearing, and wrist pin bushings.
☐ Check crankshaft counterweights for tightness.
☐ The same procedure should be followed for annual inspection as described in daily, monthly and semi-annual inspection information.

Above are intended as a guide only. Refer to manufacturer's instruction book for more specific information.

Figure 5-6. Double-Acting Compressors—Maintenance and Service

Double Acting Compressors

TROUBLE-SHOOTING CHART

NUMBERS IN SYMPTOM COLUMN INDICATE ORDER IN WHICH POSSIBLE CAUSES SHOULD BE TRACED

POSSIBLE CAUSES	SYMPTOMS: Failure To Deliver Air	Insufficient Capacity	Insufficient Pressure	Compressor Running Gear Over Heats	Compressor Cylinder Over Heats	Compressor Knocks	Compressor Vibrates	Excessive Inter-cooler Pressure	Intercooler Pressure Low	Receiver Pressure High	Discharge Air Temp. High	Cooling H_2O Dis-charge T High	Motor Fails To Start	Motor Over Heats	Valves Over Heat
Restricted Suction Line		4							4						
Dirty Air Filter		3							3						
Worn or Broken Valves L. P.	2	1	2		2				1		1	3	4	3	3
Worn or Broken Valves H. P.	2	1	2		2			1			1	3	4	3	3
Defective Unloading System L. P.	1	2	1		7		4		2	1	2	4	3		4
Defective Unloading System H. P.	1	2	1		7		4	2		1	2	4	3		4
Excessive System Leakage		5	3												
Speed Incorrect		6	6	3	3		7								
Worn Piston Rings L. P.		7	4						5						
Worn Piston Rings H. P.		7	4					3							
System Demand Exceeds Compressor Capacity			5												
Inadequate Cooling Water Quantity					4						4	1			
Excessive Discharge Pressure				4	1	11	9			2	3	5		2	1
Inadequate Cylinder Lubrication					6	10	8				8				
Inadequate Running Gear Lubrication				1		1								5	
Incorrect Electrical Characteristics													2		
Motor Too Small													5	1	
Excessive Belt Tension				2									7	7	
Voltage Low													6	6	
Loose Flywheel or Pulley						7	2								
Excessive Bearing Clearance						5									
Loose Piston Rod Nut						4									
Loose Motor Rotor or Shaft						9	6								
Excessive Crosshead Clearance						3									
Insufficient Head Clearance						2									
Loose Piston						6									
Running Unloaded Too Long															2*
Improper Foundation or Grouting						8	5								
Wedges Left Under Foundation							10								
Misalignment (Duplex Type)							3								
Piping Improperly Supported							1								
Abnormal Intercooler Pressure											7	7		4	
Dirty Intercooler											6	6			
Dirty Cylinder Jackets					5						5	2			
Motor Overload Relay Tripped													1		

* Inlet Valves

Figure 5-7. Compressor Trouble-Shooting Chart

in all lubricated compressors using petroleum base lubricants. While the condition occurs infrequently, violent explosions can result where heavy carbon deposits have formed. One manufacturer has said that one ounce of oil per 50HP hours is acceptable while one ounce per 25HP indicates repairs are necessary. Of course, this is a rule of thumb and each machine is different.

"Proper" oils and viscosities depend on conditions. A suggested lubricant for single-acting machines would be a straight mineral oil with rust and oxidation inhibitors. For double-acting machines, a guide for cylinder lubricants would be as follows:

- Flash point—400°F
- Viscosity SSU at 100°F—245-420
- Viscosity SSU at 210°F—45-90
- Pour point—30° maximum
- Conradson carbon residue, percentage—1.0

In double-acting machines, the frame or crankcase is lubricated by constant circulation. This may require changing once every six months. Generally, the cylinder is fed from an external lubricator that is driven by a mechanical linkage. The lubricator is adjusted to supply the cylinder with the proper amount of oil. While this is measured in drops, it is an unprecise and unreliable measure. Valves and cylinders that are properly lubricated will have a light film of oil. Pools of oil or concentrations of oil in valve pockets indicate too much lubrication. Inadequate lubrication will be evidenced by dry cylinders and valves. Once the proper rate has been established, drops can be counted and used as the measurement—providing the oil characteristics remain constant and the lubricator is in good condition. If the oil consumption rate changes, inspect the lubricator for malfunctioning. A formula for estimating the amount of oil to inject into a cylinder is:

$$\frac{B \times S \times N \times 62.8}{10{,}000{,}000} = Q$$

B = bore (inches)

S = stroke (inches)

N = rpm

Q = quarts of oil per 24-hour day

Synthetic lubricants have been used more extensively in the past few years. The primary objection has been the cost and the possible incompatibility with materials used in the compressed air system. If petroleum base lubricants have been used, synthetics may dissolve the deposits and form a viscous tar, possibly causing damage to the intercoolers, aftercoolers, and valves. Complete cleaning should be done before changing from petroleum base to synthetic lubricant. In addition, be certain the synthetics are compatible with all materials used in the compressor and in all components downstream, especially polycarbonates used in filter bowls. Ask the compressor manufacturer to provide recommendations on any modifications that may be necessary *before* changing to a synthetic lubricant.

ROTARY COMPRESSORS

The early rotary compressors were water-cooled, injection-lubricated, sliding vane machines. (See Figure 5-2.) They could be used for single-stage low pressure service to 50 psig or in tandem with an intercooler and second stage to pressures of 125 psig. Compact design, low vibration, inexpensive initial cost, and simple design were among the advantages offered versus large reciprocating compressors. The chief disadvantage cited by most plant engineers was the total oil consumption. An external lubricator driven by a V-belt from either the compressor or motor shaft force feeds oil into the air stream and would have to be separated after it left the compressor. In this design, inspection and possible replacement of the vanes was recommended at yearly intervals.

Later designs used the sliding vane compressor with an integral oil reclamation system as part of the design. In this system the oil is used to lubricate, absorb the heat of compression as air is being compressed, and provide a seal between rotors, vanes, cylinder and end plates. Today's rotary machines are primarily of the twin helical screw, oil-injected design consisting of two rotors meshing in a dual bore cast cylinder (Figure 5-2). The primary configuration is a four-lobe helical male and six-groove helical female screw. The main rotor (usually the male) is driven through a shaft extension by the electric motor while the secondary rotor (usually the female) is driven by the main rotor. There is no metal to metal contact because of the oil film developed by oil injection. The heat of compression is absorbed by the oil, which is water cooled with a shell and tube heat exchanger or air cooled with a fan and radiator assembly. This has been referred to as a flood lubricated compressor and is available as a pre-engineered package (see Figure 5-8) with all components mounted and wired to simplify installation. The oil is recirculated and mixed with the air in the compression chamber. The oil and air are separated after compression in an efficient combination receiver and air/oil separator.

> **Note:** The twin screw compressor is also available in oil-free design where the rotors are driven through timing gears. Multi-stage units are used for pressures above 50 psig. The advantage of this design is the lack of oil in the compression chamber and in the compressed air. The discussion on rotary compressors is limited primarily to the oil-flooded design.

Air- and water-cooled units perform equally well and have the same efficiencies. The selection of air- or water-cooled is completely a function of ambient conditions, location, convenience of installation and availability, and economics of water supply versus air cooling. In both air- and water-cooled units, a temperature regulating valve is used to maintain optimum oil temperatures. It is important that oil temperatures are maintained above the pressure dewpoint, which is approximately 125°F in

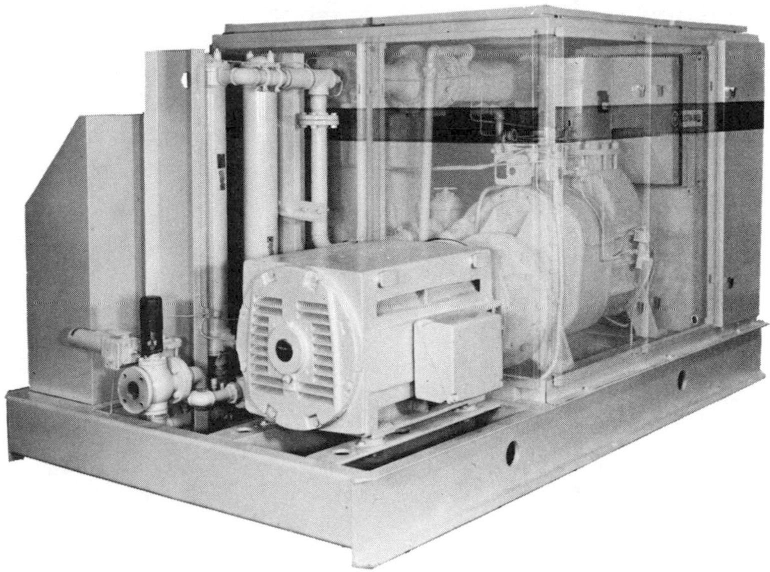

Courtesy of Gardner-Denver Division of Cooper Industries

Figure 5-8. Packaged Water-Cooled Rotary-Screw Compressor Available in Air-Cooled or Water-Cooled Through 500HP

most installations. The temperature valves are generally set to provide oil at 140°F, which will provide air discharge at approximately 180°F. This is low compared to reciprocating compressors at 100 psig, and thus, carbon and varnish deposits are virtually eliminated. It is important to maintain oil temperatures above the pressure dewpoint to avoid condensation of the air in the oil reservoir. In very humid areas this temperature may be as high as 145°F. However, it is also important to ensure that oils are not subjected to high temperatures, and a high air temperature switch is used to shut the compressor down if discharge air temperatures reach approximately 225°F.

Rotary compressors are designed to run at 100 percent capacity. The method of controlling the capacity is to modulate or to gradually choke the intake as pressure rises, thus reducing the intake air. With this arrangement, the volume is reduced but the pressure ratio across the machine increases, resulting in a relatively flat horsepower curve over the capacity range of the compressor. Most plants should operate a rotary flood lubricated compressor at full load and use a reciprocating unit to provide for peak loads. Some manufacturers of air compressors have introduced a power-saving device built into the rotary screw casing which effectively shortens the rotor lengths at part load. This will reduce the horsepower requirements proportionately to capacity and more closely approximate the reciprocating compressor power characteristics, while retaining the features of the rotary.

Primary Components of the Rotary Compressor (see Figure 5-9).

Intake Filter

This very essential item is especially important in the rotary machine where air and oil are mixed and the oil is then removed in the separator. If the air entering is contaminated, the oil and air mixture is dirty and the separator will require frequent replacement.

Oil Separator

The separator is basically a large submicronic filter generally located in the air/oil receiver. Much of the oil is separated in the receiver and the air/oil mixture is passed through the separator to remove most of the remaining oil, and returned to the oil system through a small scavenger line while the air goes to the discharge line. Oil carryover varies from 2.25 ounces per 100,000 cubic feet of free air (and according to some manufacturers) down to perhaps ¼ of an ounce per 100,000 cubic feet.

Oil Filter

The oil filter is an essential component that ensures foreign materials do not enter the compressor.

Note: *The above three filters are the primary maintenance items in a flood-lubricated compressor. Proper maintenance of the intake filter and oil filter will extend separator life and will reduce compressor maintenance.*

Safety and Control Components

- Thermostatic valve—regulates water flow in water-cooled machines or bypasses oil around the radiator in air-cooled units to maintain 140°F oil temperature.
- High air temperature switch—will shut down the machine at approximately 225°F if air temperature rises.
- Pressure relief valve—mounted in the air/oil receiver and acts as a safety valve against excessive pressure.
- A check valve—used after the air/oil receiver to prevent air from coming back to the compressor when the compressor is unloaded or shut down.
- An automatic solenoid valve—will "blow down" or relieve the pressure in the air/oil receiver on shutdown. Some manufacturers also operate this valve each time the compressor fully unloads to reduce the unloaded horsepower.
- Inlet valve—As explained, this valve is operated by a pilot valve (also called a pilot differential, subtractive pilot, modulating valve, and so on), which gradually opens or closes the inlet valve as pressure rises or falls. Some manufacturers allow the compressor to modulate to zero. Others will modulate to 40

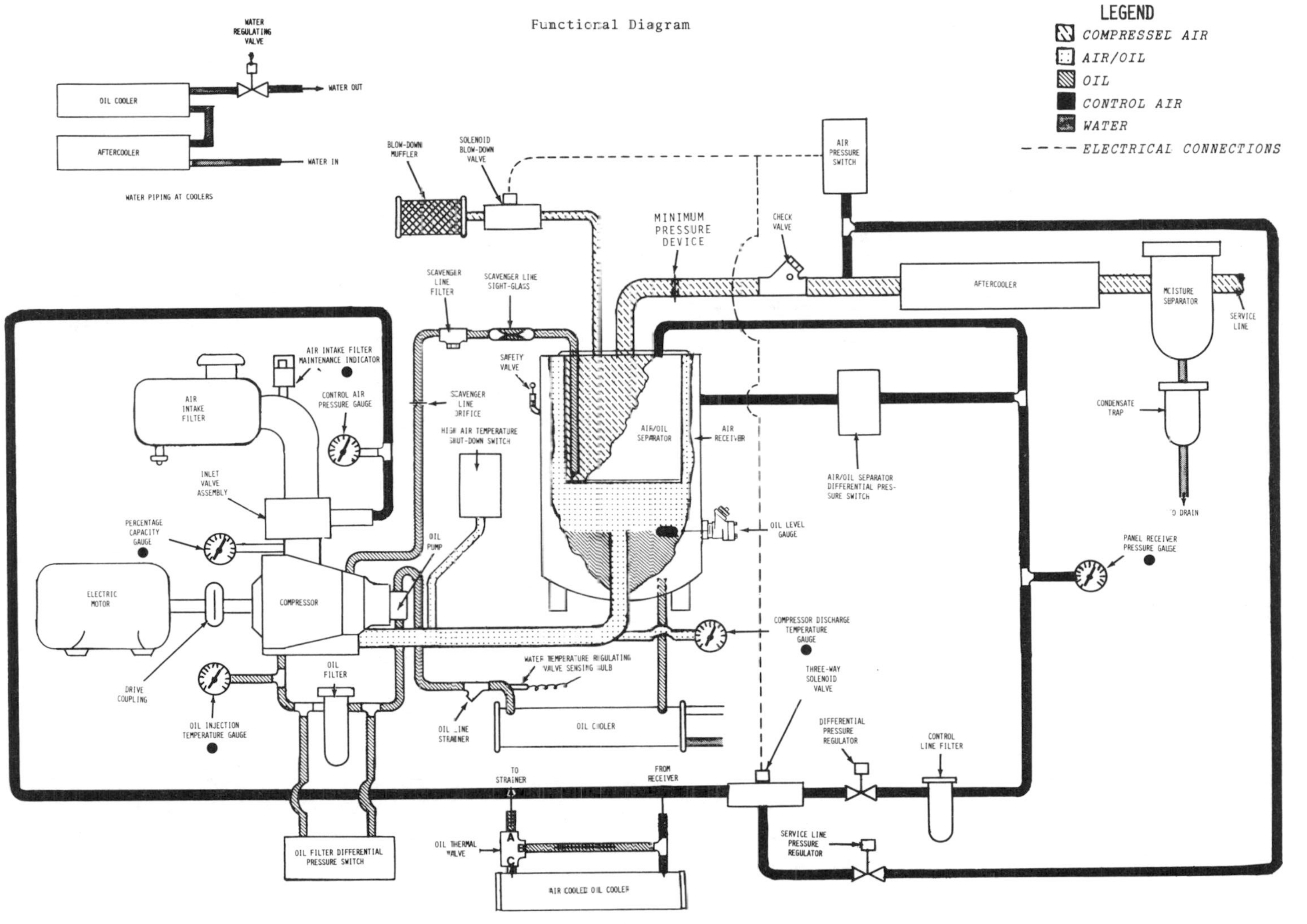

Figure 5-9. Rotary Compressor Diagram

percent of full load capacity and, at that point, will fully unload by closing off the intake completely and simultaneously blow down the receiver. This is done to reduce the no-load horsepower.

- Instrumentation—There are many devices, gauges, lights, and so on, used to indicate performance of a rotary screw compressor. These include, but are not limited to, air and oil temperature gauges, pressure gauges, filter and/or separator warning lights, and air filter indicators. Each manufacturer has different standards and options. Familiarity with the instrumentation will enable the operator to avoid major problems. Record readings periodically and investigate any deviations.

Lubricating Rotary Compressors

The selection of a proper lubricant must be the responsibility of the manufacturer. Because of the wide variance in requirements and opinions among manufacturers, no recommendations can be given that will cover all rotary compressors. Since manufacturers change their specifications and new oils are being developed, call the manufacturer annually to get the most recent recommendations.

The primary synthetic lubricants which have been used in air compressors are diesters and polyalphaolefins (PAO). The diesters have been used primarily in reciprocating compressors, where minimizing valve deposits is essential. Most air compressor manufacturers have selected the PAO lubricants, which are synthesized hydrocarbons with a selected additive package for rotary compressors. Some manufacturers have specified the diester fluids for rotary compressors, while others have chosen polyglycols or even semi-synthetics. The advantage of synthetics in rotary compressors is the extended life and drain cycle, which is often three to four times that of standard petroleum base lubricants.

Some suppliers have received FDA authorization to use their PAO lubricants in compressors, where the compressed air may have incidental food contact. However, the limitations of the lubricant additive package reduces the life expectancy to about half that of the normal synthetics.

Important:

- The basic rule is **NEVER MIX LUBRICANTS,** or change from one type to another, unless complete draining of the old oil and a thorough cleaning process precedes the changeover. This applies to petroleum base and synthetics equally.
- When using synthetic lubricants and when extended drain intervals are sought, it is essential that a testing program be instituted, so that samples of the lubricant are analyzed periodically by a reputable laboratory or the lubricant supplier.
- Obtain the air compressor manufacturer's approval before changing lubricants. In some cases, the lubricant may not be compatible with materials used in the compressor.

Routine Inspection and Service of Rotary Compressors

Note: These maintenance and inspection routines are guides. In all cases the manufacturer's recommendations are to be given preference over routines listed above.

Servicing the Air Filter

Frequency of servicing depends on the environmental conditions under which the compressor is operating.

Service the air filter element under any of the following conditions, whichever comes first:

- As indicated by the air filter service indicator (located at the instrument panel)
- Every six months
- Every 1,000 hours of operation

In some machines filter elements can be cleaned by washing in a warm water solution of household detergent. Discard the element after a maximum of six washings. Keeping a spare filter element on hand is recommended to keep machine downtime to a minimum. Safety element (where applicable) should be changed once a year. Discard the used element.

Servicing the Air/Oil Separator

Change the separator element under any of the following conditions, whichever comes first:

- As indicated by the separator warning light (located at the instrument panel) or indicator
- Once a year
- Every 4,000 to 8,000 hours (refer to manufacturer's instruction book)
- If excess oil appears in the service line as a result of a faulty separator (due to pin hole, loosened plastisol, and so on)

Servicing the Oil Filter

Change the oil filter element(s) under any of the following conditions, whichever comes first:

- As indicated by oil filter warning light (located at the instrument panel) or indicator
- Every six months
- Every 1,000-2,000 hours or as directed by instruction book
- Every oil change

Changing the Oil

Change oil after the first 1,000 hours of operation to make sure the initial contaminants in the system are removed. After that, the oil should be changed every 2,000 hours or in accordance with instruction book; check manufacturer's recommendations if synthetic oil is being used.

Servicing the Scavenger Line Filter and Orifice (where applicable)

Clean scavenger line filter element and orifice:

- when no oil flow is visible through the sight glass
- every 2,000 hours

Servicing the Control Air Line Filter (Manual)

Condensate collected in the bowl should be drained periodically. Excessive condensate built up in the bowl could adversely affect the performance of the capacity control system.

Servicing the Oil Cooler

If air-cooled, keep cooling fins clean. If water-cooled, periodic inspection and chemical cleaning may be required.

Trouble-Shooting of Rotary Compressors

Problem: *Failure to Start*—Possible causes:

- Power failure or power not supplied to starter
- Improperly wired
- High temperature switch shutdown
- Overloads tripped
- Control circuit malfunction

Problem: *Unscheduled Shutdown*—Possible causes:

- High air temperature caused by:
  —low oil level
  —clogged oil cooler or filter
  —thermostatic valve malfunction
  —poor ventilation (air cooled only)
- Overload Relay Tripped
  —discharge pressure too high
  —low voltage

—incorrect thermal relay
—poor contacts
—faulty pressure switch
—separator clogged

Problem: *Low Capacity*—Possible causes:

- Clogged air filter
- Inlet valve not fully open
- Pilot valve adjustment required
- Too much air demand

Problem: *High Oil Consumption*—Possible causes:

- Oil level overfilled
- Faulty air/oil separator
- Scavenger tube clogged
- Incorrect oil
- Oil leaks at seals, fittings or gaskets

CENTRIFUGAL COMPRESSORS

Whereas rotary screw and reciprocating compressors are positive displacement machines, the centrifugal compressor is a dynamic type of compressor. (See Figure 5-10). It uses a series of impellers and diffusers to impart velocity into pressure. The mass flow of air determines the horsepower requirements and, therefore, the inlet air temperature affects the operation of the compressor. The inlet air flow to the compressor is controlled down to a certain point (turn-down) by throttling the inlet through an automatic butterfly valve or adjustable guide vanes.

For many years, centrifugal compressors were installed only in larger plants, where great quantities of processed air were required. However, recent developments have introduced units as small as 100 HP. Before purchasing a unit of this type, it is important to consider the energy consumed at full load and at part load, and to provide a comparison versus other types of equipment under similar conditions. The centrifugal compressor at 100 psig will normally require three stages for most efficient operation, but some manufacturers have been producing two-stage units, which will reduce the initial price.

Maintenance is extremely important, especially as it relates to intake filters and the intercoolers. Proper operating temperatures must be maintained, if the compressor is to provide trouble-free service. The centrifugal compressor does not have any oil in the compression chamber and, thus delivers oil-free air to the system. Proper start-up and shutdown procedures must be followed, and good control of discharge pressures maintained to avoid a condition known as "surge." This condition normally occurs because of inadequate air flow from the compressor, which will in-

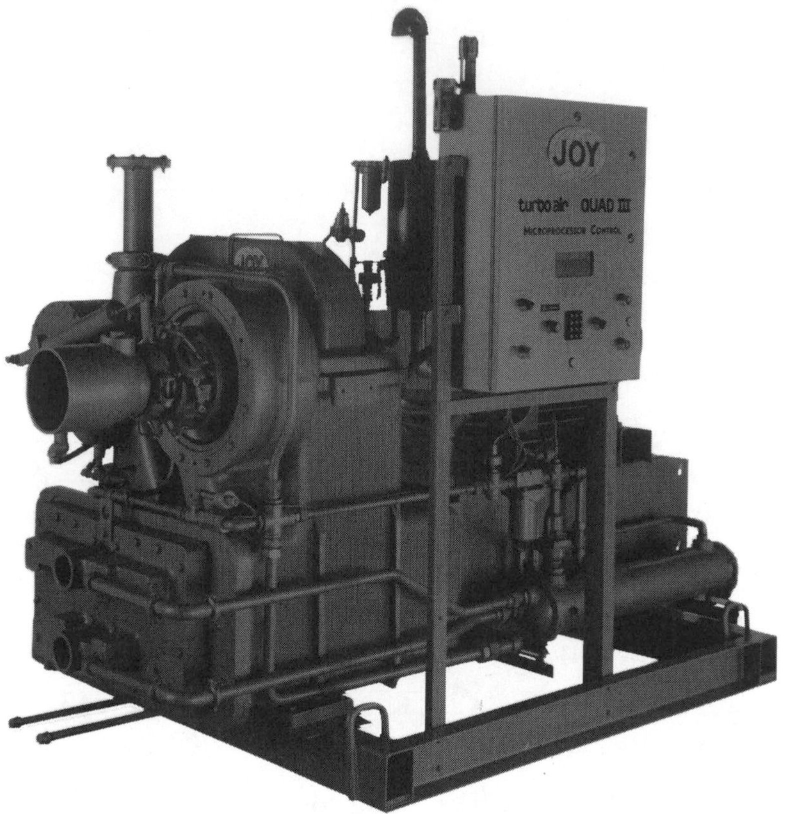

Courtesy of Cooper Industries Turbocompressor Division

Figure 5-10. 3-Stage Centrifugal Compressor

duce backflow from the line and then oscillate the flow between forward and reverse flow.

Instructions on the use of throttle and blow-off valves, proper pressure and temperature settings and good maintenance of filters and coolers will help to avoid this undesirable condition.

Figure 5-11 provides a checklist for maintaining centrifugal compressors.

CHOOSING ACCESSORIES

While the trend is towards packaged units with accessories mounted and piped as part of the package, selection of the proper accessories will reduce maintenance problems. In addition, while manufacturers do standardize on the accessories that are suitable for most applications, some problems have been eliminated by changing accessories.

For example, one plant had many lubricated reciprocating compressors and was constantly replacing aftercoolers with the original type provided. After examination

CENTRIFUGAL COMPRESSOR CHECKLIST

Daily
(*Record* pressures and temperatures)

Lube Ol Level
Lube Oil Pressure
Lube Oil Temperature
Water Temperature: Entering
Leaving
Water Pressure: Entering
Leaving
Air Inlet Differential Pressure
Inlet Air Temperature
Control Filters Drained

Interstage
1. Air pressure
   Air temperature
2. Air pressure
   Air temperature

Discharge Air Pressure
Discharge Temperature
Vibration Monitor(s)
Control Panel Alarms
Blowdown Strainers for Traps
Drain Traps Working
Leaks: Air, Water, Oil

In addition to the daily recordings and checks, it is recommended that the following *minimum* maintenance be performed:

				Comments		
	3 mos.	*6 mos.*	*yearly*	*Inspect*	*Clean and/or replace*	*Record*
Inlet controls	X			X		
Surge controls	X			X		
Main motor current amperes	X					X
Automatic traps & strainers	X			X	X	
Oil mist filter		X		X	X	
Oil sample-analyze		X				
Inlet air filters		X		X	X	
Intercoolers			X	X	X	
Oil cooler			X	X	X	
Aftercooler			X	X	X	
Change oil		X	(X)		X	
Motor			X	X		
Coupling			X	X		
Lube pump			X	X		
Gear box			X	X		

It is important that all components and accessories be kept clean.

Figure 5-11. Centrifugal Compressor Checklist

it was determined that the cooler had small tubings with an internal fin that would quickly plug with carbon deposits. Changing to an aftercooler made by the same manufacturer, but with larger unrestricted tubes, solved the problem. Another plant had continuous problems with the separators in the oil-flooded screw compressors. Upon investigation it was noted that a very fine cosmetic dust was getting through the intake filter and collecting in the separator, causing frequent costly replacement. Replacing the intake filter with one that had a finer filtration and a larger element solved the problem.

Aftercoolers

Reducing moisture in compressed air is a problem most plant engineers face at some time during their careers. The basic accessory for moisture removal is the aftercooler. By reducing the temperature of the air, the water vapor is condensed and can be separated and drained from the system. In addition, the aftercooler provides a degree of safety in preventing explosions. If carbon accumulated in the compressor discharge of a reciprocating compressor and it became sufficiently incandescent, the fire would probably be quenched as it entered the aftercooler, precluding the possibility of a major explosion if the fire reached the air receiver.

Aftercoolers are rated in cfm at a given pressure and the approach temperature. In the case of a water-cooled shell and tube aftercooler, the approach temperature is the difference between the incoming water temperature and the outgoing air temperature. Thus, a cooler with water entering the shell at 85°F and air leaving the tubes at 100°F would have a 15°F approach. Check the approach temperature with the manufacturer and the amount of water required to cool. This may vary from one gallon to four gallons per 100 cubic feet. Also, check tubing size and ease of cleaning the tubes when ordering aftercoolers. Install indoors to prevent freezing and leave room to pull the tube bundles for cleaning and maintenance. Clean, soft water is essential; and the cooler, the better. Warm water results in warmer air with more vapor left in the air, and also results in accelerated scaling in the jacket. Check the tubes and jackets yearly and clean as required in accordance with the manufacturer's instructions.

Air-cooled aftercoolers are generally radiator-type units that cool the air to an approach temperature from 10°F to 25°F of the ambient temperature. In a high ambient temperature an air-cooled aftercooler may be unsuitable. Some air-cooled aftercoolers are cooled by the flywheel of an air-cooled compressor while others have a separate fan. Be sure to check the approach temperature of the aftercooler and do install it in a cool, clean location. Maintenance consists of cleaning the fins with a compressed air blowgun. If the fins are oily, a nonflammable solvent may also be necessary.

Separators should be purchased with aftercoolers; and as the name implies, they separate the water from the air. An automatic drain trap is installed at the bottom of the separator to drain accumulated moisture.

Tips for installing aftercoolers:

- On installations where large aftercoolers are not purchased as part of a package, a three-valve bypass will enable maintenance to be done on the aftercooler while the compressor continues to operate. Be certain to relieve all pressure in the aftercooler and shut off and drain all water before attempting repairs or maintenance.
- A safety valve before the aftercooler is useful if carbon buildup could be a problem. If the cooler begins to plug, the safety valve will pop.
- Place a manual shutoff valve before any drain trap so that it can be repaired without shutting down the system.
- The outlet from an automatic drain trap should be visible to the operating engineer. Visibility is necessary to determine if the trap is working. Either pipe to an open funnel or have gauge glasses installed on the separator to see if water accumulates. There should not be water in the separator.
- Check automatic drain traps manually to be certain they are working.

Air Receivers

- Be certain the air receiver is of ASME construction and that the stamped working pressure is well above the operating pressure. An ASME safety valve should be installed and set at least 5 psi below the working pressure of the receiver.
- Receivers should be drained frequently to be clear of condensate, oil, scale, etc.
- As a rough rule of thumb, receivers are sized so that the capacity of the air compressor in cfm divided by three is the size in cubic feet of the receiver for automatic start-stop control. The cfm divided by seven is generally used for constant speed control. For rotary compressors operating on modulating controls, the receiver can be very small and in some applications is not required.

The proper selection of one or more tanks is based not only on the capacity of the compressor itself, but also on the shop load cycle. The following formula can be used to determine the correct size:

$$V = \frac{T(C\text{-Cap})\,(P_b)}{(P_1 - P_2)}$$

T = Time interval in minutes, during which a receiver can supply air without excessive drop in pressure

V = Volume of tank in cubic feet

C = Air requirement in cubic feet of free air per minute

Cap = Compressor capacity

P_b = Absolute atmospheric pressure, psia

P_1 = Initial tank pressure, psig (compressor discharge pressure)

P_2 = Minimum tank pressure, psig (pressure required to operate plant)

If Cap is greater than C, the resulting negative answer indicates that the air compressor will supply the required load. If the compressor is unloaded or shut down, Cap becomes zero and the receiver must supply the load for "T" minutes.

Intake Filters

Air entering any compressor must be filtered. Foreign matter drawn into the machine will cause rapid wear and will also contaminate the lubricant.

For nonlubricated compressors or where extremely fine filtration is necessary, a good dry-type filter with either a felt cloth or special paper media is recommended. The felt type can be cleaned, whereas the paper type is replaced when dirty.

For lubricated compressors, a viscous impingement-type or oil-bath filter may be acceptable. The viscous impingement type has woven wires packed into a cylinder or cell that is coated with oil to hold dirt. This should not be used in a dusty area. The oil-bath filter is preferred over the viscous impingement and removes dirt by washing it down a fine strainer into a sump from which it is manually drained. The oil must be changed and the filter cleaned as conditions require. Clogged filters are responsible for many problems besides wear on the compressor. A dirty filter reduces the volume of air compressed, increases the pressure differential across the machine, causes overheating, and may even collapse and be drawn into the compressor.

Some filters are manufactured with a silencing chamber to reduce noise level. There are also separate intake silencers available which, when installed close to the compressor intake, will often reduce noise levels. When air is drawn from outside the compressor location, increase the pipe one inch for every 10 feet of length and use a full-size filter with a rain hood if the filter is outdoors. Check with compressor manufacturer on lengths to avoid resonance in intake pipe.

There have been cases where intake filters were located near soot blowers and steam exhausts, and compressor life was substantially reduced. In other cases the intake was near paint fumes, which was not only detrimental to compressor life, but posed a definite safety hazard as well. Be certain that wherever the intake is located, it will draw clean, cool, dry air.

Control Panels

In multiple compressor installations, the proper control of the air compressors can result in significant energy savings and maintenance costs. The panels are usually designed to:

- Shut off unnecessary air compressors at any time.
- Base load the rotary or centrifugal compressors to maximize their effectiveness at full load, and retain the "turn-down" characteristics of centrifugals

down to the surge control point. Reciprocating compressors are often set to operate below the maximum setting of the rotary compressors.

- Alternate the lead machines.
- Monitor the system for faults.

It is very important that the maintenance engineer understand the operation of the control sequence fully and to be instructed on the proper troubleshooting.

CHECKLIST FOR REDUCING COMPRESSOR MAINTENANCE BY PROPER INSTALLATION & STARTUP PROCEDURES

- Read the instruction book thoroughly.
- Is the intake piped to a clean location?
- Be certain there is adequate cooling.
  - Air cooled—Adequate ventilation and cooling
    —Clean, dry air.

  Remember, an air compressor will dissipate 2545 BTU HP-Hour. On an air-cooled unit, this will add to the ambient temperature already in the area unless adequate ventilation is available.
  - Water-cooled—Is there adequate water at the proper pressure?

    Can the water be drained to a sewer or must a cooling tower be considered?

    Must the water be treated?
- Is there adequate power at the compressor location?

  Does the voltage on the motor nameplate agree with the supply?

  Is the correct magnetic starter installed to meet inrush current limitations?
- Is a foundation necessary? If yes, what soil bearing pressure can be tolerated?
- At start-up:
  - Proper lubricant and at the correct level?
  - Machine level?
  - Coupling aligned or belts adjusted properly?
- Turn machine over by hand.
  - Press button to jog compressor and check rotation.
  - Listen for any unusual noises.
  - Run the compressor—unloaded for period of time, if recommended.
- Measure and record:
  - Full load current and voltage.
  - Time to pump air receiver to cut out pressure.

For evaluations and considerations on what type of air compressor should be purchased, see Figure 5-12.

The basic types of compressors and some of the parameters to evaluate are as follows:

Type	Initial Purchase Price	Installa-tion	Repairs	Mainten-ance	Power Cost (Full Load)	Power Cost (Part Load)
Reciprocating						
1. Air-cooled	1	1	4	3	4	2
2. Water-cooled single-acting	2	2	3	3	2–3	2
3. Water-cooled double-acting:						
Single-stage	3	4	2–3	3	2–3	1
Two-stage	4	3*	2–3	3	1	1
Rotary Screw	1–2	1	1	1	2–3	4**

KEY: 1. Very Good 2. Good 3. Fair 4. Poor

It is well to consider the packaged concept where the manufacturer supplies all components prepiped and prewired to facilitate installation. Most rotary screw compressors and smaller air-cooled reciprocating compressors are sold packaged.

*A two-stage unit generally requires smaller foundations than single-stage. However, rigging costs may be higher.

**Some rotary screw compressors with a power-saver feature can have good part load characteristics and can be rated 2.

Other factors to consider that can influence purchase include:

- Floor space
- Noise level
- Ventilation requirements
- Quality of air required
- Subsoil conditions
- Availability of qualified maintenance

Figure 5-12. Types of Compressors and Parameters of Their Evaluation

CHOOSING MAINTENANCE AND SERVICE CONTRACTS FOR AIR COMPRESSORS

Air compressors are often called "the fourth utility," and many manufacturing operations will cease when air compressors malfunction. Service and maintenance contracts are available for almost all equipment. Most daily maintenance and routine repairs can and should be done by qualified in-house personnel; however, there is a trend towards purchasing maintenance and service plans which are intended to:

- avoid costly downtime
- save money on emergency repair costs
- ensure the air compressor is operating at peak performance and receiving the regular required maintenance to prolong the life of the equipment
- budget the costs associated with maintenance

Maintenance logs must be kept by the plant, and some basic charts appear as part of this chapter. It is recommended that maintenance people write the numbers for all readings, rather than merely providing a check mark. The manufacturer or factory authorized service depot knows the important points and the frequency of checking each item. The servicing of these components will be offered as part of the contract. Often, an extended warranty with annual inspections and renewals and free or discounted emergency service provisions or parts purchases may be included as part of the agreement.

There are many variations of the three basic types of maintenance agreements which follow:

- *Preventive Maintenance Agreement*—This provides for:
  - Scheduled maintenance visits (during regular business hours) as established by use and environment
  - Emergency visits at a discount
  - Parts at service list price (or discount)
- *Emergency Maintenance Agreement*—This provides for:
  - Preventive maintenance is done by the plant personnel.
  - Emergency repair calls are answered within an agreed time.
  - Minor parts are included in the cost, but maintenance items, such as filters and lubricants, are not included.
- *Total Coverage*—This provides for:
  - Preventive maintenance and emergency service provided by contracts.
  - All minor parts and maintenance items included.

All of the above types of agreements are covered at the agreed price. If the air compressor service company and the plant both profit, the price for the contract was correct.

When evaluating the supplier of the service, ask the following questions:

- Are the technicians "factory trained" (at the manufacturer's plant)?
- Are the service vehicles fully equipped and stocked with genuine parts?
- What is the emergency response time, and is it available twenty-four hours per day?
- Is the contractor's plant properly stocked with genuine replacement parts? What are you expected to keep in inventory?
- Will the contractor offer rental machines in the event of major breakdowns? At what price?
- What is the contractor's reputation for service with other plants?

Also, keep in mind the following considerations:

- Do not assume that the nearest contractor offers the "best" service; nor should you assume that size of the company (large or small) will guarantee the best service. The "best" service is the result of an attitude and philosophy of doing business.
- Is there a need to bring the equipment up to a standard performance before a maintenance contract can begin? How much will it cost?
- Higher priced synthetic lubricants have extended drain intervals. This should reduce labor costs. Who pays for replacement lubricant? Is a periodic oil analysis included in the price?
- How about the accessory equipment in the plant air system, such as dryers? filters? Should they be included?
- For a plant with various types of compressors, is each individual machine and the system operated, so that optimum performance at the lowest energy cost is realized? Are rotary screw or centrifugal compressors base loaded and recips loading and unloading as required? Is this the best mode for your equipment? Is there a central control panel to insure maximum system performance? Who maintains it? Monitors it?

All of the variables and questions relating to air compressor contracts cannot be covered. For each machine, the manufacturer's instruction book is the primary source of developing a maintenance schedule. The charts which appear in this chapter may assist in planning the correct program.

AIR DRYERS

There are three basic types of air dryers—deliquescent, regenerative desiccant, and refrigerated. (See Figure 5-13).

Typical Refrigerated Dryer

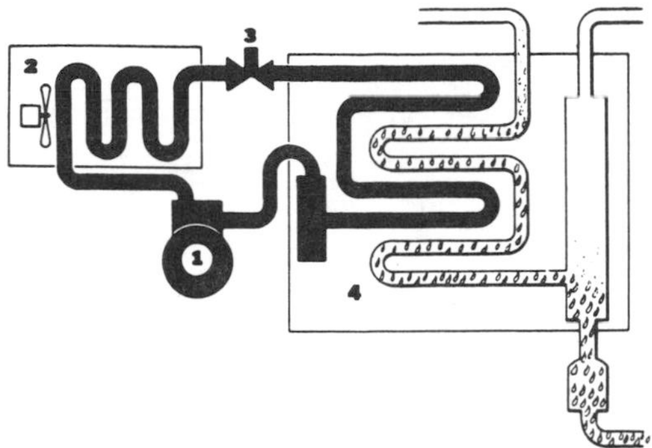

MAJOR COMPONENTS
OF THE REFRIGERATED AIR DRYER

1. Refrigeration Compressor: A hermetically sealed motor driven compressor operates continuously. It generates a high pressure refrigerant gas.

2. Hot Gas Condenser: The high pressure refrigerant gas enters an air cooled condenser where it is partially cooled by a continuously running fan.

3. Automatic Expansion Valve: The high pressure liquid enters an automatic expansion valve where it thermodynamically changes to a subcooled low pressure liquid.

4. Heat Exchanger: A system of coils where dry air is produced.

Typical Regenerative Dryer
[Heatless type - also available with heaters for desiccant regeneration]

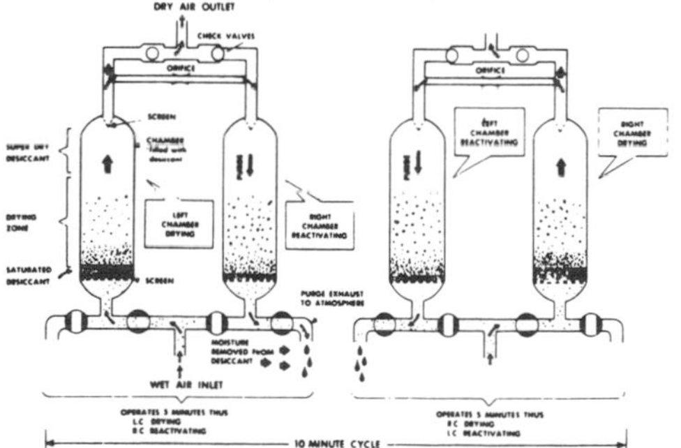

ACTUAL OPERATION — **As shown in the above flow diagrams, the heatless Dryer cycles between two desiccant chambers, one serving as a drying medium, while the other is undergoing a reactivation process. The diagram at left shows the wet gas entering at the bottom of the left-hand chamber, passing upward through the desiccant where it is dried to an extremely low dewpoint. The dry air passes through the check valve to the dry air outlet. Simultaneously,** a small percentage of the dry gas is expanded through the orifice between the chambers, and flows down through the right-hand chamber, reactivating the desiccant, and passing out through the purge exhaust. At the end of the cycle, the chambers are automatically reversed, the right-hand chamber serving as the drying medium, while the left-hand chamber is being reactivated, as shown in the diagram at right.

Typical Desiccant Dryer

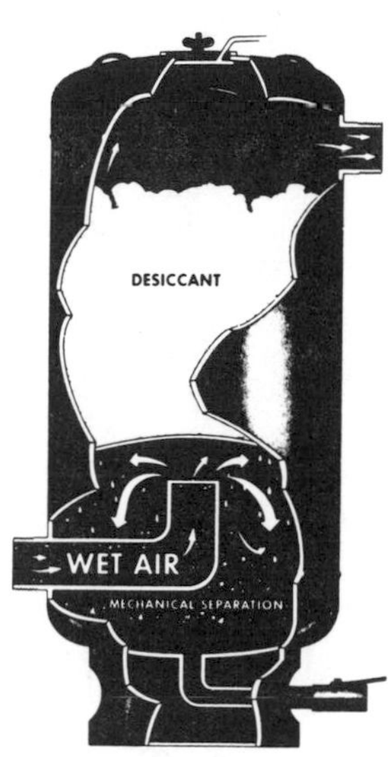

ATMOSPHERIC DRYNESS

DEW POINT °F: 110, 100, 90, 80, 70, 60, 50, 40, 30, 20, 10, 0, −10, −20, −30, −40, −50, −60, −70, −80, −90, −100

PARTIAL PRESS MM HG: 24 0, 22 0, 20 0, 17 5, 15 0, 12 5, 10 0, 9 0, 8 0, 7 0, 6 0, 5 0, 4 0, 3 0, 2 0, 1 0, 5, 32, 178, 095, 050, 026, 013, 0062, 0028, 0012

GRAINS MOISTURE # AIR: 400, 300, 200, 150, 100, 90, 80, 70, 60, 50, 40, 30, 20, 10, 8, 6, 4, 2, 1, 8, 6, 4, 2, 1, 08, 06, 04, 02, 01

MOISTURE # AIR: 0600, 0500, 0400, 0300, 0200, 0150, 0100, 0090, 0080, 0070, 0060, 0050, 0040, 0030, 0020, 0010, 0008, 0006, 0004, 0002, 0001, 00008, 00006, 00004, 00002, 00001, 000008, 000006, 000004, 000002

GRAINS MOISTURE CU FT AIR: 25, 20, 15, 10, 9, 8, 7, 6, 5, 4, 3, 2, 1 5, 1, 8, 6, 4, 2, 1, 08, 06, 04, 02, 01, 008, 006, 004, 002, 001

PPM: 60.000, 50.000, 40.000, 30.000, 20.000, 15.000, 10.000, 9.000, 8.000, 7.000, 6.000, 5.000, 4.000, 3.000, 2.000, 1.000, 800, 600, 400, 200, 100, 80, 60, 40, 20, 10, 8, 6, 4, 2

% R H a RM TEMP: 100 0, 91 7, 83 3, 72 9, 62 5, 52 1, 41 7, 37 5, 33 3, 29 2, 25 0, 20 8, 16 7, 12 5, 8 3, 4 2, 2 1, 1 3, 74, LESS THAN 5% R H

VOL %: 9, 8, 7, 6, 5, 4, 3, 2, 1 5, 1, 9, 8, 7, 6, 5, 4, 3, 2, 1, 08, 06, 04, 02, 01, 008, 006, 004, 002, 001, 0008, 0006, 0004, 0002

Figure 5-13. Types of Air Dryers

Deliquescent Dryers

Deliquescent dryers consist of a receiver filled with a chemical desiccant that reacts with moisture in the air and absorbs it by dissolving the surface desiccant. The water-salt solution drains to the bottom of the receiver where it must be drained. This type of dryer requires the desiccant to be replenished. The inlet temperature must be limited to 100°F or less to avoid increased consumption. The effectiveness of this dryer is questionable under certain conditions and while it does have a low initial cost, it is no longer considered the prime method of eliminating moisture in a compressed air line. A dewpoint suppression of 20°F is obtainable.

Regenerative Desiccant Dryers

Regenerative desiccant dryers consist of two towers, each containing a charge of a solid desiccant such as silica gel, activated alumina, or molecular sieve. The air flows alternately through each tower so that air is being dried or regenerated either by heaters or dry air is being purged back through the desiccant.

The regenerative dryers can produce dewpoints of −40°F to −100°F depending on the desiccant material selected. With purged air being used for regeneration, the cycle is very short (5 to 10 minutes) and the purged air can be 10 to 15 percent of the total supplied, so that this reduces the total air available. With a heater type dryer, the cycle is four to eight hours and electric heaters (or steam or hot air) are used. Consider total energy required for regeneration when evaluating this type dryer.

Regenerative dryers should be sized quite closely to the requirements to avoid wasteful purging or excessive heater operation and replacement. The desiccant will probably have to be replaced yearly in most cases.

The deliquescent and regenerative dryers have desiccant that will not remove the moisture if it is coated with oil. Therefore, in a lubricated system, put an oil removal filter before the dryer. Since desiccant dusting can occur, a fine filter after the dryer will prevent desiccant from entering the system.

Refrigerated Dryers

The refrigerated dryer has almost become the standard to eliminate moisture in manufacturing plants where the dewpoint required is above 35°F. Most manufacturers use a noncycling type of dryer that has an air-to-air and refrigerant-to-air heat exchanger packaged to include instrumentation, automatic drain trap and proper controls, to match the compressed air load to be dried to available refrigeration capacity. Larger chiller-type dryers use a closed circuit water chiller which cools a mass of water which is circulated through an aftercooler. The compressed air passes through this aftercooler and water vapor is condensed as it would be in a standard aftercooler.

Some dryers require pre-filters because the heat exchangers have very small passages that may easily plug with contaminants such as rust, scale, or carbon from a compressed air system.

The primary consideration in selecting a refrigerated dryer to remove moisture is to understand the design criteria so that an evaluation can be made that will prevent problems from occurring when the dryer is placed in service. The parameters to consider include:

- inlet air temperature
- ambient air temperature
- cfm to be cooled at pressure
- pressure drop
- pressure dewpoint

There have been cases where a dryer purchased with particular design criteria did not operate on a hot summer day when it was needed most.

For example, most dryers are rated at 100°F inlet air temperature operating in a 100°F ambient. At temperatures other than 100°F, the capacity is increased or decreased as follows:

Inlet Temperature, °F	Capacity Multiplier
90	1.23
95	1.13
100	1.0
105	.9
110	.83
120	.69

Ambient air temperatures have a similar effect but not as marked.

Ambient Air Temperature, °F	Capacity Multiplier
85	1.09
90	1.05
100	1.0
110	.91

A dryer rated for 100°F ambient air and 100°F inlet air would have far greater capacity than one rated at lower temperatures. Conversely, and more importantly, the purchase of a dryer rated at 1,000 cfm with air entering at 90°F in an ambient of 85°F would be capable of drying only 746 cfm at 100°F inlet air and 100°F ambient temperature.

Pressure drop is normally 5 psi maximum, but remember that this will vary as the square of the cfm through the dryer. Thus, increasing the cfm by 10 percent will increase the pressure drop by 21 percent. Also, changes in operating pressure affect the pressure drop inversely as the ratio of absolute pressures. If the pressure drop is 5 psi with 1,000 cfm flowing at 100 psig (114.7 psig), the pressure drop at 50 psig (64.7 psig) would be $\frac{114.7}{64.7} \times 5$ or 8.9 psi.

The pressure dewpoint of 35°F is one of the standards most commonly accepted, but 50°F is also used. This becomes a plant engineer's choice as to which is best suited for his plant.

Basic maintenance of dryers consists of monitoring refrigeration gauges to insure operation of the refrigeration compressor, checking the automatic drain trap and repairing when necessary, and keeping the condenser clean. Cleaning the compressed air side of the heat exchangers can be done in many dryers by back flushing air in reverse direction through the dryer, or in some cases, by pumping an acceptable solvent through the dryer. Check the manufacturer's instructions for detailed instructions. If the heat exchangers become dirty, the pressure drop across the dryer will increase. Always install a three-valve bypass around the dryer to facilitate maintenance when required and to ensure uninterrupted air supply.

Oil removal filters used in conjunction with air dryers have helped produce "instrument quality" air where lubricated compressors have been installed. Where no oil can be tolerated, such as use of air in a food product, it is generally best to use non-lubricated compressors. It is recommended that pressure gauges be installed before and after dryers and filters to monitor pressure drop that when excessive, will indicate a maintenance problem.

INCREASING THE ENERGY EFFICIENCY OF YOUR AIR COMPRESSOR SYSTEM

- Fix all leaks.
- Operate the air compressor at the lowest possible pressure that will still allow all equipment to be run. Some compressors will save 1 percent in power for every 2 psi pressure reduction so that a 10 psi drop could result in a 5 percent power savings. A second benefit is that compressors generally last longer at lower pressures. (Rotary flood lubricated compressors have a pressure below which they should not operate unless specifically designed for low pressure.)
- Use pressure reducing valves at all equipment which consume air, and set at the lowest possible pressure. Shut air off when machines are not in use.

- Pipe intake to a clean, cool location. Reciprocating compressor capacity can be increased 1 percent for every 5°F drop in temperature with no increase in energy consumption. (Rotary compressor efficiency is basically unaffected by changes in inlet temperature. Centrifugal compressors are "mass flow" machines, and scfm and horsepower are both affected by inlet air temperature.
- Heat recovery from air compressors has become an excellent method of conserving energy.
  - In water cooled compressors, water discharged from an air compressor at 110-125°F can be used to feed processes that had been heating water from municipal water supplies. One food processor required huge quantities of hot water for cleaning and was able to use the hot discharged water through a heat exchanger to heat pure cleaning water.
  - In air-cooled compressors, rotary screw compressors have become especially adaptable to use the dissipated heat from the oil cooler and aftercooler to heat an area for many months of the year. During warmer months, the hot air is vented to the outside through duct work. Don't forget to consider the cost of water in computing water-cooled costs and the fan horsepower in considering air-cooled energy costs.

DESIGNING AN AIR COMPRESSOR SYSTEM

Figure 5-14 provides friction losses through pipelines. This should be used to size a distribution system. Some hints that have helped plant engineers design a good system include:

- Size main lines for maximum flow with a maximum 2 percent pressure drop.
- Size branch lines for average flow with 2-3 percent pressure drop.
- Pitch all lines to a low point at one inch per 20 feet. Place drains at the low points.
- Take all compressed air outlets from the top of the pipe so that any condensation, scale or contaminants remain in the pipeline and can be drained off at the low points.
- Where threaded pipe is used, use tees with the branch plugged rather than couplings to join two pieces of pipe. This allows future expansion and branches if necessary without the necessity for cutting into a pipeline.
- Consider expansion when designing a system so the pipelines are large enough to supply additional equipment.
- If a certain piece of equipment uses a large volume of air suddenly, intermittently but repetitively, a surge receiver at the point of use will prevent pressure fluctuations in the line.

AIR TRANSMISSION – Friction Losses Thru Pipe

Delivery in Cu. Ft. of Compressed Air per Min.	Equiv. Delivery in Cu. Ft. of Free Air per Min.	SIZE OF PIPE, INCHES 1/2	3/4	1	1-1/4	1-1/2	2	2-1/2	3	3-1/2	4	4-1/2	5	6	8	10	12
		LOSS OF AIR PRESSURE DUE TO FRICTION (in psi in 1000 ft.* of pipe)															
		At 100 Pounds Gauge															
5.12	40		16.0	4.45	1.03	.46											
8.96	70		49.3	13.7	3.16	1.40	.37										
12.81	100			27.9	6.47	2.86	.77	.30									
15.82	125			48.6	10.2	4.49	1.19	.46									
19.23	150			62.8	14.6	6.43	1.72	.66	.21								
22.40	175				19.8	8.72	2.36	.91	.28								
25.62	200				25.9	11.4	3.06	1.19	.37	.17							
31.64	250				40.4	17.9	4.78	1.85	.58	.27							
38.44	300				58.2	25.8	6.85	2.67	.84	.39	.20						
44.80	350					35.1	9.36	3.64	1.14	.53	.27						
51.24	400					45.8	12.1	4.75	1.50	.69	.35	.19					
57.65	450					58.0	15.4	5.98	1.89	.88	.46	.25					
63.28	500					71.6	19.2	7.42	2.34	1.09	.55	.30					
76.88	600						27.6	10.7	3.36	1.56	.79	.44					
89.60	700						37.7	14.5	4.55	2.13	1.09	.59					
102.5	800						49.0	19.0	5.89	2.77	1.42	.78					
115.3	900						62.3	24.1	7.60	3.51	1.80	.99					
126.5	1000						76.9	29.8	9.30	4.35	2.21	1.22					
192.2	1500							67.0	21.0	9.80	4.90	2.73	1.51	.57			
256.2	2000								37.4	17.3	8.80	4.90	2.72	.99	.24		
316.4	2500								58.4	27.2	13.8	8.30	4.20	1.57	.37		
384.6	3000								84.1	39.1	20.0	10.9	6.00	2.26	.53		
447.8	3500									58.2	27.2	14.7	8.20	3.04	.70	.22	
512.4	4000									69.4	35.5	19.4	10.7	4.01	.94	.28	
576.5	4500										45.0	24.5	13.5	5.10	1.19	.36	
632.4	5000										55.6	30.2	16.8	6.30	1.47	.44	.17

*For longer or shorter pipes the friction loss is proportional to the length, i.e. for 500 feet 1/2 of the above; for 4,000 feet four times the above, etc.

LOSS OF PRESSURE THROUGH SCREW PIPE FITTINGS

(Given in equivalent lengths (feet) of straight pipe)

Nominal pipe size, (inches)	Actual inside diameter, (inches)	Gate valve	Long radius ell or on run of standard tee	Standard ell or on run of tee reduced in size 50 per cent	Angle valve	Close return bend	Tee through side outlet	Globe valve
1/2	0.622	0.36	0.62	1.55	8.65	3.47	3.10	17.3
3/4	0.824	0.48	0.82	2.06	11.4	4.60	4.12	22.9
1	1.049	0.61	1.05	2.62	14.6	5.82	5.24	29.1
1-1/4	1.380	0.81	1.38	3.45	19.1	7.66	6.90	38.3
1-1/2	1.610	0.94	1.61	4.02	22.4	8.95	8.04	44.7
2	2.067	1.21	2.07	5.17	28.7	11.5	10.3	57.4
2-1/2	2.469	1.44	2.47	6.16	34.3	13.7	12.3	68.5
3	3.068	1.79	3.07	6.16	42.6	17.1	15.3	85.2
4	4.026	2.35	4.03	7.67	56.0	22.4	20.2	112
5	5.047	2.94	5.05	10.1	70.0	28.0	25.2	140
6	6.065	3.54	6.07	15.2	84.1	33.8	30.4	168
8	7.981	4.65	7.98	20.0	111	44.6	40.0	222
10	10.020	5.85	10.00	25.0	139	55.7	50.0	278
12	11.940	6.96	11.0	29.8	166	66.3	59.6	332

Figure 5-14. Friction Losses Through Pipe

RECOMMENDED ARRANGEMENT OF TOOL STATIONS

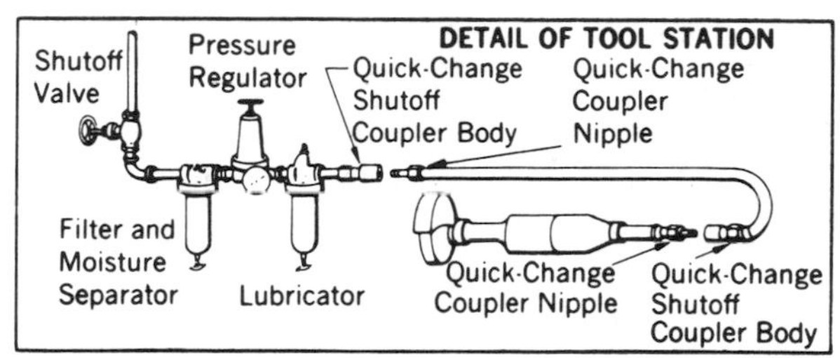

Courtesy of Scales Air Compressor Corp.

Figure 5-15.

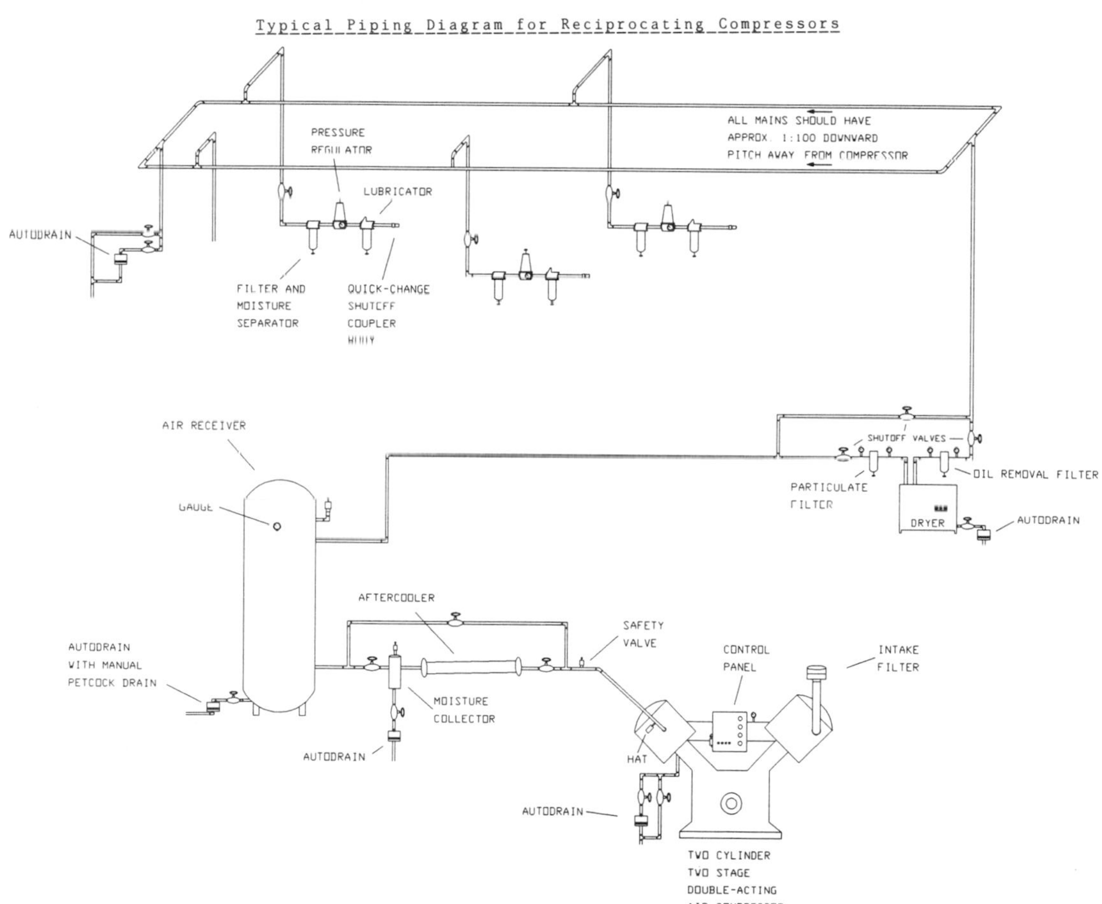

Courtesy of Scales Air Compressor Corp.

Figure 5-16. Typical Piping Diagram for Reciprocating Compressors

- Remember that pressure drop in a pipe varies as the square of the volume flow. For example, 1,000 cfm of free air in a four-inch pipe has a pressure drop of 2.21 psi in 1,000 feet of pipe at 100 psig. Two thousand cfm will have a pressure drop of 8.80 psi in the same pipe. Also, if the pressure drop at 50 psig (64.7 psia) is required in the same four-inch pipe with 1,000 cfm of free air, it would be inversely proportional to the absolute pressures. Therefore, the pressure drop would be $\frac{114.7}{64.7} \times 2.21 = 3.92$ psi.

A proper compressed air piping system is shown in Figure 5-15 and 5-16.

If good selection and installation procedures are followed, many problems with air compressors, accessories and the compressed air system can be eliminated and maintenance substantially reduced. Where repetitive problems occur, it is very possible that the equipment selected is not correct for the application.

Operating logs and records are essential in establishing a maintenance program that will pinpoint problem areas and reduce operating costs. This can be accomplished with a commitment to evaluate problem areas and work towards eliminating them.

6

Effective Motor and Automatic Control Maintenance

Evans J. Lizardos, P.E.
Douglas J. Pavone, P.E.

An understanding of motor and automatic controls is essential to provide proper maintenance to building equipment. This chapter covers the common motor and automatic control systems currently in use:

- Electric
- Electronic
  - Solid State Controllers
  - Programmable Controllers
  - Computer Controls

In the early part of the chapter, material is presented to enable you to:

- Identify typical automatic control components, including motor controllers and the different control systems.
- Read ladder diagrams. These diagrams illustrate the control logic which is fundamental to understanding how systems operate.

With a foundation established in electric controls, you can easily understand electronic controls. The latter part of this chapter describes electronic control systems currently in use.

AUTOMATIC CONTROL SYSTEMS—MAINTENANCE FUNDAMENTALS

The elementary logic contained in the ladder diagrams in this chapter is the key to identifying the function and sequence of operations for motor controllers and other types of automatic control components. For those in maintenance, it is also the key to successful trouble-shooting and repairs. Once understood, servicing of control components and overall system circuits is significantly simplified. Therefore, this chapter is dedicated to developing a basic understanding of the logic underlying simple and complex control systems.

First, however, a few sound maintenance program fundamentals for automatic control systems are included in the following elements:

Planned Maintenance

- On a regular periodic basis, inspect all system control components including all system connections, wire insulation, contact surfaces, basic parts, and other devices.
- Test all system components in accordance with manufacturer's instruction and at recommended frequency:
  - At least once a year, clean terminals, control boxes, and other parts with compressed air or by vacuum cleaning.
  - Simultaneously, check for loose connections.
- Adjust or fine-tune system components to reduce the chance of malfunction in vibration environments. Normally adjustment is unnecessary until the control circuit malfunctions.
  - Replace components in accordance with manufacturer's recommendations. Promptly overhauling the removed components will help ensure an inventory of readily usable units.
  - Run through logic of the entire system.

Failure Response

Failure frequently occurs during emergency situations. Consequently, it is best to prepare with an adequate inventory of spare parts, relays, contacts, spare printed circuits, switches, and so on. It is also prudent to have readily available the manuals for operation, service, maintenance, and spare parts. In addition, keep a posted list of current telephone numbers of key maintenance personnel and emergency service companies, as well as the telephone numbers of manufacturers and spare parts suppliers.

UNDERSTANDING CONTROL SYSTEMS

There are three basic techniques used to describe control circuits: wiring diagrams, written operational sequences, and logic diagrams.

Wiring Diagrams

Wiring diagrams illustrate, by the use of technical symbols and drawing conventions, the physical arrangements and complete electrical wiring requirements. All wiring connections to and from controls, devices and terminals on equipment are located and identified in these diagrams.

Wiring diagrams are required for a complete and proper installation of control equipment. These diagrams enable electrical personnel involved with system design, installation, cost estimates, etc., to determine location, type, and number of devices and the extent of wiring to be provided. Essentially, these diagrams are shop or field installation drawings. They supply the mechanical and electrical contractors with information required to locate and install equipment in accordance with manufacturer's requirements or engineer's design. However, these wiring diagrams assist operating and maintenance personnel only partially in understanding how the overall system operates or how the individual control device functions. For a more complete understanding there is need for a description of the control system operation and sequence.

Operational Sequences

An operational sequence can best be described as a written technical explanation of the established operating sequences and prescribed performance of the control and monitoring system. Usually an operational sequence includes lengthy and extensive descriptions to define properly the intended needs and logic of the system. As these descriptions extend into long and complex explanations, comprehension becomes difficult for some people and impossible for others. As a result, there is a need for a third control explanation technique to convey technical understanding. This is the logic diagram that is often available but seldom used, except by those involved in the original control system design.

Logic Diagrams

Logic diagrams have been in existence as long as wiring diagrams. Nevertheless, they have received little use because few operating and maintenance personnel are aware of their advantages. Wiring diagrams and operational sequences satisfy most maintenance needs, especially when control circuits are simple. In recent years, though, these circuits have become more complex and increasingly difficult to understand. Control equipment manufacturers have increased the number of operating modes and introduced more safety features in their equipment. As a consequence, manufacturers have recognized the value of logic diagrams, and are now actively promoting their use.

The logic diagram is extremely valuable in understanding how a control circuit operates. Its greatest attribute is the ease with which one gains an understanding of the step-by-step logic. It is especially valuable as programmable controllers and computers become more widely used.

Logic diagrams are also called elementary, fundamental, or ladder diagrams. In this chapter, they are referred to as ladder diagrams, because each step of the logic builds on the previous steps, like rungs in a ladder.

The application of ladder diagrams, for understanding control logic, permits maintenance personnel to readily grasp the systems logic and intent, thereby enabling them to understand the sequence of operation. Trouble-shooting a control circuit is considerably simplified when maintenance personnel understand how the system functions.

The underlying principles used in logic diagrams are the same for electric, pneumatic, and other control circuits. However for brevity and convenience, this chapter explains logic primarily in electric control circuits. The approach to understanding other automatic control systems is identical. This includes, but is not limited to, electric and electronic (including solid-state). This commonality is illustrated in the application section of this chapter.

The basic source of ladder diagram information is the equipment manufacturer who supplies a variety of pertinent information including:

- Sales literature and data sheets
- Engineering details and specifications
- Equipment manuals
- Wiring diagrams (frequently mounted somewhere on the equipment)
- Written description of operations
- Manufacturers' ladder diagrams

Other sources of ladder diagrams and related information can be obtained from engineers or others responsible for design and installation of control equipment or systems. Their inputs would include ladder diagrams, shop drawings, contract drawings, system manuals, and other related material.

TYPICAL CONTROL EQUIPMENT HARDWARE AND EXPLANATIONS OF SELECTED RELEVANT TERMS

Primarily for economic reasons, most electric motors over ½ horsepower are powered by three-phase alternating current; and motors under ½ horsepower are usually served by single-phase circuits. Fig. 6-1 summarizes alternating current characteristics most frequently used to control and power equipment in the United States.

SUMMARY OF ELECTRICAL CHARACTERISTICS IN THE U.S.

Function	Voltage	Amperes	Phase	Hertz
Control	24 to 120	5 to 15	Single	60
Power for Small Equipment	120 208 240 277	10 to 40	Single	60
Power for Large Equipment	208 240 480 575 or 600	30 to 400	Three	60

Figure 6-1. Summary of Electrical Characteristics in the U.S.

Figure 6-1 indicates control circuitry is always at 120 volts or less. Consequently, control equipment and devices are designed to be powered at relatively lower voltages. To facilitate explanation and understanding of control systems, the more common types of control equipment, components and devices are identified in the following pages. Where appropriate, selected photographs are used to identify and aid in the description of the equipment. In addition, terms necessary to understand control circuit logic are also defined in this section. These descriptions and equipment identifications are arranged in a building block sequence. The explanations begin with the simplest control terms and equipment, followed by definitions of the progressively more complex components.

Control Voltage, Control Power, Control Circuit

All three power source terms mean the same thing. They denote a 120 volt (or less) single-phase electric power source. Most regulatory agencies are adopting 120 volts as a maximum control voltage. They also require the electric source to be grounded.

Line Voltage (Power)

This term identifies the electric power source that is 120 volts or greater. It may be single- or three-phase. Line voltage is usually associated with the voltage supplied by the utility for equipment and systems found throughout the plant or building.

Switches, Disconnects, and Manual Starters

Switches and disconnects are used to interrupt manually a single- or three-phase electric supply to equipment motors and systems. They can be used in control or power circuits. Manual starters are similar to switches and disconnects, except that starters contain a motor thermal overload control device. This overload device automatically interrupts power to the motor when the current draw exceeds the motor's "must trip" ampere rating.

Courtesy of Square D

Figure 6-2. Manual Motor Starter

Figure 6-2 shows a simple manual starter with thermal overloads for motor protection. This three-pole starter is shown with an enclosure. The left side screw terminals connect the starter to the motor circuit. Figure 6-3 shows a motor pushbutton station housing switch pushbuttons that control the motor circuit. Figure 6-4 shows a three-phase manual motor starter that contains start-stop pushbuttons, a three-pole switch, and a thermal overload device.

Courtesy of Allen-Bradley

Figure 6-3. Motor Start-Stop Station

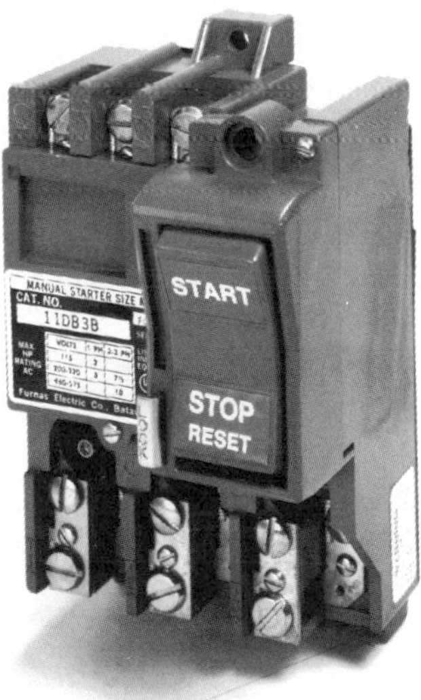

Courtesy of Furnas Electric Company

Figure 6-4. Three-Phase Start-Stop Manual Motor Starter

Magnetic or Solid-State Type Relays, Contactors and Starters

Magnetic-type relays and contactors function in the same way. They consist of a magnetic coil that, when energized, pulls together pairs of contacts to complete a circuit. This permits the flow of line power to equipment.

Magnetic starters are contactors equipped with thermal motor overload protection devices. These are properly referred to as magnetic motor starters. Figures 6-5 and 6-6 picture representative magnetic motor starters. Figure 6-7 illustrates a cutaway view of a typical thermal overload relay found in a magnetic motor starter.

Relays and contactors are similar devices. Relays are frequently used in control circuits at 120 volts or less while contactors are identified with line voltage circuit applications. Consequently, the current capacity rating of control relay contacts is usually less than 10 amperes at 120 volts. On the other hand, contactor ratings match the ampere rating of the power-consuming equipment on the line.

Solid-state relays, contactors, and starters perform the same functions as electro-mechanical relays, contactors, and starters. However, instead of contacts moving mechanically, solid-state circuits are employed to engage and disengage the controlled equipment, component, or device.

Time-delay relays are either magnetic, pneumatic, or solid-state type.

They have an added built-in mechanism to delay the opening or closing of contacts when a relay is energized.

Solid-State and Magnetic Type Multi-Speed Motor Starter

Solid-state and magnetic multi-speed motor starters (usually two-speed) operate similarly to the preceding description. Their operation requires energizing one or the other contactor coils to select the desired motor speed. A time-delay device is normally used between the high and low speeds to avoid high ampere draw at initial low motor speeds.

Solid-State and Magnetic Type Reversing Motor Starters

These are very similar to multi-speed starters, except when an alternate coil is energized in lieu of the initial contactor coil, the motor rotation is reversed. This is accomplished by changing two of the three phase legs supplying power to the motor. A time delay between rotation change is usually required to allow the motor to come to a full stop mode before it is reversed in rotation.

Solid-State and Magnetic Type Reduced Voltage Starters

Energizing a contactor in a motor starter, with its associated instantaneous closing, is referred to as an across-the-line starter. Often it is desirable to limit the maximum motor horsepower size for across-the-line starting use.

A reduced-voltage starter like an across-the-line starter energizes a coil that closes the power contacts to allow electricity to flow to the motor. However, a reduced-voltage starter through the use of an auto-transformer or similar ampere-

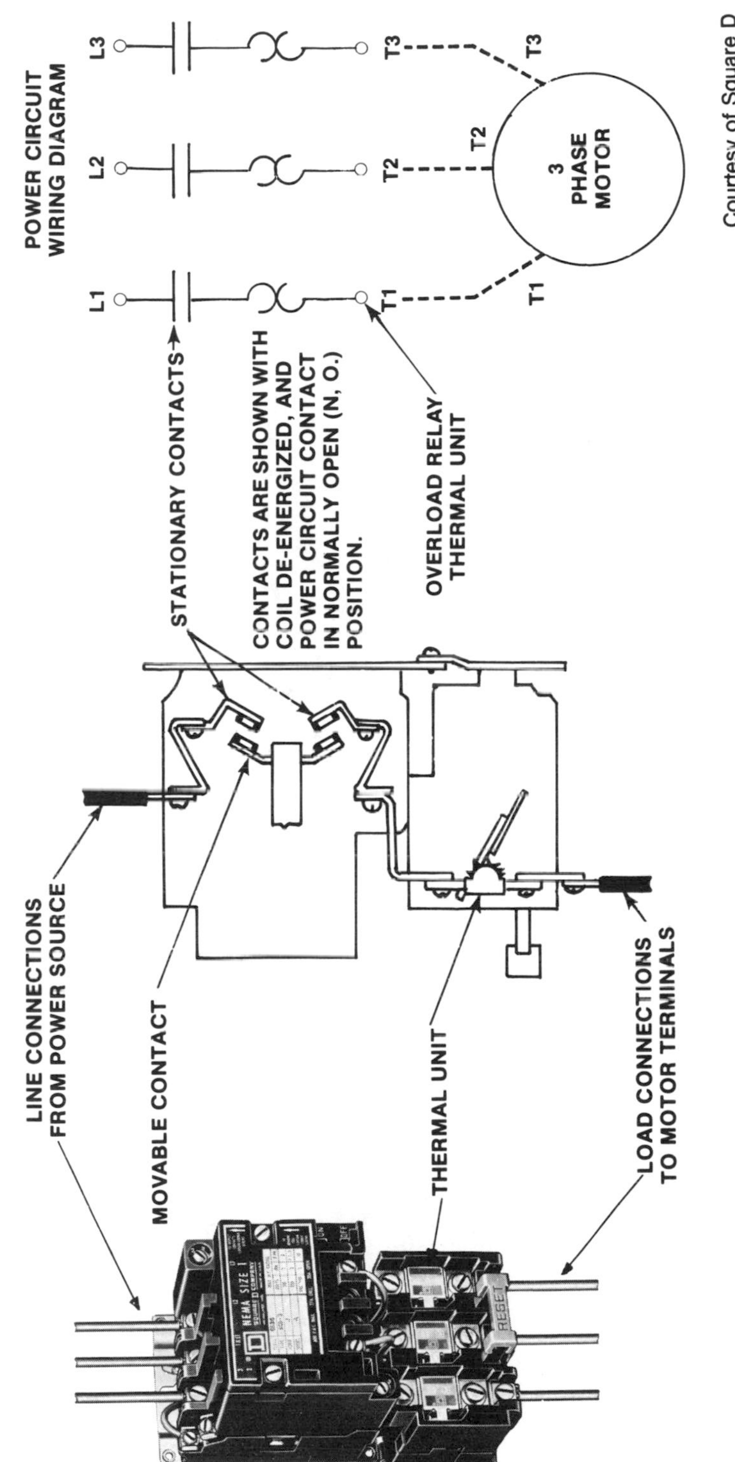

Courtesy of Square D

Figure 6-5. Magnetic Starter

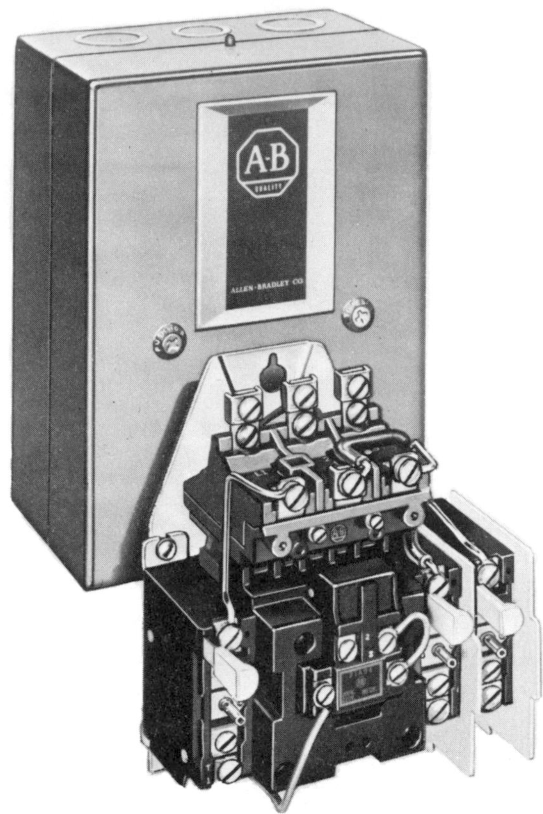

Courtesy of Allen-Bradley

Figure 6-6. Magnetic Motor Starter with Nema Type 1 Enclosure

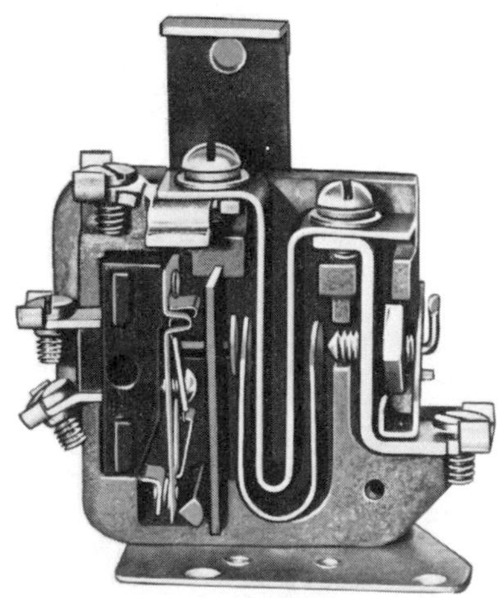

Courtesy of Square D

Figure 6-7. Thermal Overload Relay

limiting components can prevent an unwanted high initial current surge by reducing the input voltage to the motor being started.

Reduced voltage starters are frequently used for large motors to control the starting current (ampere) surge so that momentary voltage dips in the electric distribution system of the plant or building will be minimized.

Combination Starters

When a solid-state or magnetic motor starter is incorporated into an enclosure with a circuit breaker, disconnect switch, or fused disconnect switch, it is termed a combination motor starter.

Motor Control Center

In some applications, it is more economical to connect many motors from a central power and control source. In lieu of mounting individual combination motor starters for each motor, a multiple modular package called a Motor Control Center can be used.

Pilot Devices, Safeties and Accessories

An important component of the electric control system is the primary pilot (sensing or triggering) device. Also important are the safeties and accessories. In general, pilot devices are used to govern the on-off current flow in control circuits. This includes controlling relays, contactors, and starters. Pilot devices can take many forms, including: pushbuttons, thermostats, float switches, time clocks, pressure switches, mechanical and electrical alternators, etc.

A control device used to shut down a system and to prevent injury or damage to life and property is referred to as a *safety device.* This is usually accomplished by opening (stop) the control circuit. Some typical safety devices are:

- high or low water cutout devices as used in boilers,
- high-limit thermostats,
- fire thermostats,
- freeze thermostats,
- low-limit thermostats
- smoke and fire detectors.

To visually indicate the on-off condition of equipment, pilot lights are frequently incorporated into control circuits.

Control transformers are used to change the line voltage power source (generally over 120 volts), to the equipment to be controlled, to a control voltage of 120 volts or less. In this manner pilot lights, safety devices, and other components are energized at 120 volts or less, thereby avoiding high voltage control circuits that require larger devices with heavy duty components.

Whenever the line voltage to the equipment to be controlled is 120 volts, the on-off devices (pilots) and safeties are often wired directly to the controlled equipment. This integrates the control and power circuits as one. However, when a three-phase power system supplies the controlled equipment, a 120-volt control circuit is used.

Multiplexing, Interfacing, and Programmable Controllers

System control and monitoring of remote components from a central location can involve significant numbers and lengths of cable. This could add appreciably to initial installation costs.

Multiplexing provides one method of avoiding large quantity of wiring. An example of this technique involves the hard wiring of several closely located or common control devices to a nearby input/output controller. From such a remote input/output controller, one pair of wires is installed to a centrally located programmable controller (PC) or computer.

The single pair of wires carries pulsating electric signals that identify an individual control device. These signals are generated and interpreted by the programmable controller (PC), and/or computer. This method enables many remote devices to be controlled simultaneously via a minimum number of wires.

The combining of two distinct systems to operate as one is called interfacing. When controlled devices are hard wired to an input/output controller for transmission to the PC or computer, they are termed as *interfaced* into the PC system. Similarly, when a pneumatic system is activated by opening an electrical solenoid air valve, interfacing occurs.

UNDERSTANDING ELECTRIC CONTROL LADDER DIAGRAMS

The logic for understanding (elementary) ladder diagrams is explained in this section. The technique used includes an explanation of the construction and logic of 12 diagrams. This procedure starts with the simplest of preliminary diagrams and progresses to more complex ladder diagrams by adding additional features. In the simpler diagrams components are enclosed in square or rectangular outline boxes for easier understanding.

The predetermined steps of control logic are usually displayed as inverted steps in a ladder. They start from the top with first step (Rung No. 1) and work down on the ladder diagram. The first step usually starts with the power source for the control system. Each subsequent step adds one or more elements to the control circuit in the sequence they would normally occur, or in the sequence that is easy to understand. To conserve space, standard symbols and abbreviations are used on each rung of the ladder; however, they convey a complete control message. A written explanation of the control logic for one complex rung could require several paragraphs of text. The symbols and notations on the rung convey the same information.

Ladder diagrams have several applications, but the three most common uses are identified as follows:

- Accurate trouble-shooting: The maintenance worker who understands the control logic can pinpoint control system problems faster than one who does not understand the function. On a production line, this can often make the difference between minutes and hours of downtime; long delays in production operations can be disastrous. A knowledgeable maintenance technician avoids erroneous assumptions about where the control fault is located. The ladder diagram assists in localizing the fault area and also permits the trouble-shooter to recognize the cause of the malfunction.
- Add to existing system: Additions to existing control systems require more than adding wire and components. Additions must be totally compatible with system logic and the physical installation constraints. Since there are several ways to wire a complex circuit, the best arrangement can be identified on paper through modifications and extensions of the original logic diagram.
- Understanding sophisticated solid-state controllers: Without a logic diagram, the complex circuits inherent in solid-state controllers would be nearly impossible to follow and comprehend. Since programmable controllers are rapidly becoming popular, it is incumbent on those in maintenance to be prepared to service an increasing number of these sophisticated systems.

Single-Phase Line Voltage Control

Figure 6 8 shows a simple wiring diagram (6-8a) and the corresponding ladder diagram (6-8b) for a single-phase 120 volt motor control circuit. The wiring diagram identifies the 120 volt supply, a control thermostat in the "hot leg," and a manual motor starter near the agitator motor AG-1. This starter consists of a manual switch and a thermal motor overload that protects the agitator motor AG-1. With this wiring diagram, an electrician would have no difficulty wiring to the system components. The ladder diagram depicts the logic involved when the thermostat, which is the primary controller in this example, closes the circuit. If the manual switch is closed, the agitator motor AG-1 will start. Either the manual switch or the thermostat can open the circuit and stop the motor. Of course, the thermal motor overload can also interrupt the circuit.

Single-Phase Pilot Control

Figure 6-9 shows a control condition similar to that described for Figure 6-8, except the control circuit is reduced from a 120 volt supply to a 24 volt control voltage. The wiring diagram shows each element in its respective position for effective wiring. The five-step ladder diagram (6-9b) shows the logic of the control circuit via the rungs:

1. The power source is 120 volts.
2. It supplies the high voltage side of the transformer in the control circuit.

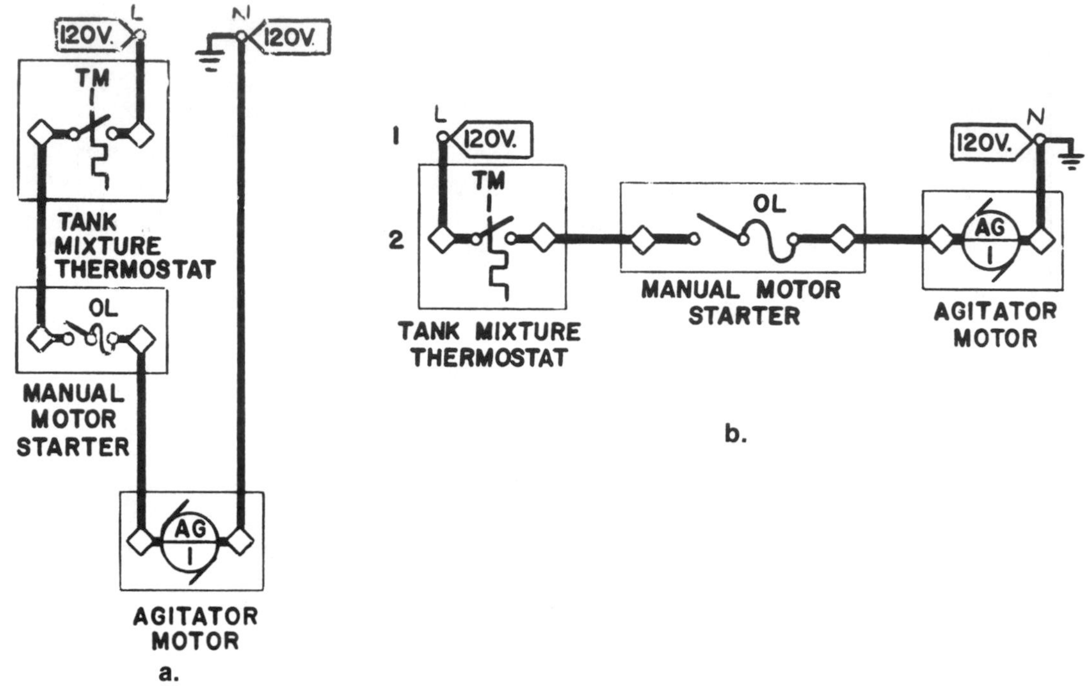

Figure 6-8.

Single-Phase Line Voltage Control
a. Wiring Diagram
b. Ladder Diagram

3. The low voltage side of the transformer supplies the control voltage at 24 volts. This pilot control voltage is more sensitive than higher voltage and therefore controllers are easier to fine adjust. This circuit is grounded and a fuse protects the 24 volts control circuit from overload.
4. On Rung 4 the thermostat is the main controller. When the thermostat closes, the relay R-1 in the control circuit is energized. The notation "5" to the right of the fourth rung indicates Rung 5 contains a device that is activated when the relay R-1 is energized.
5. When this occurs, the contacts R-1 in the hot leg of the electrical supply to the motor are closed. If the switch in the manual motor starter is closed, the agitator motor AG-1 will start. The motor will continue to operate until the circuit in Rung 5 is interrupted. This can be accomplished by opening the manual switch in the motor starter, a motor overload triggering the thermal overload device, or when the relay contacts are opened. The contacts are opened only when the 24 volt control circuit is interrupted, usually by the thermostat; but they can also be opened by a short circuit or an overload causing the fuse to open the control circuit. Note: The "4" under contacts R-1 indicates this device is controlled by an element in Rung 4.

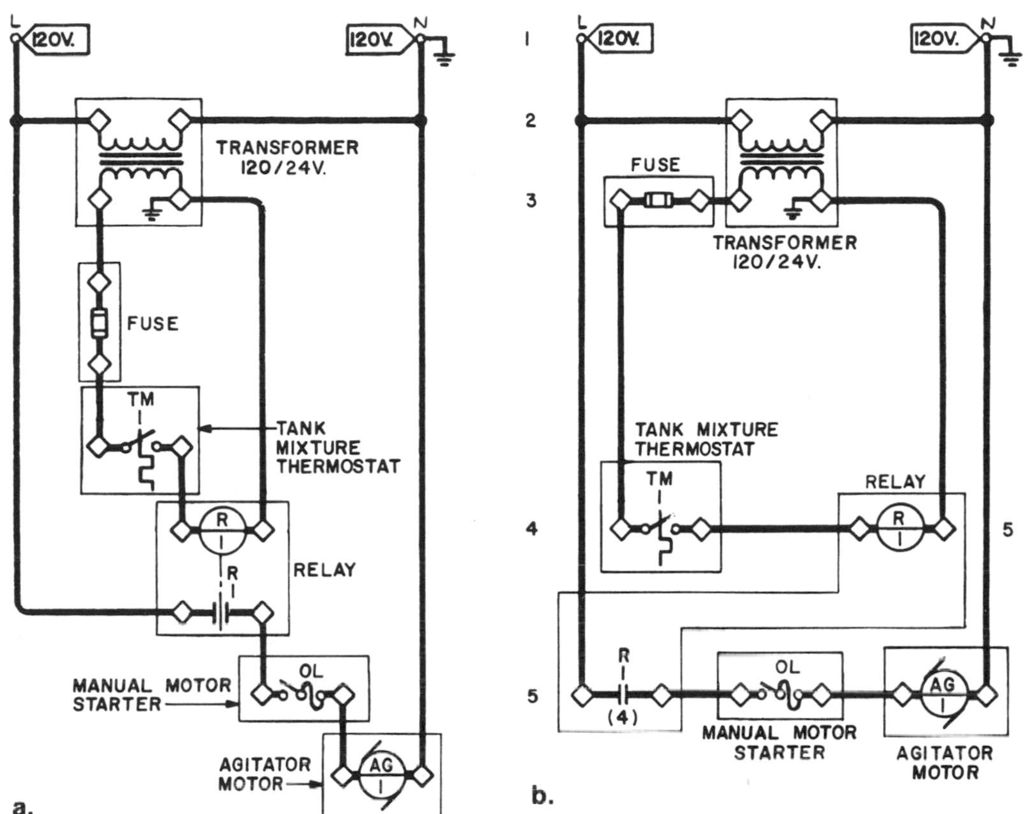

Figure 6-9.

Single-Phase Pilot Control
a. Wiring Diagram
b. Ladder Diagram

Three-Phase Motor Control

Figure 6-10 shows the wiring and ladder diagrams for a motor control circuit which is similar to Figures 6-8 and 6-9. The major difference in this example is the three-phase 480 volt supply to the agitator motor. The wiring diagram (Figure 6-10a) shows all components and connections in both the 480 volt supply and the 120 volt control circuit. The ladder diagram (Figure 6-10b) shows only the elements in the control circuit.

For example, the manual disconnect switch at the motor, normally used as a maintenance safety feature, is shown in the wiring diagram. It is excluded from the ladder diagram, however, because it has nothing to do with the control circuit. In the four-step ladder diagram, all aspects of the control circuit are shown in their logical sequencc, as follows:

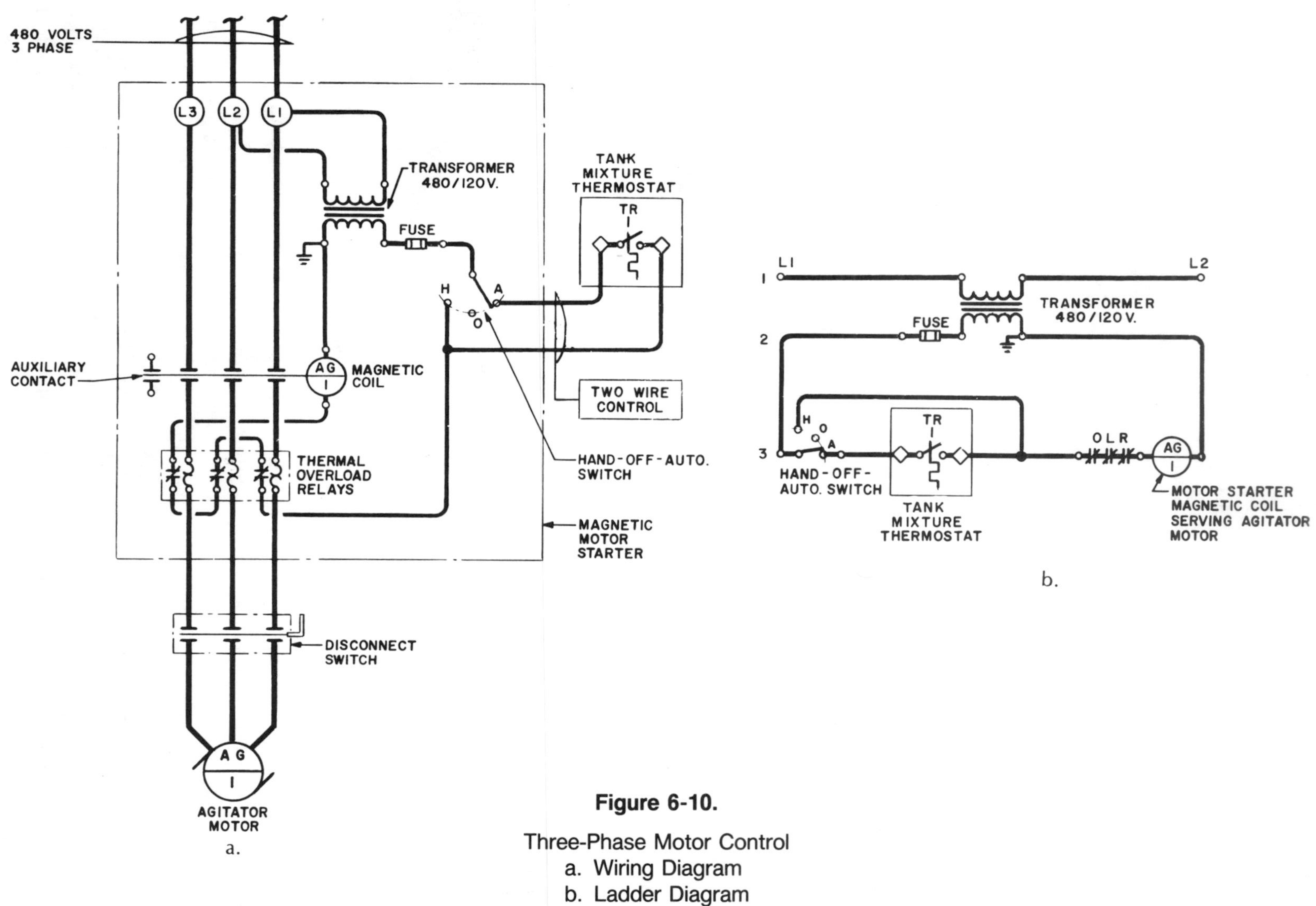

Figure 6-10.

Three-Phase Motor Control
a. Wiring Diagram
b. Ladder Diagram

1. Line voltage at 480 volts is taken from any two hot leads of the three-phase, three-wire supply.
2. The high voltage (480 volts) side of the transformer is reduced to the control circuit low voltage (120 volts) side, where the hot leg has a protective fuse and the neutral side is grounded. This could be reduced to 24 volts or any other control voltage.
3. The interlocking controls are in series. They include a three-position "hand-off-automatic" selector switch, a thermostat that is effective only when the switch is on automatic, and thermal overload relay devices to protect the agitator motor AG-1. With this type of control circuit, the motor after power restoration will start up unattended; provided the three-way switch (H,O,A) is in the automatic mode and the thermostat closes the circuit, or if the three-position switch is in the "hand" mode.
   This type of control is commonly known as a two-wire control circuit because two wires are required to connect the thermostat to the circuit. *Note:* the magnetic coil and motor are both identified as AG-1, but their symbols are slightly different. In addition, the wiring diagram shows auxiliary contacts that are used to control auxiliary equipment. The use of the auxiliary contacts is demonstrated in the next ladder diagram.

Manual Three-Phase Control

Figure 6-11 contains a ladder diagram that is similar to Figure 6-10, except a manual reset is required after a power failure. This motor control circuit is commonly referred to as a three-wire control circuit because three wires are required to connect the remote stop-start pushbutton station feature to the circuit. Also, auxiliary contacts, shown in the wiring diagram, are used to supply constant control voltage to the control circuit. The five-rung ladder diagram contains the following logic:

1. Line voltage is supplied to the high voltage side of the transformer at 480 volts.
2. The 120 volts control voltage has a fuse in the hot leg and is grounded in the neutral leg.
3. The two stop pushbuttons, one at a remote location, are in series with one another and both are normally closed. The start pushbuttons are normally open. One start pushbutton must be manually (momentary push) closed to energize the motor. The thermal motor overloads, which protect the motor, are normally closed. When the magnetic coil AG-1 is energized, it closes the main motor power contacts and simultaneously the auxiliary contacts AG-1 also close.
4. The start pushbutton that is remotely located is parallel to the local start pushbutton and with the auxiliary contacts AG-1.

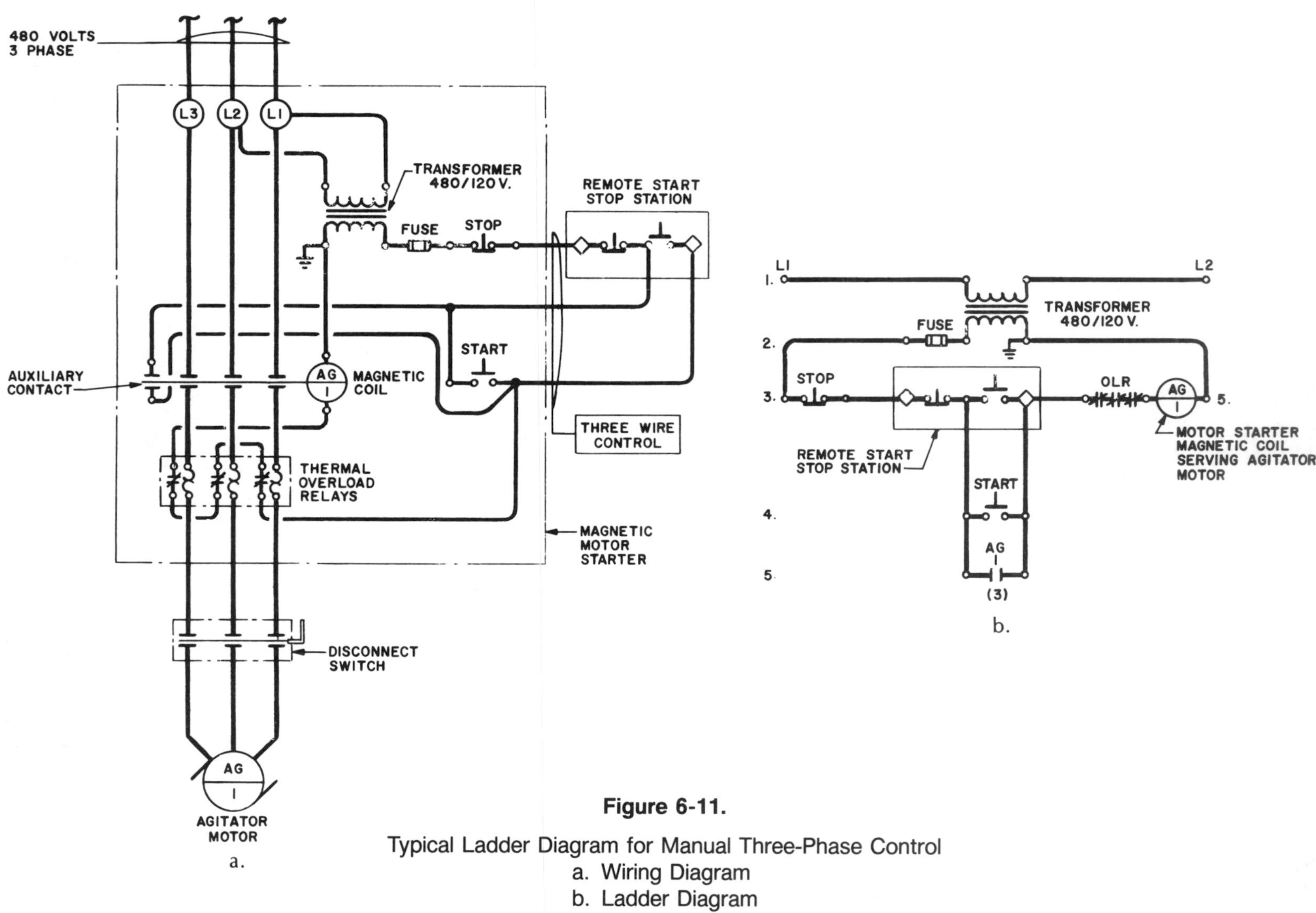

Figure 6-11.

Typical Ladder Diagram for Manual Three-Phase Control
a. Wiring Diagram
b. Ladder Diagram

5. When the magnetic coil energizes auxiliary contact AG-1, which is parallel to the two parallel-connected start pushbuttons, the motor control circuit will remain closed when the manual start pushbutton is released. However, if a power failure occurs, the magnetic coil AG-1 will be de-energized and the entire control circuit will open. When power is restored, the motor will not start until a start pushbutton is manually closed.

Single-Phase Automatic Start-Up Control

Figure 6-12 is very similar to the control wiring diagram in the previous example, except controls are in parallel in Figure 6-11 and in Figure 6-12, they are in series. Accordingly, this is a two-wire circuit containing the automatic start-up feature following a power failure. *Note:* wiring diagrams are excluded from the remaining figures because ladder diagrams are the primary subject of this chapter. The control features in this three-step ladder diagram are described as follows:

1. The 120 volt power source is attainable directly from the line service; or from a transformer.
2. An aquastat is normally closed, but will interrupt the circuit when the water temperature in the pipe is below a predetermined level. The three-position selector switch, shown in the automatic position, also has off and hand (manual) positions. The hand position enables the circuit to bypass the main controller, which is a room thermostat. The thermostat is normally open and closes the circuit to engage the unit heater motor. Finally, the manual motor starter contains the typical switch (manual, normally closed except for maintenance) and the thermal motor overload.
3. A green pilot light is in parallel with the motor, when the motor is operating, the light is on.

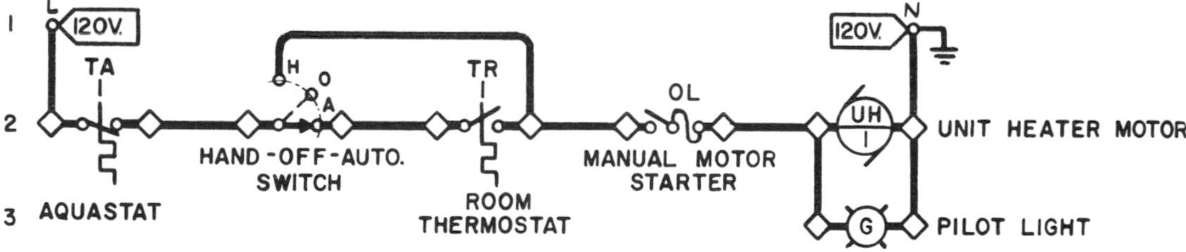

Figure 6-12. Ladder Diagram for Single-Phase Automatic Start-Up Control

Automatic Control of Single-Phase Exhaust Fan

Figure 6-13 contains the ladder diagram for controlling a typical single-phase 120-volt exhaust fan. The control circuit includes several safety features as noted in the following ladder diagram explanation.

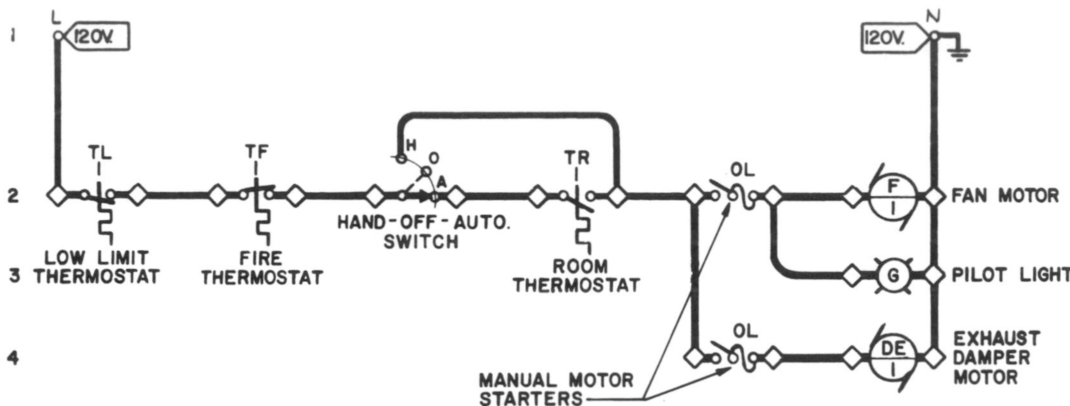

Figure 6-13. Ladder Diagram for Automatic Control of Single-Phase Exhaust Fan

1. The 120-volts power source supplies the control circuit.
2. This is the key control rung in the ladder diagram, and it contains the following elements.
   - A room-type low-limit thermostat TL-1 is normally closed. When the room temperature drops below 55°F, it will open, thereby interrupting power to the fan.
   - The fire thermostat TF-1 is located in the exhaust fan suction plenum. It is normally closed, but will open at the high temperature of 125°F, which can be produced by a fire. If smoke detection and control is required, a fire and smoke detector would be substituted for the fire thermostat and wired in series on this rung.
   - The three-position selector switch is shown on automatic, but the hand position H enables the circuit to bypass the main controller TR-1.
   - A room-type thermostat TR-1 automatically starts the fan when the room temperature reaches a predetermined temperature. At that point, the circuit is closed and the fan starts up. When a preset lower temperature level is reached, the thermostat opens the circuit stopping the fan.
   - The manual motor starter is depicted by a manual switch and thermal motor overload. This manual switch is normally in the closed position and primarily used for maintenance purposes.
   - Finally, a fan motor is shown on this 120 volt single-phase power and control voltage line.
3. A green lamp illuminates when the motor is operating.
4. The normally closed exhaust damper DE-1 is opened when the fan is operating. The damper motor is equipped with a manual motor starter (switch and thermal motor overload). The manual switch is normally closed.

In this two-wire control circuit the fan will start up automatically following a power failure, provided the room temperature is higher than the room thermostat setting.

Manual Control of Three-Phase Exhaust Fan

Figure 6-14 exhaust fan motor control is the same as Figure 6-13 except for manual control (start-stop) and in the use of a three-phase fan motor.

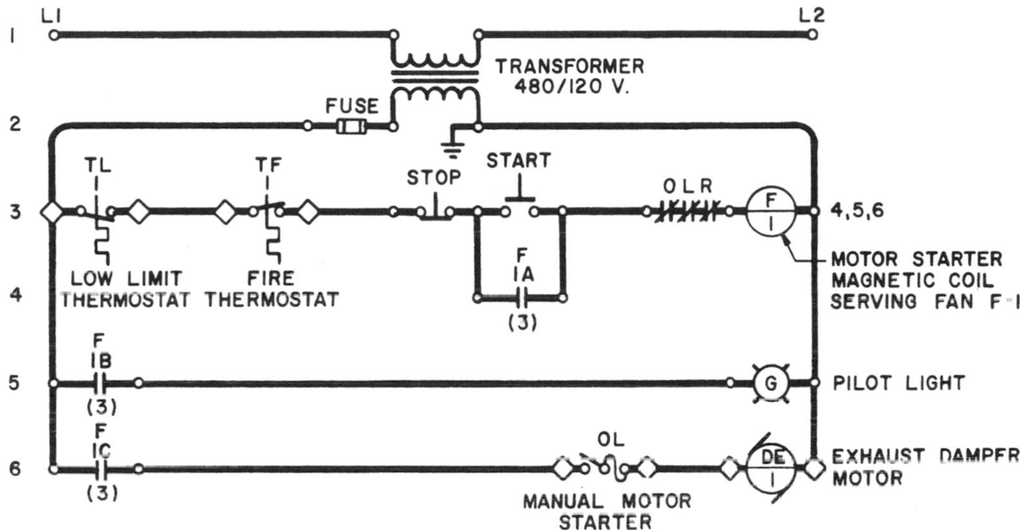

Figure 6-14. Ladder Diagram for Manual Control of Three-Phase Exhaust Fan

1. Power is taken from two hot legs of the 480 volt supply. Therefore, the high voltage side of the transformer is at line voltage.
2. The transformer secondary provides control voltage at 120 volts. The hot leg of the control circuit is fused to protect the components, and the neutral leg is grounded.
3. The third rung of the ladder diagram contains the principal controller and safety devices. These operate as follows:
   - The normally closed low-limit thermostat TL-1 will open the circuit when the room temperature drops to 55°F.
   - The fire thermostat TF-1 is located in the exhaust fan suction plenum. It is normally closed, but will open at the high temperature of 125°F, which can be produced by a fire. If smoke detection and control is required, a fire and smoke detector would be substituted for the fire thermostat and wired in series on this rung.

- The stop pushbutton is normally closed. However, when depressed, the circuit is opened, causing the exhaust fan to stop.
- The start pushbutton is normally open. When closed, the circuit is complete and the fan motor starts. However, when the start pushbutton is released, the switch opens that part of the circuit. Therefore, the maintaining contact in Rung 4 is necessary to keep the fan motor-operating. This is further explained below:
- The motor starter thermal overload relays are normally closed, and are disengaged only when line current to the motor is too high. At overload conditions, these contacts open the control circuit and the fan motor stops.
- The fan motor magnetic contactor F-1 is energized when the third rung circuit is closed. When energized, this coil closes the line voltage contacts to the motor, thereby starting it. The magnetic contactors also close the auxiliary contacts shown on Rungs 4, 5, and 6.

4. When Rung 3 circuit is closed, the contact F-1A is closed and remains closed until the fan circuit is interrupted. The purpose of the contact is to lock in the start pushbutton and enable the fan motor to continue operating (circuit remains closed) after the start pushbutton is released. Once the control circuit on Rung 3 is interrupted, auxiliary contact F-1A is returned to its normally open position. The motor will not restart until the start pushbutton is depressed.
5. Auxiliary contactor F-1B is normally open. However, when the fan operates, it is closed and the green pilot light is illuminated.
6. Auxiliary contactor F-1C closes the circuit in Rung 6, when the fan operates. As a result, the exhaust damper motor is energized and the damper opens. When the fan shuts down, auxiliary contactor F-1C is opened and the damper is closed. A manual switch and thermal motor overload, both normally closed, are located near the damper operating device.

Automatic Control of Three-Phase Exhaust Fan

In Figure 6-15, the control circuit for this exhaust fan motor is similar to Figure 6-14, except it has an automatic control feature. This means the room thermostat is the prime control when the three-position selector switch is in the automatic position.

1. The line voltage is supplied to the primary high voltage side of the control circuit transformer.
2. The control voltage is at 120 volts. The hot leg has a lockout switch, which, although not an approved disconnect means, is a good maintenance feature. Rung 2 also has a circuit fuse. The neutral side is grounded.

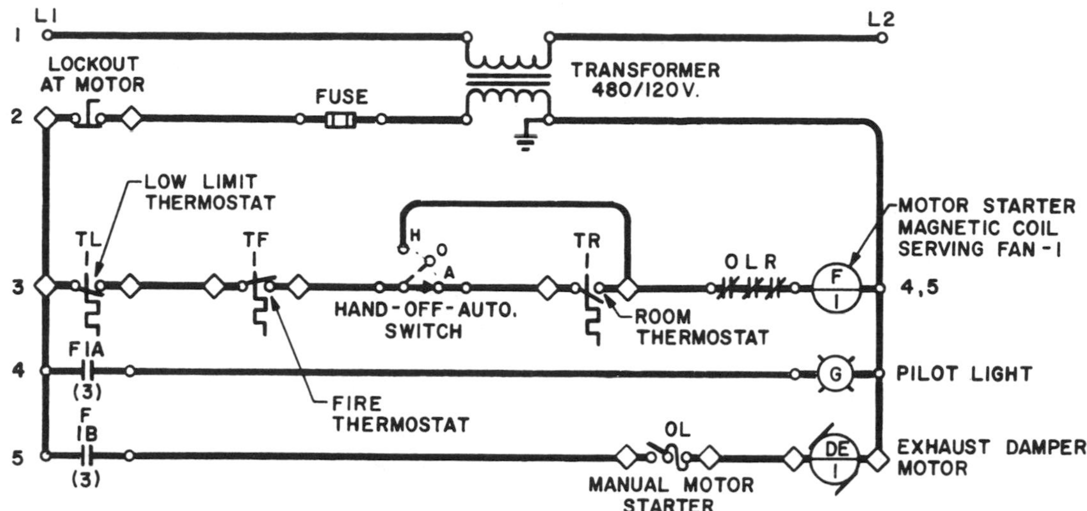

Figure 6-15. Ladder Diagram for Automatic Control of Three-Phase Exhaust Fan

3. Rung 3 of the ladder diagram contains the principal controller and safety devices. These operate as follows:
   - The low-limit thermostat TL-1, the fire thermostat TF-1, the motor starter thermal overload relays OLR, and the magnetic coil F 1 operate the same as outlined in the Figure 6-14 description (see Rung 3). The three-position selector switch and the room high-limit thermostat TR-1 operate the same as outlined in the Figure 6-13 description (see Rung 2).
4. The normally open auxiliary contacts F1A and F1B are closed whenever the exhaust fan motor starter magnetic coil is energized. The contacts operate the same as explained for Figure 6-14.

Mechanical Alternation of Two Single-Phase Motors

Figure 6-16 shows the control circuit ladder diagram for a single-phase mechanical alternator that is alternating the use of two motors. This is typical for controlling the fluid level in a vessel with two independently driven single-phase pump motors. The control circuit logic for single-phase motors is explained as follows:

1. The 120 volts power source is grounded on the neutral leg and has a fuse for circuit protection in the hot leg (not shown).
2. The mechanical alternator contacts are normally open. They are closed only when the alternator selects pump motor P-1 for operation. The manual motor starter switch and thermal motor overload OL are normally closed. The motor P-1 is energized when the circuit is closed and the motor operates

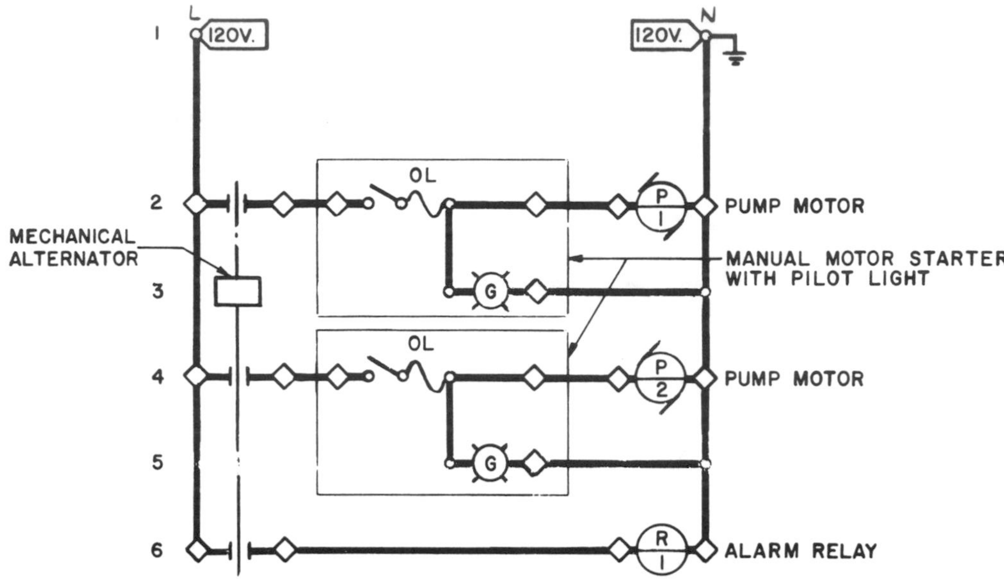

Figure 6-16. Ladder Diagram for the Mechanical Alternation of Two Single-Phase Motors

the pump. When the alternator contacts are opened, the motor stops running.

3. The green indicator light is on when motor P-1 is operating.
4. The mechanical alternator is a mechanical device that alternates the use of two motors, each capable of meeting operating requirements. When the float device (presumably in a tank) reaches a primary level, the alternator calls for motor P-1 to operate by closing the contacts in Rung 2. Pumping is completed when the float lowers to a predetermined lower level and the alternator disengages the contacts in Rung 2. The next time the float reaches the primary control level, the alternator closes the contacts in Rung 4, thereby energizing pump motor P-2. This cycle is repeated sequentially, alternating the use of and wear on the two-pump motors.
   The alternator also has two other features. First, when the water rises to a secondary control level (above the primary level), the alternator closes the contacts on both Rung 2 and Rung 4, thereby operating both pumps. This higher level would be reached when the inflow to the tank exceeds the capacity of a single pump. Second, if the water level continues to rise above the secondary level the alternator will then close another set of contacts shown on Rung 6 and an alarm will sound. This will occur when the inflow exceeds the capacity of both pumps or a malfunction of the pumps.
5. Rungs 4 and 5 are controlled exactly as explained for Rung 2 and Rung 3.
6. As explained in Rung 4, when a high level is reached, the contacts will be closed by the alternator and alarm R1 will be energized. This will signal the imminent overflow of the tank.

Mechanical Alternation of Two Three-Phase Motors

In Figure 6-17, this control circuit is similar to Figure 6-16 except the alternator, which is the primary controller, is controlling two condensate pump motors each powered by three-phase current. The logic of each rung in the ladder diagram is as follows.

1. The line voltage is supplied from two of the three-phase feeding condensate pump motor CP-1. This line voltage supplies the primary high voltage side of the transformer.
2. The low voltage side of the transformer is at 120 volts, which is the control circuit voltage. The hot leg of the circuit is fused and the neutral side is grounded.
3. Auxiliary contact CP-1 is normally open, but when magnetic coil CP-1 is energized, this coil is closed. The circuit is complete, and the green pilot light lamp illuminates, indicating motor CP-1 is operating.
4. This is the main control rung for motor CP-1. When the mechanical alternator closes the contacts in this rung, the circuit is completed. Magnetic coil CP-1 is energized and the motor supply line voltage

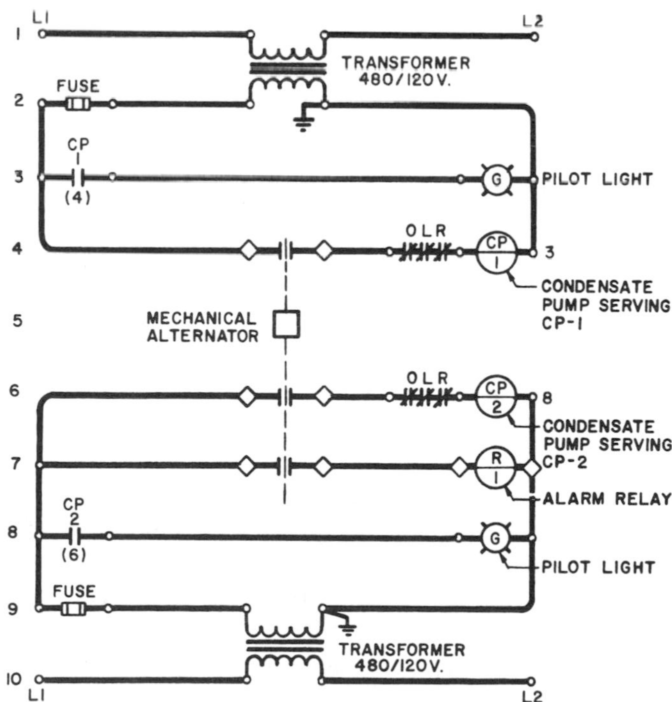

Figure 6-17. Ladder Diagram for Mechanical Alternation of Two Three-Phase Motors

contacts are closed. The motor starts and continues until the mechanical alternator opens the contacts on this rung.

5. The operation of the alternator is exactly the same as described at Rung 3, Figure 6-16.
6. This rung functions exactly the same as Rung 4 above; except this rung controls condensate pump motor CP-2.
7. As explained for Figure 6-16, Rung 6, an alarm R1 is sounded when the liquid level in the tank reaches a very high level. This circuit is activated when the alternator closes the contacts in this rung.
8. This rung operates the green pilot light the same as Rung 3 above.

9. & 10. The two rungs perform the same function for motor CP-2 as Rung 1 and Rung 2 perform for motor CP-1. Each motor has its own line voltage supply to its control circuit. With this arrangement, each motor (and its control) is completely independent from the other. This is advantageous for maintenance and emergency repairs.

Manual Control

In Figure 6-18, this ladder diagram identifies the control logic for a manually operated supply air unit powered by three-phase current.

1. The line voltage is taken from two phases of the three-phase power supply to the motor. The line voltage input to the control circuit is through the primary high voltage side of the transformer.
2. The 120 volt control circuit is fused and grounded.
3. This is the main control circuit that is protected by a normally closed smoke detector SD-1 and a normally closed freeze thermostat TZ-1 in Rung 6. The start-stop switches, the thermal motor overload relays OLR, and the motor magnetic starter F-1 are operated exactly the same as described for Rung 3 in Figure 6-16. The magnetic coil F-1 operates auxiliary contacts in Rungs 5, 7 and 9.
4. By its parallel arrangement with the local start button the remote start pushbutton can energize magnetic coil F-1.
5. Auxiliary contact F-1A is energized by magnetic coil F-1. It remains closed while the motor operates. When the motor stops (Rung 3 circuit is interrupted) coil F-1 is de-energized and contacts F-1A are opened (the normal position). The motor cannot be restarted until a start pushbutton is engaged.
6. The freeze thermostat is located in the fresh air supply duct. When the inlet temperature (downstream of the heating coil) drops to 36°F, the freeze thermostat opens the contacts in Rung 3. It also en-

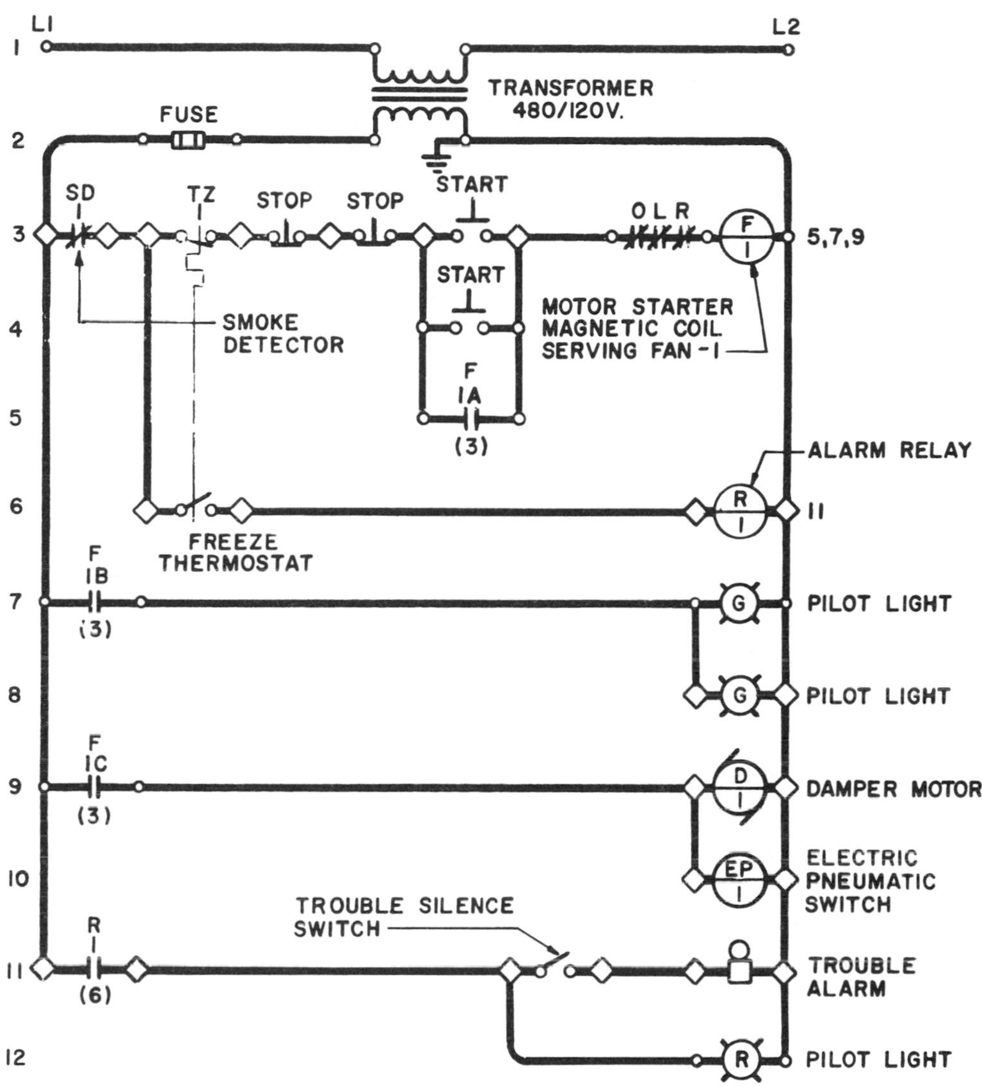

Figure 6-18. Ladder Diagram for Manual Control

ergizes relay coil R1. The TZ-1 control device opens the main control circuit, and the air supply fan motor stops (if operating) and will not restart until the condition is corrected. The relay coil R1 closes contacts in an alarm circuit described below.

7. Auxiliary contact F1B is closed when magnetic coil F-1 is energized. This closes the circuits in Rungs 7 and 8. A green pilot light is illuminated at the motor starter.
8. When contacts F-1B are closed, a remote green pilot light will also illuminate. Both lights indicate the air supply fan motor is running.

9. & 10. When magnetic coil F-1 is energized, the auxiliary contact F-1C is closed and the fresh air intake dampers are opened. These are usually opened by electric damper motors D-1 or an electric pneumatic switch.

11. When relay coil R1 is activated by the freeze thermostat in Rung 6, the contacts at R1 are closed and the circuits in Rungs 11 and 12 are completed. The alarm sounds and a manual switch is available to silence the audio alarm.

12. The red pilot light alarm remains on as long as the contacts at R1 Rung 11 are closed. The red pilot light will not go off until the freeze thermostat TZ is above 36°F and reset.

Automatic Control of Three-Phase Supply Air Unit

Figure 6-19 is a ladder diagram very similar to Figure 6-18, except this one is automatically controlled. It also includes a fire thermostat TF safety device.

1. & 2. These rungs are exactly the same as the explanation in Figure 6-18.

3. The main control circuit has the following features:
   - A freeze thermostat and a fire thermostat are two safety devices in this primary control circuit. They are normally closed, but they interrupt the circuit when either set of contacts are opened by their respective thermostat.
   - The three-position selector switch (hand, off, automatic) is usually in the automatic position. The manual position (sometimes called the hand position) bypasses the primary control, which is called the pilot device.
   - For a supply air system, the pilot device would normally be a thermostat, but in some process operations it could be a time clock or similar control device. A time clock is shown in this example.
   - The thermal overload relays OLR and fan motor coil F-1 are the same to those described in previous figures. In this application the magnetic coil F-1, when energized, closes the auxiliary contacts in F-1A and F-1B on rungs 6 and 8, respectively.
4. The fire thermostat functions similarly to the freeze thermostat except for its operating temperature and location. Fire thermostats are usually activated at about 125°F. When activated, the normally closed contact in Rung 3 would be opened. In this application, the fire thermostat would probably be located in the inlet duct to the fan.
5. This freeze thermostat rung is exactly the same as described in Rung 6 Figure 6-18. When the freeze thermostat is closed, relay coil R2 will also be activated. This relay controls the alarm circuits in Rungs 12 and 13.

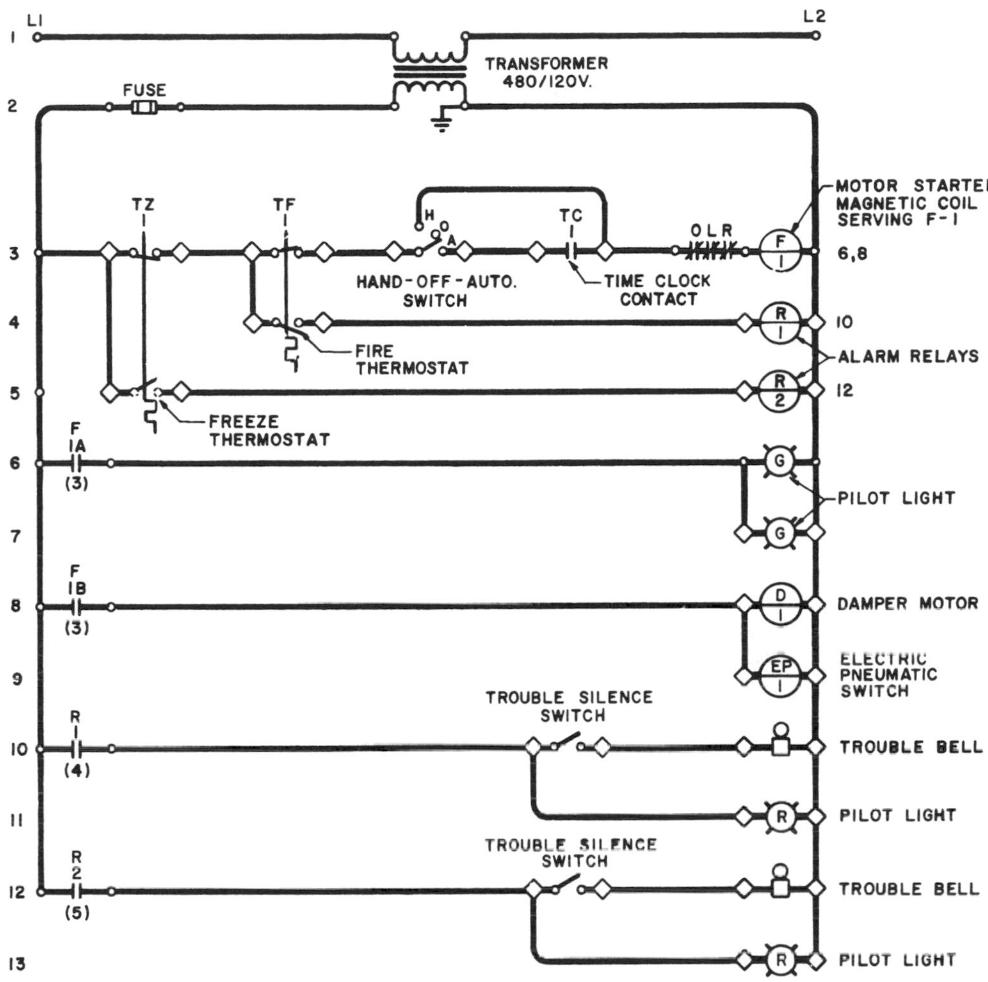

Figure 6-19. Ladder Diagram for Automatic Control of Three-Phase Supply Air Unit

6. & 7. When magnetic coil F-1 is energized, the auxiliary contact at F-1A is closed. This completes the circuit and green pilot lights in both rungs signal that the motor is operating. One of the lights is located at the motor starter. The other pilot light is remotely located (possibly in a master control panel).

8. & 9. When magnetic coil F-1 is energized, the auxiliary contacts at F-1B are also closed. This completes the circuits in Rungs 8 and 9 and the air inlet dampers are opened. These are usually actuated by electric motors or an electric pneumatic switch. When coil F-1 is de-energized, the contacts at F-1B are opened and the dampers closed by a spring drive.

10. When relay coil R1 in the freeze thermostat Rung 4 is energized the contacts at R1 are closed. This completes the circuits in Rungs 10 and

11. The audio alarm (bell) sounds, indicating the presence of cold air in the intake duct. A silencing switch is located in this circuit to silence the audio portion of the alarm.

11. The red trouble pilot light also goes on when the audio alarm sounds. However, when the silencing switch turns off the audio alarm, the trouble light remains on until the air temperature in the intake duct is increased to above 35°F and the freeze thermostat is reset.

12. The contacts at R2 are closed when the relay coil R2 in the fire thermostat Rung 5 is energized. This completes the circuits in Rungs 12 and 13. The audio alarm sounds, indicating the presence of hot air (over 125°F) in the outlet duct. A silencing switch is located in this circuit to turn off the audio alarm.

13. The red trouble pilot light also goes on when the alarm bell sounds. However, when the silencing switch turns off the alarm, the trouble light remains on until the discharge air temperature is decreased below 125°F and the fire thermostat is reset.

APPLYING LADDER DIAGRAM LOGIC TO MAINTENANCE

The confidence gained by understanding the 12 ladder diagrams in the previous section should provide the maintenance trouble-shooter with the analyzing skill necessary to take on control circuit maintenance. However, control circuits may be electrical, pneumatic, electronic, solid-state, or a combination of these elements. Accordingly, the typical application explained in this section includes all of these control elements. As mentioned before, the logic is the same regardless of the control medium.

Figure 6-20 contains a floor plan and an air flow diagram of a complex heating-ventilating system for three laboratory rooms. Each room has its complete and independent air handling system with corresponding discharge ducts and fans. The heating-ventilating (HV) units for Room No. 1 are electrically controlled. In Room No. 2, the HV control is electric/pneumatic; and, in Room No. 3, the HV controls are electric/solid state. This third system contains remote control capability using a programmable controller (PC) centrally located in a control room. It is made more complex by multiplexing the signals between the PC and the remote local control stations.

Consolidating these three different control mediums into one system is complicated and difficult to comprehend without a ladder (logic) diagram. However, with a properly prepared ladder diagram, and a meticulous step-by-step analysis, the control system logic can be understood. Figure 6-21 provides a ladder diagram for this example.

The structure of the ladder diagram depicted in Figure 6-21 has been designed for ease of following the logic. Each of the four vertical groupings apply to a major segment of the control system, as follows:

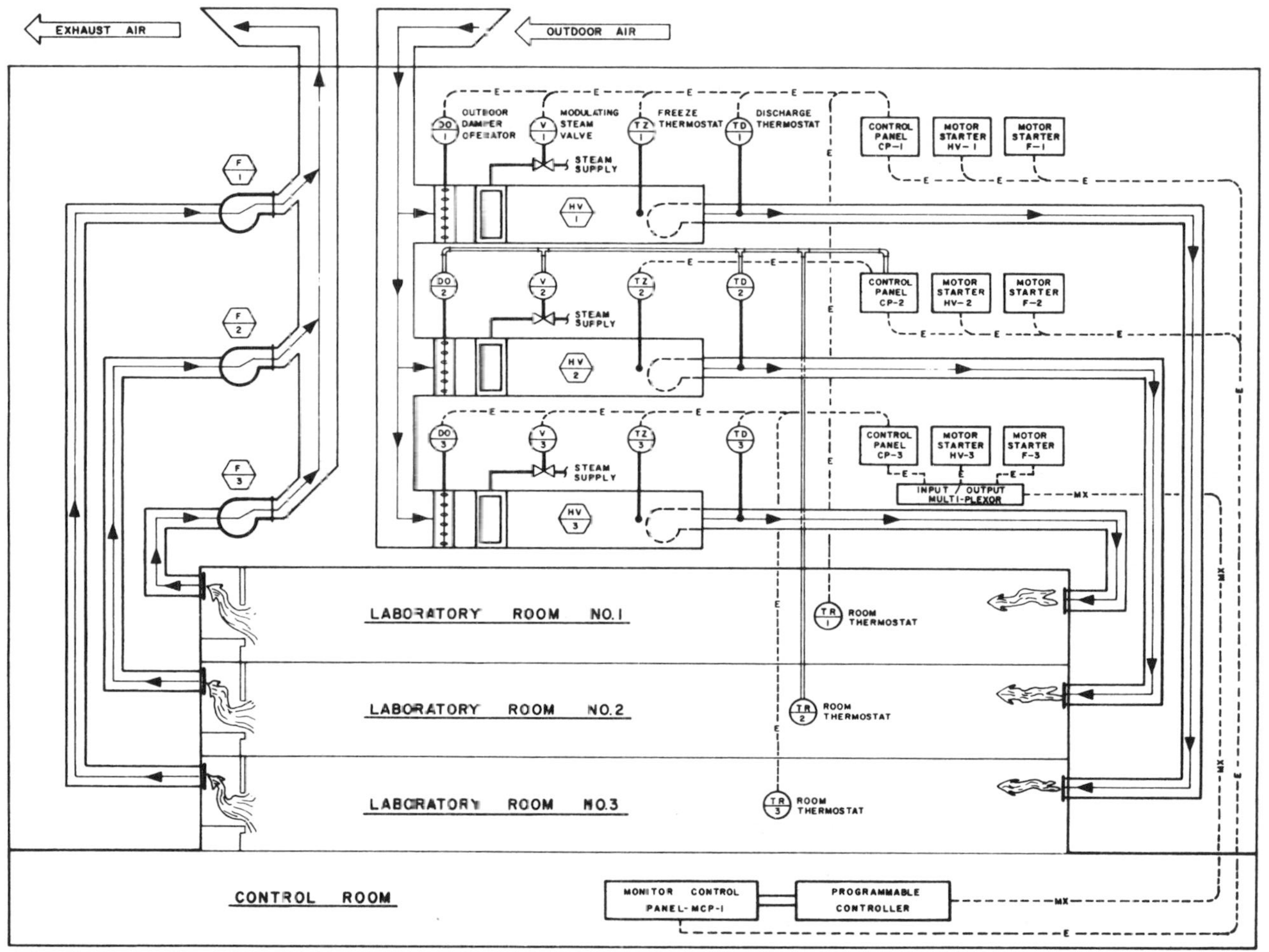

Figure 6-20. Air Flow Diagram for Heating and Ventilating Three Laboratories

- Rungs 1 through 17 identify those circuits that are centrally located in monitor control panel CP-1 for systems HV-1, HV-2 and HV-3. (Note: The HV-3 system is interfaced to the PC.)
- The second vertical ladder groupings, Rungs 18 to 34, contains the logic for the electrical control of system HV-1.
- The third ladder grouping, Rungs 35 to 52, identifies the logic of the electrical/pneumatic control circuit for system HV-2 serving the heating ventilating system for laboratory room No. 2.
- Finally, the fourth grouping, of this ladder diagram, Rungs 53 to 70, contains the logic circuitry in the electric/solid-state used in conjunction with the PC.

The following explanation of this complex control system will parallel the analysis of the four-part ladder diagram. However, first a clarification of minor variations from ladder diagram explanations used in the previous section is in order. These include the following:

- Selected rungs in this final ladder are left open to provide separation between key component segments in the logic.
- Rungs in the four vertical ladders diagram groupings are conveniently numbered for ease of identification and explanation.
- In this complex ladder diagram example, following the logic requires a path that varies from the rung numbering sequence commonly used.
- For convenience and brevity, symbols and terms that are explained earlier are not repeated in the following ladder diagram explanation.

Understanding a Representative Ladder Diagram (Figure 6-21).

- Rung 1 indicates the central control circuit is powered at 120 volts.
- Rung 2 contains the 120 volt power supply to drive the program clock.
- When the program clock contact PC-1 is closed, the Master Relay RM is energized. This action closes contacts in Rungs 20, 25, 37 and 42. It also sends an impulse to Rung 16. In following the logic, the effect of RM on the second vertical ladder grouping at Rungs 20 and 25 are reviewed first. Subsequently, the third and fourth vertical groupings are analyzed in their turn.
- In Rung 20, RM contactors are closed and the entire circuit on this rung is energized. R-1 is an override described in the Rung 33 explanation. The overload relays open the circuit whenever the motor overheats. When magnetic coil HV-1 is energized, contactors are closed in Rungs 21, 22 and 29. Also, the magnetic motor starter contactors are closed starting the motor which drives the heating-ventilating fan for laboratory room No. 1 (system HV-1).
- In Rung 21, the closing of these contactors turns on a green pilot light located near the magnetic motor starter.

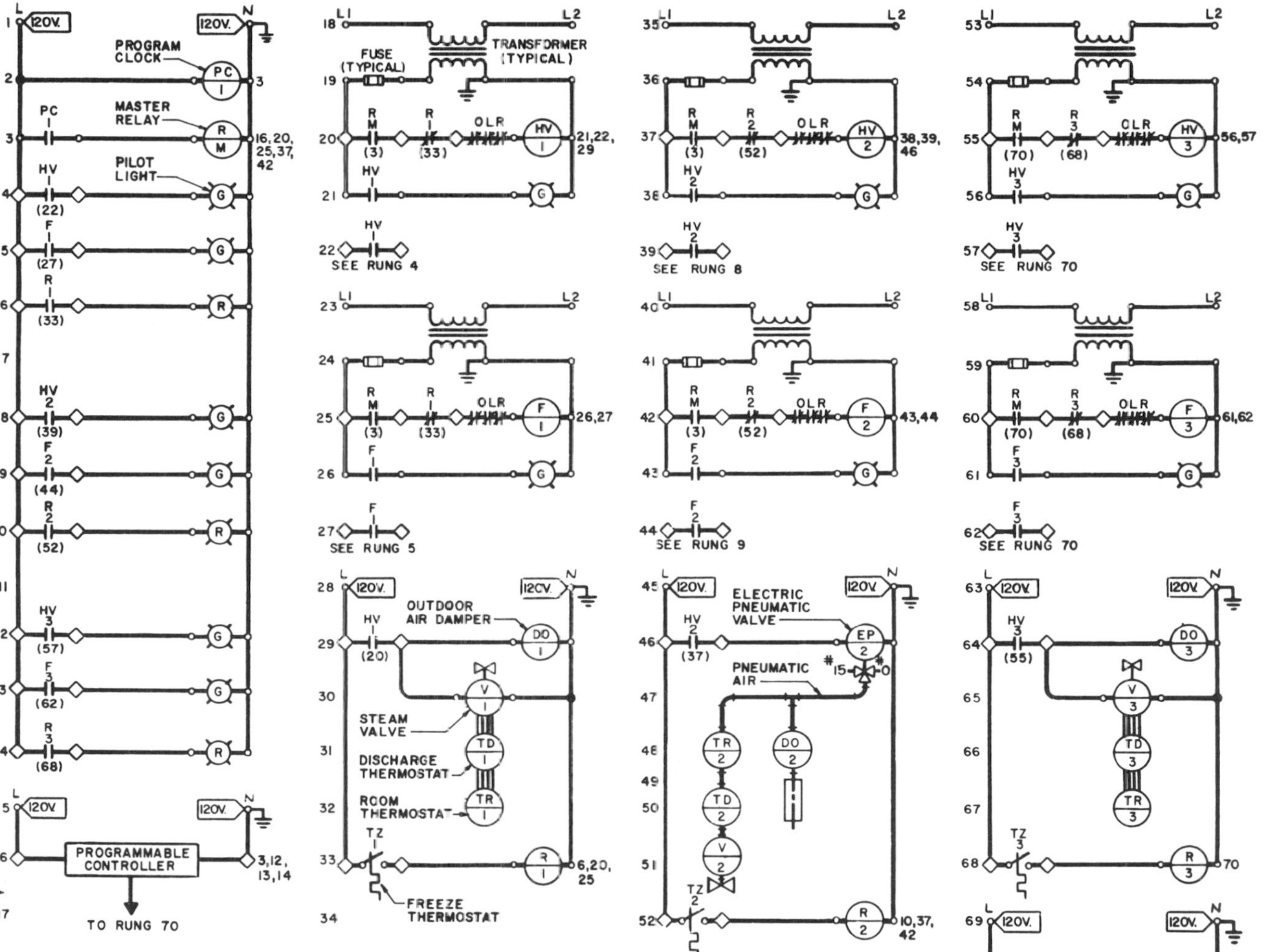

Figure 6-21. Ladder Diagram for Three Labs in Figure 6-20

- Rung 22 indicates HV-1 in Rung 4 closed the circuit, thereby lighting a green pilot light at the central control station. These green pilot lights indicate that fan motor HV-1 is operating.
- Rungs 18 and 19 identify control circuit powered from two legs of the HV-1 motor power supply source. This power is stepped-down to a control voltage (usually 120 volts) that is properly fused and grounded.
- Rungs 29 to 32 contain the heating coil control circuit. The heating coil is located in the outside air intake duct. When magnetic coil HV-1 Rung 20 is energized, the magnetic coil contactors HV-1 in Rung 29 are closed. This energizes the circuit and the outside dampers DO-1 are opened. Simultaneously, the steam valve V-1 is energized and capable of modulating to a controlled position. The valve modulated position is electrically-controlled by two thermostats. TR-1 is the room thermostat in room No. 1; and TD-1 is the thermostat in air discharge duct of system HV-1.
- Rungs 23 and 24 perform the same power source control for Exhaust Fan F-1 as described for motor HV-1.
- Rung 25 is also energized when the master relay RM closes the contactor RM. The contacts at R1, controlled by Rung 33, and the motor overload relays OLR can interrupt this circuit. The energized coil at F1 closes the contactors on the fan motor starter and also on Rungs 26 and 27.
- When contactor F-1 is closed on Rung 26, the local green pilot light is illuminated indicating the fan motor is operating.
- Rung 27 identifies the contacts that are also at Rung 5. When closed, a green pilot light at the central control panel indicates the exhaust fan F-1 is operating.
- Rung 33 contains the freeze thermostat located in the inlet air duct. When inlet air is too cold, the freeze thermostat closes and relay coil R1 is energized. This action causes the contacts in Rungs 20 and 25 to open, thereby stopping both motors. Simultaneously, the contacts R1 in Rung 6 are closed and a red warning pilot light is illuminated on the control panel.
- Rung 28 shows power supply at 120 volts controls the freeze thermostat circuit Rung 33 and the steam coil circuit Rungs 29 to 32.
- The control logic for system HV-1 is explained in the previous 14 steps. The discussion continues by returning to Rung 3 and explaining control logic when the master relay RM closes the contacts at Rungs 37 and 42. These rungs are located in the third vertical ladder grouping, which is an electrical/pneumatic control system.
- The control logic in Rungs 35 to 45, Rung 52 and Rungs 8, 9, and 10 is identical to the explanations for Rungs 18 to 28, Rung 33 and Rungs 4, 5, and 6.
- The interface between electric and pneumatic control exists at Rungs 46 and 47. When the system HV-2 magnetic motor starter coil HV-2 on Rung 37 is

energized, auxiliary contactor HV-2 closes the circuit on Rung 46. The three-way electric/pneumatic valve EP-2 is energized and the 15 psi control air supply enters the pneumatic control system. When the coil is de-energized, the valve is closed and the pneumatic system is opened to atmosphere. The three-way pneumatic valve is normally opened to the atmosphere. It closes when coil EP-2 is energized. At that moment, the 15 psi control air supply enters the pneumatic control system.

- At Rung 48, the outside damper DO-2 is pneumatically operated and the intake dampers are opened.
- The steam valve V-2 modulates the steam flow to the heating coils. However, this valve is controlled by pneumatic thermostats TR-2 and TD-2 shown on Rungs 48 and 50.
- As stated above, the all-electric and electric/pneumatic systems perform the same functions and the logic is identical. Only the conveying medium is different. In the fourth and final ladder diagram grouping of this complex control system (electric/electronic) the conveying medium (electricity) is the same as the all-electric control circuit, but two modern technological improvements are introduced. These are the programmable controller and multiplexing. It is the intent here only to introduce you to these two control devices.
- The explanation of control logic for the fourth vertical ladder, grouping the electric/electronic control, once again returns to Rung 3. Here, the master relay RM transmits a signal to the PC located on Rung 16. The PC in turn, analyzes the signal, determines what should be done, and transmits an appropriate signal by multiplexing to the input-output (I-U) multiplexer at Rung 70. The multiplexer directs the signal to the proper rung in this fourth vertical ladder grouping. This procedure is also reversed from component to multiplexer to PC to the proper rung in the control center in the first vertical ladder grouping. There are many advantages to these types of electronic controls, but the more important ones involve the following:
- Multiplexing enables many different signals to be transmitted long distances (from central control to remote station) over two wires. This is accomplished through variations in signal impulse and intensity.
- PC systems enable remote components to be controlled automatically from a central location. They also permit a limited number of people to control an almost limitless number of control devices in many control systems. Finally, PCs have the capability of rearranging a control circuit from a control panel key board, hence the term *programmable.*
- The logic of control for Rungs 53 to 68 and Rungs 12 to 14 are identical to control for Rungs 18 to 33, and Rungs 4 to 6, with two exceptions. These are the multiplexing between the first and fourth vertical ladder groupings, and the PC centrally located in a control room.

This representative example of a system ladder diagram highlights the following salient points to the maintenance operator:

- The basic start-stop control and actuating of control systems is no different in electric, electronic, solid-state, or pneumatic systems.
- The description of sophisticated modulating controls can be shown in a ladder form and maintain the system logic.
- A ladder type diagram can be used to describe and understand a programmable controller or a computer. Their logic is the same. Their components may disguise it from the basic hard wire concept. However, the ladder diagram sequentially links them together.

The use and understanding of control systems with ladder diagram logic is an important tool that will not become obsolete. It has its place with new programmable controllers and control computers.

Study of the previous ladder diagrams should provide you with the information necessary to understand the discussions on solid-state electronic and computer controls which follow.

ELECTRONIC CONTROL (SOLID-STATE CONTROLLERS)

With HVAC control several standard control applications exist which are common to the industry. These applications include:

- Discharge air temperature control in air handling units
- Economizer operation control of return, outdoor, and mixed air dampers
- Sequencing of cooling stages in chiller applications by either starting and stopping of compressors, or unloading of compressor cylinders
- Discharge water temperature control for hydronic systems

While the relay logic for each of these applications is not complex, many control relays and contacts are required in the circuitry. With the development of electronic circuitry, the relays and contacts can be replaced by electronic circuit boards contained in solid-state controllers.

Operation of Solid-State Controllers

These controllers act much like electric circuits in their operation. Input signals are received by the controller, interpreted, and output signals are sent to control devices.

The temperature input signals received are generally from either **resistance temperature devices (RTDs)** or **thermistors.** These sensors change resistance as a function of temperature. When a 2 to 20 vDC voltage is imposed on the sensor, the temperature measured is transmitted as a specific device resistance.

Input signals may be temperature, pressure, position, level or any number of other inputs. The input signal is received by the controller where electronic components in the circuitry perform basically the same relay logic as done with control relays and contacts in electric circuits.

The output signal from the solid state controller is typically either a **digital** or **analog** signal. A digital signal is a discrete signal such as the opening and closing of a relay contact or the switching action of a single-pole double throw switch (SPDT). An analog signal is a variable output such as a 4 to 20 milliamp current signal, or a 2 to 20 vDC voltage signal. The output signal positions the controlled device.

Solid state controllers can be found integrated into the electric circuitry of air handling unit and packaged rooftop unit controls, boiler control panels, or chiller controllers (See Figure 6-22). Used in combination with other relays and contacts in

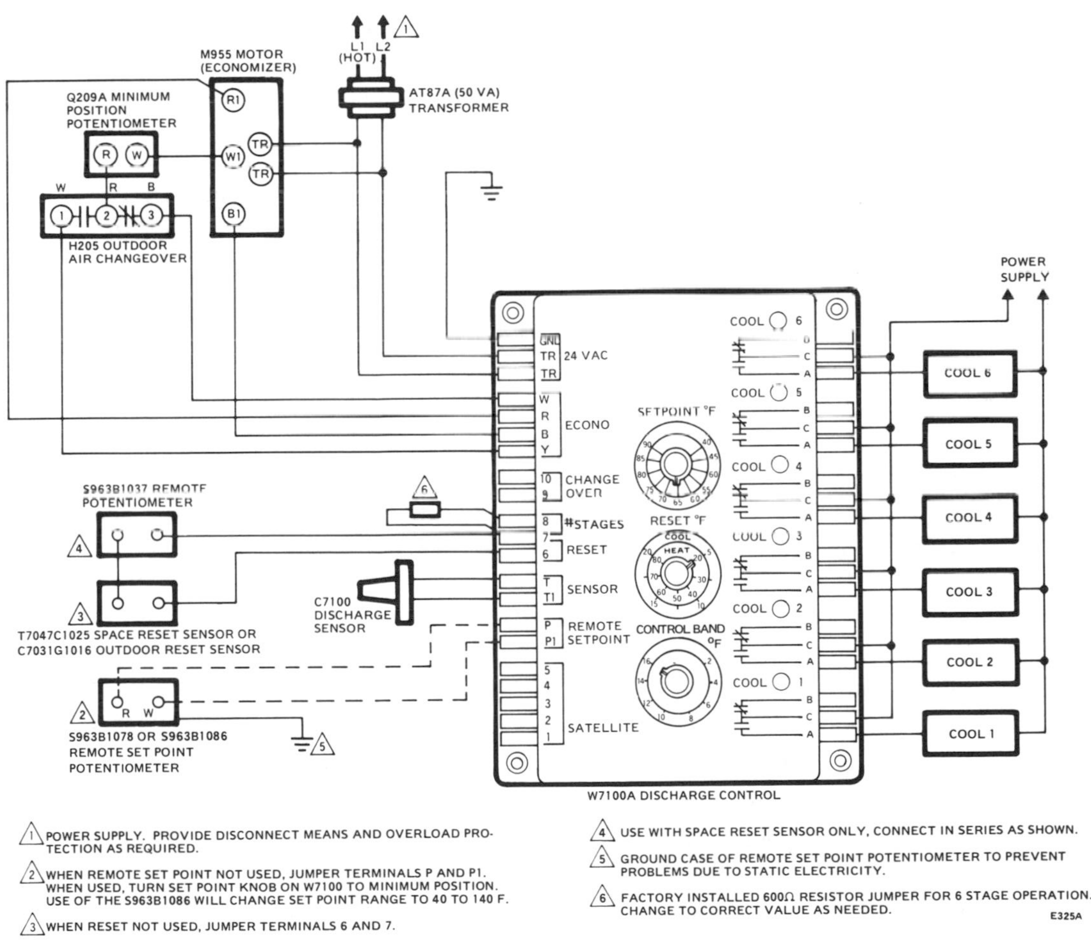

Courtesy of Honeywell Inc.

Figure 6-22. Honeywell W7100A Discharge Air Controller used to control 6-stages of cooling

a control panel, they can provide excellent operation where fixed logic control is required.

Generally, the solid state controllers have anywhere from 1 to 4 input signals and from 1 to 8 output signals. Frequently the controllers are provided with **light emitting diodes (LEDs)** to indicate the status of output commands.

Maintenance and Troubleshooting of Solid-State Controllers

Maintenance of a solid state controller is generally limited to preventing dust and dirt from building up on circuit boards or terminal blocks. Frequently the controller is encased in a plastic enclosure sealed to the outside. With these units no maintenance is required. In the event of a failure, the controller is merely replaced rather then repaired.

Trouble shooting of a controller is similarly limited. Efforts should be made to diagnose whether the problem is in the controller or in the field wiring from sensors or to controlled devices. A multimeter can be used to measure voltages on the terminal blocks where the field wiring is connected. If the input and output voltages are correct, and the controller is malfunctioning, the controller should be replaced.

The solid state controller is the simplest of the electronic control systems discussed in this chapter. While limited in capability, solid state controllers will eventually be found in most HVAC control panels where fixed logic is required.

PROGRAMMABLE CONTROLLERS

In the preceding sections of this chapter, the discussion concentrated on ladder diagrams used in hard-wired relay logic circuits. In recent years with the development of microelectronics, programmable controllers have been used in place of the hard-wired relay circuits. These controllers are increasingly being used as unit controllers for roof-top packaged air conditioning equipment, air handling units, chillers, boilers, and packaged controls for duplex fuel oil pump sets and water treatment systems. In the future it is expected that most HVAC control panels will use programmable controllers in place of today's hard-wired relay logic circuits.

Programmable Controller Operating Principles

Basically a **programmable controller** is a solid-state device used to control equipment by means of a stored program and input from field devices. The programmable controller differs from hard-wired relay circuits in that the controller's logic is easily altered, while the hard-wired relay circuit is fixed and unchangeable.

The controller can generally be readily reprogrammed to alter the way in which the controlled equipment operates. The hard-wired relay circuit can only be changed by rewiring.

A programmable controller is composed of 2-basic sections: a **central processing unit (CPU)** and **input/output (I/O) interface** devices. The CPU reads input data from various sensing devices, executes the stored program from memory, and sends appropriate output commands to control devices. The process of reading inputs, executing the program, and controlling outputs is done continuously.

The control logic used in the programmable controller is based on 3 basic logic functions (**AND, OR** and **NOT**). These functions are used either singly or in combination to form instructions. The instructions are entered into the memory of the programmable controller by a programming **language.** The languages in use in programmable controllers include:

- Ladder diagrams
- Mnemonic statements
- Boolean equations

However, ladder diagrams are the most commonly used language. The ladder diagrams used are called **contact ladder diagrams** and are very similar to the relay ladder diagrams previously covered in this chapter. Figure 6-23 illustrates the differences.

The Central Processing Unit

The central processing unit can be thought of as the brain of the programmable controller. It is composed of a **power supply, memory,** and **programming device.**

The **power supply** typically accepts a 120 Volt AC power input and using a transformer converts the input power to the 24 Volt DC power generally required for printed circuit boards.

The **programming device** is the communications device used to enter the program, to allow reprogramming, and to indicate the status of I/O points. Typical programming devices in use include:

- Mini-programmers, including:
  - Hand-held programmers
  - Limited memory programmers
- Program loaders, including:
  - Cassette recorders
  - Floppy disks
  - Hard drives
- Memory chip burners
- Computers, including:
  - Desktop personal computers
  - Laptop or notebook computers

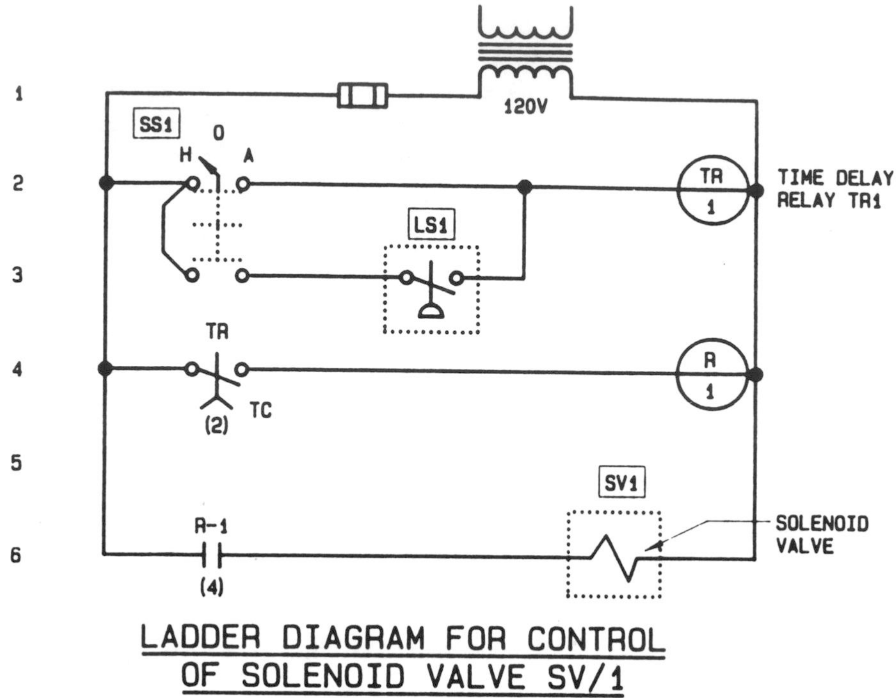

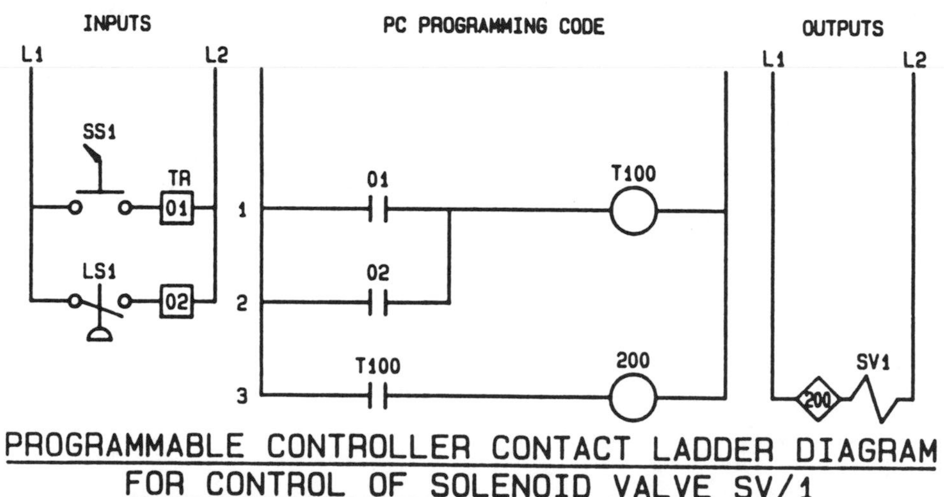

Figure 6-23. Illustration of Ladder and Contact Ladder Diagrams

The **memory** stores the program which reads the input signals and executes the output commands. The memory consists of printed circuit boards containing of memory chips and solid-state devices.

Figure 6-24 illustrates the use of an Allen-Bradley PLC-5/250 Programmable Controller in a control system.

Memory Overview

Several kinds of memory are used in a programmable controller. **Read Only Memory (ROM)** is a permanently fixed instruction set generally not alterable under ordinary circumstances. Executive programs are typically stored in ROM memory chips. In smaller, dedicated programmable controllers, ROM memory may be the only memory used.

Random Access Memory (RAM) or Read/Write memory is memory that can be written or read. Two kinds of RAM exist:

- Volatile RAM which loses its contents upon loss of power. Use of this memory in a programmable controller requires battery backup power so that the controller will not lose its program or data set upon a power failure.
- Nonvolatile RAM retains its contents upon loss of power and requires no backup power.

Programmable Read Only Memory (PROM) is nonvolatile memory which can be programmed. However once programmed, it cannot be reprogrammed. New chips must be used if PROM memory must be altered. Erasable Read Only Memory (EPROM) is a specially designed PROM that can be reprogrammed after being erased.

Memory Structure

Information contained in memory is stored in a binary array. The smallest measurement of memory is the **BIT or BInary digiT**. This represents one single item of information stored as either a 1 or 0. To store numbers or codes requires a grouping of BITS. A grouping of 8 BITS is called a byte.

Memory containing 1024 bytes is called a kilobyte, or Kb. Typical microprocessors memory capacities include 4K, 8K, 16K, 32K, 64K, 128K, 256K 512K or 640K. Above this sizing the megabyte (Mb) or 1,000Kb sizing is used as for 1MB, 2Mb, 4Mb or 16Mb ratings.

Input/Output Interface

The second key component of a programmable controller is the Input/Output Interface. This device consists of a series of racked input and output modules. Typical inputs sensors are:

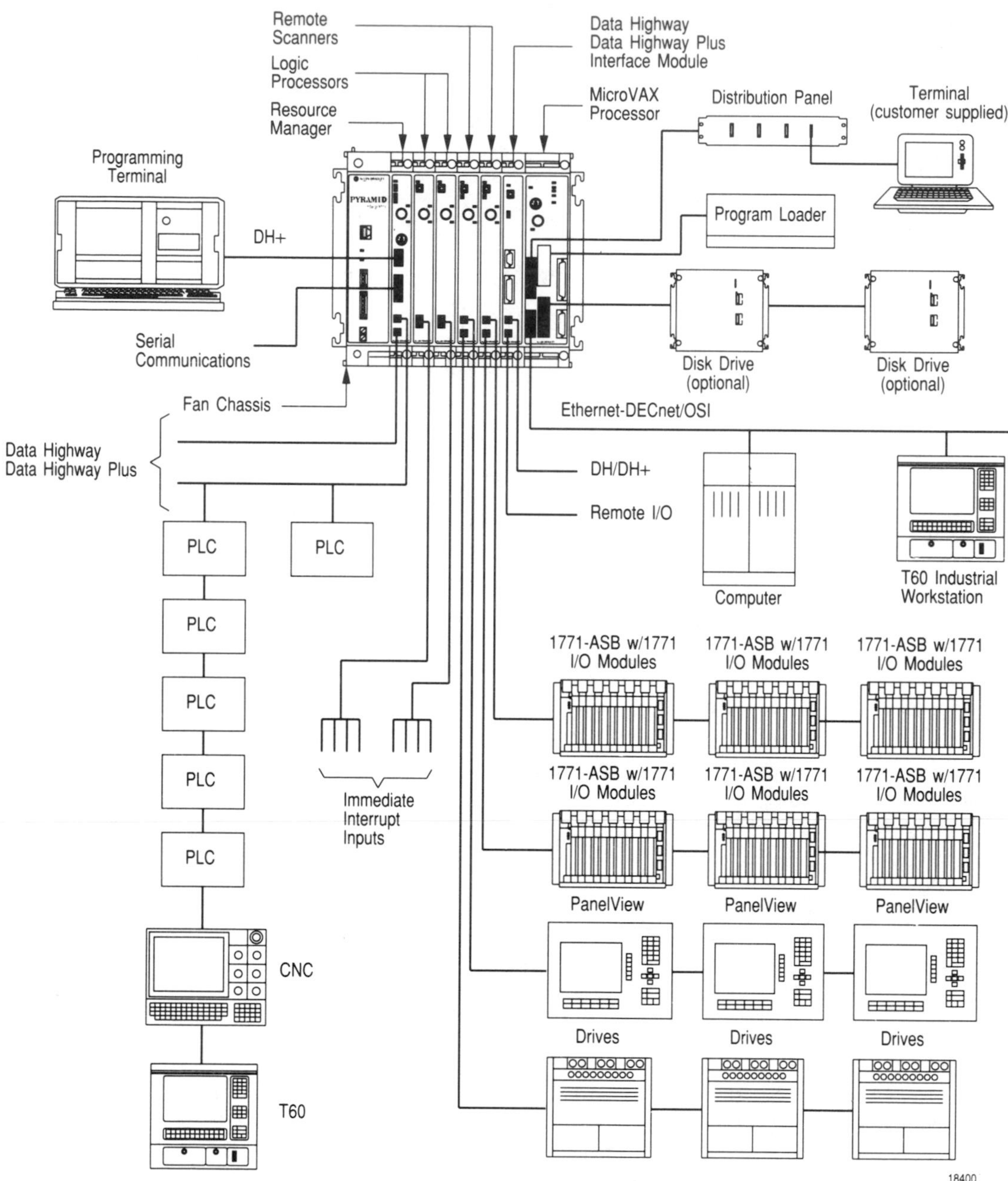

Courtesy of Allen-Bradley, a Rockwell International company

Figure 6-24. Allen-Bradley PLC-5/250 Programmable Controller Integrated into Control System with I/O Modules and Computer Control

- Selector switches
- Pushbuttons
- Photoelectric eyes
- Limit switches
- Circuit breakers
- Level switches
- Flow switches
- Pressure sensors
- Temperature sensors
- Motor starter controls
- Relay contacts

The information on the status of the input sensors is encoded in different signals such as:

- 24 Volt AC/DC
- 48 Volt AC/DC
- 120 Volt AC/DC
- 4-20 Milliamp current
- Pneumatic signals
- Contact closures

These signals are interpreted by the input modules on the I/O interface which communicates to the memory of the CPU. The CPU then sends signals through the output modules to operate the controlled equipment. The controlled equipment might include:

- Alarms
- Control relays
- Fans
- Lights
- Horns
- Valves
- Motors
- Motor starters
- Solenoid valves

Typically 1, 2, 4, 8 or 16 points are controlled by each input or output module.

Deciding Between a Programmable Controller or Hard-Wired Relay Circuit

Frequently, service personnel are in a position to influence the decision whether a programmable controller or hard-wired relay circuit panel is to be provided to control equipment. In this circumstance, the serviceperson should consider the following questions:

- Is there a need for flexibility in control logic?
- Is high reliability required?
- Are future increased capability and output required?
- Is data to be collected?
- Will there be a need for rapid logic modification?

If flexibility, future requirements, data collection, or rapid logic modifications are required, the programmable controller should be specified. If high reliability and future needs are expected to be fixed, the hard-wired relay circuits are the better choice.

Maintaining Programmable Controllers

Maintenance of a programmable controller generally involves diagnosis and replacement of modular, plug-in type components. Frequently, the circuitry of the controller contains built-in diagnostic and fault detection circuits to aid in the process. On occasion, the preprogrammed logic of the controller may be faulty and require reprogramming. Several preventive maintenance procedures exist to reduce operating troubles. These procedures include:

- Clean or periodically replace air filters used to protect the controller enclosures.
- Prevent dust and dirt from accumulating in controller enclosures.
  - If dust accumulates, heat dissipation off the electronic circuits may be hindered resulting in circuit malfunctions.
  - The build-up of conductive dust may result in short circuits permanently damaging controllers.
- Shield or remove controllers from noise-generating electric equipment such as large motors or transformers. The resulting fields from such devices can disrupt memory or cause malfunctions of electronic circuits.
  - Check plug-in and terminal block connections for tightness. Loose connections can result in improper I/O signals.
  - Check to see that air flow passages are not blocked.
  - Stock replacement spare parts, particularly power supplies, I/O modules, and replacement controller boards.

Troubleshooting Programmable Controllers

On occasion, service personnel may be required to troubleshoot a malfunctioning controller. Generally the first place to start is to check diagnostic or fault finding circuitry. Frequently controllers are provided with LED (Light Emitting Diodes) which indicate the trouble or defective component.

If the problem involves I/O circuitry, the serviceperson should attempt to isolate the problem to determine whether the fault is with an I/O module, or with the field wiring. If power and logic LED's are provided, the module failure should be readily apparent. If not apparent, a multimeter can be used to measure the voltage level at the field terminals. In the case of an input module, if the voltage is correct, the input module should be replaced. For an output module, if the voltage is correct, the field wiring should be checked. If the voltage is incorrect, the module should be replaced. All connections and solder joints should also be checked for tightness.

COMPUTER CONTROLS

Programmable controllers are used to provide electronic control for a limited number of devices. As the number of devices grows, a computer control system is required. Typically a programmable controller will control up to 40 devices, generally in the immediate vicinity of the controller. A computer control system may control hundreds or thousands of devices spread throughout a facility.

A computer control system can provide all the capability of the programmable controller, plus additional capability for programming, data acquisition, and communications. The full range capability of a computer control system is the reason such systems are often called building automation systems.

System Architecture

The hardware configuration of a computer control system is called the system architecture. The first computer control systems used a central processing unit (CPU) based architecture. This configuration consisted of a CPU with local control panels. In the CPU resided all memory. The local panels were basically "dumb" panels lacking memory or control capability. Field devices were wired to the local panels, with all input and output functions controlled by the CPU. All information from the local panels was transferred back to the CPU. Transfer of information was slow given the amount of information going back and forth on the communications network.

With the development of circuitry, distributed processing systems have begun to predominate. The architecture for these systems utilize a series of local field panels with each panel having memory and control capability. Field devices are wired to the local panels, with control done at the local panels. Generally, an operator's station is provided at a central point for operator interface and data management.

Hardware

The operator's station, sometimes called the headend, generally consists of a computer terminal with CRT (cathode ray tube), console, printer and frequently a mass storage device. The operator's station may also contain a communications controller if the system is extensive.

The field panels function much like individual programmable controllers with a processing unit containing memory, input/output interface device, and frequently a programming device.

In a distributed processing system, programming is generally done at the operator's station, with the program downloaded to the local panel where control is required. However, frequently the local panels can also individually be programmed.

Communication between the local panels and the head end is done on a data highway. The highway may be a twisted pair of small gauge wires or coaxial cable. The rate of communication is generally expressed as a baud rate. Typical baud rates are 1200, 2400, 4800 and 9600 bits of information transferred per second.

Software

The software contained in a computer control system can be divided into 3 types: system, operator, and applications.

System software is the operating system for the computer. The software may be written in a number of different programming languages including Assembler, Pascal, Fortran, or Basic. Many systems using IBM compatible computers utilize the Disk Operation System (DOS) for their system software. System software includes the bootstrap program used to initiate system operation, and to reload the system program after a loss of power.

Operator's software consists of the programs used for operator interface with the system. These programs include graphic interface, database, spreadsheet, report generation and trending of data. The Operator software provides for operator interaction with the system by allowing values and status to be displayed, resetting of setpoints, and providing for reprogramming.

Applications software consists of those programs used for device control. These programs, sometimes called software algorithms, include:

- Equipment Start/Stop
- On/Off Status
- Economizer Control
- Demand Limiting
- Duty Cycling
- Optimal Start
- Chiller Optimization
- Temperature, Pressure, Level, and Position Monitoring

- Night Setback
- Hot Water Temperature Reset
- Control Point Adjustment
- Static Pressure Control
- Damper Control

Applications software is modified to the individual installation by establishing setpoints, schedules, control limits, and feedback values required for control operation.

Maintenance of Computer Control Systems

Maintenance of a computer control systems is more difficult than for a programmable controller due to the greater complexity of the system. It is generally best for an organization to have 1 or 2 key service technicians factory trained in the operation and maintenance of the system.

It is also a good idea to retain the services of the computer control system manufacturer on a service contract to periodically maintain the system. By providing a service contract, the manufacturer maintains an interest in the continued operation of the system, and the Owner obtain updates of system software. The manufacturer is also available for debugging of software, which generally is not attempted by a system Owner.

The preventive maintenance procedures presented with programmable controllers are valid for the computer control systems. The computer system can be thought of as a series of programmable controllers linked together.

Troubleshooting Computer Control Systems

Trouble shooting of a local panel is performed the same as for a programmable controller. An effort should be made to diagnose whether the problem exists on the field wiring or on the I/O module. If the field wiring is correct, and replacement of the I/O module does not correct the situation, the problem rests with software.

If a software problem is suspected, the setpoint values in the particular program should be displayed to check for proper settings. If values are correct, the program should be erased and reprogrammed from the backup. If a bug is suspected in the program, a factory trained technician may be able to correct the problem, or the problem can be dealt with under a service contract.

Computer control systems provide for complete monitoring of an entire building from a central point. When integrated with lighting controls, fire, and security systems, the entire operation of a building may be controlled from the central point. With proper operation and maintenance, computer control systems can provide the service manager with the information and control needed to provide comfort to hundreds or thousands of building occupants.

7

Structure and Maintenance of Central Air Conditioning Systems

Central A/C systems, long the dominant cooling scheme for large buildings, now play a diminishing role in new construction. In spite of this trend, there are still major challenges and opportunities in maintaining and upgrading central A/C systems in existing structures.

The challenge for maintenance professionals is to develop a preventative maintenance program that meets the following objectives:

- Durability—This equipment represents a substantial capital investment. The expected service life can be achieved only by preventative means because irreversible damage can occur before any symptoms become obvious.
- Reliability—Ultimately, the purpose of this equipment is to provide comfort to the occupants when and where it is needed.
- Safety—The basic concern is operator and occupant safety and there is the additional impact of insurance costs on the building's financial health.
- Efficiency—The increasing cost of energy mandates equipment be in top operating condition to avoid energy waste.

Opportunities exist due to the increased need for property owners to remain competitive in today's difficult real estate market. The maintenance professional provides a valuable service to the building owners by identifying the need to upgrade or replace existing equipment or systems. Also valuable to this process is keeping abreast of new developments such as modular water chillers and off-peak cooling systems, as discussed in this chapter.

The following sections on refrigeration machines and absorption chillers, heat pumps, off-peak cooling systems, and modular chillers provide a knowledge base to help you structure your maintenance program.

Michael Heit

Section 1: ABSORPTION AIR CONDITIONING

William B. Birkel

In the 1990s, there are several reasons for the increasing popularity of absorption chillers:

- Since a depletion of the ozone layer has been discovered, absorption chillers play a major role in eliminating refrigerant losses to the atmosphere, because they do not use conventional refrigerants such as R-12, R-11, etc. Instead, absorption chillers use a lithium bromide salt solution and their refrigerant is ordinary H_2O (water).
- Absorption chillers have low electrical demand, since power is only needed for solution and refrigerant pump motors, not a large compressor motor.
- Absorption chillers can use free steam usage from an existing steam turbine, waste steam, or high-temperature water from building sources.
- Rebates are offered by many local utility companies on the purchase of replacement absorption chillers.
- There is no need for an operating license, as needed with other types of large chillers.
- Absorption chillers offer quiet, vibration-free operation so that this machine can be installed on rooftops or in close proximity of office space.

Currently, there are thousands of absorption machines in use throughout the air conditioning industry. The major manufacturers of this equipment are the Carrier, Trane, and York Corporations. The operating cycles of all three manufacturers are basically the same, although machine construction and component design do vary. Carrier and York house the cycle components in a two-shell machine, whereas Trane employs the use of a single shell constructions. (See Figures 7-1 and 7-2.)

UNDERSTANDING THE BASIC ABSORPTION CYCLE

The refrigerant in an absorption machine is water. This low-cost refrigerant is desirable for the absorption cycle because it can absorb energy by being made to boil

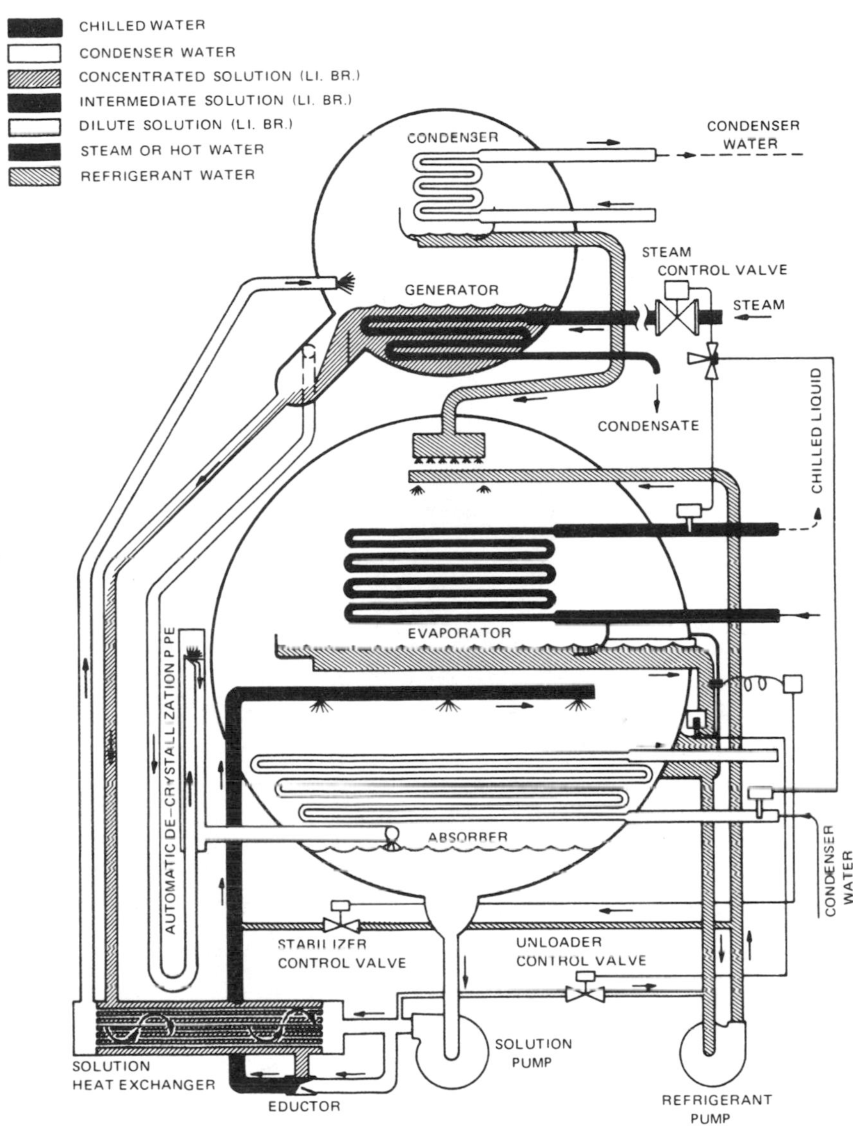

Courtesy of York Division of Borg-Warner Corp.

Figure 7-1. York Absorption Cycle

at low temperatures, when subjected to an atmosphere that is maintained at low pressure or a high vacuum. For example; boiling point is 40 degrees F at a pressure of 0.25 inches of mercury absolute.

The low pressure necessary for the refrigerant is maintained through the use of an absorbant, which in all cases is lithium bromide. Lithium bromide is a chemical that is used in an absorption machine in a crystalline form dissolved in water. The

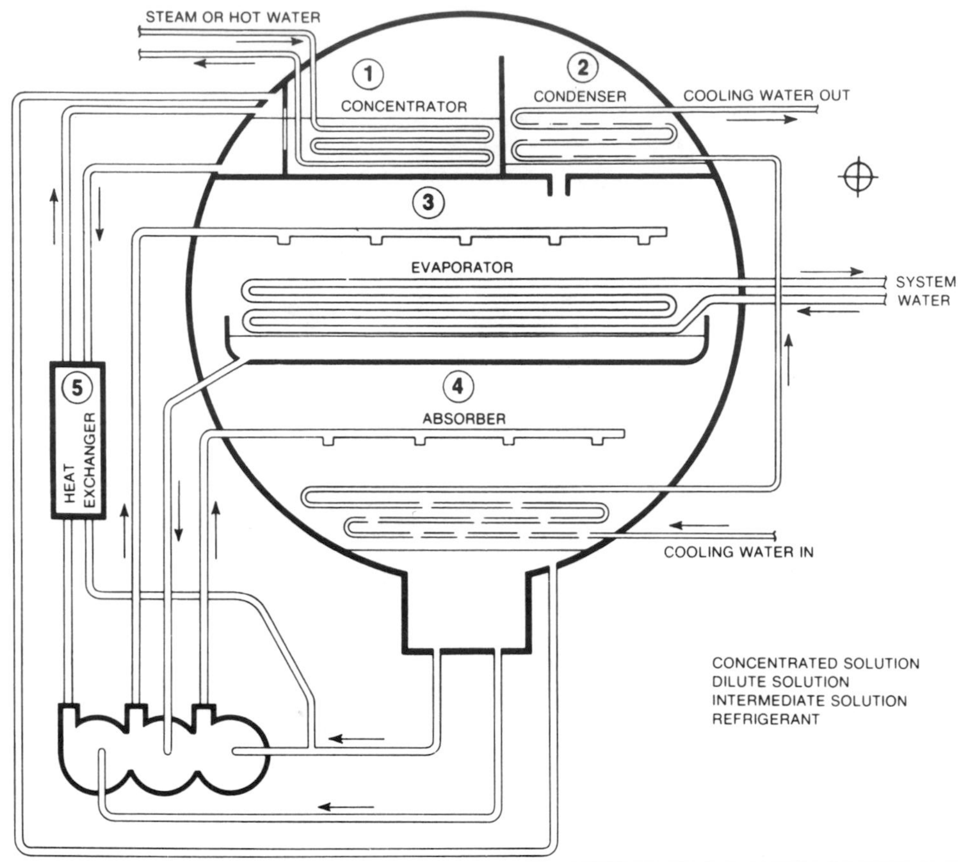

© The Trane Company 1969. Used by permission

Figure 7-2. Trane Absorption Cycle

lithium bromide solution has a great affinity for absorbing water vapor, thus making it a good absorbent for the cycle.

Note: The absorption machine cycle is based on two principles:

- Water will boil at a low temperature when subjected to a high vacuum.
- Lithium bromide solution has a great affinity for absorbing water.

Beginning with the evaporator section, the refrigerant (water) is subjected to a high vacuum, initially obtained by the use of a vacuum pump. The refrigerant is pumped through the spray header over the heat source, in this case the air conditioning load of the chilled water tube bundle. From there it absorbs heat energy, causing the refrigerant to boil and form water vapor, while chilling the water in the chilled water tubes. The water vapor must be removed immediately from this area to ensure that the high vacuum and low boiling temperature for the refrigerant can be maintained. Next, the absorbent, lithium bromide, which is housed in the absorber section (as seen in Figures 7-1 and 7-2) is used. The water vapor in the evaporator section now flows to the lower pressure absorber area where it is absorbed by the lithium bromide, which is being sprayed through a spray header.

As the lithium bromide absorbs water vapor, it becomes more diluted, thus reducing its capacity to absorb. It is therefore necessary to reconcentrate the lithium bromide by removing the water. For this purpose, the third section, called the generator or concentrator is utilized. The weak lithium bromide is pumped through a shell and tube heat exchanger (to increase heat efficiency) to the generator section, where it is subjected to a heat source (steam or hot water tube bundle). This provides the heat energy to boil excess water out of the lithium bromide, thus concentrating the solution. The excess water flows in the form of vapor to the fourth section, which is the condenser. This is provided with a cooling medium that, in most cases, is condensing water from a cooling tower. The water vapor is condensed and then flows back to the evaporator section, thus completing the cycle. The concentrated lithium bromide returns to the absorber section of the machine via the heat exchanger and is further cooled in the absorber section by the use of the cooling tower water, which is pumped through the absorber tubes prior to its use in the condenser. The cooling tower water flowing through the absorber tubes also removes the heat of condensation, which is heat created when the lithium bromide solution absorbs the water vapor.

The four basic components of the cycle—the evaporator, absorber, generator (or concentrator), and the condenser—can be seen as they are actually housed in Figures 7-1 and 7-2.

CONTROLLING THE ABSORPTION CYCLE FOR GREATER EFFICIENCY

It is universally agreed that the best method of controlling the absorption cycle is to use a modulating steam inlet valve which is controlled by a thermostat, whose temperature sensing element is located in the chilled water line leaving the machine. Older methods of control, such as modulating condenser water flow or solution flow to the generator, have given way to the use an automatic steam valve.

From the description of the absorption cycle, you can see how heat energy is transported from the inner spaces of a contained area, such as a building, via the closed chilled water system, through the absorption cycle, to the condensing water system, and on to a cooling tower, where it is ejected to the atmosphere. The absorption machine, which is a series of heat exchangers, is used as an intermediate transfer device.

Certain essential maintenance functions became obvious from an understanding the absorption cycle:

- Services to the machine, such as adequate steam volume, proper condensing, and chilled water flow, are necessary for proper operation.
- Heat transfer surfaces, both on the water and lithium bromide sides of the cycle, must be kept clean.

- The machine must be kept airtight. Noncondensables that gather at the lowest pressure area of the cycle at the absorber cause an increase in vapor pressure, thus raising the boiling point of the refrigerant and subsequently lowering the capacity of the machine.

PERFORMING ESSENTIAL MAINTENANCE

As with any heat exchanger, the heat transfer surfaces must be kept clean. This begins with a proper chemical water treatment program for both the closed chilled water system and the open condensing water system. Use strainers in both lines to protect these bundles, and keep these strainers clean to ensure adequate water flow. Clean the condenser and absorber tubes with nylon brushes before each operating season to remove loose mud and dirt. At the same time, inspect tube bundles and tube sheets for hard scaling and signs of corrosion or erosion.

You can clean the closed chilled water system less frequently. Because the absorption machine condenser tubes are made of copper, and absorber tubes are made of cupro-nickel, avoid acid cleaning these tubes unless absolutely necessary, in which case, it should be done only by qualified personnel.

Nondestructive or eddy-current electronic tube testing is becoming a very common procedure for checking the physical condition and life expectancy of all types of machine tubes. These tests can measure tube wall thickness, indicating pitting, cracking, or tube wear, thus showing up potential *problem* areas and perhaps avoiding costly emergency tube replacements. You should test machines that are more than five years old every two to three years, and maintain accurate records for comparison. REMEMBER: the tests should be performed only by trained, experienced personnel using the proper equipment.

No area of maintenance is more essential for the extended life of the lithium bromide circuit than the assurance of machine tightness. It is imperative that air be kept out of the vacuum side of the machine to avoid lithium bromide overconcentration during operation and to avoid corrosion at all times. During machine off-season shutdown periods, you should replace any areas of possible wear such as valve diaphragms, bellows assemblies, seals, and so on, in accordance with the manufacturer's recommendations. Use only dry nitrogen to break the vacuum in the unit during repairs. Perform pressure leak testing after repairs, using an electronic leak detector and Freon-12 as a tracer gas to test all gasket joints, sight glasses, welds, etc. Then evacuate; the machine subsequently, it should be able to maintain a standing vacuum during a prolonged shutdown period.

During operation, the need for frequent purging is one indication of a possible air leak, and you should take immediate steps to effect repairs. Lithium bromide in the presence of oxygen is highly corrosive, even if air is being purged through the machine in a small enough quantity not to affect cycle operation. The byproducts of corrosion clog internal passages (such as spray nozzles, heat exchanger tubes, and baffles) enough to decrease machine efficiency—sometimes to the point of impairing total operation. Shell surfaces and tubes can become so adversely affected as to destroy the entire machine long before its normal wearing life.

One simple test that you can use as a general indicator of machine condition, (in addition to physically taking cycle temperatures and pressures as recommended by the manufacturer) is a chemical lithium bromide analysis. In addition to measuring the corrosion byproducts washed in the bromide, the test measures the level of corrosion inhibitor still remaining within the solution, which can then be treated for maximum protection. The machine manufacturer or a qualified independent testing laboratory should be contacted to perform this test at least once per operating season.

All manufacturers that currently produce machines use some type of hermetic circulating pump for lithium bromide and refrigerant water. Most of these pumps have replaceable internal bearings that should be changed periodically. Again, when changing pumps, follow the manufacturer's recommendations to protect against a mechanical failure. You should megger-test electric motors at least once a year and regularly check all protection devices to avoid motor burnouts. Most of these motors cannot be rewound quickly in the field and can be replaced only at great cost and only where equipment is available.

PREVENTATIVE SERVICE RECOMMENDATIONS

The **best advice** that can be given for an optimum running absorption machine is: keeping a vacuum tight 29.62 inches absolute. Keep all tubes clean, which will also save considerable energy costs. Here are some maintenance guidelines:

- Annually brush clean all condenser tubes, utilizing a nylon brush.
- Annually brush clean all absorber tubes, utilizing a nylon brush.
- Check all diaphragm valves for operation. Renew diaphragms every two (2) years.
- Clean water side of evaporator tubes every three (3) to five (5) years.
- Check pump starters, clean contactors, overload relays, magnetic starters, and safety controls.
- Clean magnetic strainers in machine pump motor cooling coils, where applicable.
- Check efficiency of purge units in machines where vacuum pumps are used.
- Clean and flush purge vacuum pump and charge with fresh oil.
- Megger-test all motors.
- Check hermetic pump bearings where applicable every two (2) years.
- Change seals on open pumps where applicable every two (2) years.
- Change seals on open pumps where applicable every two (2) years.

Table 10-1 provides a troubleshooting guide for absorption chillers.

PROBLEM	PROBABLE CAUSE	REMEDY
Machine will not start	• No power at panel	• Check building circuit breaker and machine fuses
	• Water flow switches open	• Check flow switches
	• Condenser or chill water pumps not started	• Start pumps; check contractors
	• Machine solution and refrigerant pumps starters open	• Reset pump starters
Chill water temperature fluctuates	• Chilled water flow or building load cycling	• Check chill water system, (coils) controls and load
	• Condenser water temperature fluctuating	• Check cooling tower
	• Steam pressure or high-temperature water fluctuating	• Check steam or hot water source
	• Steam condensate not leaving machine, possible water hammer build-up in generator	• Check condensate pump, steam traps
	• Steam valve not opening	• Check calibration of steam valve actuator
Chilled water temperature high	• Noncondensables (air) in machine	• Check absorber loss (described in previous section of chapter)
	• Fouled tubes	• Brush-clean tubes
	• Machine depleted of octyl alcohol (tube wetting agent)	• Add octyl alcohol
	• Solution solidification	• Noncondensables in machine (check for high absorber loss); Heat machine, stop water flow, add refrigerant water

Table 7-1. Absorption Troubleshooting Guide

PROBLEM	*PROBABLE CAUSE*	*REMEDY*
Chilled water temperature high (cont.)	• Low hot water temperature or steam in generator	• Raise temperature
	• Steam condensate not leaving machine, possible water hammer build-up in generator	• Check condensate pump, steam traps
	• High condenser water temperature	• Check cooling tower water level and fan operation
Excessive machine purging	• Air leakage in vacuum side of machine	• Analyze solution, leak test and repair leak
Inadequate machine purging	• Inhibitor depleted	• Analyze solution, adjust alkalinity, and add inhibitor
	• Purge lines solidified	• Heat Solution Lines

Table 7-1. Absorption Troubleshooting Guide (*continued*)

COMMONLY USED TERMS OF ABSORPTION TROUBLESHOOTING

Absorber Approach—The temperature difference between the solution at the generator pump discharge and the condenser water leaving the absorber.

Normal 8-12 degrees	**Trouble 15+ degrees**
PROBABLE CAUSES	*REMEDY*
• Dirty tubes	• Clean absorber tubes
• Low water flow GPM	• Check condenser water strainers
• Plugged solution spray nozzles	• Remove absorber spray header

Absorber Loss—The temperature difference between the refrigerant and pump discharge and the solution saturation temperature at the gen. pump discharge.

Normal 0-2 degrees	**Trouble 4+ degrees**
PROBABLE CAUSES	*REMEDY*
• Non-condensables	• Purge machine; if problem persists: leak test and repair leak

Condenser Approach—The temperature difference between the condenser water leaving the condenser and the vapor condensate temperature.

Normal 10-12 degrees	**Trouble 15+ degrees**
PROBABLE CAUSES	*REMEDY*
• Dirty condenser	• Clean condenser tubes
• Low water flow GPM	• Check condenser water strainers

Evaporator Approach—The temperature difference between leaving chilled water and the refrigerant pump discharge temperature.

Normal 1-2 degrees	**Trouble 3+ degrees**
PROBABLE CAUSES	*REMEDY*
• Fouled chilled water tubes	• Clean chilled water tubes
• Excessive GPM	• Slow water flow
• Contaminated water	• Clean tubes and strainers; apply water treatment

Generator Approach—The temperature difference between the steam and the strong solution leaving the generator.

Normal 18-25 degrees	**Trouble 30+ degrees**
PROBABLE CAUSES	*REMEDY*
• Low steam volume	• Increase steam supply poundage
• Defective trap	• Clean or replace steam trap

RUNNING A VACUUM TEST AND CHECKING FOR LEAKS

1. Make certain chilled water is fairly stable.
2. Make certain there is no refrigerant overflow.
3. Isolate purge for system.
4. Fill thermometer wells with oil and insert thermometers.
5. Remove a sample of solution from the generator service valve. Measure and record specific gravity and temperature.
6. Record the solution temperature in the generator pump discharge line.
7. Record the temperature of the refrigerant water in the refrigerant pump discharge line.
8. Note and record position of the capacity control valve.
9. Plot the specific gravity and temperature of the solution sample on the equilibrium diagram.
10. Note the percent of concentration and move vertically along this percent of concentration line to the actual solution temperature found in step 6.
11. Project this point (step 10) to the saturation temperature scale on the right of the diagram to determine solution vapor temperature.
12. Subtract this reading from the actual refrigerant temperature taken in step 7. This difference is the absorber loss. (NOTE: if the loss is greater than 3 degrees, the machine must be purged until the loss is less than 3 degrees. If a large leak is suspected, take readings before the 24-hour period has elapsed.)
13. When taking the second set of readings, make sure the control valve is within 15% of valve setting in the first set of readings.
14. Compare the second set of readings with the first set. If the increase in the absorber loss is greater than allowed for that particular machine, a leak is indicated.

OPTIONS FOR ENERGY CONSERVATION

This is the most important thing to keep in mind when operating absorption machines. Earlier machines operated at a steam rate of 18 pounds per hour per ton of refrigeration; currently, the rate is 12 pounds per hour per ton. Needless to say, recently, there have been many improvements in their design.

Recently introduced is a natural-gas-fired absorption chiller manufactured by numerous manufacturers. For example, the Trane Company has introduced the **Thermachill Direct-Fired Absorption Chiller™** which is available from 100-1100 tons. The chiller uses the proven combination of water as the refrigerant and lithium bromide as the absorbent solution. Advanced inhibitors are used for corrosion protection.

The double effect of the cycle is achieved with a two-stage concentrator. The high-temperature concentrator is fired by an integral burner. The hot refrigerant vapor boiled from solution is used to fire the low-temperature concentrator.

The reverse flow cycle improves chiller efficiency and allows a more compact design. Absorbent solution flows in reverse through the two-stage concentrators, as compared to the traditional series flow. Dilute absorbent solution flows to the low temperature concentrator first. Then the solution is split, with half returning to the absorber, while the rest is pumped to the high temperature concentrator. An economizer heat exchanger further increases unit efficiency by recovering thermal energy from exhaust gas. (See Figure 7-3.)

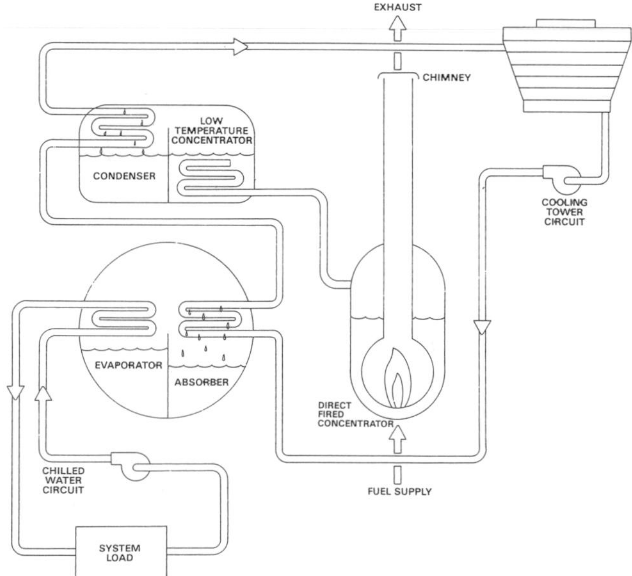

The Trane Company—Used by Permission

Figure 7-3. Typical Thermachill Direct-Fired Absorption Chiller™

Section 2: CENTRIFUGAL AIR CONDITIONING

William B. Birkel and
Frank L. Phillips, P.E.

HERMETRIC COMPRESSOR MACHINES

Hermetic centrifugal machines consist of two heat exchangers, the cooler located on the bottom and the condenser on top. The cooler cools water which flows through its tubes by means of evaporation of refrigerant which is inside the vessel. The condenser liquifies refrigerant vapor form the compressor. There is a float chamber that regulates the refrigerant flow between the cooler and condenser.

Function of Component Parts (See Figure 7-4)

Chiller: Heat exchanger that cools water in tubes by evaporation of liquid refrigerant in shell surrounding the tubes.

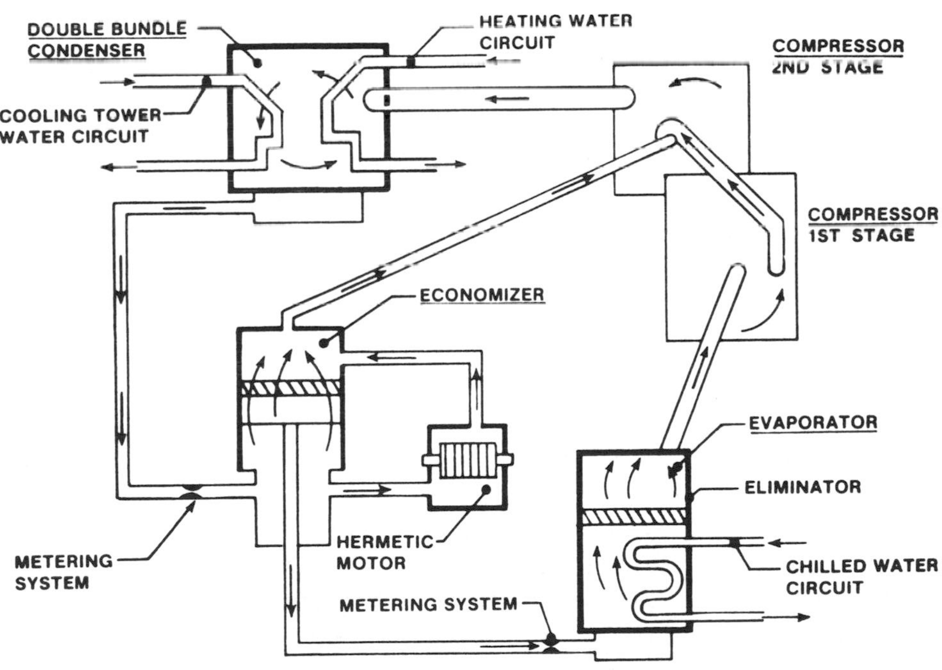

Figure 7-4. Diagram of Typical Centrifugal Machine with Compressor Chiller-Condenser

Compressors: Centrifugal pump that reduces cooler pressure to facilitate evaporation of refrigerant and compresses gas and delivers it to the condenser at high pressure and high temperature.

Condenser: Heat exchanger that cools the high temperature gas from the compressor to condense it and change it to a liquid state. Condenser cooling water from a cooling tower passes through tubes, and refrigerant remains in the shell of the condenser surrounding the tubes.

STRUCTURE OF OPEN DRIVE CENTRIFUGAL COMPRESSORS

These compressors are driven by steam turbines or variable-speed, wound rotor electric motors through self-aligning couplings. Steam turbine drive is usually a direct drive requiring no increasing gear, because turbine speed matches the optimum speed of the centrifugal compressor and can be controlled.

Electric motor drive usually requires a speed increasing gear assembly, because the synchronous speed of a four-pole or a six-pole electric motor—1,800 rpm and 1,200 rpm respectively—are too low for the centrifugal compressor, and the full load speeds of these motors would be even lower: 1,725 rpm and 1,125 rpm.

Electric motor drives are usually wound motors controlled by resistance inserted in the rotor circuit through a drum-type controller to allow efficient, variable speed control of the centrifugal compressor to suit load conditions. Drum control can be either manual or automatic.

CAPACITY CONTROL

Controlling compressor speed is the most efficient method of capacity control for power conservation. There are 4 methods:

- By means of a speed controller and turbine governor to vary turbine speed.
- By a variable-resistance drum controller for wound rotor induction motor to vary rotor speed.
- By magnetic or hydraulic coupling, where synchronous and squirrel-cage, constant-speed electric motors are used to vary compressor speed.
- By throttling compressor suction by means of a thermostatically controlled suction damper operating in response to chilled water temperature. Variable inlet guide vanes are sometimes used in the first-stage position of the compressor for control below 10 percent capacity. Capacity of a centrifugal machine is controlled to maintain the correct water or brine temperature to suit load conditions.

Therefore, the best control from the standpoint of power input is the speed control method.

These methods of controlling compressor output are effective from 30 percent of full capacity to full load. Below 30 percent capacity, a hot gas bypass is usually provided, except on a machine using the variable-inlet, guide vane control.

MAINTAINING THE PURGE UNIT

- Periodically clean all strainers.
- Periodically clean float valve or metering device.
- Periodically check and adjust controls.
- Keep sight glass clean. If presence of water is indicated, drain the water, measure the quantity drained, and record the amount for future reference.

MAINTAINING THE CONDENSER

- Annually brush clean condenser tubes.
- Periodically check the purge condensing chamber and clean the chamber.
- Periodically check the oil level of a motor driven purge unit.
- Provide water treatment if the condenser water is recirculated, to a cooling tower.

OPERATING THE PURGE UNIT

Both types of purge units return refrigerant to the cooler or an area in economizer where temperatures are above freezing, the back pressure regulating valve in the purge return line will operate to prevent the purge unit from freezing. Adjust this back pressure valve to the manufacturer's specifications (furnished with the unit). Should condensation of refrigerant occur in the suction line between the purge compressor and the condenser with compressor-type purge units, the condensate may enter the crankcase, causing oil to "foam" and resulting in loss of oil. Therefore, readjust the pressure-reducing valve in accordance with the manufacturer's recommended setting (this should be furnished with the unit).

KEEPING DAILY LOGS

The most important part of any operation and maintenance schedule is the method of recording data. A daily log should be started with the start-up of the machine and should record the following information:

- Whether or not there is purge discharge immediately after start-up that could indicate leaks during shutdown period.
- Frequency of air discharge by the purge unit—once or twice a day is normal. More frequent discharges while machine is in operation indicate leaks.
- Amount of air discharge, which can be observed by directing into a container of water. (The manufacturer's manual will list several methods of testing for air and water.)
- Water in the purge separation chamber, which indicates refrigerant has reached saturation. Water must be removed manually by the operator, and this operation makes the operator aware that water is present. The operator must then take the necessary steps to determine the source of water leaks and effect repairs.

Purge units should be operated during long periods of shutdown to determine if there are air leaks and/or water leaks.

The operator should be aware that the automatic exhausting of air and noncondensables by the purge unit also causes some refrigerant loss; some refrigerant gas is always present, because condensing of the refrigerant is not always complete. The operator should therefore check the refrigerant level from time to time.

DAILY INSPECTION AND MAINTENANCE

Make sure that the machine is clean and tight. Immediately investigate the presence of foreign material or evidence of leaks to determine the source and cause, and take remedial action to effect a clean, tight machine.

- Check for proper purge operations as a further check on machine tightness.
- Check all safety controls and keep them in first-class working order.
- Check lubrication and be sure to maintain a clean lubrication system.
- Keep adequate operating records for a day-to-day comparison that will indicate changes in the efficient operation of the equipment.

PERIODIC OR ANNUAL MAINTENANCE

- Clean heat exchange tubes.
- Check refrigerant level.

Because some refrigerant is lost when relief pressure devices operate to relieve pressure, set the relief pressure high enough to minimize refrigerant loss.

LUBRICATION

Bearings last longer when a good, clean lubricating system is maintained. The lubrication system is affected by heat, dilution (with refrigerant), and moisture. The operator should keep the lubrication system clean at all times and use only the lubricant specified by the manufacturer. This is usually a good quality turbine oil especially treated and fortified with inhibitors to minimize oxidation and foaming according to the type of compressor, evaporator, temperature, etc. Never mix oils or use oils that have been used in another machine unless their characteristics are similar in all respects.

Change oils once a year or every time the centrifugal is opened for inspection or repair. Seals on the oil container should not be broken until the operator is ready to refill the oil container with new oil. Have oil sample analyzed at each oil change by an experienced laboratory to detect any unusual conditions which might indicate excessive wear, contamination, improper lubricant, or defects in the lubrication system.

MAINTAINING OIL PRESSURE AND TEMPERATURE

Never maintain oil pressure higher than the manufacturer's specifications. Oil pressure should be noted in the daily log and the oil-regulating valve adjusted if necessary to maintain correct oil pressure.

Make sure you maintain the correct oil temperature during both machine operation and shutdown, to minimize the amount of refrigerant absorbed by the lubricating oil. If the lubricating oil heater or the controlling thermostat does not function correctly as per the manufacturer's instructions, a high concentration of refrigerant can occur during shutdown, which will cause a very high release of refrigerant at start-up. The result is oil foaming, decreased oil pressure, loss of oil, and unnecessary "trips out." Such conditions, including the frequent restarts, cause excessive bearing wear and deterioration.

REPLACING OR ADDING OIL COOLERS AND OIL FILTERS

Oil coolers are usually furnished with centrifugal machines but, if not, they can be added. Such devices consist of a shell and tube heat exchanger utilizing a small quantity of cooling water to maintain normal lubricating of oil temperatures.

Replace oil filters whenever you change the lubricating oil. Daily logs taken at start up and at least once during the daily operating cycle will record machine operation, tightness, changes in power input, purge operation, refrigerant loss, oil temperature, and other pertinent data, which will indicate to the operator the general operating condition of the equipment and help analyze machine performance. Conditions that may cause trouble can thus be corrected before major breakdown occurs. This method serves as an excellent guide for planning preventive maintenance.

PROBLEM	*PROBABLE CAUSE*	*REMEDY*
Machine fails to start	• Power failure or interruption	• Restore power, reset and start
	• Condenser or chill water pumps not started	• Start pumps
Machine runs but does not chill water	• High ambient temperature (excessive load)	• Close off building building to outside air
	• Short of refrigerant charge	• Add refrigerant
	• Warm condenser water	• Check cooling tower
Discharge pressure too high	• Air in condenser water circuit	• Vent air
	• Dirty condenser tubes	• Clean tubes
	• Cooling tower not operating correctly	• Check cooling tower
Discharge pressure too low	• Too much condenser water	• Adjust flow
Suction pressure too low	• Low refrigerant in machine	• Add refrigerant
	• Liquid line float stuck	• Check float and repair or replace
Suction pressure too high	• Excessive load on machine	• Check for outside air entering building
Purge will not build up pressure	• Purge condenser float stack open	• Clean or repair purge float valve

Table 7-2. Centrifugal Trouble Shooting Guide

MAINTAINING ROTARY SCREW TYPE COMPRESSOR CHILLERS

In the 1990's, a number of manufacturers have introduced a rotary screw type **compressor chiller**. The York Codepak Rotary Screw Liquid Chiller is comonly applied to large air conditioning systems. The Codepak consists of an open motor connected to a rotary screw compressor, a condenser, a cooler, and a MicroComputer Control Center.

The Codepak is controlled by a modern state-of-the-art microComputer control. Automatic timed start-ups and shutdowns are also programmed to suit night time, weekends, and holidays. The operating status, temperatures, pressures, and other information pertinent to the operation of the Codepak Chiller are automatically displayed.

In operation, a liquid (water or brine to be chilled) flows through the cooler, where Refrigerant-22, boiling at low temperature, absorbs heat form the liquid. The chilled liquid flows through pipes to fan coil units or other air conditioning terminal units, where it flows through finned coils, absorbing heat form the air. The warmed liquid is then returned to the cooler to complete the chilled liquid circuit.

The refrigerant vapor produced by the boiling action in the cooler is compressed by the rotary screw compressor and discharged into the oil separator section which removes the oil before the high pressure gas flows into the condenser tube bundle. Water flowing through the condenser tubes absorbs heat form the refrigerant vapor causing it to condense. Condenser water is supplied to the chiller from an external source, usually a cooling tower. The subcooled refrigerant drains form the bottom of the condenser into a pipe connection, where the flow restrictor orifice expands the flow of liquid refrigerant to the cooler to complete the refrigerant cycle.

The major components of a Codepak are selected to handle the refrigerant which is evaporated at full load design conditions. However, most systems will be called on to deliver full capacity for only part of the time the unit is in operation.

The major components are selected for full load capacity, therefore capacity must be controlled to maintain a constant chilled liquid temperature leaving the cooler at run and low load conditions. A slide valve arrangement located on the rotary screw compressor compensates for variation in load. The slide valve arrangement is controlled by the microComputer control center and unit controls that sense the building load conditions. The control center sends signals to the solenoid valve that loads and unloads the slide valve with the use of compressor oil under hydraulic pressure. A cylinder located on the inlet end of the compressor houses a spring and pistons which is fed through its ports by pressurized compressor oil. The flow of the oil is controlled by the equalizing solenoid valve which modulates to load and unload the slide valve that increases or decreases the amount of refrigerant flowing to the compressor, thus controlling the chiller capacity. (See Figure 7-5.)

A real asset to owning a rotary screw compressor chiller is they run on few moving parts, unlike your average centrifugal chiller. Maintenance is drastically cut by only requiring annual oil & filter changes, refrigerant analysis, leak testing, tube cleaning of the condenser, and a dry run of the machine safety every year.

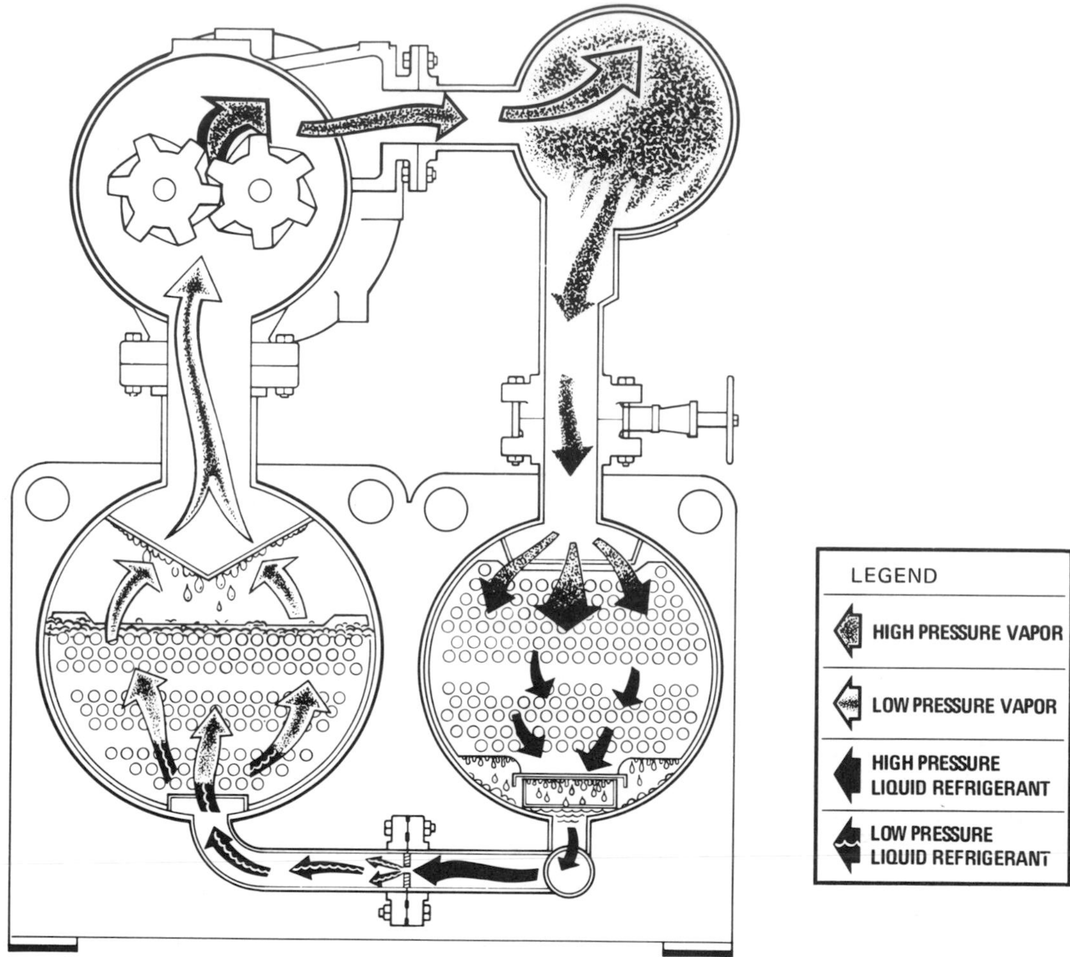

Used by Permission of York International Corporation

Figure 7-5. Refrigerant Flow Through Codepak

Section 3: RECIPROCATING AIR CONDITIONING

Robert Barr

Reciprocating air conditioning has changed vastly over the last 40 years. The building boom (which lasted for nearly four decades after the second world war) itself brought many changes, the most common being the rooftop unit, which became ubiquitous from the late 1960s onward. Due to these changes in the equipment itself, both the size and type of compressors changed. The open compressor of the

1940s and '50s became semi-hermetic—and the average horsepower dropped from 50-60 to 20-40. The driving forces in these changes were size and cost.

With the changes in compressors came new ways of servicing. Open compressors had largely been field-repaired. But the smaller semihermetics didn't lend themselves to such repair; and, in addition, the mechanics who knew how to do that work were retiring. Moreover, motor burnout was unheard of in the days of open machines; with semihermetics, motor burnout became commonplace, because the same gas was used for refrigerant and for cooling the motor. That meant motors could be made smaller (and cheaper) because return gas came back at a maximum of 65° F. (instead of 105° ambient on an open motor) and at between 2 and 3 atmospheres of pressure instead of at one.

But such a motor is often friction-fitted into the cast iron compressor body. It can't be removed in the field. Therefore, replace the compressor—don't try to fix it in the field.

And compressors do fail, bringing the entire air conditioning system to a halt. The semi-hermetic compressor is literally twice as likely to fail as its open predecessor. Insurance premiums verify the fact that the cost of failure is twice as high on semi-hermetic machines. But a word of caution is necessary. The term *fail* implies some deficiency on the part of the compressor itself, as if its design were inadequate. Such an implication is unfair and untrue. Most compressors manufactured today have a realistic life span of seven to eight years, when used in air conditioning applications in the northern United States. In heavier usage, such as in Florida, only a year or so would need to be deducted, assuming the same use. In an application such as a computer room, where the machine operates virtually continuously, the actual number of anticipatable hours would rise, though the number of years would decline.

Only in a very few cases does the compressor simply *fail*. Much more frequently, it has been subject to some abuse or lack of care or misapplication or failure of some other control or component, which causes failure of the compressor. The fault, almost always, is in the *system* rather than in the compressor. This section of this chapter deals with such abuses and how to avoid them.

The purpose of the refrigeration compressor is simply to move gas from a low pressure region to a high pressure region. The term "gas" is used advisedly, because when we do the same thing with a liquid, we use a pump and not a compressor.

PROTECTING THE COMPRESSOR FROM ABUSE

There are two stages that require protection. The first is in the original installation. The second stage is start-up and operation.

Installing the Compressor

The following is a simple checklist to help sidestep major installation problems. Of course, follow the manufacturer's instructions for a particular unit in detail.

Sizing

Avoid continuous operation of a refrigeration compressor at minimum load. Oversizing leads to a variety of problems in the system and with the compressor itself. Among the more obvious are:

- poor oil movement at low load, (low refrigerant velocity leading to poor oil return from other parts of the system)
- coil icing, (due to suction temperature dropping too low)
- frozen coolers, (for similar reasons)
- compressor overheating (low gas volume at high suction temperatures often causes motor failure in semi-hermetics despite the presence of thermal overloads).

Most manufacturers provide some form of capacity control accessory to accommodate load variation, and one of these should be employed in the original system design.

Vibration and Mounting

Follow the manufacturer's recommendations carefully. Vibration causes annoying noise and can also lead to serious mechanical problems.

Space and Ventilation

Make sure sufficient space is available all around the compressor to permit work to be done without undue restriction. The maintenance manager should be consulted at the time of the original design; and the possibility that load and/or equipment may be added as time goes by should be taken into consideration.

In the case of water-cooled systems, take care to provide enough end space for cleaning and replacing condenser tubes.

Ventilation is often mandated by local codes, but imagination and common sense will often suggest ways of relieving problems. For example, building exhaust air might be passed through the compressor area; or, conversely, the equipment room might be used as the air intake plenum. Forced ventilation may be used where natural ventilation is impractical. In any case, excessive temperatures will shorten the operating life of open compressor motors. Since semi-hermetic compressor motors are cooled by suction gas, they pose fewer problems in this area.

Alignment

Alignment is a problem peculiar to open compressors, whether direct coupled or belt driven. It is critically important that the installer align coupled machines with a dial indicator. It simply cannot be done by visual inspection. Misalignment can result in rapid wear and deterioration.

Exposure to Atmosphere

The compressor leaves the factory sealed against moisture or other atmospheric contamination. Take care not to remove these seals until piping is to be connected. (Later discussion will clarify the importance of eliminating moisture.)

Interlocks

Set up the control system so that the auxiliary contacts of the system fan (or of the chilled water pump starter) on the one hand, and the condenser water pump starter on the other hand, are *closed* before the compressor can be started. Such an interlocking circuit will provide that, should either starter fail to energize or become de-energized by any of its safety controls, its auxiliary contacts will open, thus stopping the compressor. Provide the same kind of interlock for the exhaust system—so that (typical of a hospital) in the event of a fire alarm, which stops all exhaust activity, the compressor is also de-energized.

Separate Wiring for Crankcase Heaters

A crankcase heater is designed to minimize the collection of refrigerant in the crankcase by warming the oil enough so that refrigerant boils off. But it is, typically, only about 100 watts—similar to an average light bulb. It requires, therefore, a fairly long time to heat up the reservoir of oil; it is simply not possible to turn it on and to expect that the crankcase will be freed of refrigerant quickly. The safest course is to permit the crankcase heater to stay on all the time, except when the compressor is actually running. The electric consumption is quite small and the additional margin of safety quite large. It should be confirmed that the crankcase heater is wired independently of the on-off switch used to shut off the compressor for the winter.

Note: a crankcase heater is required to protect *every* compressor, whether it is indoors or outdoors. Although it is not generally recognized, refrigerant will dissolve even easier in warm oil than in cold (much like sugar in tea), so that the object of the crankcase heater is to raise oil temperature above the ambient.

The Air Filters

They are generally of the throwaway type and quite inexpensive. If there is the slightest question of their being dirty from construction activity, change them to assure good air flow through the direct expansion air cooling coil.

Suction Line Drier

A temporary suction line drier (replaceable core type) in the system at the original installation is good extra insurance against compressor damage. Check the pressure drop, however, to be sure it does not exceed the design. Cores should be removed after the system has been proved out.

Wiring of Controls

This would apply especially to such items as oil pressure switches and flow controls on chilled water systems. A few minutes to check that they are properly wired may save much time and expense later.

Suggestions on Soldering

Fluxes containing acids should never be used to make R12 or R22 refrigeration joints; the acidic ingredients will contaminate the system and can damage the compressor severely.

Flux is applied only to the outside of the tubing on the male end of the fitting—the end is never dipped into the container of flux. Such fluxes, when heated, become hard and brittle as glass. If allowed on the inside of the joint, particles would break off and eventually find their way into the compressor.

Good soldering technique usually requires the use of an oxyacetylene torch and silver solder. When extensive soldering is to be performed, the passage of an inert gas, such as dry nitrogen through the tubing, will prevent the formation of damaging oxides on the inside of the joint.

Testing for Integrity

Once convinced that the system is leak free, conduct an overall test for integrity. Pull a vacuum to below 250 microns and blank-off the system. Then close the valve and watch the pressure on the gauge. Either it will increase for 15 to 30 minutes and then stop (which means the pressure in the system has equalized and there are no leaks), or it will continue to increase (which means that leaks still exist and must be found and corrected). *Never charge a system with Freon while there is still a leak in it: That will only compound the problem.*

Start-up and Operation of the Compressor

The second stage of protecting the compressor has to do with start-up and operation, not only the initial start, but with each seasonal start as well. Use the following checklist to inspect all components:

Air Handling Equipment

- Are all belts in place and in good shape?
- Are they aligned and properly tightened?
- Are bearings lubricated?
- Are air passages clear—unblocked by cartons and trash?
- Have filters been checked and replaced if necessary?

On Water-Cooled Systems

- Has the sump of the tower or evaporator condenser been cleaned, leak checked, and filled with water?
- Has water treatment been initiated?

Valves

- Are they tagged and have they been properly opened? This applies not only to the shut-off valves of a water-cooled condenser, but to the service valves of a compressor, both suction and discharge. Operating the compressor with valves closed will result in rapid overheating and seizure.

Controls

- Are they set to automatic, where this is required? This applies to liquid line solenoids and interlock circuits.

Electrical

- Have controls been visually inspected for pitting or welding on starters and/or for obviously worn components?
- Is the crankcase heater working?
- Has it been on long enough to heat the oil and drive out the refrigerant?

After Start-up

Once the system has operated for 15 to 30 minutes, it may be assumed to be in equilibrium. Then check the compressor oil level at the sight glass; check the oil pressure with a gauge; and check the liquid line sight glass visually.

It is obvious that, once the system is on line, periodic checks must be taken. But how frequently? Given the trade-off between cost and safety, it is suggested that a thorough monthly check is better than more cursory weekly checks. To that end, the following checklist should be followed.

Preventive Maintenance

This monthly checklist *must be logged* and kept as a permanent record. It is simple, its cost is negligible and its potential benefits are large.

- Read and record refrigerant gauge pressures.
- Read and record oil pressure gauge reading.
- Check compressor oil level at sight glass and record.

- Take sample of compressor oil and conduct a "one time" acid test. Record result. (This costs only a few dollars, but it is the best early warning signal of moisture.)
- Check and record refrigerant charge at liquid line sight glass.
- Check and record suction gas temperature for proper superheat. (It is suggested that superheat setting be locked down and painted over with red nail polish. If it is ever necessary to change it, it will also be necessary to replace the expansion valve.)
- Check crankcase heater and all pressure switches to be sure they are working. Record.
- Check air filters and change if necessary. Record.
- Check and tighten or lubricate, if necessary, bearings, belts, fans. Record.
- If you find a leak of any kind—refrigerant, water, steam—stop and repair it. Record. Leaks are costly and may be dangerous.

THREE ULTIMATE FAILURE MODES

When compressors stop running, either one, two or all three of the following will be found to have occured:

- The compressor tried to compress the incompressible.
- The bearing surfaces did not receive proper lubrication.
- The motor burned out.

Because *any* of the seven common "causes of failure"—slugging, flooding, flooded starts, loss of lubrication, contamination, overheating or electrical problems—could bring about any of the three failure modes above, the whole system requires checking in order to pinpoint the real culprit.

The following sections summarize some of the more common events and the ways to avoid them.

The Compressor Tries to Compress the Incompressible

In the refrigeration cycle, it is essential for the flow of refrigerant to the evaporator to be controlled. Too little will provide inadequate cooling; if there is too much, some will pass through unevaporated and enter the compressor suction line. The compressor is designed to pump *gas,* not refrigerant in liquid form. Liquid, being noncompressible, can easily destroy the compressor. The metal parts will break upon impact with a noncompressible liquid.

Oddly, while valves often break first, there are instances when valves remain intact while pistons, rods, and rings break up. In any case, this type of failure is often accompanied by considerable noise as bits and pieces fly around inside the compressor; in extreme cases, the casing itself breaks or a crankshaft snaps, or (in semi-hermetics) the motor burns out because of a stray piece of metal lodging in the windings.

The most common cause of liquid slugging is improper setting of, or malfunction of, either the superheat or the thermostatic expansion valve. (See Figures 7-6 and 7-7.)

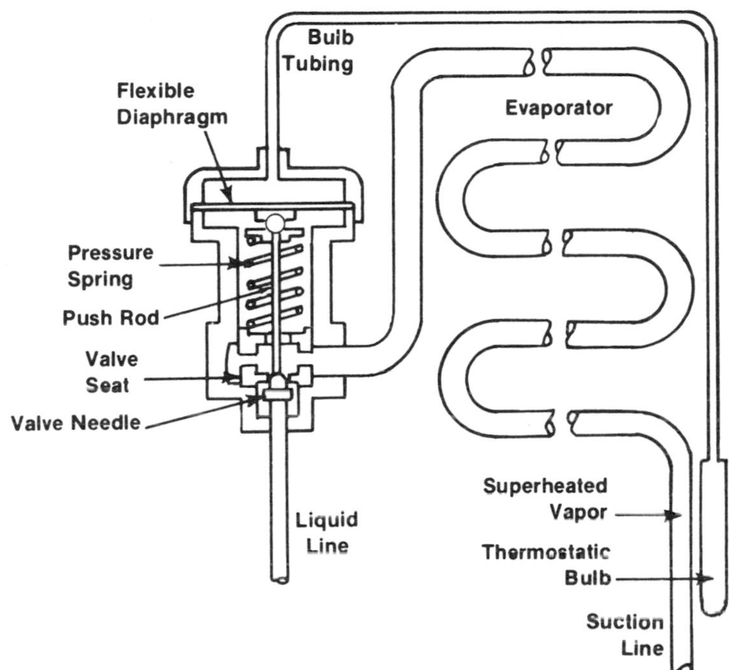

Figure 7-6. The Principal Parts of a Thermostatic Expansion Valve

An expansion valve is a relatively simple mechanical valve. The bulb and capillary tube contain a gas or liquid, usually the same refrigerant used in the system. When heat is applied to the remote bulb, the pressure within increases and is transmitted to the area above a diaphragm. The diaphragm flexes downward against the tension of a spring, pushing a pin away from its seat and causing the valve to open. The adjustment changes the tension of the spring; this variability is referred to as the *superheat adjustment.*

Note: The remote bulb location is important. Usually it is placed on the suction piping, near the outlet of the evaporator coil.

Setting Superheat

The function of superheat is to protect the compressor from either of the possible extremes: too much liquid refrigerant coming back in the suction (with resultant damage as noted) or not enough cooling for the motor. For many years, superheat

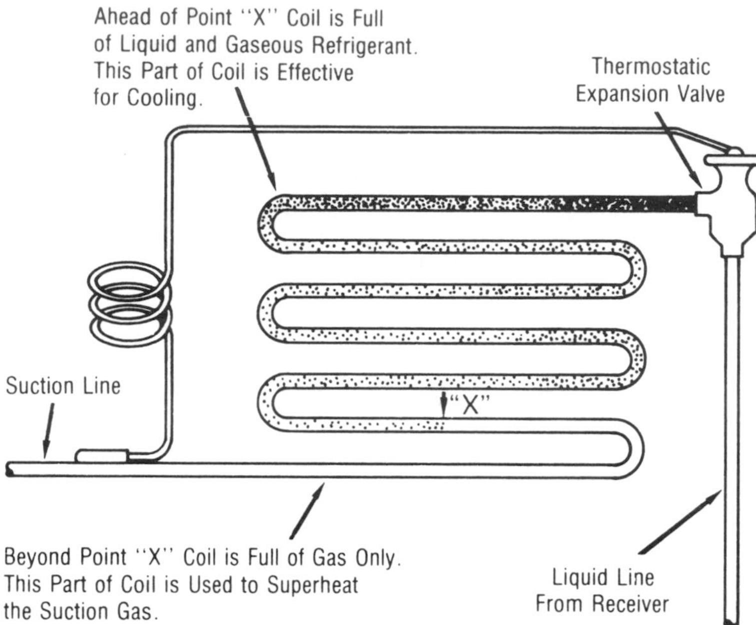

The Trane Company 1977. Used by permission.

Figure 7-7. Diagram of an Evaporator Coil

was regarded as the difference in temperature between the evaporating temperature and the gas temperature at the remote bulb. This definition is useful, in that will provide an efficient evaporator, but it has the drawback that it does not take into account the significant amount of superheat that can be added where a long suction line exists (as, for example in a split system).

A better definition of superheat is: the difference between the saturated temperature (taken from the appropriate table and equivalent to the suction pressure) and the *actual* temperature of the suction gas (taken with an accurate—prefereably electronic—thermometer). *Both of these measurements are taken as close to the compressor as possible.*

It is important that you observe the maximum and minimum settings established by the manufacturer. They are always expressed as absolute outside figures. All manufacturers provide this information for each model compressor, but some make the information more accessible than others. For example, the Worthington/Climatrol/Fedders 3V model lists a minimum of 10° F and a maximum of 25° F, measured at the compressor.

On the other hand, Copeland's instructions are: "A minimum of 15 degrees F. superheat at the compressor must be maintained at all times to insure return of dry gas to the compressor suction chamber and a minimum of 20° F superheat is recommended."

> **Note:** It is absolutely essential that superheat be checked and properly set, according to the manufacturer's instructions, at the time a compressor is replaced.

It is also necessary to recognize that—once it is properly set up—it should not be necessary to change a superheat setting. If something happens that requires a superheat adjustment, it is probable that something is wrong with the expansion valve itself. It may have sprung a leak in the bulb, it may have been plugged up with wax from the oil, or it may have hung up on copper plating as a result of some acid in the system.

A Good Rule: If you have to change the superheat setting, replace the expansion valve.

Inadequate Lubrication

A second major cause of failure is low, or no, oil pressure. Occasionally a compressor will leave the factory without oil, or an accident in transferring it from place to place drains it. Remember that some oil always moves around the system with the refrigerant. But if there was a correct oil level in the system originally, it should return to the compressor unless something is wrong, such as:

- *Oil pump failure*—This can be anything from a broken tang to the reversing gear being stuck. Sometimes just reversing the direction of compressor rotation by switching the motor leads will do the trick.
- *Oil relief valve blocked open*—A little piece of dirt can cause big trouble.
- *Worn bearings, worn unloader assembly, broken oil tube*—A number of internal problems of this kind can occur. In most cases, the most economical solution is to exchange it for a remanufactured compressor.
- *Excessive refrigerant in the crankcase*—This can occur during layoff periods or in swing seasons because of migration. Migration is the tendency of a refrigerant to move to and condense in the coldest part of the system. Since *cold* is a relative term and the difference may be only a few degrees, this phenomenon often occurs after an overnight layoff. For example, the compressor, when operating, tends to run at a fairly high temperature; the heads will literally feel hot. Since it is made of cast iron, it is also quite heavy in relation to its surface. Therefore, by definition, it is a heat sink: it is the last part of the system to cool off at night, but once down to ambient temperature, it is equally slow to heat up. As the ambient rises (for example, on a sunny spring morning) the compressor warms along with the rest of the system, but lags behind due to its mass, and it can become the coolest part of the system just at the time when the thermostat is calling for cooling. For this reason, noise and vibration from the compressor during a morning at start-up can usually be attributed to refrigerant in the crankcase.
  Check the crankcase heater for proper operation. Check the liquid line solenoid valve for leaks and be sure the system is not overcharged.

The Basic Rule Is: refrigeration oils absorb refrigerants at all temperature ranges; this absorption must be controlled. Oil will move around the system, but *must* be able to return to the compressor.

The Motor Burns Out

This section refers to semi-hermetic compressor failures only. Open motors just do not burn out due to compressor failures. Even if an open compressor seizes up, the coupling is likely to break first, before the motor fails. There are 6 principal causes of burnouts, described in the following sections.

Low Voltage

Sometimes this occurs because power companies have had to cut back on voltage in peak periods. When the voltage drops, the amperage goes up. Within a few percent leeway, the motor can probably take it; beyond that, failure is likely. Overloads built into the winding—whether simple temperature-operated switches or fancier solid state devices—are only a partial answer. The overload covers a relatively small part of the winding and unless the overheating takes place in that particular area, the motor may be burned before the sensor can function. Sometimes, just a bad contactor gives low voltage at the compressor itself, although the voltage being delivered by the power company is all right.

Voltage Imbalance

The high incidence of customizing new buildings or of converting older ones to new uses has led to a situation in which a dozen separate tenants may hire electricians to supply single phase power in individual sections of a large building.

In turn, each electrician pulls his single-phase wiring from the three-phase power entering the building. But which phase? If too many use the same leg, voltage imbalance is likely. To check for this condition, follow these steps:

The three voltages in Step 1 are 215, 221 and 224. The average, therefore, is 220V. The worst leg is the first one—with a difference of 5 volts from the average. When we divide that 5V by 220V (Step 4) and multiply the result by 100, we find that the voltage imbalance is 2.27%.

That's a problem by definition. The maximum allowable imbalance is 2%—which on a 220V average hookup is only 4.4V instead of the 5V we actually have.

This difference doesn't look critical, but it is—because the increase in temperature at the stator is two times the square of the imbalance.

Therefore the motor temperature increase (heat gain) is two times 2.27 × 2.27, or 10.3%.

Taking this computation further, if the voltage imbalance were 3%, the temperature increase would be 18% (two times 3 × 3). And on a 4% voltage imbalance, the temperature increase would be 32% (two times 4 × 4).

It gets worse. Keep in mind that the temperature increase isn't uniform throughout the stator. Instead, it is concentrated in one phase. As the voltage imbalance increases, so does the chance of a motor burnout.

What can be done on the job to avoid this hazard?

While it is possible that the problem originates with the power company, it can also be caused inside the building—and corrected there.

Keep in mind that neither a power supply nor a stator is perfectly balanced. Each of the three phases entering a building is connected to a different phase in the stator.

Thus it is possible to connect a different power phase to a different stator and thereby correct minor problems. This is similar to mixing and matching mechanical parts in order to achieve the closest possible tolerance.

After determining that the imbalance is too high, take these steps:

A. Roll the terminals one forward. Connect the wire now on Terminal 1 to Terminal 2, 2 to 3, and 3 to 1.

B. Repeat Steps 1 to 5 above, remeasuring and recalculating with the new hookup.

If the imbalance is within 2%, stop. If not, proceed.

C. Roll the terminals one more forward—Terminal 1 to 2, 2 to 3, 3 to 1.

D. Repeat the remeasuring and recalculating procedures outlined in Steps 1-5 above. If the imbalance is within 2%, stop.

If the imbalance is still excessive, you (or a representative of your company) should contact the owner. Tell him you have turned off the equipment to protect it from a voltage imbalance problem which could cause the compressor to fail. He should contact an electrician to analyze the building's power supply and load distribution.

Here is how you calculate voltage imbalance on the job. This easy step-by-step procedure follows the example. Keep in mind that the **maximum allowance voltage imbalance is 2%.**

As shown in this drawing, there are three possible hookups on every three-phase job:

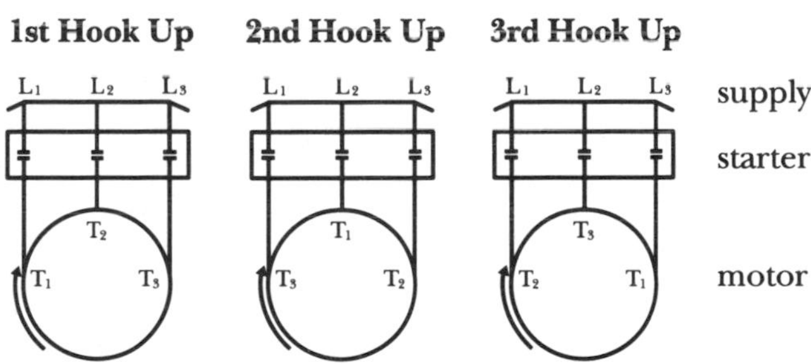

There's no theoretical way or shortcut to find which is best—you must take current readings in volts on each possible hookup. Then you must calculate which hookup has the least imbalance.

1. Start by adding up all three readings on Hookup #1:

$$T_1 = 215$$
$$T_2 = 221$$
$$T_3 = 224$$

Total: 660V

2. Divide the total by 3 to obtain the average:

$$3 \overline{)660V} = 220V$$

3. Calculate the greatest voltage difference **from** the average:

$$\begin{array}{r} 220V \\ -215 \\ \hline \mathbf{5V} \end{array}$$

4. Divide this difference **by** the average to obtain the percentage of imbalance:

$$220V \overline{)5V} = \mathbf{.022}$$

5. Multiply this total by 100 to establish the percentage of imbalance:

$$.022 \times 100 = \mathbf{2.2\%}$$

6. Repeat the above steps for Hookups 2 and 3. Move the terminal wires **forward** one leg for each hookup.
7. **Select the hookup which has the smallest percentage of imbalance.** 2% is the maximum allowable imbalance. If the "best" hookup still exceeds this percentage, you have done everything you can do on-site. It's best to shut down the equipment and notify the electrician.*

Failure to Clean Up a System from a Previous Burnout

This happens surprisingly often, considering that so much attention has been given to clean-ups. When there have been several burnouts in a row on the same unit, it seems obvious that the fault lies not with the compressor, but with the system.

Not only must a suction line drier be put on (and, preferably, the cores changed several times) but the compressor oil should likewise be analyzed each time the drier core is changed. Change the oil when indicated by analysis or when its condition is doubtful.

Simply blowing out a system with Freon will not do. The carbon and acid associated with a burnout will find its way into every crack and crevice in the interior of the compressor body and into every bend and valve in the piping system. If not cleaned out thoroughly, the Freon (one of the world's great solvents) will eventually dissolve one of these clumps and carry the material back into the stator where it will again eat through the insulation and burn out the new windings.

Cleaning out acid and carbon can be done effectively with a liquid line drier *and* a suction line filter-drier but the clean-up should be confirmed by an oil analysis.

The concept of flushing out with R11 is outdated and is not recommended for the following reasons:

---

*JobScope 1990. Used by permission.

- it does not get out all the contaminants, but simply shifts their location from one spot to another in the system,
- it is difficult to remove all the R11 from the system, and
- even small amounts of R11 left in the system may cause adverse chemical reactions.
- under new regulations, it is illegal.

By the same token, it must be recognized that just putting in a drier core and walking away is not enough. Oversized drier shells can and will trap oil. The oil level in the compressor should be checked carefully while the clean-up driers are in place. If a drop in oil level is noted, it tells the craftsman to remove the drier shells and correct the oil return problem. See detailed instructions in the section on "Cleaning Systems Properly." **Every manufacturer and remanufacturer requires that a proper clean-up be performed in order to validate the warranty on an exchange compressor.**

Spot Burnout

In this case, the damage is usually confined to a small area. It might happen as a result of a short between windings, or even between wires within a winding. It might be from windings to ground. Several causes are possible: faulty manufacture, mishandling (stator dropped, for example, which will leave a flattened spot), or deterioration of insulation caused by age. But, it might also come about as a result of a metal particle broken loose during a liquid slug and subsequently shorting the winding.

Single-phase Burn

This happens when two phases of a three-phase motor burn because they receive the full current flow. Most frequently, this results from malfunction of contactors or other electrical failure. It is characterized by a "striping" effect at the end of the stator.

General Burnout

Often it is difficult to pinpoint the cause because literally nothing is left. Generally, a strong smell accompanies a burnout. Remember: many burnouts are caused by moisture in the system.

ELIMINATING MOISTURE

Much has been written about moisture in an air conditioning or refrigeration system, but it all boils down to one thing: *Moisture is an absolute enemy and must be eliminated.* The reason is simple. Moisture (even in very tiny amounts) combines with Freon to make hydrofluoric acid. For example, in a typical refrigeration system, damage from moisture in R12 can occur from solubility as low as two ppm (two parts per million).

In short, take all the normal precautions during installation to ensure that moisture does not enter the system. Normally, this presents no problem—the problems tend to arise later, when the system is in service. Even though moisture will not usually enter a system while it is pressurized, it can enter in the process of repairing a leak when it is necessary to pump down the compressor and remove the system gas. When the system is at atmospheric pressure, moisture is very likely to enter.

HOW TO PREVENT BURNOUTS

There are two ways to prevent burnouts: check for electrical problems and clean the system properly. These prevention techniques are described in detail in the following sections.

Check for Electrical Problems

Electrical problems generally break down into four categories:

- Abnormal supply
- Malfunction of some electrical component or accessory
- "Jumping" a control
- Improper wiring

Abnormal Electrical Supply

Generally, this is a voltage problem—either high or low. The motor winding temperature will tend to rise on either side of the normal line voltage, but low voltage tends to be the more prevalent cause of motor failure.

Three-Leg Motor Protection. Frequently, only two legs of a three-phase motor have had overload protection. This leaves a motor vulnerable to damage by single-phasing, due to failure of the unprotected phase. It is good practice to use phase failure protection *on all three legs* of the power supply.

Malfunction of Electric Component

Visual inspection can often detect possible failures—for example, pitting, welding, or corrosion on starter contacts. If a starter doesn't look perfect, replace it—it's far cheaper than replacing a compressor. The same is true of relays; contrary to popular opinion, they do wear. A good rule of corrective maintenance is to replace starting relays whenever you suspect deficient performance. Another good rule is to replace the starting contactor any time you replace the compressor. Be sure that contactors, controls, fuses, and size of wires used are correct for the nameplate and the locked rotor amperage (LRA) of the motor. Chattering controls can produce high transient voltages, which may cause motor winding breakdown.

"Jumped" Controls

One of the more common failures in air conditioning is caused by bypassing a control that is *known* to be inoperative. The craftsman jumps a defective oil safety switch, with the intention of coming back the next day with a replacement control. But he is delayed and ultimately forgets. Subsequently, a condition arises that should have caused this control to shut down the system; but, it can't do so, and a major failure results.

Improper Wiring

This often shows up immediately on installation, although damage sometimes occurs quickly, as well.

Any of the following signs would lead to further checkout:

motor drawing excessive amperage

motor cycling on overload protection

motor will not start, or runs below speed

Note: Be sure to carefully compare the old and new terminal pattern and wiring when replacing a compressor. Manufacturers change number and position of terminals, which can result in miswiring.

Cleaning Systems Properly

Original Installation

Special care should be taken to be sure that all contaminants are removed from the system at time of original installation. Anything except oil and refrigerant is a contaminant—e.g., dirt, filings, shavings, solder, flux, steel wool, metal chips, grit from sandpaper or emery paper, bits of wire, air and moisture. Some are so fine that they will pass through a suction screen; others are sharp enough to cut such a screen.

A PROPERLY SIZED, HEAVY-DUTY SUCTION LINE FILTER SHOULD BE INSTALLED AT THE ORIGINAL INSTALLATION, ALONG WITH A PRESSURE FITTING AHEAD OF THE FILTER TO AID IN CHECKING PRESSURE DROP. The benefits are: elimination of hard-to-remove foreign matter and prevention of contamination of the entire system through the suction line, should a burnout occur.

System Cleaning

Many years ago, common practice involved flushing out the system with R11. The difficulty lay in applying this practice properly and in getting out all of the R11. The practice is now regarded as out-of-date.

Instead, current practice *requires* the installation of a suction-line cleanup kit of the recommended size. Strongly recommended is the installation of an oversized filter-drier in the liquid line as well. The sizing should be such that the maximum pressure drop under normal operating conditions is within the limits shown on the curves on the reverse, with connections of the filter-driers the same sizes as the connecting lines.

Evacuation

Proper evacuation with a properly sized vacuum pump is essential, since the presence of either air or moisture is detrimental. Purging with refrigerant, either by blowing out lines or feeding from the top of the condenser and receiver is not adequate, since it cannot remove the air trapped in the compressor during installation. Both new and replacement (remanufactured) compressors are shipped either with a dry air holding charge or a dry nitrogen charge; in any case, the compressor must be evacuated.

The industry standard is triple evacuation of the system or compressor—twice to 1500 microns and the third time to 500 microns, breaking the vacuum each time with the refrigerant to be used and building a positive pressure to at least 2 psig. Vacuum pump must be connected to both high and low sides through connections of proper size.

System Cleanup After Burnout

After closing service valves, remove burned-out compressor and install replacement. Save a sample of the oil in the replacement compressor in a sealed glass bottle. This will be used later for comparison.

It should be necessary only to evacuate the compressor. Open compressor service valves, close liquid line valve and any other shut-offs to minimize the amount of refrigerant to be handled. Pump system down. While some contaminants will come back to the compressor during pump down, the compressor will not be harmed in this short period. The contaminants will be removed by the filter-driers.

Inspect, clean or replace all system controls—expansion valves, solenoid valves, check valves, etc. Remove and replace any filter-driers previously installed, as well as any filters or strainers. Install the recommended size filter-drier in the suction line and an oversized filter-drier in liquid line. Start the compressor, observing pressure drop across the filter-drier after a *minimum* of 4 hours. Compare to curves shown on the Filter Drier Pressure Chart on adjacent page; replace as necessary. Take oil sample and check with acid test kit. After 24-48 hours, recheck pressure drops and oil acidity. Replace oil if acidity persists. After 7-10 days, check same factors again. Compare oil color and odor to original sample to be sure it compares favorably. At this time, replace liquid line filter-drier with normally recommended size. Remove the suction line filter-drier and replace with a permanent type suction line filter.

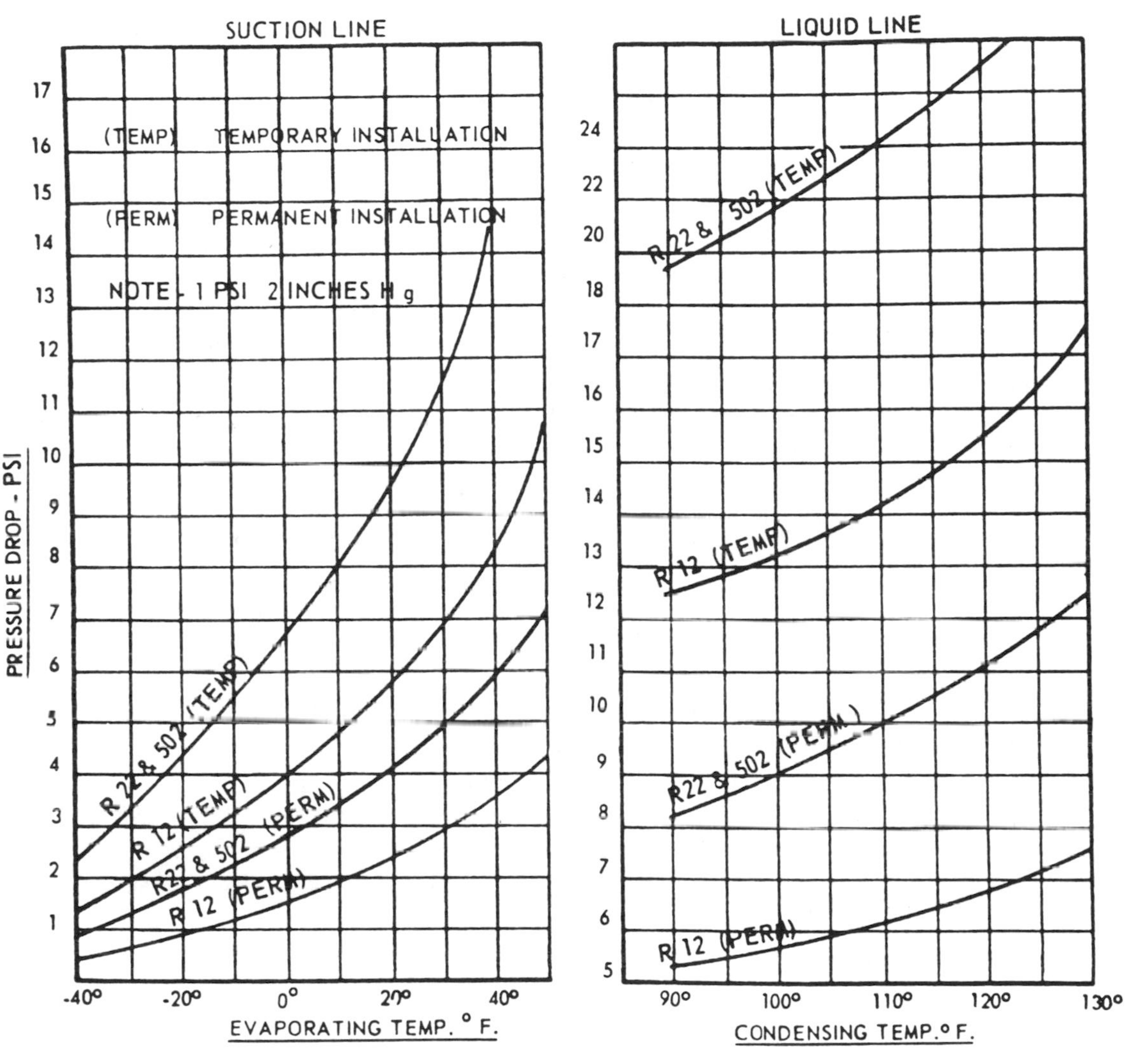

CFCs AND THE ATMOSPHERE

Much has appeared in the press toward the end of the 1980s about the effect of chlorofluorocarbons (CFC) refrigerants on the ozone layer. Congress under Title 7 of the Clean Air Act has established a number of restrictions which affect production of CFCs, resulting in a complete phaseout of CFCs by the year 2000. State and local governments have also established restrictions.

While much remains unclear about the subject, clearly it will no longer be possible to vent refrigerants to the atmosphere. Reclaiming is becoming the mode and, in late 1990, ASHRAE released its new Guideline.

In part, that Guideline states, "Before topping off any system that is short of refrigerant, a diligent effort must be made to discover and repair the cause of the refrigerant leak. When the cumulative loss in 12 months exceeds 10% of the operating charge of 50 lbs. or more, the system should be promptly taken out of service and repaired."

It goes on to say: "Refrigerant in any system that has failed may be contaminated with moisture, metal particles, acid, carbon, sludge, etc. Depending on the severity of the contamination, the refrigerant may be saved and reused or it may be removed in suitable containers for later use after being processed by a reclamation procedure.

"Once a decision has been made regarding the status of the refrigerant, the procedures to be followed should be as indicated in the ASHRAE Handbook, 1986 Refrigeration Volume, Chapter 7, 'Moisture and Other Containment Control in Refrigerant Systems', except that no deliberate discharge of refrigerant to the atmosphere should occur."

HOW TO HANDLE A FAILURE OF THE COMPRESSOR

Air conditioning stoppage, according to one of Murphy's many laws, always occurs about 5 p.m. on Friday evening or, alternately, on a holiday immediately before the day of the big meeting attended by all the "top brass."

As we have noted, the initial cause may have been electrical or mechanical—either designed in or the result of a component breakdown—or even of simple abuse of the equipment. But the net result will often be the breakup or burnout (or both) of the compressor.

Now what? The answer depends on a number of factors. If a big meeting is indeed scheduled, then budgetary considerations may well be ignored and speed may become the most important item. In such a case, a replacement compressor from a competent remanufacturer or from the manufacturer is probably the quickest, most reliable solution. Together with a good bit of sweat and a lot of overtime, the crisis may be overcome without the "brass" ever becoming aware of it.

If time is less pressing and labor availability and skill are sufficient, you may want to try rebuilding in place—*if* the compressor is an open design; *if* parts are readily available at reasonable cost; *if* you have literature that shows the manufacturer's recommendations of wear tolerances and procedures; *if* the physical location of the compressor gives enough light, reasonable protection from the elements and sufficient space in which to work; and *if* the savings to be enjoyed offset the disadvantage of having no warranty.

Considerations of experience and skill aside, it pays to understand the structure and philosophy of the industry that produces the equipment and supplies replacements.

The Manufacturer

Virtually all of the leading manufacturers supply replacement compressors for current designs through their parts' department or through their distributors. Obviously, stocks are not always local or complete, but all such compressors, whether open or semi-hermetic, are remanufactured. They are seldom, if ever, *new* in the strict sense of the word.

All replacement commercial and industrial compressors are sold with a one-year warranty.

On the whole, the manufacturers (either directly or through authorized shops) do an entirely satisfactory job of remanufacturing their own compressors.

The Independent Remanufacturer

This classification has only emerged in the marketplace within the last ten years, as the need for more local stocks of reliable replacement compressors has grown.

At this writing, only a handful of independent companies produce compressors which meet the ultimate requirements of remanufacturing:

- Compressors are restored to the *new* specification, without any allowance for wear.
- Compressors are supplied complete with all components which are supplied by the manufacturer himself. That would include such items as terminal boxes, phase separators, solid state modules, and so on.
- Compressors are updated to current design.

Typically, such compressors would also be furnished with a complete set of installation/startup instructions.

These compressors are available in the US through a network of wholesalers or distributors, who keep common models in stock. In some cases such distributors handle a particular manufacturer's own brand and utilize independent remanufacturers to round out their stock to include all other brands as well.

The Rebuilder (Repair Shop)

For many years, local rebuilders have existed all over the country, usually confining their activity to a small area or region. Such rebuilders often maintain their own inventory. Many satisfactorily repair compressors, but none restores according to the requirements above. For that reason, quality varies from shop to shop and from compressor to compressor and local experience may be the only guide available.

SUMMARY

Many compressor failures can be avoided by using common sense and taking care to set up a regular system of inspection, checkout, and component change. The methodical monthly checkout (including an oil analysis) will, by itself, reduce compressor failures by two-thirds.

The enemies of reciprocating compressors are moisture, liquid slugging, and electrical problems.

A compressor failure—whether the result of abuse or not—brings the system to a complete standstill. How quickly that particular system has to be back on line is usually the determining factor in how the failure is corrected. If it is a critical system, the fastest, most reliable and, therefore, least expensive solution is usually a fully warranteed remanufactured exchange compressor.

Section 4: HEAT PUMPS.

Alvin Soffler, P.E.

It is not the intent of this section to necessarily explain all the details and highbred adaptations that can be used with a heat pump system. The intent is rather to show how this relatively new type of air conditioning and heating system (approximately 30 years old) has many applications not formerly available by the older but still current and reliable systems of heating and cooling.

Fossil fuel heating systems require a pipe line to the heating equipment for either natural gas, oil, or propane. Heating and cooling systems utilizing fossil fuels for the heat also need electricity. A heat pump serving both heating and cooling requires only electricity.

Fossil fuel systems require a flue for combustion exhaust and provisions for combustion air intake located within the structure. Heat pumps have no such need.

The potential of fire from open flame is eliminated with heat pumps. In some areas, the fire insurance rates could be somewhat reduced if the building has no source of open flame or combustible products. Where packaged rooftop heating and cooling equipment utilizing fossil fuels is not practical in cases where the location of the equipment is far from the space to be conditioned, the split heat pump system may be more practical and economical to install.

The material that follows attempts to explain the basic principles of the heat pump and the various types of systems that are available and commonly used, and the various ways in which the heat pump system can be applied to commercial and industrial as well as residential structures.

UNDERSTANDING HOW THE HEAT PUMP WORKS

The heat pump is basically one system which gives cooling when operated in one direction or manner and by "reversing the cycle" becomes the heating system. The basic components can be utilized for both purposes.

By way of illustration, the operation of a simple room air conditioner can display the operation of a heat pump. In summertime, the unit supplies cool air by means of a compressor system that removes heat from the space and discharges it to the outdoor air. Consider now if we took this same room air conditioner and installed it backwards. The cold air emanating from the grille would be discharging to the outside, and the hot air from the condenser side of this unit would be discharging heat into the space. Instead of removing and reversing the air conditioner for the respective heating and cooling seasons, heat pumps have a "reversing valve" to reverse the cycle depending on the season and requirements of the space to be conditioned.

Heat will always flow from a high temperature source to a lower temperature source—"sink." However, by a mechanical device such as a compressor system, a refrigeration machine (heat pump) is a device whereby heat can be made to flow in the reverse direction. This equipment "pumps" heat from a "source" at a lower temperature to a "sink" at a higher temperature. It is the compressor in the heat pump that is used to reverse the natural flow of energy.

We usually associate the term *heat pump* with a heating device, but cooling equipment also functions as a heat pump. For example, a room air conditioner is a heat pump because it makes heat flow from a lower temperature area (inside the room) to a higher temperature area (the outside air). The terms *refrigerator* and *air conditioner* are commonly associated with cooling equipment. These devices are not normally called *heat pumps,* but when a refrigeration machine is used for heating also it is usually called a heat pump.

Although it might seem confusing at first as to how you can extract heat from cold air and produce warm air, remember that any substance contains heat if it has a temperature above absolute zero or minus 460°F. Similarly, water at various temperatures also contains heat to be liberated or pumped to a "sink".

In the heating mode, certain conditions of temperature and humidity cause the formation of frost or ice on the outdoor coil. If this should happen, a defrost control is provided in the system which at outdoor coil temperatures below 37° F will energize the outdoor fans and close the circuit to the defrost relay. At this point in the operation, the defrost relay closes, which in turn energizes the reversing valve placing the system into the cooling mode. In the cooling mode, the compressor will discharge hot gas into the outdoor coil thereby melting any frost or ice formation on the coil. During this period, the system will energize the first bank of electric (or other) supplemental heaters to provide heat to the space during this defrost operation. When the outdoor coil temperatures reach the higher temperatures (approximately 70° F), the system will again change back to the heating mode.

TYPES OF HEAT PUMP SYSTEMS

The various types of heat pump systems are classified by the source from which the heat is extracted and the fluid which is used to distribute the heat to the occupied space.

Although heat pumps are most often identified with their operation in the heating mode, note that heat pump are also used to provide summer cooling. When the equipment cycle is reversed to provide cooling, the source and sink are also reversed. For example, a building that is heated with a water-to-air heat pump in winter will be cooled by an air-to-water heat pump in summer. The equipment will be referred to as a "water-to-air" heat pump because it operates in this mode when it is used as a heating device.

Air-to-Air Heat Pumps

This system extracts heat from one air stream (the source) and delivers the heat to the air stream that is circulated through the occupied structure (the sink). See Figure 7-8a. Air-to-air systems are the most common heat pump systems because the atmosphere is so readily available as a heat source.

One drawback associated with the air-to-air heat pump is that the efficiency of this equipment is reduced as the source air temperature becomes colder and thus requires more supplemental heat energy. This efficiency is noted by the term "coefficient of performance" (COP). For Heating and Cooling cycles, see Figures 7-9 and 7-10.

Water-to-Air Heat Pumps

Water-to-air heat pumps extract heat from a water source and deliver heat to the indoor supply air stream. See Figure 7-8b. Usually, a well provides the water source. Wells provide consistent performance during all seasons having a more constant sink temperature year round.

Within limits, this type of system is especially versatile when used for heat recovery applications in that it can extract heat from other fluid sources. It can efficiently recover heat from water or other fluids discharged from building water systems used for manufacturing processes. In some applications, buildings may have a closed water loop system. This might consist of a cooling tower which can maintain a constant source water temperature by means of the evaporative cooling of the tower in summer or heating of the tower water in winter with steam or other forms of heat. This relatively constant water temperature source can be circulated to various heat pumps throughout the building, which can alternately be used for cooling and regenerating heat to the loop system for furnishing heat to a given space and rejecting the cooling back to the loop system.

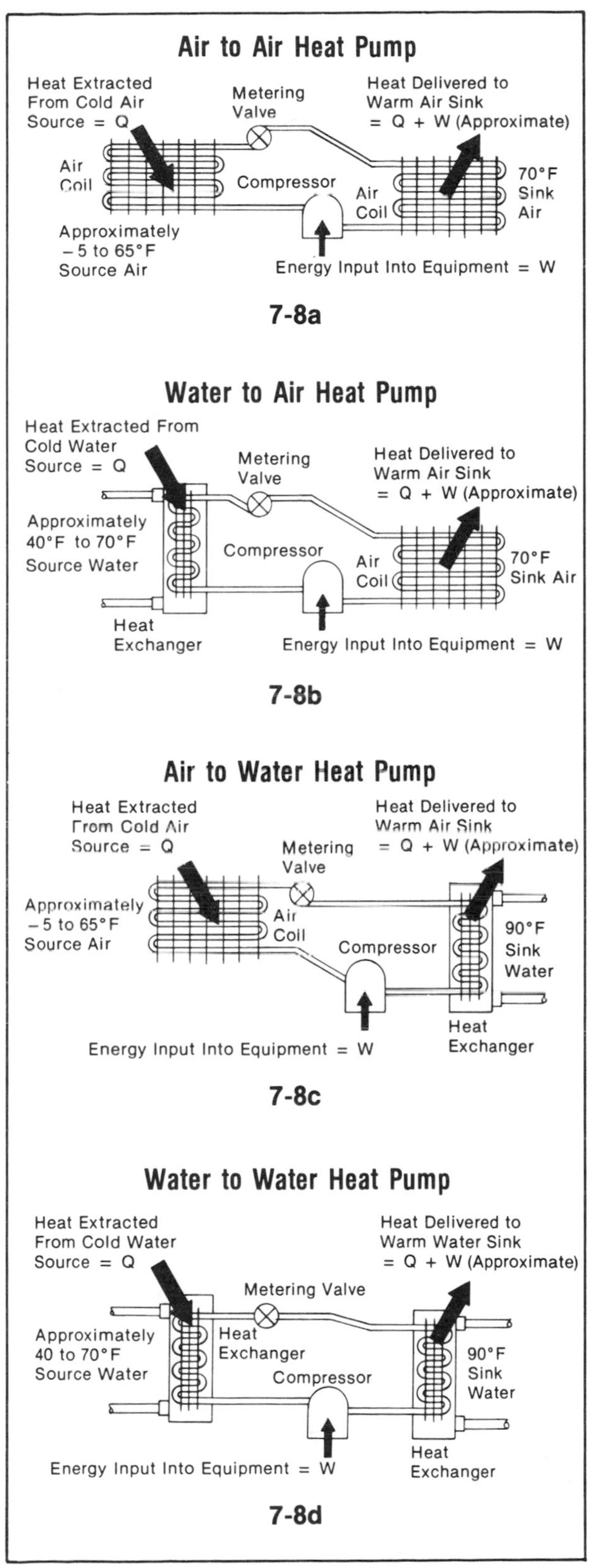

Courtesy of Air Conditioning Contractors of America

Figure 7-8. Types of Heat Pump Equipment

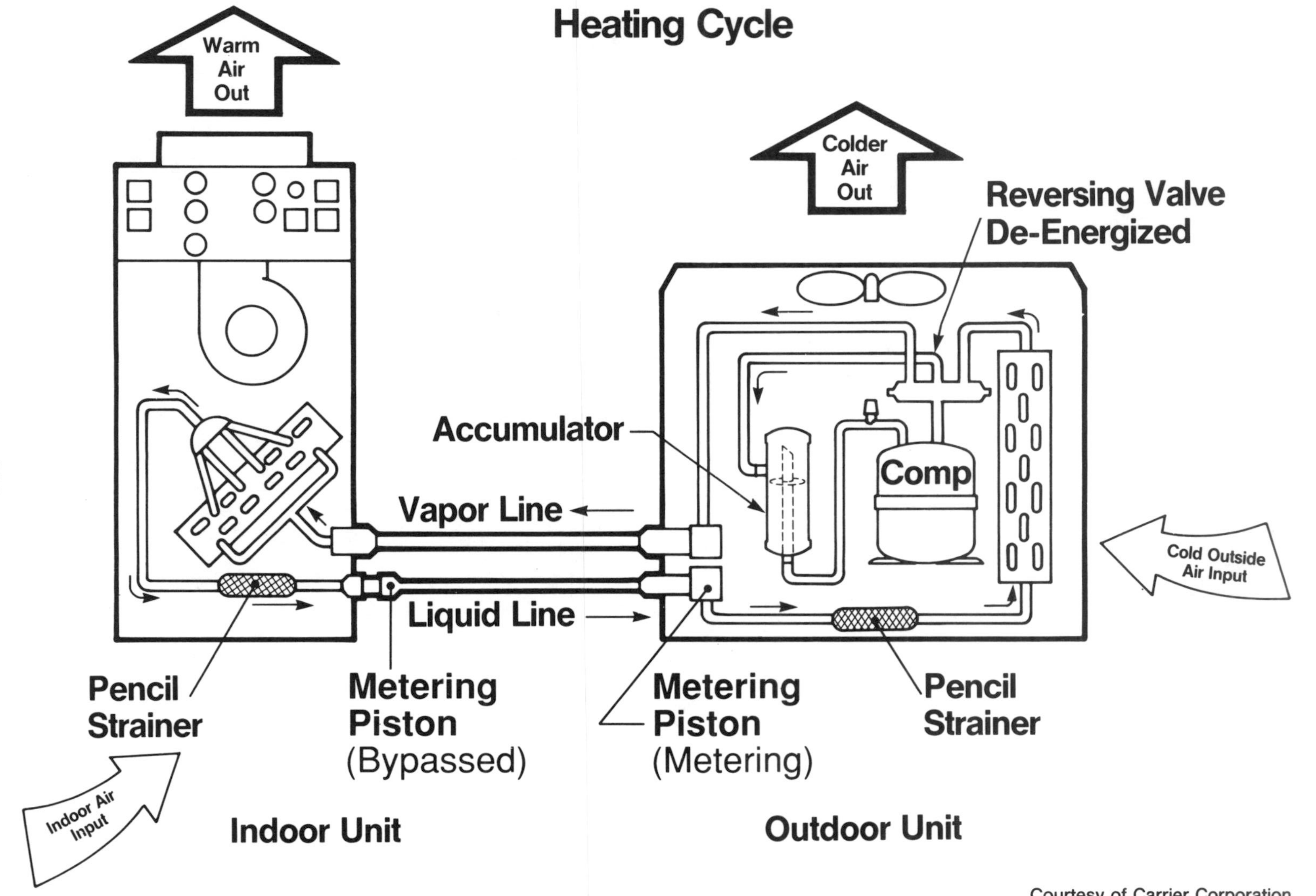

Courtesy of Carrier Corporation

Figure 7-9. Heating Cycle, Air-to-Air Heat Pump

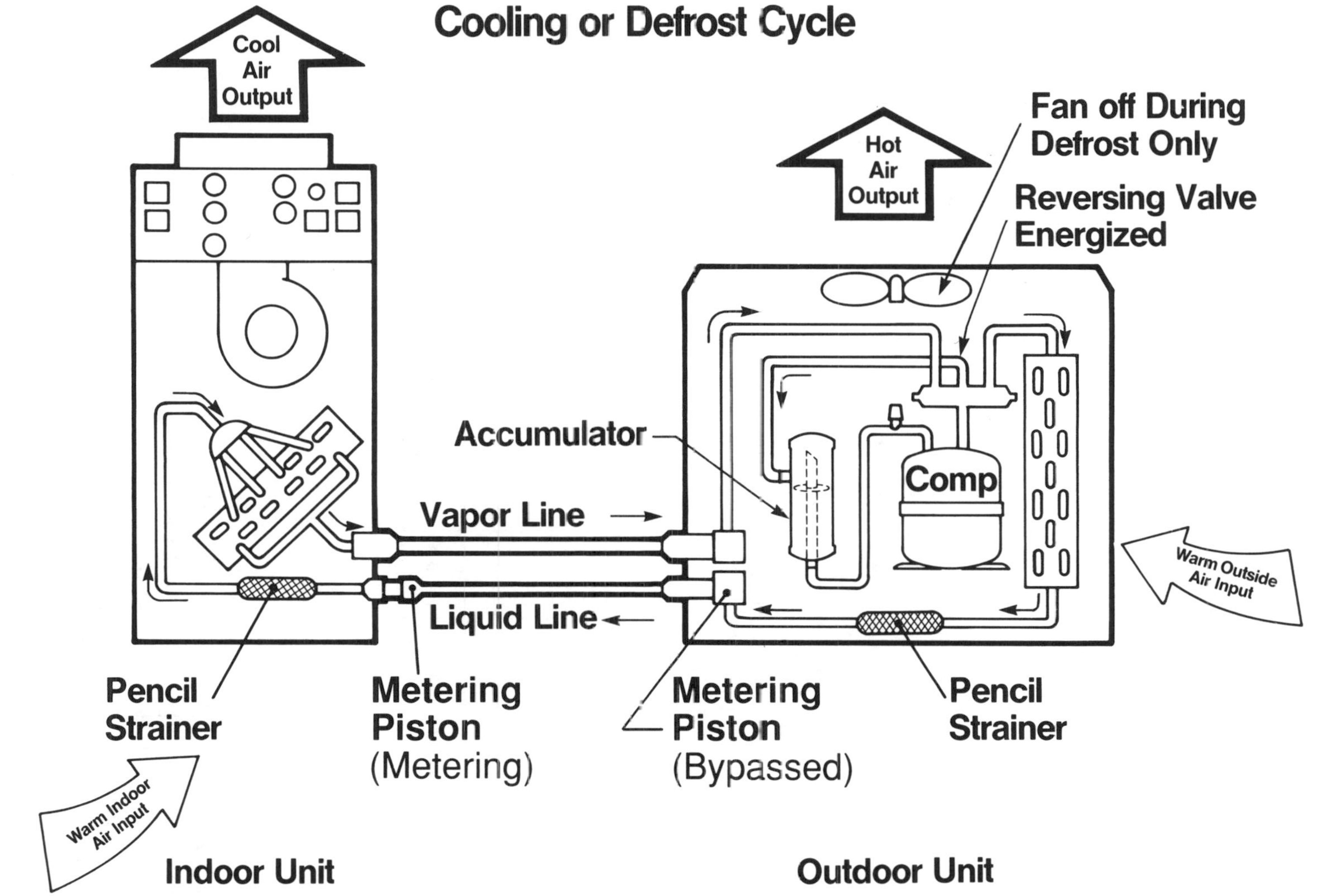

Courtesy of Carrier Corporation

Figure 7-10. Cooling or Defrost Cycle Air-to-Air Heat Pump

Where ground aquifers permit, the groundwater temperature is at least 40° F. Groundwater heat pumps have a definite place in this type of heat pump system. The water source is plentiful and the temperatures relatively constant for year-round operation of heating and cooling. In many cases, the groundwater temperature is such that pre-cooling groundwater coils placed in the air stream can offer free cooling and thus save compressor power energy.

In such groundwater applications, a diffusion well that will return the water back to the ground is mandatory, in order to conserve ground water.

Air-to-Water Heat Pumps

These devices extract heat from the source air stream and deliver heat to water that is circulated through the hydronic space heating system. See Figure 7-8c. The source air may be provided by the atmosphere or by a building or process exhaust system. Although this type of heat pump is not as common as air-to-air or water-to-air equipment, it is available in sizes for many commercial applications.

Because the occupied space is ultimately heated or cooled by hydronic devices, the air-to-water heat pump lends itself to designs where zone control is required. This is accomplished by allowing the zone thermostats to operate the various hydronic devices while the heat pump maintains a constant water supply temperature.

Water-to-Water Heat Pumps

This equipment extracts heat from the source water and delivers the heat to water that is circulated through hydronic devices. See Figure 7-8d. The source water may be provided by a well or fluid discharge from a process or ground water.

Water-to-water systems are not as common as air source or water-to-air equipment, but this equipment is available in sizes suitable for commercial applications. This equipment also finds application in situations where water-to-water reclaim is desired.

DESIGN AND INSTALLATION CONSIDERATIONS

For a given building envelope, the heating loss or BTUs required to maintain building temperature decreases as the outside temperature increases. The capacity of the heat pump to deliver heat increases as the outside, or source temperature, increases. Where these two curves cross (if plotted) is the theoretical balance point of the system. If the design outside temperature is below the balance point, then supplemental heat is required to maintain the design envelope temperatures

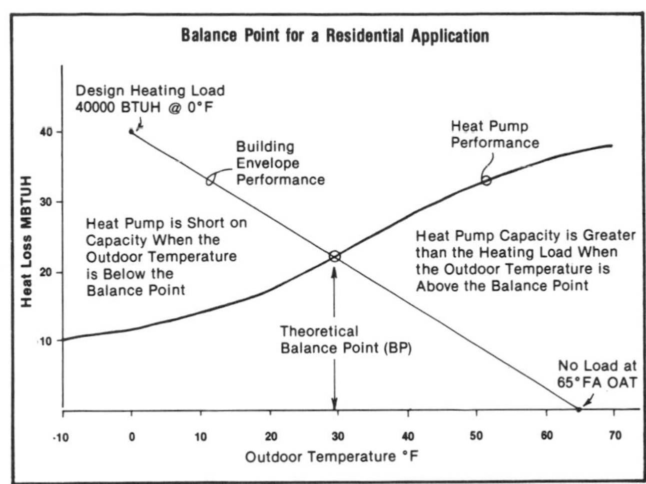

Courtesy of ACCA—Air Conditioning Contractors of America

Figure 7-11. Heat Pump Balance Point

at outdoor design temperatures. In general, for commercial and industrial buildings, the heating loads are smaller than the cooling requirements because of the high internal load. Accordingly, little if any supplemental heat is required in these types of structures as compared to a residence or structures with low internal heat.

Figure 7-11 indicates the heat loss of an envelope having a heat loss of 40,000 BTUH at 0° F outdoor while trying to maintain an indoor temperature of 65° F. The heat pump performance curve represents the heat delivery curve of the specific heat pump selected.

Electric resistance heaters may be used to add supplemental heat as the outside temperature drops and the heat produced by the heat pump diminishes. In buildings where there is steam or hot water sources of heat, this supplemental heat can be accomplished by means of hot water or steam coils. This supplemental heat can also be considered a back-up system: if properly sized, it can act as a source of heat, albeit not totally adequate to maintain design conditions, but as a means of emergency heat if the compressor or the heat pump malfunctions or fails. This supplemental heating can maintain some semblance of comfort until the compressor or equipment is repaired.

From the standpoint of energy conservation, supplemental heat acts as a second stage. Multiple staging of supplemental heaters, whatever the source, is a form of more uniform comfort control and energy conservation during fluctuating outside temperatures and indoor requirements.

Checklist for Installing Heat Pumps

Some additional considerations to the proper application of a heat pump system follow:

- Have you calculated the cooling and heating loads precisely, using proper design temperatures which stipulate the outdoor design conditions for which certain indoor temperature conditions are to be maintained?
- Have you plotted or calculated the balance point of the system and sized the unit for the supplemental heating required for winter use?
- Have you taken the individual rooms CFMs from load form (recommended CFM being 450 CFM per ton)?
- Have you designed the supply and return duct systems for low velocity systems on the basis of .10 and .08 inches loss per 100 feet respectively?
- Have you selected the proper diffusers and registers and are they located properly?
- Does the fan section of the system provide enough available static pressure to overcome all system losses, i.e., ductwork, filters, heaters, coils and outlets?
- Is the ductwork insulated and vapor sealed?
- Have you used flex connectors at either side of the unit at junctions with ductwork to minimize transmission of vibration and noise?
- For split system applications, have you located properly the outdoor units and have you properly sized tubing lengths or piping runs and insulated them accordingly?
- Does the building electrical service meet the equipment requirements in voltage, ampacity, and phase?
- Are the thermostatic controls staged to implement the use of supplemental heaters when required?
- Is there automatic emergency heat provision for back-up for emergency heat operation?

CHECK LIST FOR TROUBLESHOOTING A SELF-CONTAINED, AIR-TO-AIR HEAT PUMP

Trouble	Potential Cause	Investigate
Unit won't start.	No electrical power.	Check circuit breaker or fused disconnect switch.
	Contacts welded or defective switch or thermostat.	Check contactor & wiring. Take voltage readings to test. Replace contactor if defective.
No cooling while in the cooling mode.	Thermostatic control defective.	Operate to test or jump out to test. Replace if defective.
	Loss of refrigerant.	Check liquid refrigerant line sight glass for bubbles. Repair leak & recharge refrigerant system.
	Low ambient temperatures causing pressure switch to trip out on low pressure.	Add low-ambient control kit and/or damper if unit must operate under conditions of low outside air temperatures.
	Outside condenser coil dirty or clogged, pressure switch causing tripping out on high head pressure.	Clean outside condenser coil. Check, clean & lubricate fan motor.
	Drop in power source line voltage.	Verify incoming line voltage. If more than 10% below nameplate rating, contact utility company or electrician.
	Motor-compressor defective.	Verify starting & operating compressor current & compare against rated values. Listen for unusual noisy operation or none at all. Replace or rebuild compressor accordingly.
	Unit is in heating mode.	Check control setting. If outside discharge air is cold, check reversing valve. May be stuck.

(continued)

CHECK LIST FOR TROUBLESHOOTING A SELF-CONTAINED, AIR-TO-AIR HEAT PUMP (*continued*)

Trouble	Potential Cause	Investigate
Low air flow.	Evaporator fan scroll dirty or damaged.	Clean blower & lubricate fan & motor or replace as necessary.
	Filters are dirty.	Replace filters.
	Inside evaporator coil dirty or clogged.	Examine & clean evaporator coil.
	Formation of ice on evaporator coil. Possibly caused by insufficient evaporator air flow or low refrigerant charge.	Verify free air flow thru evaporator section. Operate unit on "fan" only to melt ice. Veryify refrigerant charge & test
No heat although in heating mode.	Thermostatic control defective.	Operate to test or jump out to test. Replace if defective.
	Loss of refrigerant.	Check liquid refrigerant line sight glass for bubbles. Repair leak and recharge refrigerant system.
	Motor-compressor defective.	Verify starting & operating compressor current & compare against rated values. Listen for unusual noisy operation or none at all. Replace or rebuilt compressor accordingly.
	Inside coil dirty, pressure switch causing trip out on high pressure.	Clean inside coil & louver. Check, clean & lubricate fan motor.
	Low line voltage.	Verify incoming line voltage. If more than 10% below nameplate rating, contact utility company or electrician.
	Reversing valve stuck in "cool" position for defrost cycle.	Possible defective defrost sensor & reversing solenoid valve.
	Low outdoor temperatures.	Verify operation of auxiliary electric heater (if existing). Add heater if required.

Trouble	Potential Cause	Investigate
Water leaks inside.	Condensate water is not being drained or discharged. Drain pan leaking or improper pitch.	Check pan & piping for free passage of condensate water.
	Wind-driven rain penetrates through to inside.	Check drain pan for proper pitch. Check outside louver. Check seal gasketing.
Compressor cycles excessively.	Internal compressor protector overheating, due to high head pressure or loss of refrigerant or low line voltage or defective sensor.	Verify compressor voltage & current readings & compare with rated values. Check condenser coil & fan. Test for refrigerant leaks. Purge & recharge refrigerant, add line filter/dryer.
Excessive cooling.	Thermostatic control defective.	Operate to test or jump out to test. Replace if defective.
	Contacts or defective switch.	Check contactor & wiring. Take voltage readings to test. Replace contactor if defective.
Excessive noise.	Fan not balanced or hitting housing or worn bearings.	Inspect & clean fan housing & motor. Lubricate fan & bearings. Replace fan assembly if necessary.
	Sheet metal vibration.	Tighten all loose fasteners of supports & covers. Add fasteners & gasketing if necessary. Brace ductwork. Verify flexible connections.
Conditioned space air uncomfortable	Insufficient air distribution; no or inadequate fresh air.	Verify supply air delivery to space; verify if fresh air louvers closed or blocked.

Section 5: OPAC—OFF-PEAK AIR CONDITIONING

Calvin D. MacCracken

Off-peak air conditioning can save utilities from needing new generating plants by using their idle night capacity. This approach to air conditioning requires certain changes:

- Thermal storage must change from refrigeration to HVAC, totally different contractors and equipment (package, not built-up).
- All kinds of buildings must be adapted to thermal storage because the utilities' peak loads are broad based.
- Cost of equipment must decrease because much longer cooling periods are required (i.e., offices: 10 hours, malls: 14 hours).
- Off-peak kW hour energy rates and kW demand rates must be structured to provide incentive to the customer.
- Low maintenance and high reliability and predictability must be assured.

This has all been done in a variety of ways for commercial applications, but industrial and residential uses (with their 24-hour loads) have not been generally practical. Manufacturing and installations are growing rapidly worldwide wherever utilities have been forced to provide incentives due to inadequate capacity. The name *thermal energy storage* has been widely used by engineers, but *off-peak air conditioning* (OPAC) is now preferred by many because it is more easily understandable and describes what the system is in recognizable terms.

TYPES OF SYSTEMS

Chilled Water Storage

The first system to be established in the HVAC field was the storage of chilled water in large tanks, usually underground. Dallas and Toronto were the leading cities. Originally there were problems in keeping the warmer water returning from the fancoils separate from the chilled water being pumped out, so diaphragms, multiple tanks, and empty tanks were tried. Later research sponsored by the Electric Power Research Institute (EPRI) proved that separation (known as a "thermocline") of layers of warm and chilled water differing by 15°F to 20°F could be maintained by careful low velocity layers of water entering and leaving the tank.

Chilled water tanks require 5 to 10 times the volume of equivalent ice storage and thus must be site built, sometimes in basements made of concrete or constructed of steel round above-ground tanks. Flow must reverse direction between charging with chilled water and discharging when warmer water returns. This warmer water is very slightly lighter than the 40°F to 42°F chilled water and thus is kept on top requiring reversal of flow with valves. Chilled water storage is primarily used on very large jobs. An example is a 3,000,000 gallon tank for the Stanford University campus. Standard chillers provide the chilled water off-peak. The tank is open so a heat exchanger is often used to avoid pump losses in multistory buildings. Water treatment is required.

Brine Solid Ice

The most widely used system is modular plastic tanks, either round or rectangular, with a uniformly distributed plastic tube heat exchanger with extensive tube surface. These tanks serve as an accessory to a chiller which is designed to handle ice freezing temperatures. The brine is 25% ethylene glycol (like car antifreeze) and 75% water to protect it from freezing above 12°F. The chiller cools the brine to about 25°F during the charging freeze cycle with return from the tank heat exchanger at about 31°F. The ice freezes about 0.5″ thick on a tube before joining ice frozen on the adjoining tube. Expansion of the water vertically upward takes place as ice is built uniformly throughout because of counterflow temperature averaging.

Ice is melted from the tube side by the warmer fancoil return liquid passing through the tube heat exchanger, cooling down to 32°F and rising a few degrees as the discharge is completed. (See partial storage under the section on "Sizing and Strategy.") There are no moving parts, the polyethylene and polypropylene are noncorrodible, the water that freezes and melts never moves, and the brine system is closed and handles 90 PSI. The heat exchanger takes up 10% to 15% of the tank volume. It is highly efficient, reliable, and uncomplicated. The tanks are externally insulated and hold from 100 to 500 ton-hours capacity.

Brine Encapsulated Systems

Water may be encapsulated in round plastic balls or flat trays with cold brine pumped around them within a pressured tank. The balls are factory filled and trays are usually site filled. Different methods are used to account for the 9% expansion when ice freezes.

Brine External Melt

This system is similar to the brine solid ice system, except metal tubes are used and the ice is melted from the outside since only half the water is frozen. The remaining water is pumped to the fancoils instead of the brine. A heat exchanger may be used if the brine system must supply cooling at the same time. The water side is open to atmosphere and thus may need a heat exchanger also. A finned tube heat exchanger also is used to assure low temperature throughout the discharge. The system may incorporate a packaged chiller and controls.

Harvesters

Vertical evaporator plates or tubes suspended over an open top cylindrical ice tank are used to freeze part of a stream of water falling down over the evaporator plates. When the ice builds up to about ¼″ thick, it is dropped into the tank by releasing the ice/plate bond by heating or chemicals. The ice is made on the evaporator but stored in the tank. Uneven distribution of ice in the tank has been a problem causing rising discharge temperatures. A heat exchanger is usually required to act as a closed system.

The plates are harvested in series so as to keep a balanced ice production. Each plate has its own refrigeration DX valve and solenoid controls. Several manufacturers are in the field.

Water/Eutectic System

Encapsulated stackable trays are filled with a Glauber Salt eutectic that freezes and melts in the 45°F to 48°F range with a capacity of 41 Btu/lb and 1.5 specific gravity compared to water at 144 Btu/lb. and 1.0 gravity. These eutectic trays are hand stacked in swimming-pool-shaped rectangular tanks with unpressurized water flowing around them. Brine is not usually needed because of the high fusion point. Flow velocities are low with high residence time.

A key application is retrofitting old centrifugal compressors that cannot reach ice-making temperatures. Costs are high because of increased volume of eutectic trays to reach capacity.

Ice Builders (Ice on Coil)

The dairy type ice system used for over 60 years has refrigerant in metal evaporator pipes in a rectangular water tank with ice stored in place on the pipes. The 50% water not frozen is pumped to the fancoils during cooling. This is rarely used in the HVAC field.

WORKING WITH THE UTILITIES

The utilities have hesitated so long to construct new generating plants that in many areas they will be out of power before they can build new ones. Orders for small gas turbine generators running on natural gas are mounting but this will not fill the bill and will raise energy cost. Therefore, load shifting, peak cutting, and conservation are their options. Only shifting keeps them from losing business while reducing their peak.

Therefore, the utilities have adopted the carrot-and-stick approach. Recently, state public utility commissions are getting more active and other options will occur, such as selling large blocks of peak power to be used at night at low rates to financial investors who will pay a penalty if not done.

For example, utilities have offered the following incentives to encourage off-peak air conditioning:

- Seminar—publicity
- Demonstration job
- Avoided KW subsidy
- Engineering fee support
- Marketing
- Lease program

When these "carrots" fail, utilities have tried the following measures:

- Off-peak KWH, required
- High KW rate
- Full storage window
- Storage rate—ratchet
- Storage mandate
- Connected KW limit

Urban utilities are all summer peaking as are southern rural producers. Northern rural generators are winter peaking but see the summer peak growing. All over the world air conditioning is growing rapidly. Only with OPAC can the problem be handled.

The Electric Power Research Institute (EPRI) has financed over 35 research contracts to aid OPAC paid for by member utilities and more are coming. (See the section on "Cold Air Systems.")

SIZING AND STRATEGY OF OPAC SYSTEMS

Sizing of OPAC systems is very different from non-storage commercial building cooling systems. It is affected by the shape of the load curve hour-by-hour, how much of the load is to be shifted, and, to a lesser degree, what temperature is needed at the coil in the duct system. If a building has 10 hours of required cooling like most office buildings, the size of the storage and the chiller is much different than with 14 hours of cooling, as in shopping malls.

Computer programs are available that provide sizing of storage, chiller (compressor), liquid flow rate, and ice consumption hourly on several different arrangements. Full storage involves the compressor only operating during unoccupied (or off-peak) hours, shifting the full compressor demand and energy load. For office buildings with 10 hours of cooling, the ice chiller will be of conventional (non-storage) size, and storage to cool for the full 10 hours must be added causing higher cost.

Partial storage is far more popular because smaller size with lower cost is achieved by arranging for the chiller to assist storage during peak cooling periods. Often the chiller is about half size or less with storage making up the balance. In mild weather storage may not be used at all, being only a back-up. Figures 7-12 and 7-13 illustrate partial and full storage.

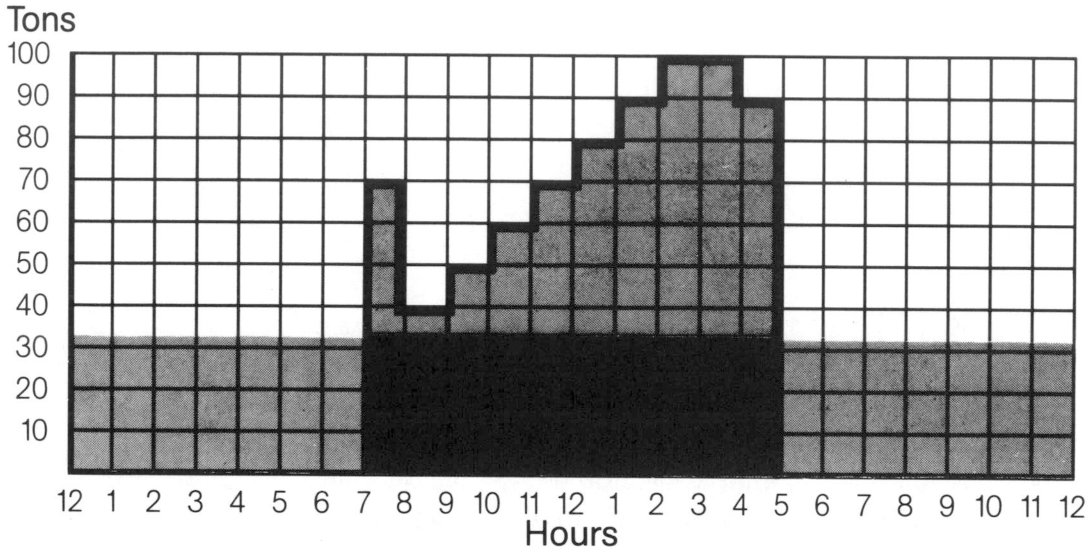

Figure 7-12. Partial-Storage OPAC System

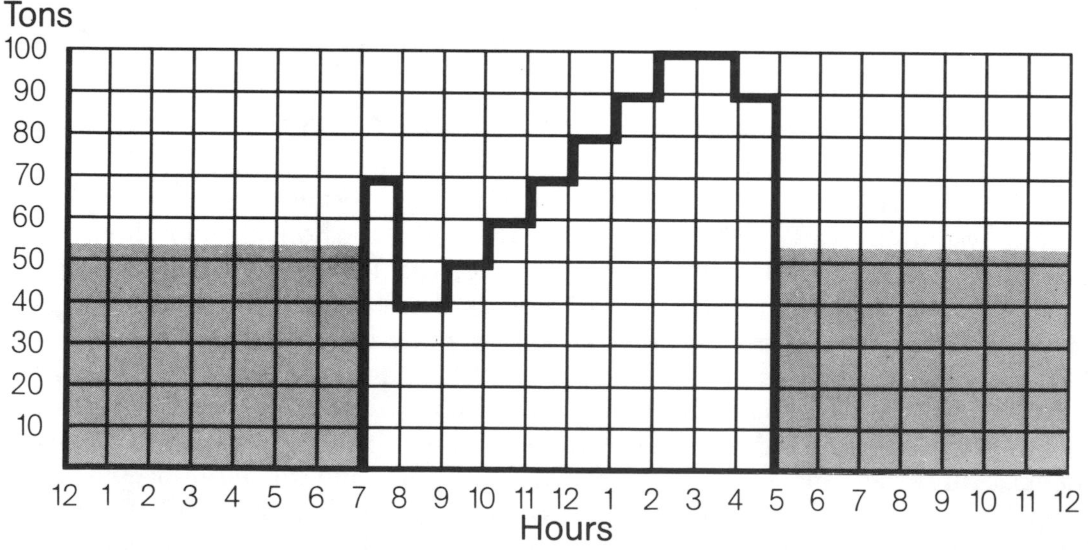

Figure 7-13. Full Storage OPAC System

When ice is used as the back-up, it is called chiller priority strategy and is the most efficient, because the ice making is minimal. When the chiller is the back-up, it is called ice priority and, although less efficient, it is often used when off-peak energy rates are more than 25% lower than day rates. Setting and selecting the controls for these options (partial, full or window storage, and chiller or ice priority) requires special care. For larger systems, an automatic control is available and will be provided by chiller manufacturers in the near future.

INSTALLATION AND MAINTENANCE

There is so much difference in the three major types of OPAC systems that you must obtain installation instructions from the manufacturer. Some require a great deal of site work, some relatively little. Some can be located on roofs, whereas others must be on ground level. Some have their own combined refrigeration system, and others must be combined with chillers supplied by a different vendor. Some are unitary packages, and others require filling and assembly of thousands of parts on site. And finally, there are fully buried storage units where maintenance is almost prohibitive and installation is a very final step and is split among several trades.

Regardless of what type or make of system—whether ice, chilled water, or eutectic—a final complete check-out, called commissioning, is essential. Far too many systems of all types do not operate as intended because of installation errors that were not caught, often for years. Commissioning should be done by a third party not involved in the construction.

Piping, valves, pumps, and pipe insulation are a major part of the installer's responsibility, and later the maintenance work. Temperature modulating valves which control the bypass mixing and blended temperature to the load coils must operate correctly and should be checked for proper operation during early runs. The piping and insulation are not much different from a standard chilled water system except that insulation needs to be thicker and vapor tight because of the lower temperatures and condensation caused by ice, and it often needs to be weather proof when used at outdoor locations.

Water Treatment to Prevent Corrosion

Often pipe can be PVC, particularly the outdoor runs, covered by closed-cell insulation and an outer aluminum wrapper. Stainless steel is rarely used both because of cost and also it being in most cases a closed system. Where open systems are used, such as with brine external melt ice builders, eutectics, and harvesters, water treatment is necessary to prevent corrosion. Water treatment is seldom done with cooling towers of 100 tons or smaller, so water must be continuously replaced by make-up water in a process called blowdown. You are accustomed to seeing that this blowdown is being used, adjusting the volume of make-up water flow to keep the tower basin clean, and vacuuming sediment from the basin as needed. This same process should be followed with the cold water of open OPAC systems unless it is large enough to warrant chemical water treatment. As with cooling towers, chlorine is used for killing off bacteria and other pollutants. Chromates are no longer permitted by the EPA, so alternative corrosion inhibitors such as molybdates are being applied. This is often handled once a month by special water treatment service companies.

Ozone units for water treatment, which work electrically without chemicals, are coming on the market. They require checking twice a week to insure that there is compressed air pressure per manufacturers' instructions, that the ORP (oxidation reduction potential) reading is within indicated limits, and that the pump is running. Observe the appearance of the water to insure its cleanliness and transparency. Vacuuming from the tank bottom, or draining and flushing if vacuuming cannot be done, is needed monthly.

Brine, in the HVAC/ASHRAE world, refers to any antifreeze solution, formally called *secondary refrigerant.* Brine systems are the most popular, and ethylene glycol in a 25% solution with water is the most common. The inhibited type ethylene glycol should be used, such as Dow Chemical's SR-1 or Union Carbide's UCARtherm. This provides against corrosion or foaming problems. The freezing point is about 12°F. To measure it, use a refractometer, and check it periodically and certainly if any leak occurs with subsequent refilling of the system. It is biodegradable, so spills into the ground or sewer system are not a problem. Mixing with potable water, however, is a problem because it is toxic and can cause death if ingested in sufficient quantity. It has a distinct unpleasant taste so no deaths have been recorded except in rare cases where it was mixed with wine.

Ethylene glycol is used also as the antifreeze in automobile cooling systems, but with different additives for sealing leaks and inhibiting corrosion. The two key manufacturers both provide free inspections of your system and will give instructions or provide needed laboratory tests to insure your system performs correctly. You should take advantage of this free offer yearly.

Propylene glycol is non-toxic, is about the same price, and similar in most ways to ethylene glycol except that a 50% heavier pump is required due to higher viscosity.

The brine solid ice system has stagnant water which freezes and melts but never moves out of its plastic non-corrodible tank. However, algae can grow over a period of time. Add a small envelope or two of a chlorine based algicide, such as Algex for swimming pool treatment, at start-up, two year intervals, and any time the water is replaced. Also check the water level through the fill port on a yearly basis, and add or siphon out water per manufacturers' instructions.

In valving, motorized valves are better in large systems because of their slow closing, compared to electric solenoid valves with a fast action that can cause water hammer or pressure spiking. In general, two-way valves are preferable to three-way when possible, even if it takes two of them. Make sure in all cases that water hammer and consequent pressure spiking is not occurring.

Using a Heat Exchanger to Circulate Pressurized Water

Only brine systems are closed systems where the heat transfer fluid is pressure tight. Ice builders, harvesters, eutectic, external melt brine are open unpressurized systems. In most cases, a heat exchanger is applied so that pressurized water can be circulated with the energy saving benefits when multistories are involved. In a closed system, the excess pressure required to pump up to elevated areas is recovered when it returns back down to the heat exchanger. Open systems have used hold-back valves in the return line or pressure recovery turbine pumps to overcome the losses. Heat exchangers are an easier answer but the temperature differentials and costs are a deterrent.

Brine systems can hold enough pressure to cover ten or more stories without a heat exchanger, so it is rarely used. Concern about glycol brine circulating through the building piping has been largely overcome. In large multibuilding campuses or airports, heat exchangers are used to reduce the cost of the glycol.

Preventing Overfreezing

Because it takes 144 Btu's to freeze a pound of water to ice at 32°F and only 0.5 Btu's per pound is gained for each degree below 32°F, there is no gain to overfreezing or reducing the ice temperature below its freezing point. In fact, the compressor loses efficiency as temperature decreases. Set the thermostatic control to avoid wasteful operation at temperatures lower than necessary.

During freezing, some brine systems locate the expansion water above the heat exchanger area and do not intend to freeze it. If the return brine from the ice tank to the chiller goes down to 25°F or lower, this water may freeze and form an ice cap that will not melt during the discharge cycle—the ice being trapped above the heat exchanger. This leads to a lack of water in the heat exchange area with impaired performance. It can easily be melted off by blowing hot air from a hair dryer through the fill hole. However, be sure to set the thermostat so brine from the tank never goes below 28°F.

Measuring Ice Inventory

The amount of ice available at any time, called ice inventory, is important to the control of an OPAC system so that the ice will be utilized most efficiently. Check the measurement devices monthly and calibrate them at the fully charged or fully discharged (all ice or all water) condition.

There are two types of water level indicators: float and pneumatic. Floats can be used on ice-on-coil ice builders and brine systems. Harvesters and eutectic systems require calculation or Btu meters from the fully charged condition. Pneumatic measurement of liquid level is used on brine solid ice and couples directly to a computer.

The level is measured by the air pressure in inches of water required to bubble air from a submerged tube. Maintenance is limited to checking proper location depth of the tube-end at the water level when fully melted.

USING COLD AIR SYSTEMS

It has long been supposed that it was more efficient to operate an air conditioning system at higher duct air temperatures, such as 55°F or 60°F, because of the compressor performance. However, that was assuming that the blower or fan was giving the same flow (CFM) in either case. In commercial or industrial buildings, the CFM and liquid flow (GPM) can be reduced as the duct air temperature is reduced, with an overall savings in energy and electrical demand.

Because the duct coil will be colder, more water is condensed out and the relative humidity is also reduced. The colder air, usually around 45°F, is mixed with room air blending up to 55°F to 60°F prior to entering the conditioned space to prevent drafts and give more recirculation in the room.

Guidebooks and manuals on cold air systems such as EPRI EM5447 are available at no charge from utilities who are members of the Electric Power Research Institute (EPRI), which has done extensive study on this new technology. They explain the performance and maintenance requirements. The fan-powered VAV boxes in the ceiling spaces require inspection and occasional maintenance on the fan motors, fans, and valve operators.

Cold air systems reduce costs of fans, fan motors, pumps, piping, and air ducts. They also reduce the drop-ceiling space so that 6″ to 12″ may be saved per floor and save horizontal space due to smaller risers. Insulation should be moderately better than 55°F ducts but, because the whole building is dryer, condensation on the ducts is rarely a problem.

The lower relative humidity (about 35% instead of 60%) has been very popular with building occupants. They report improved comfort, improved perception of indoor air quality, better alertness, and increased productivity. In addition, the thermostat may be set at a higher temperature (by 3°F or 4°F) for the same comfort level, a major energy saver. It also has been found that the chillers may be started later at the morning start-up because the humidity level can be pulled down much faster than the dry bulb (wall thermometer) temperature. For example, a building in early morning at 82°F and 70% R.H. will pull down to 78°F and 35% R.H. much faster than to 74°F and 60% R.H.

To achieve much colder air such as 35°F to 38°F, you can reduce the freezing point of the water from 32°F to 28°F by adding a patented salt mixture which forms a stable eutectic (constant freezing point). The salt constitutes only 5% of the weight (2% of the volume). This requires a small stirring pump to run during the freezing operation to prevent settling. For maintenance, check the freezing point as well as pump operation. If it has gone up to 30°F or above, check the stirring pump for cleaning out and proper operation.

SOLVING PROBLEMS IN OPERATION OF OPAC SYSTEMS

Chilled Water

The melting of ice or the stratification of warm water over colder water in a tank involve careful balancing of flow. Some types are more sensitive than others.

Chilled water stored in a tank requires that the warmer water returning from the duct coils (fan coils) be kept separate from the colder storage water. If it mixes, the water to the coils is warmer and cooling is cut down with a consequent energy loss. Many methods were tried as described before, with the low velocity thermocline winning out. When charging (cooling the tank), cool water must enter the bottom of the tank and gradually fill it from the bottom with the upper level warmer water going out to the chiller. When discharging (cooling the building), cool water must leave the bottom of the tank. This obviously requires reversing valves operating each time there is a shift of function. It is important these valves operate well in response to controls.

Also, big underground concrete tanks are prone to cracking. Check water inventory, perhaps by measuring the flow of city make-up water. EPRI has a guidebook on chilled water storage and tanks.

Harvesters

When ice is harvested (shucked) into an open tank below, partially filled with water, the distribution of the ice as it accumulates may be uneven. This often is a maintenance or service problem and may call for oversizing.

This unevenness should be observed and, if not easily correctable, reported to the manufacturer. Harvesters have multiple freezing plates which are cycled in sequence requiring multiple valves and an expansion valve for each plate. Timers for defrost cycles also are sequenced and must be checked to see that all are operating. Each of the freezing plates is flooded with falling water from holes or slots in water distribution trays above. You should check these orifices weekly to make sure there is no blockage from dirt and clean them out. Also any filter should be cleaned.

Ice on Coil Ice Builders

Air bubbling is often used to facilitate uniform melting of the ice on the refrigerant pipe coils. Nevertheless bridging over the ice from one coil to another, blocking water flow, occurs. This requires full meltdown and recharging. Sometimes this is done weekly. The thermostatic control to stop ice building before bridging over should be checked as well.

Brine Encapsulated Systems

Unbalancing on the inside of pressure tanks of the flow around balls or trays of ice can occur but is difficult to identify. Analysis of performance data may be necessary, such as watching the rise in temperature during discharge versus time. This curve should be relatively flat. Also subcooling has caused problems with initial freezing points being well below normal in some ice balls. If this is suspected because of unusually low temperatures, remove a few containers and measure the freezing point. Be sure to melt it fully beforehand. If it is subcooling, contact manufacturer.

Brine Solid Ice

Because bridging over of ice is intended, this is not a problem. You can check flow balancing between modular tanks connected by parallel headering by inspecting through the cover fill cap to see that expansion of the water level above the ice is the same in all tanks.

STORAGE LOCATION

Storage tanks come in a wide variety of sizes and weights. The smallest and lightest are the brine freeze solid type, and the largest and heaviest are chilled water storage tanks. Location might include roof, basement, pad, partially buried, fully buried, double tiered, separate floors, and retrofit.

Modular brine tanks can be used in most all of these locations and have been. Full burial has a limitation on depth of soil cover of one foot. Modular tanks can be fitted into irregular spaces for retrofit but in new construction are usually located on outdoor pads, basements, roof, or partial burial. It is wise to arrange for a drainage path or berm in case a leak should occur.

Chilled water tanks are generally of two types: welded steel above-ground and concrete in-ground. Because of their very large size, they are not well adapted for retrofit, roof, fully buried or separate floors.

Harvesters are usually at ground level, indoor or out, and often require excavation for the single large tank.

Encapsulated water/ice balls or trays are usually at ground level or buried, and appear as horizontal cylindrical pressurized steel tanks, sometimes more than one per job. They are loaded on site.

Package units containing chiller and storage also can be located in many spots except buried.

Rooftop Conversions

A growing new retrofit is to convert a group of rooftop air conditioners into a central ice storage system and abandon the old corroded rooftop compressors and condenser. This requires the coils being converted from DX (direct refrigerant) to brine. The conversion can be done easily by an experienced torch user by following instructions in a guide supplied by Calmac Manufacturing Corporation.

The insulated plastic brine pipes can be run across the roof from the outdoor air-cooled chiller and storage tanks mounted on the roof or ground. A modulating valve at each rooftop provides for temperature control.

Costs are lower than removing old rooftops which requires changing curbs and ducts to fit new ones, and using a crane again to mount the new ones. Also roof insurance bonds are preserved. The ice storage and efficient central chiller provide significant energy, demand, and operating cost savings.

THE FUTURE OF OPAC SYSTEMS

Massive growth of OPAC is inevitable. Led by the demand for air conditioning, electric utilities will use up their reserve capacity within a few years, some sooner than others. Every other system of cutting the electrical load peaks in hot summer afternoons and evenings takes away business from the electric utility which forces them to raise rates.

Building the night load while the reducing peak load increases the load factor, evens out day and night operation, uses the more efficient and lower cost base load (hydro, coal, nuclear) and reduces oil and gas consumption.

The other amazing advantages to remember are:

- Downsizing to reduce cost.
- Space saving with cold air systems.
- Energy reduction in many ways.
- Infinite modulation to match any load.
- Redundancy back-up during service.
- Unanticipated loads such as special events.
- Constant optimum compressor performance.
- Free winter heating from heat recovery.
- Greater comfort from lower humidity.
- Higher reliability from less moving parts.
- Keeping up with state-of-the-art technology.

Table of Check Points & Maintenance	Recommended Solution
1. Prohibitive service on fully buried storage.	Consider partial burial with 1 ft. above ground.
2. Commissioning essential.	Carefully check that specifications are fully met and system tested.
3. Temperature modulating bypass (blending) valves require early checkout.	Check temperatures at both entering pipes and outlet blended pipe.
4. Vapor tight pipe insulation.	Insulators must provide vapor proof insulation to prevent condensation.
5. Water treatment in open systems.	Open systems, such as chilled water, harvester, and ice-on-refrigerant coil, should be properly treated by chemicals or ozonation to prevent corrosion and bacteria growth.
6. Cleaning out storage tanks and tower basins.	Mineral deposits or loose precipitants in chilled water tanks and tower basins should be cleaned manually or with swimming pool vacuum cleaners.
7. Ozonated water purification needs special attention.	Ozonation for cooling towers should be covered by a service contract.
8. Free checkout of glycol in brine systems.	The glycol antifreeze in closed systems should be sampled and analyzed by vendor every five years.
9. Corrosion of dissimilar metals in salt brine systems.	Salt brines, used to save cost, require yearly inspection of pipe interiors, particularly near joints of dissimilar metals, use of dialectric unions, and maintenance of inhibitor.
10. Algicide need in solid ice storage.	In systems where water is stationary a packet of swimming pool algicide should be added every 3 years.
11. Water level yearly checks.	Water level in stationary water systems usually does not change, but add water yearly if needed.
12. Water hammer and pressure spiking.	Occasionally water hammer may occur from air in system or sudden valve closures. A pressure spike telltale gauge can be used to check this. Air should be purged from system. Valves should be slow acting.

(continued)

Table of Check Points & Maintenance	Recommended Solution
13. Overfreezing can be a problem.	Compressors should automatically be turned off when ice is fully frozen. A 28°F return from storage is a typical lower limit.
14. Ice inventory meters require calibration.	Knowledge of available ice is often a control necessity. Meters should be checked at no ice and full ice points.
15. Cold air duct insulation.	Cold air ducts should be checked to see if any condensation appears. Additional insulation may be required.
16. Freezing point of low temperature eutectics with stirring pumps.	When freezing points lower than 32°F are utilized, salts are added to depress the freezing point. A stirring pump is used to keep the eutectic salt in suspension. These should be checked periodically for operation.
17. Chilled water storage reversing valves.	Chilled water storage tanks require water *removed* from tank near bottom when providing cooling and water *entering* the same location when storing cooling (charging). Valve reversal operation should be checked.
18. Leaks in concrete tanks.	Concrete in-ground tanks are susceptible to cracks and leaking in early stages. Water levels should be observed and the contractor called if drop in level is found.
19. Ice distribution in harvesters and oversizing.	The angle of repose of ice chips in a harvester tank may limit the storage capacity and require more tank volume. This angle should be observed and contractor notified.
20. Cleaning of harvester water trays.	Water flow distribution over harvester vertical icing plates is controlled by gravity flow through an orifice. Anything blocking these holes should be cleaned periodically.

Section 6: MODULAR LIQUID CHILLERS

Michael Heit

A new concept in water chillers has recently been developed which may be the ideal water chiller for replacement, building renovation or industrial process purposes. It is a compact modular water chiller that has the following features:

- can provide easy equipment installation plus flexibility, so capacity can be expanded or contracted to meet building or process load requirements.
- can reduce energy costs by operating at peak efficiency even under low load requirements.
- uses an environmentally safe refrigerant.
- can be installed without cranes or special rigging.
- will pass through narrow doorways and up and down standard stairways and passenger elevators.
- can be maintained without proprietary training.

In the short time since the modular chiller has been available in the U.S., it has solved virtually all types of water chiller problems associated with replacement, renovation, and retrofit building comfort systems. See Figures 7-14 and 7-15.

The nature of the modular water chiller's design, coupled with hermetic shell compressors, results in very simple and uncomplicated maintenance procedures, when compared to conventional chillers.

III MULTISTACK™ Courtesy of Multistack Inc.

Figure 7-14. Large Tonnage Modular Chiller

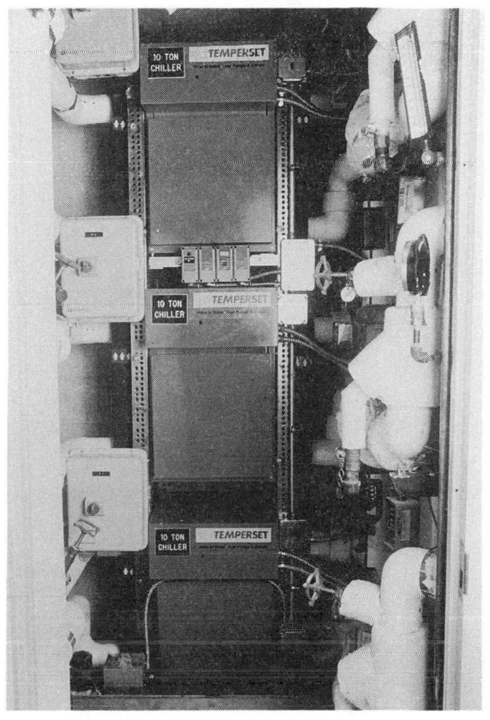

TEMPERSET Courtesy of Temperset

Figure 7-15. Small Tonnage Modular Chiller

MAINTENANCE PROCEDURES FOLLOWING INITIAL START-UP

One week after the initial start-up, back flush both the condenser and evaporator headers using the back flushing valves provided on the victaulic end caps. This back flush will clean out any residue created by the agitation caused by the start-up of water flow through the piping system.

Also check the mesh media in the condenser and evaporator strainer (or other filtration device) for blockage.

PERFORMING THE ANNUAL CHECKUP

Test all refrigeration circuit high and low pressure switches annually to determine that they function properly at the setting specified. The high pressure switch should trip at 290 psig, and the low pressure at 25 psig. Pressure can be observed on the gauges, and the faults can be induced by reducing the flow rates over the condenser and evaporator heat exchangers.

Observe the level of oil in the compressor through the oil sight glass located at the bottom of the compressor. If low, add oil until it is observed through the sight glass. Use Zerol-150 or equivalent for replacement.

Check the amount of superheat created by the thermal expansion valve. This can be done by observing the TX and suction temperatures recorded when the status of each circuit is called up on the microprocessor controller. Superheat will average about 3°F.

Condenser water treatment is important to the efficient operation of the heat exchanger. Therefore, it is recommended that a water treatment program be instituted on an ongoing basis.

SEASONAL START-UP PROCEDURES

Before starting the chiller, be sure to recheck the strainer media to insure that it is clean.

The backup battery, typically included to retain operating settings in memory, is located in the microprocessor drawer and should be checked and replaced if required (3 volt Duracell DL½N).

SEASONAL SHUTDOWN PROCEDURES

Be sure to leave the chiller disconnect "on" during the off season. This will keep the backup battery from draining.

The condenser and evaporator heat exchangers should be back flushed at shutdown.

There is no recommended maintenance action required on a 10,000 or 40,000 hour basis. The modular chiller is designed to provide long life with a minimum of maintenance and no tear-down inspection requirements.

BACK FLUSH PROCEDURE CHECKLIST

Note: The recommended back flush method is to use a cleaning solution and solution pump to recirculate the cleaning solution through the heat exchangers until they are thoroughly cleaned. An alternate method using the water system static head in lieu of a pump and without cleaning solutions follows.

On the *evaporator side,* make sure that the power to the chilled water pumps is off. This method uses the static head from the chilled water loop.

1. Close the water isolation valves.
2. Drain the evaporator heat exchangers by partially opening the lower drain valve.

3. Remove the upper end cap.
4. Close the drain valve.
5. Fully open the leaving chilled water valve for 3 to 5 seconds, then close it.
6. Repeat step 5 as many times as needed to clear the evaporator heat exchangers.
7. Replace the end cap.
8. Refill the evaporator heat exchangers by slowly opening the leaving chilled water valve and vent by partially opening the upper drain valve.

See Figure 7-16, which illustrates the back flush procedure on the evaporator side.

On the *condenser side,* make sure that the power to the condenser water pumps is shut off. This method uses the static head from the cooling tower/condenser water system.

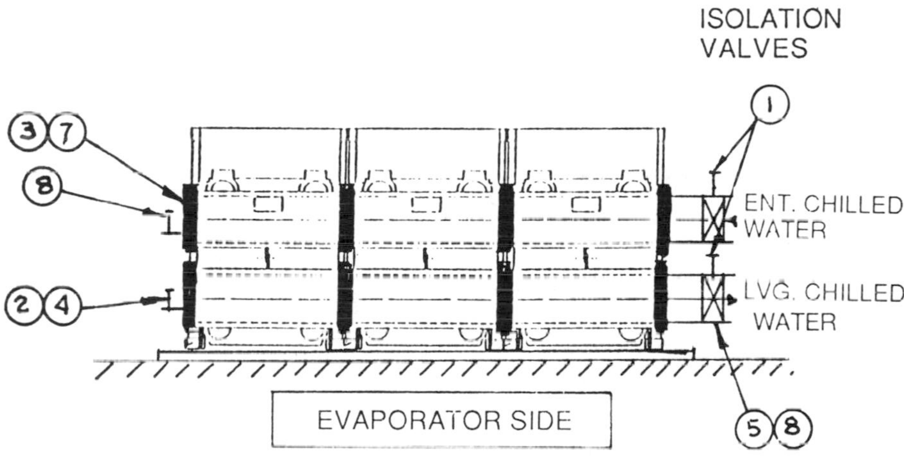

Figure 7-16. Back Flush Procedure, Evaporator Side

1. Close the water isolation valves.
2. Drain the condenser heat exchangers by partially opening the lower drain valve.
3. Remove the lower end cap.
4. Fully open the leaving condenser water valve for 3 to 5 seconds, then close.
5. Repeat step 5 as many times as needed to clear the condenser heat exchangers.
6. Replace the end cap.
7. Refill the condenser heat exchangers by slowly opening the entering condenser water valve and vent by partially opening the upper drain valve.

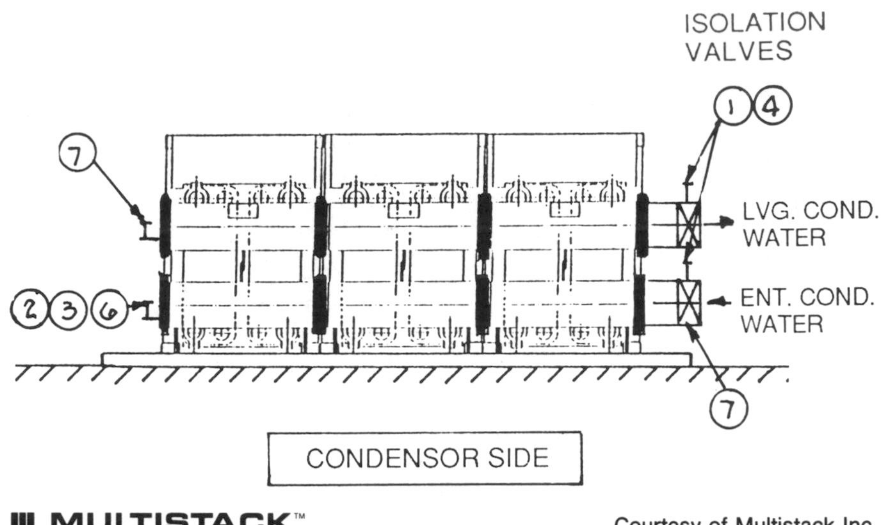

Figure 7-17. Back Flush Procedure, Condenser Side

DETERMINING WHICH DETERGENT OR ACID CLEANING MATERIAL TO USE

Flush-out 624L (manufactured by Ashland Chemical Co.), or equivalent, is a combination of iron sequestrants, dispersants and wetting agents which will remove mill scale, grease, oil, and old corrosion products. It is recommended for pre-cleaning of new systems as well as for older systems that have accumulated corrosion products and other debris from neglect or improper treatment. Any deposit is detrimental to the proper functioning of the system. It can lead to restriction of flow, under-deposit corrosion and impaired heat transfer.

Flush-out 624L or equivalent may conveniently be slug-fed through bypass feeder for closed systems and directly into tower basin for open systems. Avoid contact with skin and eyes. In case of contact, wash with clean water. If eye irritation persists, consult physician.

A dosage of 3 to 6 gallons of Flush-out 624L or equivalent per 1000 gallons of water system is recommended when used for pre-cleaning and pretreatment of new systems. Water should be circulated for a minimum of 6 hours. A longer period of 24 to 48 hours would be beneficial. Old systems with heavy encrustations should use approximately twice as much treatment with a minimum circulation period of 24 hours.

In lieu of Flush-out 624L a 10% solution of phosphoric acid can be utilized.

8

Operation and Maintenance of Boiler Room and Heating Equipment

Raymond O. Combs

Chapter 8 covers the proper preventative maintenance of burners and boilers in order to optimize combustion efficiency and to provide proper safe operation of steam and hot water heating boilers. Specific recommendations are made for the tuning up of burners and the servicing and testing of safety controls. Significant fuel savings can be achieved by following the recommendations in this chapter.

TYPES OF FUEL

Solid Fuels

Solid fuels include soft and hard coal, wood, wastepaper and cardboard, and certain waste chemical solids and liquids. Each type is described in the following paragraphs.

Coal

The burning of coal as a primary fuel is currently more commonly limited to larger installations, such as power generating stations. The primary reasons for this limitation are:

- Coal is a high carbon content fuel; therefore, many impurities (such as sulphur) may be present in higher levels than in the more volatile liquid and gaseous fuels,
- the fuel is more difficult to prepare for combustion and requires more preparation to provide for complete combustion and frequently requires "after treatment" of the stack gases to meet air quality emission requirements.
- Problems associated with volume of storage and disposal of ash residue are major considerations to be examined when dealing with these fuels,
- materials handling is usually required for both supply to the fuel burning apparatus and for ash removal.

Equipment variations are required because stokers used in the combustion of coal are usually designed for either hard or soft coal. Few (if any) of these devices are suitable for combustion of both types of coal.

Wood

The burning of wood and sawdust is largely dependent upon availability and economy. Quality of the fuel may vary greatly depending upon the source and in some cases the seasonal affect (moisture content) of the fuel.

Waste Paper and Cardboard

These fuels, which are more volatile than the wood from which they are derived, are excellent sources of energy, however consideration must be given to long term availability and transportation of available supply and disposal of residue (ash).

Certain Waste Chemical Solids and Liquids

These rather specialized fuels require significant attention with respect to environmental aspects. Specific burners, with associated higher costs, are usually required to process and burn these fuels.

The use of waste products, both liquid and solid forms, will usually result in solid particulate, in the form of ash, on the heating surfaces of the boiler. Therefore, when considering these fuels, the cost of automatic cleaning equipment should be factored into the evaluation. A typical tube cleaner for solid fuel installation is manufactured by Fuel Efficiency, Inc., Clyde, New York. This system uses air as a cleaning medium and is highly recommended for all waste fuel applications, where deposits may be anticipated on heat exchanger surfaces. Automatic cleaners should also be considered when burning coal, and the heavier grades (no. 4, 5 and 6) of fuel oil.

Liquid Fuels

Commercial fuels are defined by commercial standard 12-48 or by ASTM D396-67. These liquid fuels are classified as follows.

No. 1 Fuel Oil

This is a distillate oil intended for burners with little ability to process fuel—this fuel is generally used in smaller appliances. It is normally used where cost is not a major factor. Its primary attraction is the ease of combustion: it is the most volatile of the common liquid fuels and is frequently referred to as kerosene.

No. 2 Fuel Oil

This is usually a distillate oil for general-purpose domestic heating. This fuel is becoming more accepted in commercial installations for firing rates up to 45 gallons per hour as a primary fuel, and more significantly as a stand by fuel in conjunction with other (gaseous or waste) fuels as a standby fuel. It is most favored for this purpose because the equipment required to prepare for combustion is comparatively low in cost and the fuel is easy to prepare for burning, requiring no pre heat or special handling. In addition, the fuel remains stable for reasonably long periods of time in storage, a significant factor in its use as a stand by fuel.

No. 4 Fuel Oil

This is a "heavier" grade of fuel oil—some preheating may be required for burning. Because this fuel contains a higher carbon ratio, it is more difficult to prepare for burning, frequently requiring pre heating for combustion, and in some cases (in colder climates) may require pre heating for transportation and in the storage tank.

No. 5 Fuel Oil

In some areas, two grades of this fuel are available—light No. 5 fuel and heavy No. 5 fuel. These fuels are in between the No. 4 and No. 6 and will always require pre heating for combustion and usually require significant amounts of pre heating for handling and storage.

No. 6 Fuel Oil

Sometimes referred to as Bunker C oil, this is the heaviest commercially defined oil. It will always require pre heating for combustion and for storage and handling. This fuel must be delivered at elevated temperatures, usually 90 to 120 degrees F., in order to be handled and pumped from delivery tanks.

Selecting Liquid Fuels

See Figure 8-1 for classification of fuel oils and Figure 8-2 for gravity and heating values of commercial fuel oil.

Property	ASTM Test Method[B]	No. 1	No. 2	Grade No. 4 (Light)	No. 4	No. 5 (Light)	No. 5 (Heavy)	No. 8
Flash Point °C (°F), min	D 93	38 (100)	38 (100)	38 (100)	55 (130)	55 (130)	55 (130)	60 (140)
Water and sediment, % vol, max	D 1796	0.05	0.05	(0.50)[C]	(0.50)[C]	(1.00)[C]	(1.00)[C]	(2.00)[C]
Distillation temperature °C (°F)	D 88							
10% vol recovered, max		215 (419)	—	—	—	—	—	—
90% vol recovered, min		—	282 (540)	—	—	—	—	—
max		288 (550)	338 (640)	—	—	—	—	—
Kinematic viscosity mm^2/s[D] at 40°C (104°F)	D 445							
min		1.3	1.9	1.9	>5.5	—	—	—
max		2.1	3.4	5.5	24.0[E]			
at 100°C (212°F)								
min		—	—	—	—	5.0	9.0	15.0
max		—	—	—	—	8.9[E]	14.9[E]	50.0[E]
Ramsbottom carbon residue on 10% distillation residue % mass, max	D 524	0.15	0.35	—	—	—	—	—
Ash, % mass, max	D 482	—	—	0.05	0.10	0.15	0.15	—
Sulfur, % mass max[F]	D 129	0.50	0.50	—	—	—	—	—
Copper strip corrosion rating, max, 3 h at 50°C (122°F)	D 130	No. 3	No. 3	—	—	—	—	—

Density, kg/m³ 15 °C	D 1298							
min (°API max)		—	—	>876[G] (30)	—	—	—	—
max (°APt min)		850 (35)	876 (30)	—	—	—	—	—
Pour Point °C (°F), max[H]	D 97	− 18 (9)	− 6 (28)	− 6 (28)	− 6 (28)	—	—	—

[A]It is the intent of these classifications that failure to meet any requirement of a given grades does not automatically place an oil in the next lower grade unless in fact it meets all requirements of the lower grade. However, to meet special operating conditions modifications of individual limiting requirements may be agreed upon among the purchaser, seller and manufacturer.

[B]The test methods indicated are the approved referee methods. Other acceptable methods are indicated in Section 2 and 5.1. The table values stated in SI units are to be regarded as standard. Values in parentheses are for information only.

[C]The amount of water by distillation by Test Method D 95 plus the sediment by extraction by Test Method D 473 shall not exceed the value shown in the table. For Grade No. 6 fuel oil, the amount of sediment by extraction shall not exceed 0.50 mass %, and deduction in quantity shall be made for all water and sediment in excess of 1.0 mass %.

[D]1 $mm^2/_5$ = 1 cSt.

[E]Where low sulfur fuel oil is required, fuel oil falling in the viscosity range of a lower numbered graded own to and including No. 4 can be supplied by agreement between the purchaser and supplier. The viscosity range of the initial shipment shall be identified and advance notice shall be required when changing from one viscosity range to another. This notice shall be in sufficient time to permit the user to make the necessary adjustments.

[F]Other sulfure limits may apply in selected areas in the United States and in other countries.

[G]This limit assures a minimum heating value and also prevents misrepresentation and misapplication of this product as Grade No. 2.

[H]Lower or higher pour points can be specified whenever required by conditions of storage or use. When a pour point less than −18 °C (0 °F) is specified, the minimum viscosity for grade No. 2 shall be 1.7 $mm^2/_5$ (31.5 SUS) and the minimum 90% recovered temperature shall be waived.

[I]Where low sulfur fuel oil is required, Grade No. 6 fuel oil will be classified as Low Pour (+15 °C (60 °F) max) or High Pour (no max). Low Pour fuel oil should be used unless tanks and lines are heated.

Reprinted, with permission, from the Annual Book of ASTM Standards, copyright American Society for Testing and Materials, 1916 Race Street, Philadelphia, PA 19100.

Figure 8-1. Detailed Requirements for Fuel Oils[A]

GRAVITY		DENSITY	HEATING VALUE	
DEG. AP 1	SPECIFIC @ 60/60 F	LB PER GALLON	BTU • PER LB	BTU PER GALLON
0	1.0760	8.962	17.690	158.540
1	1.0679	8.895	17.750	157.870
2	1.0599	8.828	17.810	157.190
3	1.0520	8.762	17.860	156.520
4	1.0443	8.698	17.920	155.890
5	1.0366	8.634	17.980	155.240
6	1.0291	8.571	18.040	154.600
7	1.0217	8.509	18.100	153.980
8	1.0143	8.448	18.150	153.370
9	1.0071	8.388	18.210	152.760
10	1.0000	8.328	18.270	152.150
11	.9930	8.270	18.330	151.570
12	.9861	8.212	18.390	150.990
13	.9792	8.155	18.440	150.410
14	.9725	8.099	18.500	149.850
15	.9659	8.044	18.560	149.300
16	.9593	7.983	18.620	148.630
17	.9529	7.935	18.680	148.190
18	.9465	7.882	18.730	147.660
19	.9402	7.830	18.790	147.140
20	.9340	7.778	18.850	146.620
21	.9279	7.727	18.910	146.100
22	.9218	7.676	18.970	145.580
23	.9159	7.627	19.020	145.100
24	.9100	7.587	19.190	145.600

Figure 8-2. Specific Gravity, Density and Heat Content of Commercially Available Oils

GRAVITY		DENSITY	HEATING VALUE	
DEG. AP I	SPECIFIC @ 60/60 F	LB PER GALLON	BTU • PER LB	BTU PER GALLON
25	.9042	7.538	19.230	145.000
26	.8984	7.490	19.270	144.300
27	.8927	7.443	19.310	143.700
28	.8871	7.396	19.350	143.100
29	.8816	7.350	19.380	142.500
30	.8762	7.305	19.420	141.800
31	.8708	7.260	19.450	141.200
32	.8654	7.215	19.490	140.600
33	.8602	7.171	19.520	140.000
34	.8550	7.128	19.560	139.400
35	.8498	7.085	19.590	138.800
36	.8448	7.043	19.620	138.200
37	.8398	7.011	19.650	137.600
38	.8348	6.960	19.680	137.000
39	.8299	6.920	19.720	136.400
40	.8251	6.879	19.750	135.800
41	.8203	6.893	19.780	135.200
42	.8155	6.799	19.810	134.700

Grade Number	API Gravity
6	–2.4 to 19.7
5	6.4 to 25.7
4	9.2 to 34.0
2	26.0 to 46.7

Figure 8-2. Specific Gravity, Density and Heat Content of Commercially Available Oils (*continued*)

You should be aware that more caution is required, as more emphasis will be placed on the quality of the fuel being burned. This quality can have a dramatic impact on the stack emissions generated by the fuel burning equipment. Restrictions on the emission of sulphur dioxide are in place in most areas of the United States. Some areas, such as the southern part of California, have strict standards with regard to emissions of nitrous oxide. When purchasing new fuel burning equipment, review with the equipment supplier his capabilities to meet present and potential requirements or air quality management standards presently in effect, and those proposed in the near future.

Gaseous Fuels

These fuels are characterized by the ease with which they are burned, their higher hydrogen content (in relation to carbon), provides this advantage. Refer to Figure 8-3 for properties of the typical gaseous fuels. Just as viscosity affects the burning and use of liquid fuels, the properties of the various gaseous fuels have an important bearing on the selection of the equipment used to accomplish the burning process. Although most commercial gaseous fuels are more uniform in quality, significant differences do occur and the operator of the combustion equipment must take these into consideration. Major factors to be considered in the selection of combustion equipment and handling of the fuel are:

- specific gravity of the gas
- BTU content per cubic foot
- pressure at which the gas is available
- possibility of contaminants in the fuel supply

TYPE OF GAS	SPECIFIC GRAVITY	BTU PER CU FT
NATURAL	.60–.66	850–1250
MANUFACTURED	.40–.70	400–600
SEWAGE	.70–.80	500–900
PROPANE (LP)	1.50	2560
BUTANE (LP)	2.00	3180
LP–AIR MIXTURES	1.10–1.25	550–1400

Figure 8-3. Properties of Typical Gaseous Fuels

The first three items will have a major impact on the cost of the handling (piping network) supplying fuel to the burners. As an example, if the fuel has a low BTU content and is available only at relatively low pressures, the size of the piping and gas valves will dramatically impact the price of the equipment required to deliver the fuel to the burner. The specific gravity of the fuel will also affect the ability of the piping and valves with relation to capacity. Commercial natural gas pressures may vary from as little as 3.5″ W.C. to as much as 50 to 60 psig availability (27.71″ equals one pound). In cases where only low gas pressure is available, gas boosters (a type of centrifugal pump) may be installed in the supply piping to change the available pressure. In applications of this type, consult with the local utility, because the booster could have an adverse affect on their distribution network by reducing the main pressure to adjacent customers. Usually when boosters are allowed, the utility network is protected by the addition of a low gas pressure switch on the suction side of the booster to protect the utility network from excessive demand on the part of the customer, which would allow a dangerously low pressure in the network.

Usually, contaminants are not a problem with commercially available grades of gas fuels; however, in some areas the equipment may require the addition of an inlet strainer to protect the on site components from foreign substances. The strainer mesh should be sufficiently fine to protect the downstream equipment, but not so restrictive that it imposes a high pressure drop and impacts the flow of gas. Criteria for the selection and maintenance of inlet strainers are:

- Allowable pressure drop which may be allocated to the strainer.
- Suspected size of any contaminant that may be anticipated as being delivered from the source gas.
- Requirements of the control valve manufacturer as to the level of cleanliness required to allow safety and regulating controls to perform their respective functions.

Usually, the public utility supplying natural or manufactured gas maintains reasonably good control over contaminants, with the major problem in the distribution of these gases being the presence of moisture and small particles. The significance of quality and contamination is more critical on the liquefied petroleum gases (propane and butane), and is a very important consideration when using sewage and other sour gases. When using gases of this type, extreme care must be taken in the selection of safety valves, making certain they are suitable for the intended use. Important factors to consider when selecting safety shut off valves are:

- Physical properties of the metal of the valve body.
- Physical properties of the valve seat.

Various materials of construction of the body and seat will have the ability to withstand different materials which may be present in the gas. In extreme cases, the pipe itself may have to be specially treated to withstand the affects of corrosion.

Electric Energy

The main characteristic which makes the use of electric as a fuel attractive is convenience. In most areas of the United States, pricing structures make the use of large volumes of electric energy prohibitive. Exceptions where this may not be the case are small point-of-use devices such as would be used for hot water heating for limited usage, electric strip heaters, spot heating and small process requirements where low volume is the main consideration, and piping from a central plant may not be desirable because of distance, or because the main plant is operating at a different pressure or temperature. Further consideration may be given to providing "point-of-use" steam or hot water, which will allow the main boiler plant to be shut down during certain times of the operating year. Electric boilers are usually of low volume (small amounts of water content), because they can respond rapidly, and because the density of fuel input is usually relatively high.

Electric boilers are characterized by the type of element utilized:

- Electrode type—providing an additional safety feature—the electrode is immersed in the fluid—when the electrode is uncovered (lack of fluid) the current flow is interrupted, since the fluid completes the circuit path.
- Resistance element immersed in the fluid to be heated—this boiler has a more conventional look, although for any given output it will be smaller than a fossil fueled boiler. Because they use standard heating elements, a conventional low water cut off is employed as a means of detecting low fluid level.

When considering the use of electric energy as a fuel source, carefully consider the following factors:

- Is there adequate wiring capacity within the plant to support the additional load?
- What affect will the additional electrical requirement have on the overall rate structure for the plant operation? Certain utilities have what are called "ratchet", time of day, seasonal or peak demand charges—all of which must be carefully considered when an evaluation is made.

SAFETY VALVE OPERATION PRINCIPLES

The importance of safety valves, relief valves, and safety relief valves to the safe and proper operation of a boiler plant can not be over emphasized. For this reason, a summary of important considerations with regard to safety relief valves is presented. Full details may be obtained from A.S.M.E., Section IV for low pressure boilers and Section I for high pressure boilers. This reference is not intended to contain the complete code requirements; it is included to stress the significance of this very important safety device. Whenever a question regarding safety valves arises, you should refer these questions to one of the following:

- the state boiler inspector or other governing body having jurisdiction,
- the insurance company carrying the boiler insurance,
- the latest applicable code.

A *safety valve* is an automatic pressure relieving device actuated by the static pressure upstream of the valve and characterized by full opening pop action. It is used for steam, air or vapor service. A *relief valve* is an automatic pressure relieving device actuated by the static pressure upstream of the valve which opens further with the increase in pressure over the opening pressure. It is used primarily for liquid service. A *safety-relief valve* is an automatic pressure relieving device suitable for use either as a safety valve or relief valve, depending on application.

Safety Valve Considerations

- Each boiler shall have at least one safety valve.
- The safety valve capacity for each boiler shall be such that the safety valve(s) will discharge all the steam that can be generated by the boiler without allowing the pressure to rise more than 6 percent above the highest pressure at which the valve is set and in no case to more than 6 percent above the maximum allowable working pressure.
- One or more safety valves on the boiler proper shall be set at or below the maximum allowable working pressure. If additional valves are used, the highest pressure setting shall not exceed the maximum allowable working pressure by more than 3 percent. The complete range of pressure settings of all the saturated steam safety valves on a boiler shall not exceed 10 percent of the highest pressure to which any valve is set. Pressure setting of safety relief valves on high-temperature water boilers may exceed this 10 percent range. Because safety relief valves in hot water service are more susceptible to damage and subsequent leakage, the valve setting for high-temperature boilers should be selected substantially higher than the desired operating pressure so as to minimize the times the safety relief valve must lift.
- For high-temperature water boilers safety-relief valves must be used. Such valves must have a closed bonnet.
- A safety valve over 3 inches in size, used for pressures greater than 15 PSI gauge, shall have a flanged inlet connection or a welded inlet connection.
- When two or more safety valves are used on a boiler, they may be mounted either separately or as twin valves made by placing individual valves on Y-bases. The twin valves shall be of approximately equal capacity.
- The safety valve shall be connected to the boiler independent of any other connection and attached as close as possible to the boiler, without any unnecessary intervening pipe or fitting. Every safety valve shall be connected so as to stand in an upright position, with the spindle vertical.
- The opening or connection between the boiler and the safety valve shall have at least the area of the valve inlet. No valve of any description shall be placed between the required safety valve and the boiler, nor on the discharge pipe between the valve and the atmosphere. When a discharge pipe is used, the cross-

sectional area shall be not less than the full area of the valve outlet or of the total of the areas of the valve outlets. It shall be as short and straight as possible and so arranged as to avoid undue stresses on the valve.

- When a boiler is fitted with two or more safety valves on one connection, this connection to the boiler shall have a cross-sectional area not less than the combined areas of inlet connections of all the safety valves with which it connects.
- To ensure the valve being free, each safety valve or safety relief valve shall have a substantial lifting device by which the valve disc may be positively lifted from its seat when the valve is subjected to pressure of at least 75 percent of the set pressure. The lifting device shall be such that it cannot lock or hold the valve disc in lifted position when the exterior lifting force is released. Discs of safety relief valves used on high-temperature water boilers shall not be lifted while the temperature of the water exceeds 200°F. If it is desired to lift the valve disc to assure that it is free, this shall be done when the valve is subjected to a pressure of at least 75 percent of the set pressure. For high-temperature water boilers, the valve shall have a pressure tight packed lift lever.
- A body drain below seat level shall be provided in the valve and this drain shall not be plugged during or after field installation.
- Each steam boiler shall have one or more officially rated safety valve of the spring pop type adjusted and sealed to discharge at a pressure not to exceed 15 PSI. Seals shall be attached in a manner to prevent the valve from being taken apart without breaking the seal. A body drain connection below seat level shall be provided by the manufacturer and this drain shall not be plugged during or after field installation.
- No safety valve for steam boiler shall be smaller than ¾″ nor larger than 4½″.
- Each hot water heating boiler shall have at least one officially rated pressure relief valve.
- There shall be a lifting lever and a mechanical connection between the lifting lever and the disc capable of lifting the disc from the seat.
- Every valve shall be tested to demonstrate its popping point, blowdown, and tightness. A tightness test shall be conducted at a maximum expected operating pressure, but not at a pressure exceeding the reseating pressure of the valve. When testing on either water or steam, a valve exhibiting no visible signs of leakage shall be considered adequately tight.
- Safety valves and safety relief valves requiring repairs shall be replaced with a new valve or repaired by the manufacturer.
- Boilers cannot be operated at their design pressure, as safety valves must be set at design pressure by codes and laws. The following suggestions are made to provide proper safety valve seating and to prevent leakage:
  - For 15 psig design steam—operate at 10 psig. Operation at 12 psig occurs in special cases.

- For 30 psig design hot water—operate at 26 psig. This is for total of both static and velocity heads. For hot water above 30 psig, boiler design pressure should be 10 to 20 percent above the total of system static and velocity heads.
- For high pressure—operate at least 10 to 15 percent below the design pressure. For example, operate a boiler designed for 150 psig at a pressure no higher than 135 psig.

• Never start up a new boiler nor operate one with which you are unfamiliar without first checking the steam safety relief valve for pressure setting and relieving capacity.

• The minimum relieving capacity of a safety valve in pounds per hour shall be the greater of that determined by dividing the maximum BTU output at the boiler nozzle obtained by the firing of any fuel for which the unit is installed by 1000, or shall be determined on the basis of the pounds of steam generated per hour per square foot of boiler heating surfaces as defined in the applicable A.S.M.E. code. Both should match the design criteria for the boiler.

Installing Safety or Relief Valves

The Kunkle Valve Company, Inc., makes the following recommendations with regard to safety/relief valve installation:

• Before installing a new valve, use a pipe tap to assure clean-cut and uniform threads in the vessel opening and to allow for normal hand engagement followed by a half to one turn with a wrench.

• Avoid overtightening, as this can distort safety/relief valve seats. Remember: as the vessel and valve are heated, the heat involved will grasp the valve more firmly.

• Avoid excessive "popping" of safety/relief valves, as even one opening can provide a means for leakage. Safety/relief valves should be operated only often enough to assure that they are in good operating order.

• Avoid wire, cable, or chain pulls for attachment to levers that do not allow for a vertical pull. The weight of these devices should not be directed to the safety/relief valve.

• Avoid having the operating pressure too near the safety/relief valve set pressure. A very minimum difference of five pounds or 10 percent (whichever is greater) is recommended. An even greater differential is desirable, when possible, to assure better seat tightness and valve longevity.

• Avoid discharge piping where its weight is carried by the safety/relief valve. Even though supported separately, changes in temperature alone can cause piping strain. It is recommended that drip pan elbows or flexible connections be used wherever possible, as shown in Figure 8-4.

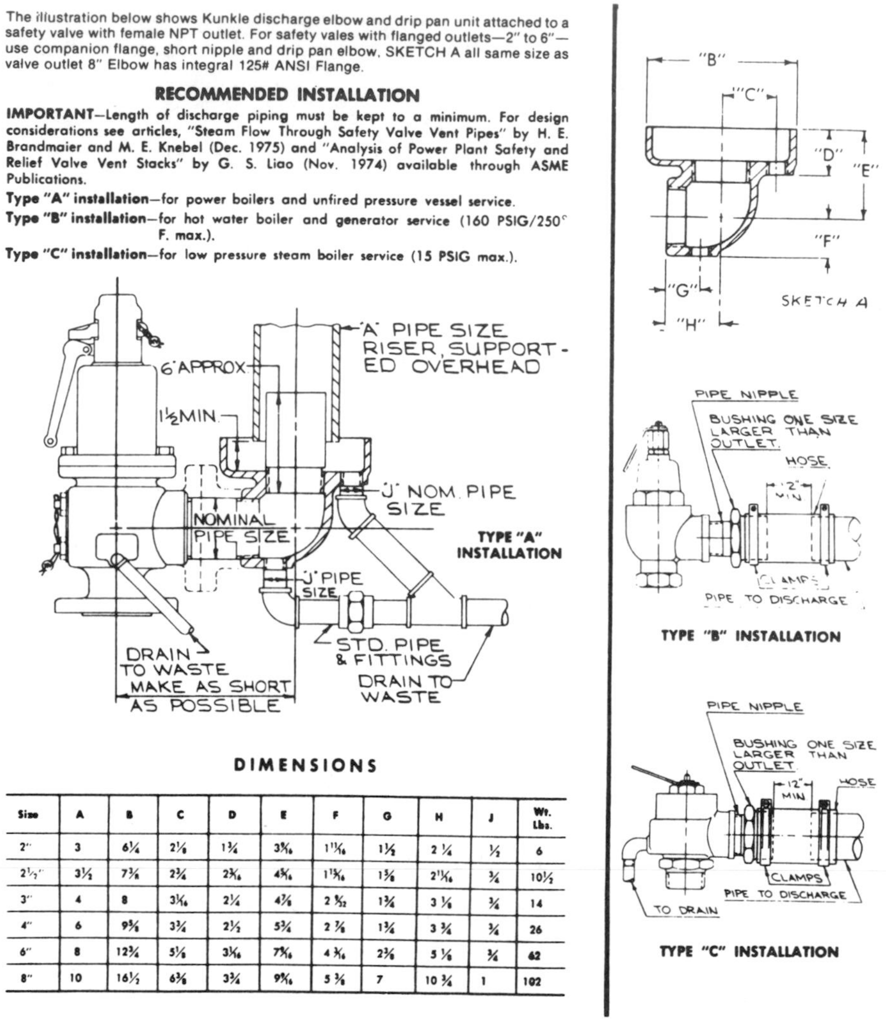

The illustration below shows Kunkle discharge elbow and drip pan unit attached to a safety valve with female NPT outlet. For safety vales with flanged outlets—2" to 6"—use companion flange, short nipple and drip pan elbow. SKETCH A all same size as valve outlet 8" Elbow has integral 125# ANSI Flange.

RECOMMENDED INSTALLATION

IMPORTANT—Length of discharge piping must be kept to a minimum. For design considerations see articles, "Steam Flow Through Safety Valve Vent Pipes" by H. E. Brandmaier and M. E. Knebel (Dec. 1975) and "Analysis of Power Plant Safety and Relief Valve Vent Stacks" by G. S. Liao (Nov. 1974) available through ASME Publications.

Type "A" installation—for power boilers and unfired pressure vessel service.

Type "B" installation—for hot water boiler and generator service (160 PSIG/250°F. max.).

Type "C" installation—for low pressure steam boiler service (15 PSIG max.).

DIMENSIONS

Size	A	B	C	D	E	F	G	H	J	Wt. Lbs.
2"	3	6¼	2⅛	1¾	3 9/16	1 11/16	1½	2¼	½	6
2½"	3½	7⅜	2¾	2 3/16	4 5/16	1 15/16	1⅝	2 11/16	¾	10½
3"	4	8	3 1/16	2¼	4⅞	2 9/32	1¾	3⅛	¾	14
4"	6	9⅝	3¾	2½	5¾	2⅞	1¾	3¾	¾	26
6"	8	12¾	5⅛	3 1/16	7 9/16	4 3/16	2⅜	5⅛	¾	62
8"	10	16½	6⅜	3¾	9 9/16	5⅜	7	10¾	1	102

Courtesy of Kunkle Valve Company, Inc.

Figure 8-4. Drip Pan Elbows

- Apply a moderate amount of pipe thread compound to male threads only, leaving the first thread clean. Compound applied to female threads or used to excess can find its way into the valve, causing leakage. Flange connections should be clean and straight, with new gaskets. Draw mounting bolts down evenly.

WATER LEVEL CONTROL MAINTENANCE

Statistics point out that the greatest number of boiler incidents involving loss and injury may be attributable to failure to maintain water level within the vessel.

This failure can result not only in high dollar cost, but can involve personnel injury and even death. The liquid level controls, together with the relief valves mounted on the boiler, are the most important safety devices in the boiler plant. Only persons thoroughly experienced in the installation and maintenance of these devices should be allowed to install, test, and maintain them.

Basically, low water cutoffs are divided into two broad classifications, those that operate on a float principle, and those employing the conductivity of the liquid as a means of determining the absence or presence of the water level. The latter are referred to as probe-type water level controls. It is our recommendation that all water cutoff devices not only function to stop the operation of the firing device, but they be connected to an external alarm to alert the operator of the problem promptly.

Maintaining Hot Water Boilers

Although the hot water system does not usually present a problem with respect to low water conditions, a number of failures take place each year with hot water systems. Because no "normal" water level exists within the boiler, the LWC may be placed at any point on the boiler above the lowest permissible water line. Systems which have all the radiation and connecting piping above the boiler rarely present a problem; however, piping networks with a large portion of piping and/or terminal units below the boiler level can result in a rapid loss of water from the system in the event of piping failure. Verifying the proper operation of the LWC in a hot water system is difficult, because it is not practical to drain the entire system to check the operation of the control, therefore, the use of Test-N-Check valves as manufactured by Mc Donnell & Miller is a valuable adjunct in connection with the testing of column mounted LWCs on HW systems. When external liquid columns are used with float type LWCs, take care to eliminate air from the column, which could result in "false" low water alarms. The probe type LWC may be inserted directly in the shell of the vessel, in which case testing normally consists of verifying the operation by means of a test button. In all cases, follow the manufacturer's instructions for testing of the device.

Maintaining Steam Boilers

Water level in a steam boiler is considerably more important, in that a rather precise level must be maintained for two reasons: both to protect the vessel from damage and to maintain the water level within boundaries which provide for maximum utilization of the steam disengaging area of the boiler chest. Operating the boiler at lower than permissible water levels may result in overheating and damage to the boiler itself. Operating the boiler with liquid levels higher than actually required, reduces the area available for the proper separation of steam from the surface of the water. This condition will result in carry over and poor quality steam.

Modulating Type Systems

These systems consist of variable inputs to the boiler. These inputs may be as a result of sensing the steam flow leaving the boiler (such as systems manufactured by Copes Vulcan), or may be of the type which respond to the actual liquid level and modulate a motorized valve (proportional systems such as Mc Donnell & Miller 7 series switches).

The advantage to these systems is that they provide continuous "metered" return of water to the steaming vessel, in lieu of the small "slug" or batch feeding which results from the normal on off action of a boiler feed or return pump. The modulating systems are more commonly used on larger boilers where it is more desirable to anticipate changes in load, and to have more uniform inputs. Obviously, the sudden action of a boiler return pump in delivering intermittent large volumes of water is not as desirable as a steady continuous varied return rate.

On-Off Type Systems

These systems consist of a boiler return or boiler feed pump, designed to return condensate from a collector or storage tank, to the boiler. The pumps are activated by means of the liquid level controller mounted on the boiler. On a drop in liquid level the pump is activated and returns the condensate to the boiler.

Sizing of Boiler Return Pumps

It is suggested that boiler return pumps have a capacity of from two to three times the evaporation rate of the boiler. Remember that a 100 HP boiler evaporates 6.9 gpm (from and at 212°F), and therefore a typical boiler return pump for this boiler would be able to deliver from 14 to 21 gpm. The boiler feed or return pump should have sufficient capacity (rated discharge pressure), to overcome the design pressure (boiler operating pressure), plus an allowance for the pressure drop of piping and valves between the pump discharge and the boiler.

As an example, a 200 HP boiler operating at 125 psig, should have a return pump with a capacity of 28 to 42 *gallons per minute* with a discharge pressure rating of 130 psig (usually a minimum of 5 psig over the boiler operating pressure is required for piping). Frequently, the manufacturer rates these pumps in "feet of head" (2.31 feet of head equals one pound). Therefore, in the above example, the discharge head could be expressed as approximately 300 feet of head.

Sizing of Boiler Return (or Boiler Feed) Tanks

This is a critical area, which unfortunately is often neglected or mistreated. Maintenance managers frequently are faced with a problem when replacing an older, existing boiler with one of the newer package variety. Although the new boiler has the ability to deliver rated output specified by the manufacturer, it may have considerably less water content, and therefore the size of the boiler feed tank is very critical, because the water formally stored in the older boiler must now be held in reserve in the boiler feed tank. The boiler feed tank should be sized so that the upper ¾ of the tank contains enough water to provide for supplying the boiler requirements for a period of 20 to 30 minutes of operation at maximum firing rate. Referring to our previous example, a 200 HP boiler would require the following:

> 200 (HP) × .069 (evaporation rate per hp/hr) × 60 (minutes in hour) divided by 3 or 2 (20 to 30 minute supply).

This volume of water should be contained in the upper ¾ of the tank—so that the make up valve on the tank will not function, except in the case of system water loss, and the water being returned, will be that which has had time to circulate the system (change from steam to condensate), and return via the piping and any collector tanks, to the boiler feed tank. Therefore the tank size in the example should be between 368 to 552 gallons. Using standard tanks, the selection would probably be 350 or 550 gallon capacity.

A reasonable "rule of thumb" is that you should allow two gallons of storage in the boiler feed tank for every boiler horsepower. This would result in selecting a tank of 400 gallons capacity. The determining factor in selecting 20 or 30 minute storage is the design of the building. A large horizontal (one story) building over considerable square footage requires more tank capacity than a building occupying a small area with multiple stories.

The boiler feed tank should be equipped with an automatic feeder to function when the level drops to approximately 25% of the tank capacity. The supply line to this feeder should be equipped with a water meter, which will indicate the amount of water "lost" from the system (that which must be made up), an important tool in preventative maintenance, and indicator of system integrity. Further, the tank should be fitted with a LWC which will shut off the pump or pumps and prevent them from operating without water and doing damage to the pumps and seals. Boiler feed tanks employed in HP operating systems will normally be equipped with deaerators, assisting in the removal of oxygen from the boiler feedwater. LP systems may not be so sophisticated as to employ deaerators; however, the benefits of partial oxygen removal by pre heating should not be overlooked. Additionally, consider protecting the boiler feed tank by means of coating the interior surfaces, the use of anode rods projecting into the tank, and the addition of chemicals.

Importance of Testing

Periodic testing of level controls is a must. Frequently, insurance companies or local policy will dictate testing low water controls on a shift basis; that is, if the boiler is operated on a round-the-clock basis, testing once each 8 hours. In less severe applications, others suggest daily, and still others weekly testing. Weekly would be a minimum, even for low pressure systems designed as minimal loss systems.

At least once each year, examine the LWC carefully. In the case of float type, disassemble it and observe all piping and sediment chambers for cleanliness. In addition to the periodic (shift, daily or weekly), testing of the LWC by the traditional method of blowing down the column, test the cutoff by the slow drain method at least twice a year. This procedure more closely duplicates the way a boiler would actually run out of water if it were in normal operation.

Remember, a sluggish LWC may function when the blow down valves are operated, but because of the physical difference, may not function to feed and/or shut down the boiler, if the drop in water is gradual. For this reason, conduct the slow drain test as follows. Shut off the pumps or water supply to the boiler, continue to operate the burner, while slowly draining water from a rear blow down (duplicating the gradual loss that would occur during normal operation), and insure that the burner stops firing. This must occur while water is still visible within the gauge glass of the boiler. If the boiler is fitted with two LWCs, test both in the manner described above. Testing of the second or lower of the two will require bypassing the first cutoff, usually by means of an electrical jumper. To prevent leaving this jumper in place, it is suggested that the jumper be made complete with a large red flag attached, plainly visible, as a reminder to remove same upon completion of the test. The code requires that water must still be visible in the water gauge glass when both LWCs have functioned to shut down the firing device.

KEEPING A MAINTENANCE LOG OF ALL BOILER ROOM AND HEATING EQUIPMENT

To control the boiler plant operation, keep a boiler maintenance log. Figure 8-5 provides a typical and comprehensive example. The following information is an item-by-item checklist of each category of the log.

1. Observe operating pressure and water level, or operating temperature. Check to make certain that the boiler is operating within its design parameters. Pay special attention to verify that the operating control set point has not been changed (increased) beyond the normal limits. Observe water level on steam boilers to make certain it is within the normal operating range and at all times above the minimum lowest permissible water line. Investigate any changes from previous settings and take corrective action as required.

2. Check and record reading of primary combustion control signal. Take reading of flame signal meter on both pilot and main flame. Record *both* readings, note any change from previous readings. If value is lower, check for possible obstructions in the scanner sight tube, weak cell or scanner, and on older installations, vacuum tubes, or changes in the pilot or main flame. Usually any degradation in signal is a sign of potential problems, therefore, determine the reason and take corrective action.

	Routine to be Performed	Freq	1	2	3	4	5	6	7	8	9	10	11	12	13	14	15	16	17	18	19	27	28	29	30	31
1	Observe operating pressure and water level or operation pressure	D																								
2	Check and record reading of primary combustion control signal	D																								
3	Observe condition of flame	D																								
4	Observe water level glass (steam systems) for proper water level & cleanliness	D																								
5	Check fuel supply if oil or LP gas	D																								
6	Inspect boiler room air intake	D																								
7	Test operation of low water cut off and/or cut off feeder combination	D																								
8	Check operation and condition of condensate return, boiler feed pumps or circulators	D																								
9	Hot water heating systems only—check expansion tanks	W																								

D—Daily W—Weekly M—Monthly A—Annually SA—Semi-Annually

Figure 8-5. Sample Boiler Maintenance Log

	Routine to be Performed	Freq	1	2	3	4	5	6	7	8	9	10	11	12	13	14	15	16	17	18	19	27	28	29	30	31
10	Operate bottom blowdown valves (steam boilers only)	W																								
11	Check flame—visibly inspect flame	D																								
12	Check condition of heating surfaces (fireside)	M																								
13	Test limit controls	A																								
14	Test operating controls	M																								
15	Inspect fuel supply system	D																								
16	Check gas piping	A																								
17	Gas burners equipped with normally open vent valves	M																								
18	Check discharge of vent lines from gas regulators high/low switches, etc.	W																								
19	Inspect fire suppression systems	W																								
20	Check condition of boiler water	W																								
21	Try lever test— safety or relief valve	SA																								

D—Daily W—Weekly M—Monthly A—Annually SA—Semi-Annually

Figure 8-5. Sample Boiler Maintenance Log (*continued*)

	Routine to be Performed	Freq	1	2	3	4	5	6	7	8	9	10	11	12	13	14	15	16	17	18	19	27	28	29	30	31
22	Testing of draft switches, draft fan proving switches, atomizing, and combustion air switches.	A																								
23	Test operation of high/low gas pressure switches	A																								
24	Perform complete combustion efficiency test	M																								
25	Perform routine maintenance on low water cut off	A																								
26	Pop test	A																								

D—Daily W—Weekly M—Monthly A—Annually SA—Semi-Annually

Figure 8-5. Sample Boiler Maintenance Log (*continued*)

3. Observe condition of flame—check for changes in flame shape, size or color, as well as possible flame impingement. Changes in flame could be due to contaminated fuel nozzles, overheating, and distortion of nozzles or gas jets, or changes in make up of fuel. Changes in fuel supply are not as likely on pipe line gas, however may vary widely on LP gases and liquid fuels (No. 2 thru No. 6 fuel oils). Cycle the burner to check for smooth light off of main flame.

4. Observe water level glass (steam systems) for proper water level and cleanliness—clean or replace glass if required. Check for any leaks in or around the gauge glass and tri cock assemblies, repair or replace glass and or gaskets or valves at once if signs of leakage are present. Remember even a small leak can affect the accuracy of the reading and small leaks get larger. In addition the leakage will usually "wire draw" the component, resulting in costly damage.

5. Check fuel supply if oil or LP gas—maintain fuel supply within limits of storage and usage. Consider storage tank size and possible delivery problems due to weather extremes. In the case of LP fuel tanks, avoid overfilling in cold weather, because if the fuel is used as a standby, the expansion due to elevated temperatures in warmer weather may lead to relief valve discharge.

6. Inspect boiler room air intake—check to make certain there is no blockage of fresh air for combustion and ventilation. In areas of snow, check especially for snow accumulation blocking the air inlet. Where leaves or other debris may be a problem, check for their presence in the air inlet. When dampers are used, check that operation is smooth, that dampers open fully and do not stick or bind. Systems employing dampers and/or fans should have interlock switches to prove that the damper is in the open position. Systems utilizing fans should have differential air switches to verify air flow. The operation of these switches should be checked periodically to make certain they shut down the firing device if air supply is discontinued.

7. Test operation of low water cut off and or cut off feeder combination—on steam boilers verify that when using the column blow down method, the firing device is shut down when the column is drained via the blow down method.

On HW boilers, the practice of draining water is discouraged, because it usually requires the release of large amounts of water, is wasteful of the water treatment, and introduces air (which causes corrosion), into the system. A practical solution to this problem is the use of properly designed valves, such as Mc Donnell & Miller Test-N-Check series valves. *Under no circumstances should a standard shut off valve be installed in the piping, which could accidentally be left in the closed position isolating the boiler from this important safety device.*

Because the loss of water from a steam boiler is usually not rapid, as would be simulated by the column blow down test, however gradual, as would occur during the steaming (evaporation) process, conduct a "slow drain" test at least twice a year. This test more nearly duplicates the occurrence which would happen during normal operation of the boiler. The makeup and feedwater to the boiler should be stopped (electrically de-energize the feedwater or return pump), slowly drain water from the boiler. Drain this by opening a valve at the bottom of the boiler, at a point as remote as possible from the water column. Observe that the cutoff functions to electrically deenergize the firing device before water leaves the gauge glass and/or before the water level drops to the lowest permissible water level (as determined by the marking provided by the boiler manufacturer). *Under no circumstances should the boiler be operated when water is not visible in the gauge glass.* If the boiler is equipped with a secondary LWC device, it should also be checked via the slow drain method. To accomplish this, it will be necessary to jumper the primary LWC. It is suggested that the jumper wire have a red cloth tag at least 6″ in length attached to it, to prevent accidentally leaving same in place. *Note, that even during the testing of the secondary LWC, control must function to stop the firing device before water leaves the gauge glass.* Upon completion of the testing, remove any jumpers, and test each control using the column blow down method. Failure of any of the above tests should result in *immediate* corrective action. A boiler with inoperative LWCs should be shut down until corrections are made.

8. Check operation and condition of condensate return, boiler feed pumps (steam systems) or circulators (HW systems)—check for excessive vibration, coupling misalignment, noise and shaft leakage. Adjust packed seals to drip at all times or damage will occur from running components dry. On the other hand, mechanical seals should operate without loss of liquid through the seal face. Corrections should be made to leaky or vibrating components at once, because if they are allowed to deteriorate, repairs will be more costly. Monthly, observe and record both operating pressure and motor amperage. Investigate any significant changes—for example, an increase in motor amperage may be due to a deteriorating condition of the motor bearing causing increased friction. Investigate significant changes in either direction, up or down, determine their cause, and where applicable, take appropriate action.

9. Hot water heating systems only—check expansion tanks to determine they are not "water logged." A good indication of water logged tanks is a rise in the system operating pressure. If tanks are losing their "air charge," determine the reason. Plain steel tanks may have a pin hole leak in the tank, changes in system capacity (additional piping), or operating at elevated temperature are all possible causes. When using hydro pneumatic tanks (diaphragm), an additional cause could be a leak in the diaphragm or a leak in the air chamber.

10. Operate bottom blowdown valves—drain sludge from bottom connection, or connections, of the boiler by operating the bottom blow down valve or valves. Amount and duration of bottom blow down required will be determined by the amount of make up water admitted to the system. The number of blow down cycles and length should be worked out in concert with your water treatment service. Excessive bottom blow down is wasteful of both chemicals and energy, insufficient bottom blow down results in accumulation of sludge and sediment along the bottom drum. It is desirable to observe the characteristics of the blow down discharge, usually the blow down should be operated until the discharge is relatively clean and free of sludge. Care should be taken when operating the bottom blow down of HP boilers, that the discharge is directed to a safe location. Where operating characteristics

and equipment is available, the blow down should be discharged to a heat recovery unit to conserve energy.

11. Check flame—visibly inspect flame. Check for appearance of flame. The flame should be concentric, not impinging on surfaces of the boiler or furnace. If the unit is equipped with smoke opacity system, check operation of the system, including alarm and shut down functions. This equipment is normally found on solid fuel, waste burners and oil fired units only. Testing of the smoke alarm should be in accordance with manufacturer's recommendations. Investigate visible flame impingement, unbalanced, uneven, or concentrated flame characteristics, as to cause, and take corrective action. Many times an offsided fire is due to dirt in the fuel tip or nozzle, and a simple cleaning is all that is required to restore normal operation.

12. Check condition of heating surfaces (fireside). Check the fireside heating surfaces of the boiler for excessive scale and/or soot build up. Frequency of this inspection may vary depending on the fuel being burned. Natural gas will rarely require examination more than once a year, but waste fuels, soft coal, and other "dirty" fuels which may contain high amounts of ash dictate a more frequent observation of the surfaces. Note that a rise in exit gas temperature also indicates the need for physical examination of the condition of the heating surfaces.

13. Test limit controls and record in the log the actual cutoff points of the limit controls (water temperature or steam pressure). To determine if the limit control (which should be a manual reset device) is functioning, shut down the system, removing all power to the controls, and place a jumper around the normal operating (steam/water temperature) control. **IMPORTANT!!** *This test should be conducted only by personnel thoroughly familiar with the operation of the boiler plant, and for obvious safety reasons, should be conducted with that person in continuous attendance during the test.* Check setting of the temperature or pressure operating device, operate the firing device and observe that the high limit control functions to shut the burner down at or prior to the set point. At no time during the test should the boiler be exposed to temperatures or pressure beyond the rated design temperature or pressure. If the pressure or temperature reaches that point, stop the test immediately, determine the reason, and take corrective action. At the conclusion of the test, reset the high limit control, remove the previously installed jumper from the operating control, and observe that the operating control functions properly to cycle the unit. It is very important that the person conducting the test perform this operation as part of the procedure, prior to leaving the boiler "on line."

14. Test operating controls, record pressure or temperature, compare with previous readings. If large differences are noted either correct the condition or determine the reason for change, such as higher operating temperature or pressure was required and control was reset. Failure of the control to "repeat" or function should be corrected by replacement, or in the case of steam pressure operating controls, check the condition of the sensing tube (connecting piping or tubing).

15. Inspect fuel supply system. Check the condition of fuel piping, supply pumps, and so on. Check for leakage, physical damage, improper pipe supports, etc. Immediately correct any leakage observed. Leaking fuel is a fire hazard, as well as a

housekeeping problem. In the case of gas supply systems, observe that physical condition of pipe is good, check for any missing pipe supports, improper supporting methods, and so on, if observed, take immediate corrective action.

16. If the installation contains gas piping and the system is equipped with gas leak detection equipment, check the operation of this equipment in accordance with manufacturer's instructions, making certain that it shuts down the plant and/or sounds an alarm. If permanently installed leak detection systems are not in place, conduct inspections of all gas piping using a portable (hand-held) leak detector.

17. Gas burners equipped with normally open vent valves (currently IRI requirements, or UL requirement over 12,500,000 BTU/hr). Check to make certain that the normally open vent valve is *not* discharging gas to the atmosphere during normal operation. Remember that this valve is normally open, so failure (for example, a coil burnout), will result in discharging gas to the atmosphere all during the firing cycle. This failure can result in tremendous increases in fuel consumption, as well as contributing to air pollution.

18. Check discharge of all vent lines from gas appliance regulators, high and low gas pressure switches, etc. Although small amounts of gas may discharge from certain types of diaphragm valves, any continuous discharge of gas is cause for *immediate* investigation and corrective action.

19. Inspect any fire suppression systems installed in connection with boiler room. This includes all types of systems, from wall mounted, portable systems up to and including permanent installed systems. If portable, make certain that all equipment is in place and that devices are currently certified, having been inspected within the prescribed time period. In the case of permanently installed systems, comply with the recommendations of the manufacturer or the requirements of the local inspection service or insurance company. Fire suppression equipment is an important part of a boiler plant, when it is required, the need is immediate and this equipment should be maintained in top notch condition.

20. Check condition of boiler water. Observe the water to determine whether any obvious changes, such as a change in color are visible. This observation does not take the place of a carefully administered boiler water treatment program, which should always be in place on any sizeable commercial installation. Larger installations will require actual boiler water tests to be made with varying degrees of frequency. In larger plants, these may include chemical tests as frequently as hourly, while in the smaller (hot water) installation, these tests may be required on only an annual basis.

21. Try lever test—safety or relief valve. The safety relief valve (or valves) is the final defense against overpressure. The safety relief valve and the LWC are the two major safety devices. It is impossible to overemphasize the proper operation of these devices. In testing safety relief valves, take care to protect personnel and property for injury or damage by escaping steam or water. As with the limit control test, this test should only be conducted by personnel thoroughly familiar with boiler operation and safety. For the try lever test, manually lift the valve or valves and hold the valve off its seat for a period of 5 to 10 seconds, by means of a pull chain attached to

the valve handle. After observing adequate discharge, release the chain and allow the spring to snap the valve to the closed position. If the valve does not seat fully, repeat the procedure two or three times to flush any sediment which may have logged in the seat area. If the valve continues to leak, it must be replaced. **IMPORTANT!!** *Never disassemble the valve or attempt to adjust the spring setting.* Repairs to safety and safety relief valves should only be attempted by the original equipment manufacturer or an authorized repair shop, who can certify the valves function and install the required seal. Depending on the size of the installation, frequency of testing may be varied.

22. Testing of draft switches. Induced draft fan proving switches, atomizing air switches, combustion (fresh air) proving switches.

23. Test operation of high and low gas pressure switches. Make certain switches are set in accordance with standards, or as a minimum that low gas pressure switch is set to a value not less than 50% of normal, and high gas pressure is set at a level not higher than 150% of normal. Record settings and trip points for future reference. Make certain that switch trips and locks out, preventing automatic reset.

24. Perform complete combustion efficiency test. Record all readings. Compare readings with previous and immediately determine the reason or reasons for any significant changes. The overall operating efficiency of the plant depends on boiler efficiency; in addition, any change in boiler operating efficiency may indicate a problem, such as scale deposits on the water side, or dirty heat exchanger surfaces. As a minimum, record efficiencies at low, medium and high fire positions. Record the following information:

- fuel flow rate
- percentage of CO_2 level in flue gas
- percentage of O_2 level in flue gas
- exit gas temperature (NET)
- smoke test (for solid and fuel oil units)
- carbon monoxide test (for gas units)
- over fire draft or pressure
- breeching draft or pressure

If changes are made to improve operating conditions (such as improvement to CO_2 level), make certain that values are rechecked, such as smoke or carbon dioxide test, to determine that they are still within proper range, after these changes.

25. Perform routine maintenance on the LWC. FLoat type—remove the head mechanism from the float chamber. Inspect the float chamber and connecting lines to the boiler for accumulations of mud, scale, or any foreign substance which would restrict water flow, and remove as required. Examine all wiring for brittle or worn insulation. Inspect the head mechanism for evidence of worn parts or corrosion, paying particular attention to solder joints on the bellows and float. Check float for evidence of collapse or other physical damage. Check mercury switches, if used, for

signs of overload. Replace any damaged or worn parts. Do not attempt field repairs to the switch and or head assembly, replace as a unit.

Probe or electronic type units—follow the manufacturer's procedure for annual maintenance and testing. As a minimum remove probes from the water column or boiler, and examine for scale or dirt buildup. Check all electrical connections and internal wiring.

When either float or probe type cut offs are placed back in service after periodic inspection, check them for performance (fuel cut off). This test should include both a column blow down and slow drain test to verify operation. (See item 7 above).

26. Pop test—operation of safety or relief valve—steam systems. Low pressure boilers only. Before conducting this test, verify that the safety valves installed have a rating (operating pressure), at or below the maximum rating allowed on the boiler name plate. The relief or safety valve is a critical part of the plant safety equipment. In addition, to try lever test conducted in item 21 above, conduct an annual test as follows:

- Isolate the boiler from the supply and return piping.
- Install a calibrated test gauge.
- Temporarily place test leads across the appropriate terminals on the operating control, to demonstrate the ability of the high limit control to shut down the firing equipment.

Upon satisfactory testing of the operating control, place test leads across the high limit (manual reset) control terminals to allow continuous operation of the burner.

It may be necessary to manually bypass the modulating control, when boilers are so equipped, to maintain the burner in "high fire" operating mode.

Observe that the safety valve operates satisfactorily when the maximum design steam working pressure is reached—that is, the set pressure of the valve, plus or minus 2 psig.

If the valve does not open within the limits specified, immediately stop the test, shut down the burner, remove the boiler from service, and do not operate until the reason for failure is corrected. Replace the required valves before the unit is allowed to operate.

Although tests for HP boilers may be somewhat similar, due to the elevated pressures and therefore higher potential energy, plan these test procedures in conjunction with your boiler insurance representative or a state or national boiler inspector.

Conduct a test of the operation of the safety valve when:

- a new safety valve or valves is installed
- a new firing device is installed on the boiler

Once the valve or valves have opened, the burner should continue to be operated at full firing rate, during which time the pressure in the boiler should not rise above 17 psig for low pressure boilers. If the pressure rises above this point, shut down the burner and replace the valve or valves as required.

IMPORTANT—*Upon completion of the above test, remove all jumpers installed for the test* and *operate the boiler through one complete normal cycle, making certain that operating controls function properly.* Compare the cutoff points with previous shut down points recorded in item 15 above. Correct any deficiencies (change from previous set points), by replacing or re setting control.

27. Test combustion air flow switches, low draft cut off switches, induced draft fan proving switches, and, if applicable proof of boiler room air delivery (combustion air for boiler room, when provided by mechanical means, or position switch if motorized dampers). Test all switches by removing one wire from the switch to determine that the control circuit is de-energized and that the combustion process is interrupted. Preferably on an annual basis, test the function of the switch itself by disconnecting the supply tubing (or other appropriate activating source) to make certain that the switch is functional, in addition to verification of the electrical circuitry. As with all tests involving combustion safety equipment, these tests should be conducted only by persons thoroughly experienced in burner controls and safety equipment.

CHEMICAL CLEANING

Chemical Cleaning of New Systems

Chemical cleaning of boilers and hot water heating systems is mandatory for all new systems to remove oils, greases, dirt, and other debris.

Steam boilers must be boiled out with an alkaline compound. This usually consists of a mixture of equal parts of trisodium phosphate, caustic soda, and soda ash.

Use one-half pound (or 1¼ percent by weight of water in boiler) of this mixture per boiler horsepower. The following procedure is recommended:

1. Fill the boiler with water, allowing enough room for the cleaning mixture.
2. Dissolve required amount of mixture in water, one pound per gallon of water.
3. Add solution directly to boiler through top manhole or other convenient opening.
4. Close stop valve in header, remove safety valve, and do not cap or plug the opening.
5. Fire boiler without generating steam for 16 to 24 hours so that water is just below boiling point.
6. Bottom blow the boiler periodically during the cleaning.
7. Drain boiler and flush thoroughly with fresh water to remove all of the boiling-out compound. This is indicated when the rinse water is no longer pink to phenolphthalein or the pH is the same as the water used for flushing.

8. Refill the boiler to its operating level with deaerated or preheated feedwater. Add required chemical treatment, replace safety valve, open stop valve in header, and fire slowly to bring boiler gradually up to operating pressure. If boiler is connected with another boiler(s) to the same header, *do not* open stop valve until boiler is at system pressure.

Cleaning Condensate Return Systems

When a steam system is put in operation for the first time or after a long shutdown, all condensate should be run to waste for 7 to 10 days, or until its runs clear. Any oil, grease, dirt, iron corrosion products, or other debris in the condensate return system is thus prevented from returning to the boiler. Problems such as deposit corrosion of the boiler tubes, foaming, and priming will be prevented by this procedure.

During this period, the boiler will require the use of 100 percent raw water for makeup. Special attention must be given to water treatment and blowdown during this period. Water treatment should be checked every four hours or more frequently.

Chemical Cleaning of Hot Water Systems

Hot water heating systems must be thoroughly cleaned with a low-foaming, detergent-type cleaner. This will remove all oil, grease, dirt, fluxes, and other debris resulting from construction. The cleaner is normally used at a concentration of two pounds per 100 gallons of water in the system (0.25 percent by weight). The following procedure should be followed:

1. Fill system with fresh water, venting where needed.
2. Open all manual and automatic water valves to avoid bypassing any parts of the system, such as coils.
3. Dissolve cleaner in water—one pound per gallon.
4. Add cleaning solution to system by means of a bypass feeder or transfer pump.
5. Recirculate system for four to eight hours at 140°F if possible.
6. Drain and flush system thoroughly with fresh water until water drained from the system has the same pH as the fresh water.

If the system circulating pump has mechanical seals, have a spare set of seals available. Dirt dislodged during cleaning may damage the mechanical seal. Where the mechanical seals are equipped for external lubrication, the seal cavity supply water should be filtered with a 20-micron filter or a cyclone separator.

CARE OF IDLE BOILERS

It is probable that more damage occurs to a boiler left idle than to a boiler in operation. The water side of a boiler will corrode whether or not the boiler is in operation, if proper precautions are not observed. Although there is no danger of scale forming in an idle boiler, severe corrosion and pitting may occur unless you take the necessary precautions while a boiler is idle or laid up, either on a short-term or long-term basis.

If a boiler is to be out of service for more than 30 days, protect it by either the dry or wet method of storage during the period of nonuse. Thoroughly clean the fireside surfaces of the boiler of any soot or other deposits (scale, adhering soot, fly ash, and so on). This is to prevent acid attack on the surfaces of the boiler if the deposits become moist from moisture (humidity) in the air. Inspect water side surfaces of the boiler, and thoroughly wash and chemically clean them if required (all scale deposits removed).

If the boiler is to be out of service for a long period of time or may be exposed to freezing temperatures, use the dry layup method:

1. Drain and clean fire and water side surfaces of the boiler thoroughly. Treat fireside surfaces with a chemical compound such as ultraspray (manufactured by Fuel Efficiency Inc., Clyde, NY) or a similar cleaner and protectorant. Clean water side surfaces to bare metal, removing all scale, sludge, rust and deposits.
2. *Completely* dry all surfaces.
3. Place slaked lime or other desicant in trays inside the water side of the boiler. Seal boiler tightly to exclude all moisture and air. Examine desicant periodically and replace or dry out as required.
4. Close all valves leading to the boiler waterside, secure all manual fuel supply valves.
5. Open boiler door at stack end of boiler to prevent flow of moist warm air through the fire side. Close breeching dampers to prevent moisture from entering the boiler via the breeching, chimney connection. If boiler is to be laid up for a considerable period of time, place a moisture absorbing compound (quick-lime or silica-gel) in the combustion area. Use approximately 30 lbs per 100 BHP. Place this material on trays in the combustion zone. Make periodic inspections and rejuvenate or replace the material as required.
6. To prevent damage to the control circuit, it is recommended that power remain applied to the control circuit components. If this is not possible, use some means of "warming" the control cabinet, such as a 100-watt light bulb which operates continuously or a small electric heater of some type (thermostatically controlled).

For shorter periods of time, use the wet layup method described here:

1. Clean fireside surfaces of the boiler thoroughly. Treat fireside surfaces with a chemical compound such as Ultraspray (manufactured by Fuel Efficiency Inc., Clyde, NY) or a similar cleaner and protectorant.
2. Flood boiler to overflowing with the hottest water possible (use deaerated water if possible). Treat water with chemical treatment to prevent oxygen pitting during idle period. Consult with your chemical treatment company regarding the type of additive to use. Treatment of water during layup period is essential, even though idle, the levels of treatment should be monitored during the layup period.
3. To prevent damage to the control circuit, it is recommended that power remain applied to the control circuit components, if this is not possible, use some means of "warming" the control cabinet, such as a 100-watt light bulb which operates continuously or a small electric heater of some type (thermostatically controlled). Regardless of the method of layup, before returning the boiler to service, drain or fill (as required) and perform a minimum of monthly and preferably an annual maintenance check, *before* the boiler is allowed to operate automatically. In addition, return water treatment levels to their normal values.

INCREASING THE PLANT'S EFFICIENCY

Using Fuel Oil Additives

Fuel oil additives are very helpful in reducing soot and vanadium deposits on the fire side of boiler tubes. They cannot, however, be expected to eliminate soot caused by improper adjustment of oil burners. These additives contain metallo-organic compounds that are soluble in oil. Iron, manganese, and cobalt compounds lower the ignition temperature of carbon (soot) to 600°F or lower by catalytic action. Combustion catalysts in a fuel oil additive will provide significant benefits such as more complete combustion, reduced soot deposits, improved heat transfer, and a cleaner stack effluent.

Vanadium slag is a product of the fusion of various inorganic impurities in the fuel oil into a molten ash that hardens at low temperatures. Fuels high in vanadium have an ash with a very low melting point. The molten ash combines with metal tube surfaces, baffles, stack, etc., reducing heat transfer and causing serious corrosion of iron and steel surfaces. The use of magnesium and manganese compounds raises the melting point of the ash or slag to 1,400°F or higher. The resultant ash can be removed easily by soot blowing or brushing.

Fuel oil additives may also contain wetting agents, emulsifiers and organic alkaline compounds. The wetting agents and emulsifiers prevent the accumulation of moisture caused by the condensation in fuel oil storage tanks. These wetting agents will take care of up to about 2 percent of water in the oil. If large amounts of water enter a fuel tank through loose fill caps or other means, it is necessary to pump such huge amounts of water from the tank because there will not be a sufficient amount of emulsifier present to cope with the water.

Organic alkaline compounds will neutralize any acidity that may be present in water lying in the bottom of the fuel tank. Tank life is thus extended by preventing corrosion caused by acidity.

Many #4 and #6 oils have high paraffin wax concentrations. The waxes crystallize from the oil even at relatively high temperatures. Waxes have been encountered with melting points as high as 140°F. Waxes clog fuel lines, causing flame outs. Fuel oil additives with wax and sludge dispersals will prevent this from occurring.

The selection of fuel oil additives on the basis of price alone should be avoided. There are many low-priced additives that are nothing more than kerosene. In some cases such additives do not even mix with the fuel oil.

There are also some fuel oil additives, many at very high prices, that are sold on the basis of saving up to 25 percent on fuel oil usage. Nothing could be further from the truth. No fuel oil additive can substitute for proper burner adjustment. If combustion efficiency is being maintained at 78 to 80 percent there is no way that a fuel oil additive can increase combustion efficiency another 25 percent!

Energy Conservation

It is in the national interest to conserve energy to preserve dwindling supplies of fuel. It is in the operating engineer's interest to conserve energy so as to lower operating costs. Everyone is looking for some dramatic way to reduce operating costs as fuel costs skyrocket. In most cases, however, significant savings will result only by saving a little here and a little there. The following fuel savings are possible in any typical operation:

- Increase combustion efficiency from 75% to 80% 6.3%
- Lower operating pressure from 125 psig to 40 psig 2.0%
- Recover 10% more condensate 1.3%
- Keep fire and water sides clean .... 5 %

Total fuel savings 14.6%

Any one of the above savings is hardly a dramatic one, but, collectively, they are dramatic. Thus, if the above operation used 100,000 gallons of fuel oil per year, annual savings of 14,600 gallons of oil could be realized. At $1.00 per gallon, annual savings of $14,600 are possible!

Is the above operation a large one? Not really. It could apply to a 200 HP boiler operating at 50 percent of capacity for only 3,333 hours per year. How to obtain the above savings is discussed below.

Combustion Efficiency

The greatest savings in fuel can be achieved by keeping the burner properly tuned. Combustion occurs when a proper relationship exists between a fuel, oxygen (from air), and heat. Because a burner can never achieve proper combustion with the theoretical fuel/air mixture, and because heat transfer is never 100 percent efficient, it is impossible to obtain 100 percent combustion efficiency. In order to obtain combustion with little or no smoke, it is necessary to provide the combustion process with excess oxygen (air). In most cases, about 30 percent excess air is required. Combustion efficiency decreases as excess air increases. Since one of the products of combustion that can be easily measured is CO_2, the greater the amount of excess air, the lower the concentration of CO_2 will be in the stack gas. Therefore, the lower the CO_2, the lower the combustion efficiency will be.

Another indication of efficiency is the stack gas temperature. The higher the stack gas temperature, the lower the efficiency. Soot formation on the fire side acts as an insulator, preventing heat transfer from the hot gases to the boiler water. Excessive heat is lost up the stack. When using the following graphs, the stack temperatures used is the *net* stack temperature. For practical purposes:

Net Stack Temperature °F = Stack Temperature °F—Boiler Room Temperature °F

Figures 8-6, 8-7, and 8-8 show the relationship between CO_2, net stack temperature, and percentage of heat lost up the stack for natural gas, #2 oil, and #6 oil respectively.

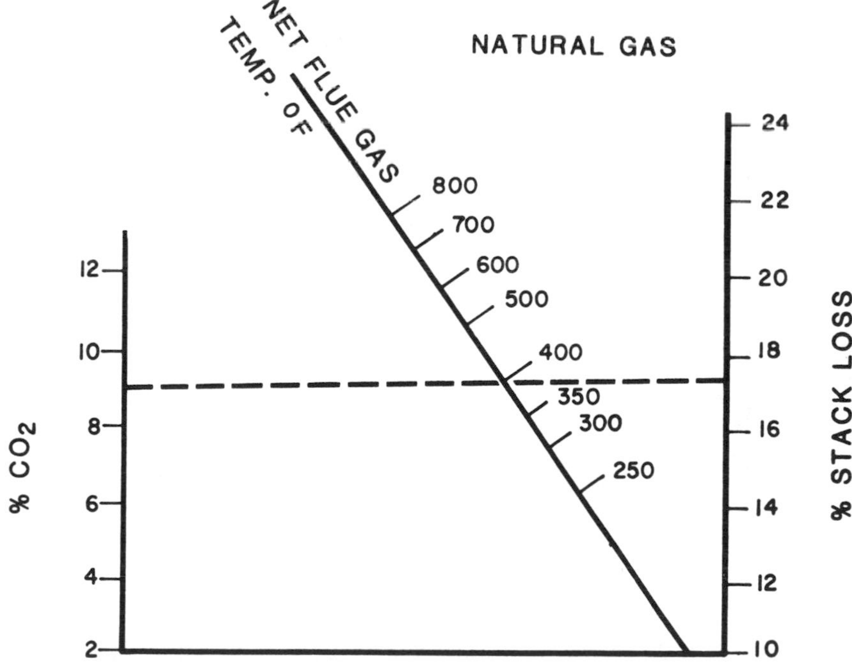

Figure 8-6. Natural Gas—Relationship Between Percent of CO_2 and Percent of Stack Loss

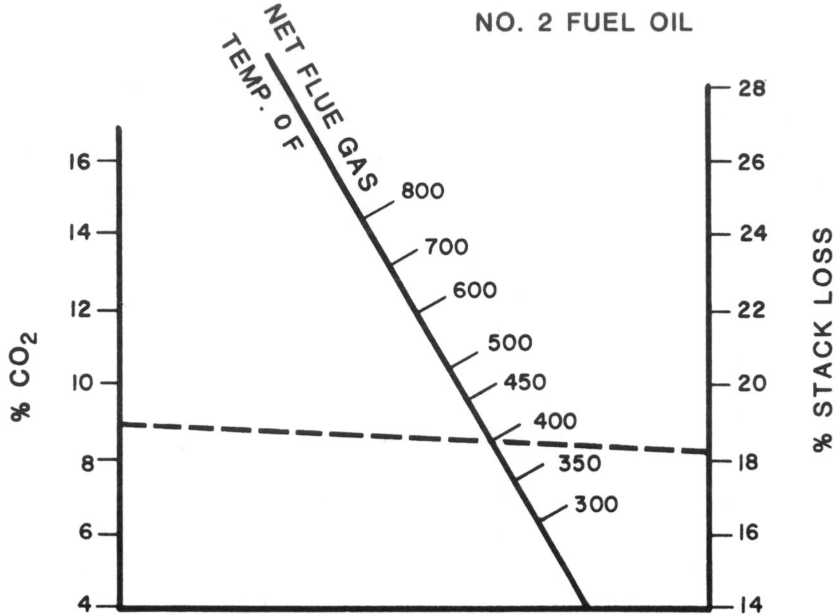

Figure 8-7. No. 2 Fuel Oil—Relationship Between Percent of CO_2 and Percent of Stack Loss

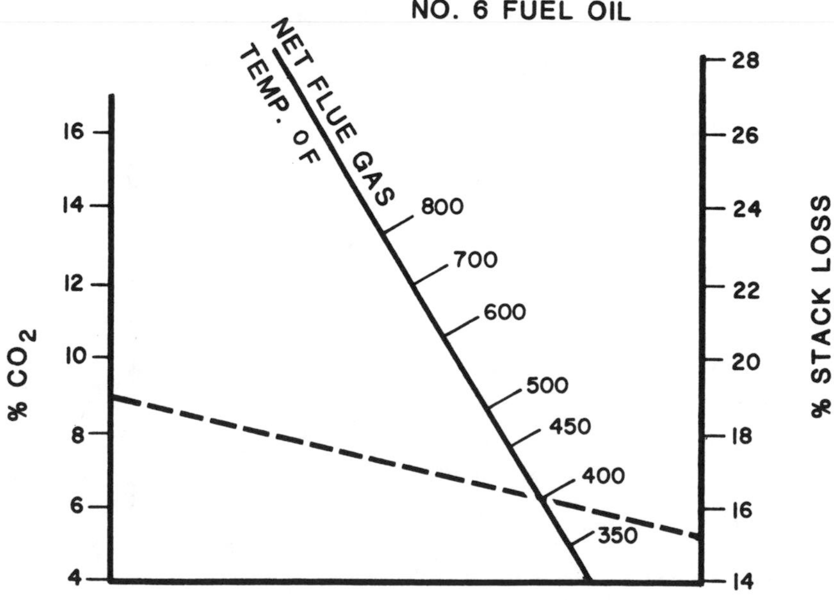

Figure 8-8. No. 6 Fuel Oil—Relationship Between Percent of CO_2 and Percent of Stack Loss

$$\% \text{ Combustion efficiency} = 100\% - \% \text{ Stack loss}$$

The dotted line in Figure 8-6 shows that with 9 percent CO_2 and a net stack temperature of 400°F, the stack loss is about 17.3 percent. The combustion efficiency is therefore about 82.7 percent.

The dotted line in Figure 8-7 shows that with 9 percent CO_2 and a net stack temperature of 400°F, the stack loss is about 18.3 percent. The combustion efficiency is therefore about 81.7 percent.

The dotted line in Figure 8-8 shows that with 9 percent CO_2 and a net stack temperature of 400°F, the stack loss is about 15.2 percent. The combustion efficiency is therefore about 84.8 percent.

It can be seen from the above that with the same CO_2 and net stack temperature readings, stack losses will vary depending upon the type of fuel used. If #4 fuel oil is being used, one can determine stack losses by averaging the stack losses for #2 and #6 fuel oils. In the above examples, if #4 fuel oil were being used, the combustion efficiency would be about 83.3 percent.

Before making any burner adjustments to increase combustion efficiency, be certain to clean the water and fire sides of the boiler. When adjusting the burner, make certain that:

- Burner is supplied with fuel at its high fire design rate when burner is operating at high fire.
- Adjust combustion air and/or fuel rate to give maximum CO_2 with zero smoke for gas and #2 oil and a maximum of two to three smoke on heavy oils as measured by the Bacharach smoke tester.

  When adjusting a gas burner, it is absolutely essential that stack gases be tested for carbon monoxide (CO). The carbon monoxide concentration must be zero.
- Check efficiency at low and high fire. Check through entire firing range with full modulation burners.

From time to time, various "gadgets" are promoted that are touted to increase combustion efficiency by 20 to 25 percent. These relate primarily to emulsifying water with oil or adding moisture to the combustion air. Remember that water and air add no BTUs to the combustion process. When such savings are observed, they usually result from proper adjustment of the fuel-air mixture. Remember also that when flue gas analyses show 80 percent or better combustion efficiency, little can be done to improve the operation significantly.

Boiler Operating Pressure

Boiler water temperature increases as boiler water pressure increases. At 15 psig, boiler water temperature is about 250°F; whereas at 125 psig, boiler water temperature is about 355°F. Net stack temperature at high fire will be about 150°F higher than the boiler water temperature. Figures 8-6, 8-7, and 8-8 show that efficiency decreases with increasing net stack temperature.

Fuel consumption can be decreased by lowering the boiler operating pressure. Figure 8-9 shows how fuel consumption decreases as operating pressure decreases for a boiler normally operated at 125 psig. Simply stated, less heat is lost up the stack. Many industrial boilers that are operated at 125 psig during the day, could operate at a lower pressure at night during the heating season and save significant amounts of fuel. Fuel savings as high as 4½ percent are possible during those hours when a lower operating pressure is permissible.

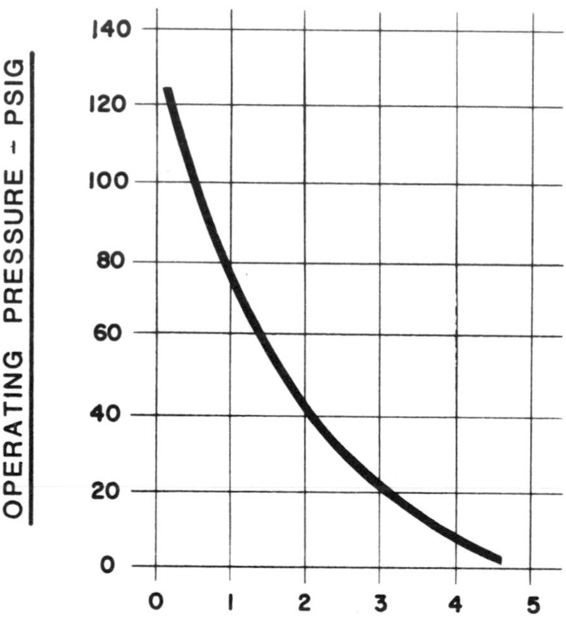

Figure 8-9. Decrease in Fuel Versus Operating Pressure for 125 Design Boiler

Condensate Losses

Another source of potentially high energy loss is condensate line leakage or other condensate losses. The loss of all returns on a 200 HP boiler operating at an average load of 50 percent means a loss of about 3,450 pounds of condensate per hour. Assuming that this condensate would have been returned at 200°F, and that the raw water used to replace it would be at 50°F, 3,450 pounds of raw water would have to be heated 150°F. The BTUs required are as follows:

$$3{,}450 \times 150 = 517{,}500 \text{ BTUs per hour}$$

Assuming the net BTUs from a gallon of fuel oil at 140,000 BTUs per gallon:

$$\frac{517{,}500}{140{,}000} = 3.7 \text{ gallons of fuel oil per hour}$$

Since at 100 HP operating load the boiler will be using about 30 gallons of oil per hour, 3.7 gallons of oil represents about 12.3 percent of the oil required for operation. Thus, every 10 percent of condensate that can be saved will save about 1.25 percent of fuel requirements. Every boiler feedwater system should be equipped with a water meter in the raw water makeup line. Condensate losses can then be measured and steps taken to minimize them. Daily water meter readings should be logged and any significant increases in raw water requirements should be investigated and corrected.

Thermal Efficiency

Operation of a boiler plant, with its related equipment, requires particular attention to economical considerations, both for expenditure of money on fuel, and to discharge a minimum of pollution to the atmosphere. Previously, portions of this chapter have discussed combustion testing and its importance to proper plant operation. We will now consider items related to thermal efficiency.

There are many older steel fire tube and CI water tube boilers having generous, and by today's standards, over size flue gas passages. The design of many of these boilers manufactured prior to the 1970s, was such that they accommodated all fuels; therefore, their heating surfaces allowed for contamination by those fuels which tended to burn leaving a residue in the tubes or on the tube surfaces. The addition of specifically designed turbulators to increase flue gas turbulence within the boiler results in lowering exit flue gas temperatures, and an increase in overall thermal efficiency. Numerous tests have been conducted proving the fuel savings which are possible when applying designed turbulators to these boilers. Fuel Efficiency, Inc. (Clyde, NY) is one of the manufacturers of this type of equipment. Investigation conducted by A.S.M.E. (their report 83-HT-44) supports the usage of this type of turbulator for improving operating thermal efficiency. The installation requires trained technicians to perform combustion testing and evaluation of available furnace draft. The proper application of flue gas turbulators will usually result in lowering the exit gas temperature to within 125 to 150 degrees of the boiler operating temperature.

Certain boilers, due to their configuration, are not suitable for installation of turbulators, however may have exit gas temperatures which are higher than desired. These boilers which are not suitable candidates for turbulators, may yield increased efficiency by installation of heat reclaiming accessories adapted to the flue gas discharge. One company which manufactures this material is Kentube (Tulsa, OK).

Plant maintenance managers may apply heat recovery equipment to new or existing boilers. The heat recovered with this type of equipment is usually used to pre heat incoming make up water to the boiler. A typical boiler economizer installation is indicated in Figure 8-10. The installation indicates a BTU meter, installed to verify savings. Once again the goal of this type of equipment is to reduce exit gas temperatures and improve the thermal efficiency of the boiler plant.

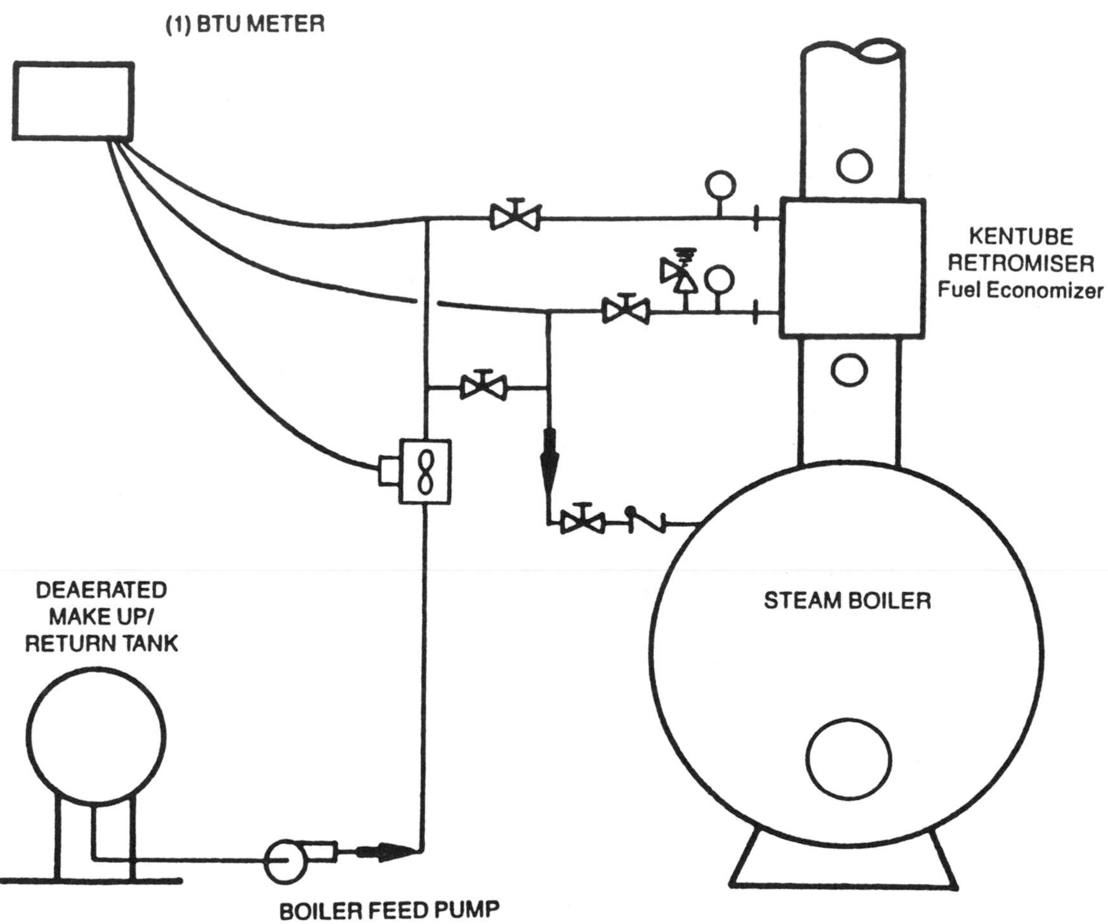

Courtesy of Kentube

Figure 8-10. Typical Boiler Economizer Installation

The maintenance of peak thermal efficiency may be enhanced by keeping fire-side surfaces of the boiler clean at all times. The best method of accomplishing this is by means of an automatic tube cleaner. The purpose of this device is to clean each tube in the boiler once each hour of operation, thereby maintaining essentially a clean boiler at all times. The automatic function and frequency (hourly), cleaning is essential to peak thermal efficiency.

TROUBLE SHOOTING CHART FOR GAS, OIL AND GAS/OIL BURNERS* (FOR GAS AND LIGHT OIL BURNERS IN THE SIZE RANGE UP TO 100 gph)

GENERAL PROBLEMS

1. Burner Fails to Start
   - A. Defective On/Off or fuel transfer switch. Replace switch.
   - B. Control circuit has an open control contact. Check limits, low water cut-off, proof of closure switch and others as applicable.
   - C. Bad fuse or switch open on in-coming power source. Correct as required.
   - D. Motor overloads tripped. Reset and correct cause for trip out.
   - E. Flame safeguard control safety switch tripped out. Reset and determine cause for apparent flame failure.
   - F. Loose connections or faulty wiring. Tighten all terminal screws and consult wiring diagram furnished with the burner.
   - G. Frozen oil pump shaft preventing blower motor operation. Replace oil pump.
   - H. Flame safeguard control starting circuit blocked due to flame relay being energized. Possible defective scanner—replace. Possible defective amplifier—replace. Scanner actually sighting flame due to leaking fuel valve—correct unwanted flame cause. Defective flame safeguard control—replace.
   - I. Defective blower motor. Repair or replace.
2. Occasional Lockouts for No Apparent Reason
   - A. Gas pilot ignition failure. Refer to pilot manufacturer's adjustment section and readjust to make certain that ignition is instant and that flame signal readings are stable and above minimum values. Use a manometer on pilot gas supply line to make certain that pressure is as recommended.
   - B. Check for proper settings on direct spark oil ignition electrodes. Make certain that gap is not too wide and that "light-off" oil pressure is as recommended by burner manufacturer.

*Courtesy of Power Flame, Inc.

C. Gas pilot ignition and direct spark oil ignition. Verify that there are no cracks in the porcelain and that transformer end and electrode end plug in connections are tight.

D. Loose or broken wires. Check all wire nut connections and tighten all terminal screw connections in panel and elsewhere as appropriate.

E. With flame safeguard controls that incorporate the air flow switch in the non-recycling circuit, ensure that when main flame lights, the air flow switch is not so critically set as to allow occasional momentary opening of the air switch contacts.

F. Occasional low voltage supply. Have local utility correct. Make certain that the burner control circuit transformer (if supplied) is correct for the voltage being supplied.

G. Occasional low gas supply pressure. Have local utility correct.

H. Air leak in oil suction line or check valve not holding. Correct as required.

GAS OPERATION PROBLEMS

1. Burner Motor Runs, but Pilot Does Not Light

A. Gas supply to burner shut off—make sure all manual gas supply valves are open. Automatic high pressure valve at meter such as "Sentry" type tripped shut due to high gas pressure—reset valve and correct cause for trip out.

B. Pilot solenoid valve not opening—listen and feel for valve actuation. Solenoid valve not being powered—check electrical circuitry. Replace coil or entire valve if coil is burned out.

C. Defective gas pilot regulator—replace.

D. Gas pressure too high or too low at pilot orifice. Check orifice size in gas pilot assembly. Replace if incorrect. Refer to manufacturers gas pilot adjustment for correct settings. Readjust as required.

E. Defective ignition transformer—replace. Incorrect ignition electrode settings—refer to manufacturers gas pilot adjustments for correct settings.

F. Defective flame safeguard control or plug in purge timing card. Replace as required.

G. Air flow switch not making circuit—check out electrically and correct pressure adjustment on switch if required. Defective air flow switch—replace. Air switch negative pressure sensing tube out of position—reposition as necessary.

2. Burner Motor Runs & Pilot Lights, but Main Gas Flame Is Not Established
   - A. Main shut off or test cock closed. Check to make certain fully open.
   - B. Pilot flame signal reading too low to pull in flame safeguard relay. Refer to minimum signal required for proof of pilot.
   - C. Defective automatic main or auxiliary gas shut off valves. Check electrical circuitry to valves. Replace valves or correct circuitry as required.
   - D. Main diaphragm shut off valve opening too slowly. Correct as required.
   - E. Defective flame safeguard control or plug in amplifier. Check and replace as required.
   - F. Butterfly valve set incorrectly on modulating burner. Readjust as required.
   - G. Main gas pressure regulator atmospheric vent line obstructed. Correct.
   - H. Defective main gas pressure regulator—replace. Misadjusted main gas pressure regulator—readjust to meet required operational values.
3. Carbon Monoxide Readings on Gas Firing
   - A. Flame impingement on "cold" heat transfer surfaces caused by excessive firing rate. Reduce firing rate to correct input volume.
   - B. Flame impingement on cold combustion chamber surfaces due to undersized combustion chamber. Refer to burner manufacturers minimum combustion chamber size requirement.
   - C. Incorrect gas/air ratios. Readjust burner to correct CO_2/O_2 levels eliminating all CO formation.
4. Gas High Fire Input Cannot Be Achieved
   - A. Gas company pressure regulator or meter operating incorrectly, not allowing required gas pressure at burner train inlet. Have gas company correct.
   - B. Gas cock upstream of train inlet not fully open. Check and correct.
   - C. Gas line obstructed. Check and correct.
   - D. Gas train main and/or leak test cocks not fully open. Check and correct.
   - E. Gas supply line between gas company regulator and burner inlet too small. Check supply pressure at meter, determine pressure drop and increase line size as required, or raise supply pressure to compensate for small line. Do not raise pressure so high that under static (no flow) conditions the pressure exceeds the maximum allowable pressure to the gas train components on the burner.
   - F. Burner gas train components sized too small for supply pressure. Increase component size as appropriate.

G. Automatic gas valve not opening fully due to defective operation. Replace gas valve.

H. Limiting orifice (if supplied) too small. Replace with correct size.

I. On modulating burner, butterfly valve not fully opened. Readjust.

J. Defective main gas pressure regulator. Replace.

K. Incorrect spring in main gas pressure regulator. Replace as required.

L. Main gas pressure regulator vent line obstructed. Check and correct.

M. Normally open vent valve (if supplied) not closing when automatic gas valves open. Check to see if valve is fully closed when automatic valves are open. Replace vent valve, if not closing fully.

OIL OPERATION PROBLEMS

1. Burner Motor Runs, but Direct Spark Ignited Oil Flame Is Not Established

A. Defective or incorrect size oil nozzle. Remove and clean or replace.

B. Low oil pressure. Check with gauge for correct "light-off" pressure.

C. Defective oil pump. Replace.

D. Defective oil solenoid valve. Replace.

E. Oil pump coupling loose or defective. Replace or lighten as required.

F. Low oil pressure switch (if supplied) defective or incorrectly set. Adjust or replace switch.

G. Ignition transformer defective. Replace.

H. Ignition electrode set incorrectly. Remove electrodes and reset.

I. Ignition electrodes cracked and grounding out spark. Replace electrodes.

J. Ignition leadwire defective and grounding spark out. Replace.

K. Ignition plug in connections at transformer or electrodes loose. Tighten.

L. Air flow switch (if provided) not making. Reset pressure or replace.

M. Defective flame safeguard control or plug in purge time card. Replace.

N. Air dampers held in high line position due to mechanical binding of linkage. Readjust linkage.

O. Loose wiring connections. Check and tighten all connections.

2. Oil Flame Ignites, but then Flame Safeguard Control Locks Out on Safety

A. Flame scanner lens dirty. Remove and clean.

B. Scanner sight tube blocked or dirty. Check and clean.

C. Flame scanner defective. Replace.

D. Defective oil nozzle causing unstable flame and scanning problems. Replace oil nozzle.

E. Fuel/air ratios incorrect, resulting in unstable or smoky flame causing scanner flame sighting problem. Readjust ratios for clean stable flame.

F. Defective flame safeguard amplifier or control. Replace as appropriate.

3. Oil Flame Extremely Smoky at Light Off or In Low Fire Position

A. Defective or incorrect size oil nozzle. Replace.

B. Fuel/air ratio incorrect. Readjust.

C. N.C. oil solenoid valve in oil nozzle return line not opening. Check electrical circuitry and replace valve if defective.

D. On two step pump—N.O. pump mounted solenoid valve malfunctioning. Replace valve or pump.

4. Light Off Oil Flame Is Established and Proven, but Burner Will Not Attempt to Go to the High Fire Position

A. Low/High/Low or Modulating burner high fire temperature or pressure control could be defective or not set to call for high fire. Readjust or replace control.

B. Loose wires or incorrectly wired. Verify wiring and tighten all connections.

C. Flame safeguard control or high fire panel switching relay (if supplied) defective. Verify and correct as required.

D. High fire 3 way solenoid valve defective. Replace.

E. Hydraulic oil cylinder defective. Replace.

F. On two step pump—N.O. solenoid valve defective (not closing). Replace pump or valve.

G. Linkage mechanically binding. Readjust linkage.

H. On modulating system—defective modulating motor. Replace.

5. Low Oil Flame Is Established and Proven, but Flame Out Occurs in Transition from Low Fire to high Fire

A. On Low/High/Off or Low/High/Low system—N.C. oil solenoid valve in nozzle return line not closing (or leaking). Check valve operation and replace if necessary.

B. On two step oil pump—N.O. solenoid valve defective (not closing). Replace valve or pump.

C. Defective or incorrect size oil nozzle. Replace.

D. High fire oil pressure too low. Readjust.

E. Air dampers set too far open at low fire, which causes flame to blow out in starting to high fire. Readjust dampers.

F. Oil pump coupling loose or defective. Tighten or replace.

G. Defective oil pump. Replace.

H. Linkage mechanically binding. Readjust.

I. On modulating systems—fuel/air ratios set incorrectly, causing flame to blow out when going to high fire. Readjust linkage.

6. White Smoke Formation on Oil Firing

A. Oil/Air ratios incorrect due to excess air, or oil flow is too low. Readjust for proper fuel input, CO_2 and smoke reading.

7. Gray or Black Smoke Formation on Oil Firing

A. Impingement on cold combustion chamber surfaces due to undersized chamber, or incorrect oil nozzle spray angle for application. This could also result in carbon formation on chamber surfaces. If chamber is the correct size, change nozzle spray angle in order to shorten or narrow the flame as required.

B. Defective or dirty oil nozzle. Replace or clean nozzle.

C. Incorrect oil/air ratios. Readjust burner to correct CO_2 and smoke levels.

D. Oil pressure too low resulting in poor atomization. Readjust.

E. Impingement of raw oil spray on the blast tube choke ring or oil nozzle air diffuser. Make certain that the diffuser is seated firmly against the oil nozzle adapter shoulder. Position the oil gun assembly fore or aft in the blast tube to assist in elimination of oil spray on the blast tube choke ring.

8. Oil High Fire Input Rate Cannot Be Achieved

A. Oil nozzle size too small. Remove nozzle and check markings. Replace with correct size nozzle.

B. Nozzle defective—replace. Nozzle mesh filter dirty—clean or replace.

C. Oil supply pressure to nozzle too low. Readjust.

D. Oil pump defective. Replace.

E. On Low/High/Off and Low/High/Low systems—N.C. oil solenoid valve in nozzle return line not closing (or leaking). Check valve operation and replace if necessary.

F. On two-step pump—N.O. pump mounted oil—solenoid valve defective (not closing). Replace valve or pump.

G. Oil pump coupling loose (slipping) or defective. Replace.

H. Linkage mechanically binding. Readjust.

I. On modulating burner, oil nozzle return line metering valve set incorrectly. Readjust to attain required nozzle bypass pressure.

J. Oil suction line too small or partially blocked. Make vacuum test while at high fire. If the vacuum is in excess of 10″ HG, consult pump manufacturer for total flow rate and line sizing data. Make line size changes if required.

K. Blocked or dirty suction line oil filter. Replace or clean.

L. Manual valves in suction line not fully open. Check and correct.

M. Suction line check valve or foot valve operating incorrectly. Check and correct.

N. Vent system on oil tank blocked creating vacuum on tank, with high vacuum and lowered oil flow to burner. Check and correct.

OIL NOZZLE SERVICING

No Oil Delivered

1. The nozzles should be removed from the nozzle adapter by use of the proper wrench. They should be disassembled and thoroughly cleaned with a liquid solvent (preferably non-flammable) and a brush.
2. Do not use a screwdriver, wire brush or similar metallic objects to clean nozzles. Damage to orifices or spray slots result in off-center or "sparky" fires.
3. The nozzle should be seated firmly in the nozzle adapter to prevent leaks.
4. If a nozzle is damaged or burned, replace it.
5. The entire oil tube and nozzle assembly (the oil drawer assembly) may be removed for ease of service.
6. When cleaning and taking the nozzle apart, do not force it.

OIL PUMP OR OIL FLOW PROBLEMS AND TYPICAL SOLUTIONS

1. Reversed pump rotation
2. Suction lift too high
3. Air leak in suction line
4. Pump not primed, or has lost prime
5. Pump coupling not installed properly
6. Pump defective
7. Line plugged
8. Valve closed

Capacity Too Low

1. Suction lift too high
2. Air leak in suction line
3. Suction line too small
4. Check valve or strainer is obstructed or dirty
5. Mechanical defects—pump badly worn or seal defective

Noisy Pump

1. Air leak in suction line
2. Pump not securely mounted
3. Vibration caused by bent shaft or misalignment
4. Pump overloaded
5. Suction line vacuum so high that vapor forms within the liquid

Pump Leaks

1. Cover bolts need tightening; gasket broken or defective
2. Mechanical seal (used on certain models) may be scratched, due to dirt
3. Inlet head pressure too high. Install a pressure reducing valve set at 3 psig or less.
4. Oil line fittings not tight

9

Testing and Maintaining Electrical Systems

Ronald Texel
Alfred T. Baker, P.E.

Commerce and industry depend upon reliable and adequate electric utility supply to function. This chapter focuses on essential elements of electric systems and building power supplies, addressing and identifying those methods that optimize electric system performance, longevity, efficiency, and service reliability. These elements include the following:

- Transformers
- Switchgear and switchboard assemblies
- Circuit breakers and switches
- Protective relays
- Cables and busways
- Grounding systems
- Emergency generators
- Batteries and DC systems
- Uninterruptible power supplies
- Automatic transfer switches
- Miscellaneous equipment: capacitors, lightning/surge arresters

- Testing and maintenance staffs: in-house versus outside testing company
- Electric distribution system engineering studies
- Maintenance/test forms and checklists.

Electrical maintenance expenditures are monies well spent. When electric utility services are well run and maintained, a reputation for performance, dependability, and organizational competence is sure to result. In addition, a measure of energy conservation will also result because properly operating equipment will be more efficient.

The inspection, maintenance and testing procedures presented in this chapter are taken largely from the *1989 Maintenance Testing Specifications* developed by the InterNational Electrical Testing Association (NETA, headquartered in Morrison, CO).

The various forms in this chapter were developed by ELEMCO TESTING CO., INC. which are used as guidelines in inspection, maintenance, and testing procedures and are similar to those used by others in the testing industry (ELEMCO, headquartered in Bohemia, NY).

TESTING AND MAINTENANCE STAFFS: CHOOSING BETWEEN AN IN-HOUSE AND OUTSIDE TESTING COMPANY

Training Your In-House Staff

You should develop and train the maintenance and testing staff in accordance with the electrical equipment needs of the systems that are in service at the facility. The major considerations associated with the development of the in-house staff are:

- Technical qualifications of staff.
- Depth and sophistication of maintenance and testing equipment.
- Complexity and importance of plant equipment.
- Annual maintenance budget.

Using an Outside Testing Company

In addition to your in-house staff, you may decide to engage the services of a recognized, independent testing company to perform maintenance, inspections and tests on your electrical systems. The decision to hire an outside company could be based on the importance and complexity of the plant's equipment, the expertise of the in-house staff, and the magnitude of your maintenance budget.

The testing company should not only provide the expertise of testing the electrical systems but should also provide the needed materials and equipment to perform such tests. The testing company should perform the needed equipment repairs or at least should arrange for a reputable outside company to repair any defective equipment.

It is to your advantage to retain the testing company on a continuing basis so that the company's personnel will become familiar with the overall operation of the facility.

Qualifications of the testing company

The following are some of the qualifications the testing company should have. The following items could be used in preparing a specification associated with hiring a testing company.

- The testing company should be a corporately independent testing organization which can function as an unbiased testing authority, professionally independent of the manufacturers, suppliers and installers of equipment or systems that are to be evaluated by the testing company.
- The testing company should be regularly engaged in the testing, inspecting and maintaining of electrical equipment.
- The testing company should use qualified technicians to provide the testing services.
- The testing company should use testing and maintenance equipment that has been qualified and, when necessary, certified for the specific task.

The testing company should also submit a test report at the conclusion of work assignments; this report should include the following, as a minimum:

- Summary of findings and recommendations as to possible needed repairs or replacement.
- Description of systems and/or equipment tested.
- Test procedures followed.
- Test results, including standardized test sheets, recording tapes, pictures of defective equipment, and so on.

Contracting for Electrical Distribution System Engineering Studies

It is strongly recommended that you retain the services of a professional engineering company to evaluate the adequacy of the electrical distribution system, as it pertains to the standards published by the various national engineering and equipment organizations. The types of engineering studies to be performed should be the following, as a minimum:

- Load studies
- Short circuit studies
- Protective equipment coordination studies
- Equipment failure analysis

Engineering Studies should be made for the following reasons and the related scheduling.

Facility Expansion

Engineering studies should be conducted *before* a major facility expansion is designed. Load, short circuit and coordination studies should be made to properly integrate the new equipment with the existing electrical distribution system.

The engineering firm should provide a current and complete short circuit study, equipment interrupting or withstand evaluation, and a protective device coordination study for the electrical distribution system. The studies shall include all portions of the electrical distribution system from the normal and alternate sources of power throughout the low voltage distribution system. Normal system operating method, alternate operation, and operations which could result in maximum fault conditions shall be thoroughly covered in the study.

Short Circuit Study

The study shall be in accordance with the applicable IEEE and ANSI Standards. The study input data shall include the utility company's short circuit single and three phase contribution, with the X/R ratio, the resistance and reactance components of each branch impedance, motor and generator contributions, base quantities selected, and all other applicable circuit parameters. Short circuit momentary duties and interrupting duties shall be calculated on the basis of maximum available fault current at each switchgear bus, switchboard, motor control center, distribution panelboard, pertinent branch circuit panelboards and other significant locations through the system.

Equipment Evaluation/Load Study

An equipment evaluation study shall be performed to determine the adequacy of circuit breakers, controllers, surge arresters, busways, switches, and fuses by tabulating and comparing the short circuit ratings of these devices with the available fault currents. Any problem areas or inadequacies in the equipment shall be promptly brought to the facility's attention.

Existing and proposed load data shall also be provided to evaluate the equipment's load rating. The load data should be provided with input regarding the time of day and day of week so that the load can be evaluated on a coincident basis.

Protective Device Coordination Study

A protective device coordination study shall be performed to select or to check the selection of power fuse ratings, protective relay characteristics and settings, ratios and characteristics of associated voltage and current transformers, and low voltage breaker trip characteristics and settings.

The coordination study shall include all voltage classes of equipment from the utility's incoming line protective device down to and including each motor control center and/or panelboard. The phase and ground overcurrent protection shall be included as well as settings for all other adjustable protective devices.

The time-current characteristics of the specified protective devices shall be plotted on appropriate log-log paper. The plots shall include complete titles, representative one-line diagram and legends, associated power company's relays or fuse characteristics, significant motor starting characteristics, complete parameters of transformers, complete operating bands of low voltage circuit breaker trip curves, and fuse curves. The coordination plots shall indicate the types of protective devices selected, proposed relay taps, time dial and instantaneous trip settings, ANSI transformer magnetizing inrush and withstand curves per ANSI C37.91, cable damage curves, symmetrical and asymmetrical fault currents. All requirements of the current National Electrical Code shall be adhered to. Reasonable coordination intervals and separation of characteristic curves shall be maintained. The coordination plots for phase and ground protective devices shall be provided on a system basis. Separate curves shall be used to clearly indicate the coordination achieved to each utility main breaker, primary feeder breaker, unit substation primary protective device, main and tie secondary breakers, substation feeder breakers, and main load protective device.

The selection and settings of the protective devices shall be provided separately in a tabulated form listing circuit identification, IEEE device number, current transformer ratios, manufacturer, type, range of adjustment, and recommended settings. A tabulation of the recommended power fuse selection shall be provided for all fuses in the system. Discrepancies, problem areas, or inadequacies shall be promptly brought to the Facility's attention.

Study Report

The results of the power system study shall be summarized in a final report. Copies of the final report shall be submitted to the facility's director.

The report shall include the following sections:

- Description, purpose, basis, and scope of the study; and a single line diagram of the portion of the power system which is included within the scope of study.
- Tabulations of circuit breaker, fuse, and other equipment ratings versus calculated short circuit duties, and commentary regarding same.
- Protective device time versus current coordination curves, tabulations of relay and circuit breaker trip settings, fuse selection, and commentary regarding same.
- Fault current tabulations including a definition of terms and a guide for interpretation.
- Tabulation of appropriate relay tap settings.

You should engage an independent testing firm for the purpose of inspecting, setting, testing, and calibrating the protective relays, circuit breakers, fuses, and other applicable devices as recommended in the power system study report. The testing firm shall strictly conform to the requirements for outside testing companies discussed earlier.

Five-Year Study Program

You should also establish a program where the overall electrical facility is investigated by an engineering firm on a minimum five-year program. Numerous minor changes could have been made to the electrical system which could impact the proper operation of the distribution system. Some of these minor changes could include:

- Re-assigning loads of different distribution panels, creating possible overloads.
- Minor capital improvements.
- Adding of emergency generators and automatic transfer systems.
- Adding of an uninterruptible power supply system.
- Change of the utility's power supply transformer.

Equipment Failure

An engineering analysis should be performed to determine the cause of a specific equipment failure. This study could determine that there were several plant electrical problems that may have caused the failure and they may not be directly associated with the specific failed device. This study could also pinpoint other systems that may be subject to a similar failure, if the cause of failure is not corrected.

INSPECTING AND TESTING TRANSFORMERS

The maintenance of a transformer begins on the day that the transformer is first delivered to the site. Data concerning the design and use of the transformer should be collected and recorded from **"DAY 1"** of the transformer's existence. This data should begin with the manufacturer's factory tests.

Transformer manufacturers provide recommended maintenance procedures and schedules for their equipment. These recommendations should be followed initially. However, the actual experience of the performance and application of a specific transformer should govern the frequency and depth of the maintenance to be performed.

The main causes of transformer failures and their relative frequency are as follows:

Manufacturer	5%
Accidents (physical, thru faults, etc.)	15%
Poor application (load, environment, etc.)	5%
Poor maintenance	75%
	100%

Inspection by Senses—A Checklist

There are numerous procedures and test methods that can and should be used in maintaining transformers; but the primary methodology should be the use of one's own senses, and that means the use of all senses.

Some of the potential problems that can be detected by your senses are:

1. *Eyes*
   - Rust or corrosion can be a sign of poor ventilation or cooling.
   - Coloration of transformer oil from a clear, olive oil shade to dark molasses.
   - Broken insulators, damaged conductors.
   - Accumulation of dust.
   - Gasket deterioration, loose manholes.
2. *Ears and Touch*
   - Vibration can be caused by loose parts.
   - Overload could cause the transformer frame to vibrate or hum.
   - Hot tank enclosure indicates poor cooling or an overload condition.
3. *Smell*
   - Insulation burning caused by deterioration or low level arcing faults.
4. *Taste*
   - High voltage arcing could cause ozone to foam which can be detected by the taste buds.

Performing Transformer Maintenance Tests

Transformers should be tested and maintained at least biannually, and if the transformer serves a vital load system, it should be maintained annually.

Test reports and pertinent data should be maintained throughout the life of the transformer. How the transformer performed during previous tests is the best tool available as to the evaluation of a transformer's current performance. Refer to Figure 9-1 for a sample transformer test report.

The following tabulations provide basic lists of inspections, recommended electrical tests, acceptance criteria and test values for

- Dry-type transformers
- Liquid filled transformers
- Small dry-type transformers

TRANSFORMER TEST REPORT

Sheet No.___of___

Customer__________ Date__________ Project No.__________

Address__________ Air Temp.______ Rel.Humidity__________

Owner/User__________ Date Last Inspection__________

Address__________ Last Inspection Report No.__________

Equipment Location__________

Owner Identification__________

Transformer Impedence__________% Location of Transformer: Indoors ☐ Outdoors ☐

Manufacturer__________ Phases______ Serial Number__________

[Type: PCB ☐ Oil ☐ Dry ☐ Label:__________] KVA__________

Winding Connections__________ Pri/Sec Volts__________ HZ__________

Class Ins.__________ Gal.Oil __________ BIL__________

EXTERNAL INSPECTION	Insp.	Cleaned/ Tightened	Reading
Connections			
Bushings			
Arrestors			
Leakage			
Finish			
Fluid Level			
Pressure Gages			
Tap Changer			
Cooling Fans			
Temp. Gage			

INTERNAL INSPECTION

Sludge Deposits__________

General Cond. of Fluid__________

Moisture Present__________

Rust or Corrosion__________

Connections__________

INSULATING LIQUID TESTS

Dielectric Strength__________

Additional Report Attached__________

Filtering Recommended: Yes ☐ No ☐

INSULATION RESISTANCE TEST

Temp. Corr. Factor to 40°C__________

Pri. Winding to Grnd. __________Megohms

Sec. Winding to Grnd. __________Megohms

Winding to Winding __________Megohms

Polarization Index__________10/1 Min. Ratio

Dielectric Absorbtion__________60/30 Sec.Ratio

LIGHTNING ARRESTORS

AØ __________Megohms

BØ __________Megohms

CØ __________Megohms

TAP INFORMATION:

TRANSFORMER TURNS RATIO AND POLARITY TEST

Name Plate Voltage Ratio__________ Turns Ratio__________

Actual Turn Ratio: AØ__________ BØ__________ CØ__________

Tap Position__________ Polarity__________ Abnormal Cond.__________

COMMENTS:

____

____

____

____

____

Figure 9-1. Transformer Test Report

Checklist for Inspecting Dry-Type Transformers

1. *Visual and Mechanical Inspection*
   - Inspect for physical damage, cracked insulators, tightness of connections, defective wiring, and general mechanical and electrical conditions.
   - Verify proper auxiliary device operation such as fans and indicators.
   - Check tightness of accessible bolted electrical joints.
   - Perform specific inspections and mechanical tests as recommended by manufacturer.
   - Verify proper core and equipment grounding.
   - Thoroughly clean unit prior to testing.
2. *Electrical Tests*
   - Perform insulation resistance tests, winding-to-winding and windings-to-ground, utilizing a megohmmeter with test voltage output as shown in Table 9-1.

Transformer Coil Rating Type	Minimum dc Test Voltage	Recommended Minimum Insulation Resistance in Megohms Liquid Filled	Dry
0 - 600 Volts	1000 Volts	100	500
601 - 5000 Volts	2500 Volts	1000	5000
5001 - 15000 Volts	5000 Volts	5000	25000

Table 9-1. Transformer Insulation Resistance Test Voltage

   - Test duration shall be for 10 minutes with resistances tabulated at 30 seconds, 1 minute, and 10 minutes. Dielectric absorption ratio and polarization index will be calculated.
   - Perform power factor or dissipation factor tests in accordance with the manufacturer's instruction manual.
   - Perform a turns ratio test between windings at as found tap setting.
   - Perform winding resistance tests for each winding at as found tap position.
   - Perform individual excitation current tests on each phase.
   - Perform tests and adjustments for fans, controls, and alarm functions.
   - Verify that the tap-changer is set at specified ratio.
   - Verify proper secondary voltage phase-to-phase and phase-to-neutral after energization and prior to loading.

3. *Test Values*
   - Insulation resistance test values should not be less than values recommended in Table 9-1. Results shall be temperature corrected in accordance with Table 9-1.
   - The polarization index should be above 1.2 unless an extremely high value is obtained initially, which, when doubled, will not yield a meaningful value.
   - Polarization Index $= \dfrac{\text{Ohms @ 10 Min.}}{\text{Ohms @ 1 Min.}}$
   - Turns ratio test results should not deviate more than one-half of one percent (0.5%) from either the adjacent coils or the calculated ratio. CH and CL power factor values will vary due to support insulators and bus work utilized on dry transformers. The following should be expected on CHL power factors.
     Power Transformers: 3% or less
     Distribution Transformers: 5% or less
   - Winding resistance test results should compare within one percent (1%) of adjacent windings.
   - Excitation current test data pattern: Two similar current readings for outside coils and a dissimilar current reading for the center coil of a 3-phase unit.

Checklist for Inspecting Liquid-Filled Transformers

1. *Visual and Mechanical Inspection*
   - Inspect for physical damage, cracked insulators, leaks, tightness of connections, and general mechanical and electrical conditions.
   - Verify proper auxiliary device operation: fans, pumps, instruments.
   - Check tightness of accessible bolted electrical connections.
   - Verify proper liquid level in all tanks and bushings.
   - Perform specific inspections and mechanical tests as recommended by manufacturer.
   - Verify proper equipment grounding.
2. *Electrical Tests*
   - Perform insulation resistance tests winding-to-winding and windings-to-ground, utilizing a megohmmeter with test voltage output as shown in Table 9-1.
   - Test duration shall be for 10 minutes with resistances tabulated at 30 seconds, 1 minute, and 10 minutes. Calculate dielectric absorption ratio and polarization index.

- Dielectric Absorption $= \dfrac{\text{Ohms @ 60 Sec.}}{\text{Ohms @ 30 Sec.}}$
- Perform a turns ratio test between windings at designated tap position. The tap setting is to be determined by the owner/user's electrical engineer and set by the testing laboratory.
- Sample insulating liquid in accordance with ASTM D-923. Sample shall be laboratory tested for:
  - Dielectric breakdown voltage: ASTM D-877 or ASTM D-1816.
  - Acid neutralization number: ASTM D-974.
  - Specific gravity: ASTM D-1298.
  - Interfacial tension: ASTM D-971 or ASTM D-2285.
  - Color: ASTM D-1500.
  - Visual Condition: ASTM D-1524.
  - Perform dissolved gas analysis (DGA) in accordance with ANSI/IEEE C57.104 or ASTM D-3612.
  - PPM water: ASTM D-1533. Required on 25kV or higher voltages and on all silicone filled units.
  - Measure total combustible gas (TCG) content in accordance with ANSI/IEEE C57.104 or ASTM D-3284.

    Measure dissipation factor or power factor in accordance with ASTM D-924.
- Perform insulation power factor tests or dissipation factor tests on all windings and bushings. Overall dielectric-loss and power factor (CH, CL, CHL) shall be determined. Test voltages should be limited to the line-to-ground voltage rating of the transformer winding.
- Perform individual excitation current tests on each phase.
- Perform winding resistance tests on each winding in final tap position.
- Perform tests and adjustments on fan and pump controls and alarm functions.
- Verify proper core grounding if accessible.
- Perform percent oxygen test on the nitrogen gas blanket.

3. *Test Values*
   - Insulation resistance and absorption test. Test voltages to be in accordance with Table 9-1. Resistance values to be temperature corrected in accordance with Table 9-2.
   - The polarization index should be above 1.2 unless an extremely high value is obtained initially, which, when doubled, will not yield a meaningful value.

Temperature °C	Temperature °F	Oil	Dry
0	32	.25	.40
5	41	.36	.45
10	50	.50	.50
15	59	.75	.75
20	68	1.00	1.00
25	77	1.40	1.30
30	86	1.98	1.60
35	95	2.80	2.05
40	104	3.95	2.50
45	113	5.60	3.25
50	122	7.85	4.00
55	131	11.20	5.20
60	140	15.85	6.40
65	149	22.40	8.70
70	158	31.75	10.00
75	167	44.70	13.00
80	176	63.50	16.00

Table 9-2. Transformer Insulation Resistance—Temperature Correction Factors To 20°C

- Turns ratio test results shall not deviate more than on half percent (0.5%) from either the adjacent coils or the calculated ratio.
- Maximum power factor of liquid filled transformers corrected to 20°C shall be in accordance with Table 9-3.
- Bushing power factors and capacitances that vary from nameplate values by more than ten percent (10%) should be investigated.

Power	Oil	Silicone	Tetrachlor-ethylene	High Fire Point Hydrocarbon (R-Temp)
Transformers	0.5%	0.5%	3.0%	0.5%
Distribution Transformers	1.0%	0.5%	3.0%	1.0%

Table 9-3. Maximum Power Factors of Liquid-Filled Transformers—Corrected to 20°C.

- Excitation current test data pattern: Two similar current readings for outside coils and a dissimilar current reading for the center coil of a three phase unit.
- Dielectric fluid should comply with Table 9-4.
- Winding resistance test results should compare within one percent (1%) of adjacent windings.

Inspecting Small Transformers—Dry Type, Air-Cooled (600-Volt and Below, Less than 100KVA Single-Phase or 300KVA Three-Phase)

- Inspect for physical damage, broken insulation, tightness of connections, defective wiring, and general condition.
- Thoroughly clean unit prior to making any tests.
- Perform insulation resistance test. Calculate dielectric absorption ratio and polarization index. Measurements shall be made from winding-to-winding and windings-to-ground. Test voltages and minimum resistance shall be in accordance with Table 9-1. Results to be temperature corrected in accordance with Table 9-2.
- Verify that the transformer is set at the specified tap.

MAINTENANCE AND INSPECTION CHECKLISTS FOR SWITCHGEAR AND SWITCHBOARD ASSEMBLIES

Switchgear and switchboards are free-standing assemblies of metal-enclosed sections containing low-voltage power circuit breakers, molded case circuit breakers and/or fused disconnect switches. They contain bus bars, power cable termination provisions, auxiliary protective devices, controls and various forms of instrumentation.

The assembly may be part of a load center, single or double ended substation or a simple distribution board.

Facility personnel should maintain up-to-date copies of all switchgear schematic and wiring diagrams.

Assembly Maintenance Tests

Switchgear and switchboard assemblies should be tested and maintained on a biennial basis. Test reports and equipment data should be maintained throughout the life of the assembly. Figure 9-2 provides a switchgear inspection checklist.

The following tabulations provide basic lists of inspections, electrical tests and acceptable test values.

	Oil	High Molecular Weight Hydrocarbon	Silicone	Tetrachlor-ethylene
Dielectric Breakdown ASTM D-877	24kV Minimum	30kV Minimum	30kV Minimum	30kV Minimum
Dielectric Breakdown ASTM D-1816 @ 0.04″ gap				
–34.5kV and below	20kV Minimum	—	—	26kV Minimum
–above 34.5kV	25kV	—	—	—
Neutralization Number ASTM D-974	0.36mgKOH/g Maximum	0.03 mgKOH/g Maximum	.01 mgKOH/g Maximum	.25 mgKOH/g Maximum
Interfacial Tension ASTM D-974 or D-2285	21 dynes/cm Minimum	33 dynes/cm Minimum	—	—
Color ASTM D-1500	3.0 Maximum	N/A	0.05 Maximum (D-2129)	—
Visual Condition ASTM D-1524	Compare to Previous Tests	N/A	Crystal Clear	Clear, Slight Pink Irre-descent

Power Factor ASTM D-924 @ 25°C	1.0% Maximum	0.1% Maximum	0.1% Maximum	2% Maximum
Water Content ASTM D-1533				
–15kV and below	35 PPM* Maximum	35 PPM Maximum	80 PPM Maximum	25 PPM Maximum
–above 15kV below 115kV	25 PPM* Maximum	—	—	—
–115kV–230kV	20 PPM* Maximum	—	—	—
–above 230kV	15 PPM Maximum	—	—	—

* Or in accordance with manufacturer's requirements. Some manufacturer's recommend 15 PPM maximum for all transformers.

NOTE: The values for oil are taken from ANSI C57.106 (IEEE Std. 64). The remaining values are acceptance test values and should be modified for service aged fluid as recommended by the manufacturer.

Table 9-4. Dielectric Fluid —Minimum Test Values

Switchgear Check List

Sheet No. ____ of ____

Customer ____ Date ____ Project No. ____

Address ____ Air Temp. ____ Re. Humidity ____

Owner/User ____ Date Last Inspection ____

Address ____ Last Inspection Report No. ____

Equipment Location ____

Owner Identification ____

Mfgr. ____ S.O. ____

Dwgs. ____ Voltage Class ____ Type ____

Switchgear Ampere Rating: ____ Service Rating: ____

External Condition: ____ Good ____ Fair ____ Poor

Consisting of: ____ Total Breakers ____ Total Instruments ____ Total Relays ____ Molded Case Breakers

1. General inspections of exterior equipment. ____
2. Check panel lights for operation-burned out or missing bulbs and lamp covers. ____
3. Check control knobs and switches for freedom of movement and contact condition. ____
4. Inspect for damaged, bent, or twisted doors. ____
5. Inspect door handles, locking bars, and mechanism. ____
6. Check door interlocks or positive operation. ____
7. Inspect for broken instrument and relay cover glass and burned out phase indicator lights. ____
8. Inspect for proper grounding of equipment. ____
9. Measure resistance to ground. ____
10. Inspect and megger power cable or bus to switchgear ____
11. Dielectric test of cables - bus work and potheads. ____
12. Inspect bus and support insulators. ____
13. Torque test bolted bus. (Exposed Connections Only) ____
14. Clean bus insulators - megger test for grounds. ____
15. Inspect control and metering transformers. ____
16. Ratio test transformers - ____
17. Check resistors - grid assemblies and space heaters. ____
18. Check condition of wiring & terminal connections. ____
19. Report unsafe conditions. ____
20. Check bus for support & spacing. ____
21. Note and report any unmarked circuits. ____
22. Remove draw out breakers. ____
23. Check rails, guides, rollers, and shutter mechanism. ____
24. Lubricate draw out assembly parts ____
25. Check cell interlocks and auxiliary contact assemblies. ____
26. Inspect breaker and cell contacts. ____
27. Vacuum and clean interior of cubicle. ____
28. Perform breaker inspection and test. ____
29. Test molded case breakers ____
30. Inspect and check instruments. ____
31. Note and record as found relay settings. ____
32. Determine correctness of settings - if improperly set - advise customer. ____
33. Restore control power to switchgear. ____
34. Check relays for positive tripping. ____
35. Test annunciator - alarm or target operation. ____
36. Operate controls - close and trip breakers electrically. ____
37. Check automatic transfer relay operation (if used) ____
38. Recheck relays for positive tripping with breakers in test position ____
39. Make final visual inspection - remove leads - tools, etc. ____
40. Prepare Switchgear inspection report. ____
41. Ground Network Resistance ____
    Res. ____ Ohms.

Comments:

____

____

____

Figure 9-2. Switchgear Check List

1. *Visual and Mechanical Inspections*
   - Inspect for physical, electrical, and mechanical condition.
   - Compare equipment nameplate information with latest one line diagram when available and report discrepancies.
   - Check for proper anchorage, required area clearances, physical damage, and proper alignment.
   - Inspect all bus connections for high resistance. Use low resistance ohmmeter, or check tightness of bolted bus joints by calibrated torque wrench method.
   - In lieu of above item, inspect the assembly for high temperature problems by using an Infrared Survey Viewer. Refer to the following section on performing a thermographic-infrared survey.
   - Test all electrical and mechanical interlock systems for proper operation and sequencing.
     - Closure attempt shall be made on locked open devices. Opening attempt shall be made on locked closed devices.
     - Key exchange shall be made with devices operated in off-normal positions.
   - Clean switchgear.
   - Inspect accessible insulators for evidence of physical damage or contaminated surfaces.
   - Verify proper barrier and shutter installation and operation.
   - *Lubrication*
     - Verify appropriate contact lubricant on moving current carrying parts.
     - Verify appropriate lubrication on moving and sliding surfaces.
   - Exercise all active components.
   - Inspect all mechanical indicating devices for proper operation.
2. *Electrical Tests*
   - Perform ground resistance tests.
   - Perform insulation resistance tests on each bus section, phase-to-phase and phase-to-ground for one (1) minute. Test voltages should be in accordance with Table 9-5.
   - Perform control wiring performance test. Use the elementary diagrams of the switchgear to identify each remote control and protective device. Conduct tests to verify satisfactory performance of each control feature.
   - Perform over-potential tests on each bus section, phase-to-phase and phase-to-ground. Test voltages shall be in accordance with Table 9-6.(Optional).
   - *Control Power Transformer—Dry Type*

Voltage Rating	Minimum dc Test Voltage	Recommended Minimum Insulation Resistance in Megohms
0 – 250 Volts	500 Volts	50
251 – 600 Volts	1000 Volts	100
601 – 5000 Volts	2500 Volts	1000
5001 – 15000 Volts	2500 Volts	5000
15001 – 25000 Volts	5000 Volts	20000

Table 9-5. Switchgear Insulation Resistance Test Voltage

Type of Switchgear	Rated kV	Maximum Test Voltage kV ac	dc
MC	4.76	14.3	20.3
(Metal Clad Switchgear)	8.25	27.0	37.5
	15.00	27.0	37.5
	38.00	60.0	+
SC	15.50	37.5	+
(Station-Type Cubicle	38.00	60.0	+
Switchgear)	72.50	120.0	+
MEI	4.76	14.3	20.3
(Metal-Enclosed	8.25	19.5	27.8
Interrupter Switchgear)	15.00	27.0	37.5
	15.50	37.5	52.5
	25.80	45.0	+
	38.00	60.0	+

* Derived from ANSI/IEEE C37.202.2, Paragraph 5.5 and C37.20.3, paragraph 5.5.
+ Consult Manufacturer.

Table 9-6. Field Over-Potential Test Voltages*

- Inspect for physical damage, cracked insulation, broken leads, tightness of connections, defective wiring, and overall general condition.
- Verify that primary and secondary fuse ratings or circuit breakers match drawings.
- Perform insulation resistance test. Measurements shall be made from winding-to-winding and windings-to-ground. Test voltages and minimum resistances shall be in accordance with Table 9-1. Results to be temperature corrected in accordance with Table 9-2.

3. *Test Values*
   - Insulation resistance test shall be performed in accordance with Table 9-5. Values of insulation resistance less than this table or manufacturer's minimum should be investigated.
   - Over-potential test voltages shall be applied in accordance with Table 9-6. (Derived from ANSI/IEEE C37.20.2).
   - Test results are evaluated on a go, no-go basis by slowly raising the test voltage to the required value. The final test voltage shall be applied for one (1) minute.

Performing Thermographic–Infra-red Survey

One of the major tools available for inspecting the condition of switchgear and switchboard assemblies is the infra-red viewer which makes use of thermography to detect overloaded cables, loose cable/bus connections and internal circuit breaker problems. Thermography is the technique of using non-contact, non-destructive scanning equipment that detects invisible, infrared heat radiation and converts this energy to visible light. An infrared viewer is a highly sensitive, hand-held infrared imager which presents, in its observer's eyepiece, a detailed thermal picture of the entire scene within the field of view.

By clearly showing temperature differences between closely adjacent objects, and between objects and background, the infrared viewer quickly and accurately locates sources of heat. It can do this in total darkness. It is an invaluable aid to utility and industrial plant inspectors.

Operation of the infrared viewer is based on the principle that all objects, whether animate or otherwise, radiate infrared energy in amounts depending on their temperatures. As the infrared viewer scans a scene, it detects and converts the levels of infrared radiation to corresponding levels of visible light; in this way it produces, on a small viewing screen, a display containing easily distinguishable temperature patterns of all objects in range. The thermal picture presented is in red.

Once a "hot-spot" is detected, the temperature is recorded at the location. Table 9-7 should be used as a guide to corrective action. Figure 9-3 provides an Infra-red scan report.

INFRA-RED SCAN REPORT

SCAN NO. ________

JOB NO. ____________

TEST REPORT

INFRA-RED STUDY

DATE: ______________

CLIENT ____

PROJECT ____

CLIENT I.D. ____

LOCATION ____

PERFORMED BY ________ WITNESSED BY ________ TEMPERATURE ________ HUM.____%

AMBIENT TEMP. ______°C

INFRA-RED RESULTS

AØ ____________________	AØ ________°C	AØ ________AMPS
BØ ____________________	BØ ________°C	BØ ________AMPS
CØ ____________________	CØ ________°C	CØ ________AMPS

REMARKS: ____

Figure 9-3. Infra-Red Scan Report

1. Up to 3°C above ambient	No immediate action necessary
2. 3°C to 7°C	Correct at next routine shutdown
3. 7°C to 15°C	Correct prior to routine maintenance
4. 15°C to 25°C	Correct as soon as possible
5. Above 25°C	Correct immediately

Table 9-7. Infrared Viewer—Corrective Action Guide

MAINTENANCE CHECKLISTS AND TESTS FOR CIRCUIT BREAKERS AND SWITCHES

The circuit breakers and switches mounted within switchgear and switchboard assemblies provide the means for opening and closing circuits. However, their principal function is to protect the electrical system against any abnormally high current that could flow during a short circuit or for an overload condition.

Because of the important part that circuit breakers and switches play in protecting electrical equipment and vital distribution systems, the maintenance of this equipment should have the highest of priorities. Circuit breakers, at a minimum, should be trip tested on a two-year cycle. Switches and their related fuses should be inspected on a similar schedule.

The following sections provide basic lists of equipment inspections, recommended electrical tests and test acceptance criteria for:

- Low voltage - Molded case circuit breakers
- Low voltage - Power circuit breakers
- Medium voltage (5kV & 15kV) - Oil/air circuit breakers
- Medium voltage (5kV & 15kV) - Vacuum circuit breakers
- Low voltage air switches
- Medium voltage (5kV & 15kV) - Metal enclosed switches
- Medium & high voltage - Open air switches

Checklists for Low Voltage-Molded Insulated Case Circuit Breakers

1. *Visual and Mechanical Inspection*
   - Check circuit breaker for proper mounting.
   - Operate circuit breaker to ensure smooth operation.
   - Inspect case for cracks or other defects.
     Check tightness of connections with calibrated torque wrench. In lieu of this test, perform infrared survey and verify with a load applied to the circuit.
   - Check internals on unsealed units.

2. *Electrical Tests*
   - Perform a contact resistance test.
   - Perform an insulation resistance test at 1000 volts dc from pole-to-pole and from each pole-to-ground with breaker closed and across open contacts of each phase.
   - Perform long time delay time-current characteristic tests by passing three hundred percent (300%) rated current through each pole separately. Record trip time.
   - Determine short time pickup and delay by primary current injection.
   - Determine ground fault pickup and time delay by primary current injection.
   - Determine instantaneous pickup current by primary injection using run-up or pulse method.
   - Perform adjustments for final settings in accordance with breaker setting sheet when applicable.
3. *Test Values*
   - Compare contact resistance or millivolt drop values to adjacent poles and similar breakers. Investigate deviations of more than fifty percent (50%). Investigate any value exceeding manufacturer's recommendations.
   - Insulation resistance shall not be less than 100 megohms.
   - Trip characteristic of breakers shall fall within manufacturer's published time-current characteristic tolerance band, including adjustment factors.
   - All trip times shall fall within Table 9-8. Circuit breakers exceeding specified trip time at three hundred percent (300%) of pickup shall be tagged defective.
   - Instantaneous pickup values shall be within values shown on Table 9-9.

Checklist for Low Voltage-Power Circuit Breakers

1. *Visual and Mechanical Inspection*
   - Inspect for physical damage.
   - Perform mechanical operational test.
   - Check cell fit and element alignment.
   - Check tightness of connections with calibrated torque wrench. Refer to manufacturer's instructions for proper torque levels.
   - Check arc chutes for damage.
   - Clean entire circuit breaker using approved methods and materials.
   - Lubricate as required.

Breaker Voltage Volts	Range of Related Continuous Current Amperes	Maximum Trip Time In Seconds*
240	15-45	50
240	50-100	70
600	15-45	70
600	50-100	125
240	110-225	200
240	250-400	300
600	110-225	250
600	50-400	300
600	450-600	350
600	700-1200	500
600	1400-2500	600
600	3000-5000	650

* For integrally-fused circuit breakers, trip times may be substantially longer if tested with the fuses replaced by solid links (shorting bars).

Table 9-8. Values for Overcurrent Trip Test (At 300% of Rated Continuous Current of Circuit Breaker)

Frame Size, Amperes	Tolerances of High and Low Settings	
	High	Low
≤ 250	+40% −25%	+40% −30%
≥ 400	±25%	±30%

Table 9-9. Instantaneous Trip Setting Tolerances

2. *Electrical Tests*
   - Perform a contact resistance test.
   - Perform an insulation resistance test at 1000 volts dc from pole-to-pole and from each pole-to-ground with breaker closed and across open contacts of each phase.
   - Determine minimum pickup current by primary current injection.

- Determine long time delay by primary injection.
- Determine short time pickup and delay by primary current injection.
- Determine ground fault pickup and delay by primary current injection.
- Determine instantaneous pickup value by primary current injection.
- Make adjustments for final settings in accordance with breaker setting sheet.
- Activate auxiliary protective devices, such as ground fault or under voltage relays, to ensure operation of shunt trip devices. Check the operation of electrically operated breakers in their cubicle.
- Check charging mechanism.

3. *Test Values*
   - Compare contact resistance or millivolt drop values to adjacent poles and similar breakers. Investigate deviations of more than fifty percent (50%).
   - Insulation resistance shall be less than 100 megohms. Investigate values less than 100 megohms.
   - Trip characteristics of breakers when adjusted to setting sheet parameters shall fall within manufacturer's published time-current tolerance band.

Checklist for Medium Voltage (5KV & 15KV)-Oil/Air Circuit Breakers

1. *Visual and Mechanical Inspection*
   - Inspect for physical damage, cleanliness, and adequate lubrication. Figure 9-4 provides a sample inspection report.
   - Inspect anchorage, alignment, and grounding.
   - Perform all mechanical operator and contact alignment tests on both the breaker and its operating mechanism.
   - Check tightness of bolted bus joints by calibrated torque wrench method. Refer to manufacturer's instructions for proper torque levels.
   - Check cell fit and element alignment.
   - Check racking mechanism.
   - Verify that primary and secondary contact wipe and other dimensions vital to satisfactory operation of the breaker are correct.
   - Ensure that all maintenance devices are available for servicing and operating the breaker.
   - Lubricate all moving current carrying parts.
   - Check for proper operation of the cubicle shutter.
   - Perform circuit breaker time-travel analysis.

Oil and Vacuum - Medium Voltage
Air Cicrcuit Breaker Test and Inspection Report

Sheet No. ____ of ____

Customer ______ Date ______ Project No. ______
Address ______ Air Temp. ______ Re. Humidity ______
Owner/User ______ Date Last Inspection ______
Address ______ Last Inspection Report No. ______
Equipment Location ______
Owner Identification ______

BREAKER DATA:

Manufacturer ______ Voltage ______ Type ______ Amps. ______ Int. Rating ______
Serial No. ______ Type Oper. Mech. ______ Age ______ Other N.P. Data ______

TEST DATA:

	A	B	C
Ins. Res., ____ kv, Megohms			
Contact Resistance, Microhms			
Closing Speed/Opening Speed			
Reference, P.F. Test Sheet No.			

ADJUSTMENTS:	Mfr's Rec.	As Found	As Left
Arcing Contact Wipe			
Main Contact Gap			
Main Contact Wipe			
Latch Wipe			
Latch Clearance			
Contact Travel			
Prop Clearance			
Stop Clearance			

INSPECTION AND MAINTENANCE:

	Insp.	Dirty	Cleaned Lubed.	See Remarks
Overall Cleanliness				
Insulating Members				
Mech. Connections				
Structural Members				
Cubicle				
Pri. Contact Fingers				
Shutter Mech.				
Relays				
Auxiliary Devices				
Racking Device				
Arc Chutes				
Blow Out Coil				
Puffers				
Liner				
Arc Runners				
Main Contacts				
Cubicle Wiring				
Breaker Wiring				
Heaters				
Panel Lights				
Bearings				
Contact Sequence				
Ground Connection				
Counter Reading				

Remarks

______

Submitted by ______ Equipment Used: ______ Test Report No. ______

Figure 9-4. Oil and Vacuum-Medium Voltage-Air Circuit Breaker Test/Inspection Report

2. *Electrical Tests*
   - Measure contact resistance.
   - Measure insulation resistance pole-to-pole, pole-to-ground, and across open poles. Use a minimum test voltage of 2500 volts.
   - Perform insulation resistance test at 1000 volts dc on all control wiring. (Do not perform the test on wiring connected to solid state components.)
   - With breaker in the test position, make the following tests:
   - Trip and close breaker with the control switch.
   - Trip each breaker by operating manually each of its protective relays.
   - Perform power factor test with breaker in both the open and closed position.

   In addition, for all oil circuit breakers:
   - Sample insulating fluid. Sample shall be laboratory tested for:
     - Dielectric breakdown voltage: ASTM D-877
     - Color: ASTM D-1500
     - Power factor: ASTM D-924
     - Interfacial tension: ASTM D-971 or D-2285
     - Visual condition: ASTM D-1524

3. *Test Values*
   - Determine contact resistance in microhms. Investigate deviations of more than 50%.
   - Minimum insulation resistance should comply with Table 9-1.
   - **For Air Breakers** - Power factor and arc chute watts loss should be compared with results from previous tests of similar breakers, or referred to manufacturer's published data.
   - **For Oil Breakers** - Power factor and capacitance test results shall be within ten percent (10%) of nameplate rating for bushings. Tank loss index shall not exceed the manufacturer's allowable value.
   - **For Oil Breakers** - Insulating liquid tests shall comply as follows (see ANSI/IEEE C57.106 for additional information):
     - Dielectric breakdown voltage: ASTM D-877, 25kV minimum
     - Color: ASTM D-1500, 1.0 maximum
     - Power factor: ASTM D-924, 1.0% at 25°C
     - Interfacial tension: ASTM D-971 or D-2285, 24 dynes/cm minimum
     - Visual condition: ASTM D-1524, clear and free of visual contaminants

Checklist and Tests for Medium Voltage (5KV & 15KV)—Vacuum Circuit Breakers

1. *Visual and Mechanical Inspection*
   - Inspect for physical damage.
   - Inspect anchorage, alignment, and grounding.
   - Perform all mechanical operational tests on both the circuit breaker and its operating mechanism.
   - Measure critical distances such a contact gap, as recommended by manufacturer.
   - Check tightness of bolted connections by calibrated torque wrench method.
2. *Electrical Tests*
   - Perform a contact resistance test.
   - Perform breaker travel and velocity analysis.
   - Perform minimum pickup voltage tests on trip and close coils.
   - Trip circuit breaker by operation of each protective device.
   - Perform insulation resistance tests, pole-to-pole, pole-to-ground, and across open poles at 2500 Volts minimum.
   - Perform vacuum bottle integrity test (over potential) across each vacuum bottle with the breaker in the open position in strict accordance with manufacturer's instructions. **Do Not Exceed Maximum Voltage Stipulated for this test.** Provide adequate barriers and protection against x-radiation during this test. Do not perform this test unless the contact displacement of each interrupter is within manufacturer's tolerance. Be aware that some dc high potential test sets are half-wave rectified and may produce peak voltages in excess of the breaker manufacturer's recommended maximum.
   - Perform insulation resistance test on all control wiring at 1000 volts dc. (Do not perform this test on wiring connected to solid state relays.)
   - Perform power factor tests on each pole with the breaker open and each phase with the breaker closed.
   - Perform power factor tests on each bushing. Use conductive straps and hot collar procedures if bushings are not equipped with a power factor tap.
3. *Test Values*
   - Compare contact resistance to adjacent poles and similar breakers. Investigate deviations of more than fifty percent (50%). Investigate any value exceeding manufacturer's tolerance.
   - Contact displacement shall be in accordance with factory recorded data marked on the nameplate of each vacuum breaker or bottle.

- Apply over-potential test voltages in accordance with manufacturer's instructions. Interrupter shall withstand the voltage applied without breakdown.
- Compare circuit breaker travel and velocity values to manufacturer's acceptable limits.
- Control wiring insulation resistance shall be a minimum one (1) megohm.
- Power factor and capacitance test results shall be within nameplate rating for bushings.

Inspection & Test Checklists for Low Voltage Air Switches

1. *Visual and Mechanical Inspection*
   - Inspect for physical and mechanical condition. Figure 9-5 provides a test and inspection report.
   - Check for proper anchorage and required area clearances.
   - Perform mechanical operation tests.
   - Check blade alignment.
   - Check each fuse holder for adequate mechanical support of each fuse.
   - Inspect all bus or cable connections for tightness by calibrated torque wrench method. Refer to manufacturer's instructions for proper torque levels. In lieu of above torquing, perform an infrared survey.
   - Test all electrical and mechanical interlock systems for proper operation and sequencing.
   - Clean entire switch using approved methods and materials.
   - Check proper phase barrier materials and installation.
   - Lubricate as required.
   - Exercise all active components.
   - Inspect all indicating devices for proper operation.
2. *Electrical Test and Test Values*

   Perform insulation resistance tests on each pole, phase-to-phase and phase-to-ground for one (1) minute. Test voltage and minimum resistances should be in accordance with:

System Voltage	*Test Voltage - dc*	*Recommended Minimum Insulation Resistance Megohms*
250 Volts	500 Volts	25
600 Volts	1000 Volts	100

MEDIUM AND LOW VOLTAGE
DISCONNECT SWITCH TEST AND INSPECTION REPORT

Fused ______ Non-Fused ______ Sheet ______ Of ______

Customer ______ Date ______ Job No. ______

Address ______

Job Site ______ Date Last Inspection ______

Address ______ Last Inspection Report. No. ______

Equipment Location ______

Owner Identification ______

______

Switch Data:

Manufacturer ______ Voltage ______ Type ______ Amp ______ Int. Rating ______

Serial No. ______ Type Oper. Mech. ______

Phase ______ Wire ______

Test Data:

	A	B	C
Ins. Res. ____KV Megohms.........			
Cont.Res.-Normal-Microhms (AF).....			
Cont.Res.-Normal-Microhms (AL).....			
Cont.Res.-Emergency-Microhms (AF)..			
Cont.Res.-Emergency-Microhms (AL)..			
Contact Resistance, Fuse..........			
Cable Size--Line..................			
Cable Size--Load..................			
Cable Size--Emergency.............			
Cable Type: Alum. ☐ Copper ☐...			
Fuse I.C. Rating..................			
Fuse T.C. Curve...................			
Fuse Manufacturer.................			
Fuse Catalog Number...............			
Fuse Size.........................			
System Voltage....................			

Other Test Data:

Cable Ins.Res/KV Megohms..........			
Contact Sequence..................			
Infra Red Inspection..............			

(Note: B & C remain the same as A unless otherwise noted.)

Inspection & Maintenance

	Insp	Dirty	Clean/ Lubed
Overall Cleanliness			
Insulating Members			
Mech. Connections			
Cable Connections			
Ground Connections			
Neutral Connections			
Bonding			
Movable Contacts			
Stationary Conts.			
Manual Operation			
Electrical Operation			
Contact Lube			
Mechanical Lube			
Contact Alignment			
Paint			
Bus Connection			
Door Interlock			
GFI			
GFI CT			
System Bus			
Metering CT's			

Remarks ______

______

______

______

Figure 9-5. Test and Inspection Report for Medium and Low Voltage Disconnect Switch

Perform contact resistance test across each switch blade and fuse holder, or perform an infrared survey.

Determine contact resistance in microhms. Investigate any value exceeding 50 microhms or any values which deviate from adjacent poles or similar switches by more than fifty percent (50%).

Maintenance Checklist for Medium Voltage (5KV & 15KV)-Metal Enclosed Switches

1. *Visual and Mechanical Inspection*
   - Inspect for physical and mechanical condition.
   - Check for proper anchorage and required area clearances.
   - Verify that fuse sizes and types correspond to drawings.
   - Perform mechanical operator tests in accordance with manufacturer's instructions.
   - Check blade alignment and arc interrupter operation.
   - Verify that expulsion limiting devices are in place on all holders having expulsion type elements.
   - Check each fuse holder for adequate mechanical support for each fuse.
   - Inspect all bus connections for tightness of bolted bus joints by calibrated torque wrench method. Refer to manufacturer's instructions for proper torque levels. In lieu of above torque tests, perform an infra-red survey.
   - Test all electrical and mechanical interlock systems for proper operation and sequencing.
   - Clean entire switch using approved methods and materials.
   - Verify proper phase barrier materials and installation.
   - Lubricate as required.
   - Check switch blade clearances with manufacturer's published data.
   - Inspect all indicating devices for proper operation.
2. *Electrical Tests and Test Values*
   - Perform insulation resistance tests on each pole, phase-to-phase and phase-to-ground for one (1) minute. Test voltage and minimum resistances should be in accordance with the following:

Voltage Rating	*Test Voltage - dc*	*Recommended Minimum Insulation Resistance Megohms*
5kV	2500 Volts	5,000
15kV-25kV	5000 Volts	20,000

- Perform contact resistance test across each switch blade and fuse holder.
- Determine contact resistance in microhms. Investigate any value exceeding 100 microhms or any values which deviate from adjacent poles or similar switches by more than fifty percent (50%).

Checklists for Medium and High Voltage-Open Air Switches

1. *Visual and Mechanical Inspection*
   - Inspect for physical damage and compare nameplate data with plans and specifications.
   - Perform mechanical operator tests in accordance with manufacturer's instructions.
   - Check blade alignment and arc interrupter operation.
   - Check fuse link or element and holder for proper current rating.
2. *Electrical Tests*
   - Perform ac or dc overpotential test on each pole-to-ground and pole-to-pole.
   - Perform contact resistance test across each switch blade.
3. *Test Values*
   - Apply overpotential test voltages in accordance with the following:

Voltage Rating	*Test Voltage Minimum - dc*	*Recommended Minimum Insulation Resistance Megohms*
601 - 5000V	2500 Volts	1,000
5001 - 15000V	2500 Volts	5,000
15000 and Larger	5000 Volts	20,000

MAINTAINING AND TESTING PROTECTIVE RELAYS AND METERING SYSTEMS

This section covers the maintenance and testing associated with protective relays and metering systems and the related instrument transformers and control circuits associated with their application.

Protective relays find their application principally in the facility's main service equipment, especially in medium voltage and low voltage switchgear. In addition, the backup and emergency generators are equipped with various forms of protective relays, with their quantity and sophistication dependent on the size of the generator. The principal function of the protective relays is to protect the electrical system against any abnormally high current or voltage or the detection of potentially destructive short circuits or system overload conditions.

Meters are used to monitor the performance of the distribution system with respect to the detection of overloads, adequate voltage levels, energy (KWHR) consumption, and so on.

Using an Outside Testing Company

Because of the importance of this equipment in protecting the electrical distribution systems, the maintenance of protective relays and meters should have a high position in the priority of systems to be maintained. The sophistication and expertise that is required for this type of equipment is such that the facility staff may not be qualified to provide the proper maintenance.

Therefore, it is recommended that you engage the services of a recognized independent testing company to inspect and test the protective relays. It is further advised that the testing company be a member of a recognized association such as the InterNational Electrical Testing Association. The use of a testing company is further advised, based on the fact that many of the switchgear suppliers are using complex solid state relays and instruments.

Maintenance and Test Checklist

The following sections provide detailed checklists of recommended inspections, tests and general maintenance procedures for:

- Protective relays
- Instrument transformers
- Metering and instruments

The electrical distribution system's protective relays and instrumentation systems should be scheduled for maintenance and testing at the same time since these systems normally require a shutdown of the electrical system for most servicing.

When and How to Test Protective Relays

1. Protective relays should be maintained and tested on a two-year cycle, as a minimum. If a major change has been made to your facility's utility service equipment or to your distribution system, it is strongly recommended that an engineer review the relay settings for the possible need to revise the settings in order to maintain proper coordination between the various protective devices.
2. *Visual and Mechanical Inspection*
    - Inspect relays for physical damage. Figure 9-6 provides a test report for protective relays.
    - Inspect cover gasket, cover glass, presence of foreign material or moisture, condition of spiral spring, disc clearance, contacts, and case shorting contacts if present.

PROTECTIVE RELAY TEST REPORT

Sheet No. _____ of _____

Customer ______________________ Date ______________ Project # ______________
Address ______________________ Air Temp. ______________ Rel. Humidity ______________
Job Site ______________________ Last Date Inspection ______________________
Address ______________________ Last Inspection Report # ______________________
Equipment Location ____
Owner Identification ____

Circuit Identification ______________________ *C.T. Ratio* ______________ P.T. Ratio ______________

Visual Inspection					Routine Maintenance					
	A	B	C			A	B	C		
Cover Gasket					Glass Cleaned					Mfr:
Glass					Case Cleaned					Type ϕ:
Foreign Material					Relay Cleaned					Type Grd:
Moisture					Connection Tight					Tap Range ϕ:
Spiral Spring					Tap Tightened					Tap Range Grd:
Bearing Condition					Contacts Cleaned					Inst. Range ϕ:
Bearing End-Play					Insulation Resistance					Inst. Range Grd:
Disc Clearance					Trip Circuit					Use:
Rust										

Remarks: ____

Relay Settings														
	ICS	Inst.Elem.Setting				Tap Setting				Time Dial Setting				
		A	B	C		A	B	C		A	B	C		
Specified														
As Found														
As Left														

Test Operations - As Found - Time in Seconds

		Time Element		Current Voltage			Inst. Element Current/Voltage		ICS AMPS			
	Zero Set	P.U.		Time			Pick Up	Drop Out	Pick Up	Drop Out	Seal In	
		Tap 1	Tap 2	P.U. X__	Tap 1 X__	Tap 2 X__						
A Phase												
B Phase												
C Phase												

Test Operations - As Left - Time in Seconds

		Time Element		Current Voltage			Inst. Element Current/Voltage		ICS AMPS			
	Zero Set	P.U.		Time			Pick Up	Drop Out	Pick Up	Drop Out	Seal In	
		Tap 1	Tap 2	P.U. X__	Tap 1 X__	Tap 2 X__						
A Phase												
B Phase												
C Phase												

Submitted by ______________________ Equipment Used ______________________

Figure 9-6. Protective Relay Test Report

- Check mechanically for freedom of movement, proper travel and alignment, and tightness of mounting hardware and tap plugs.
- Verify all settings in accordance with owner/user's provided settings.

3. *Electrical Tests*
   - Perform insulation resistance test on each circuit to frame. Do not perform this test on solid state devices.
   - Perform the following tests on the nominal settings specified by the Facility's Director:
     - Pickup parameters on each operating element.
     - Perform timing test at two (2) points on time dial curve.
     - Pickup target and seal-in units.
     - Special tests as required to check operation of restraint, directional and other elements per manufacturer's instruction manual.
   - Testing Relays

     The testing and re-calibration of relays should always be based on the recommendations of the manufacturers as stated in their instruction manuals.

Testing and Maintaining Instrument Transformers

Instrument transformers, such as current transformers and voltage transformers, should always be maintained in conjunction with their respective protective relays and/or their metering systems.

Caution: Prior to any maintenance services being performed on current transformers, the phases on the secondary side of the current transformers should each be shorted and grounded. Current transformers should each be equipped with shorting type terminal blocks, which are specifically designed to provide the proper shorting and grounding functions. Open-circuited, un-grounded secondary current transformer leads could have a potential between them as high as *1000 Volts* and are very ***DANGEROUS.***

1. *Visual and Mechanical Inspection*
   - Inspect for physical damage.
   - Verify proper connection of transformers with system requirements.
   - Verify tightness of all bolted connections and assure that adequate clearances exist between primary circuits to secondary circuit wiring.
   - Verify that all required grounding and shorting connections provide good contact on short type terminal blocks.
   - Test proper operation of transformer withdrawal mechanism (tip out) and grounding operation when applicable for medium voltage level voltage transformers.

2. *Electrical Tests - Current Transformers*
   - Perform a ratio verification test of each current transformer. This shall be performed using the voltage method or current method in accordance with ANSI C57.13.1. Figure 9-7 provides a report for testing current transformers.
   - Measure relaying circuit burdens at current transformer terminals and determine the total burden in ohms at 60 hertz.
3. *Electrical Tests - Voltage Transformers*
   - Perform insulation resistance tests on voltage transformers, winding-to-winding and windings-to-ground. Value of test voltage on secondary wiring shall be 500 volts dc for one (1) minute. Do not perform this test with solid state devices connected. Figure 9-8 provides a test record for potential transformers.
   - Perform a ratio test using a transformer-turns-ratio test set or by voltage comparison method.
   - Perform a dc dielectric withstandability test on the primary windings with the secondary windings connected to ground. The dc dielectric voltage shall be in accordance with Table 9-10.

Inspecting and Testing Metering and Instrument Systems

Since most metering and instrument systems are so integrated with the distribution system's protective relays and their respective instrument transformers, they should be maintained at the same time.

1. *Visual and Mechanical Inspection*
   - Examine all devices for broken parts and wire connection tightness.
2. *Electrical Tests*
   - Check calibration of meters at all cardinal points.
   - Calibrate watt hour meters to one half of one percent (0.5%).
   - Verify all instrument multipliers.
   - Digital solid state instruments should be tested in accordance with the manufacturer's instruction manuals.

MAINTAINING CABLES AND BUSWAYS

Cables and electrical busways should be tested during their installation. These tests should be carefully recorded and this initial data should be used as a comparison for future tests. The main function of these tests is to determine if the system's dielectric protection is deteriorating.

CURRENT TRANSFORMERS (CT) TEST RECORD

Sheet No.________
Customer________ Date________ Project #________
Address________ Air Temp.________ Rel. Humidity________
Job Site________ Date Last Inspection________
Address________ Last Inspection Report No.________
Equipment Location________
Owner Identification________

Application of CT (Metering/Relaying)________ Design Ratio ________
Drawing Number/Revision________

NAME PLATE DATA

Manufacturer________ Type________ B.I.L________
Serial No.: AØ________ BØ________ CØ________ Neutral________
Frequency________ Accuracy Rating________ CT Type________
Insulation Class________ KV________ Cont. Thermal Rating Factor________ AMB____ C
Rated Current Primary________ Sec________ Ratio________ Polarity________
Instruction Book________

I Primary Ratio & Polarity Check

CT Number	Taps	Ratio	*Full Voltage			Half Voltage			Polarity CK	Megger Reading	
			ES	IS	EP	ES	IS	EP			

*Max Voltage 1 Volt/Turn or Max of Test Equipment.

III Secondary Turns Ratio**

		Test Voltage		(Voltage X1-X5)				Test Meter # & Calib. Date
Taps	Ratio	CT1	CT2	CT3	CT4	CT5	CT6	Voltmeter:
X1-X2								Ammeter:
X2-X3								Megger:
X3-X4								
X4-X5								

**Multi Ratio CT'S Only.

SATURATION MILLIAMPS

Volts	A	B	C	Neutral

COMMENTS________

TESTED BY: ________

Figure 9-7. Sample Current Transformers (CT) Test Report

POTENTIAL TRANSFORMER (PT) TEST RECORD Sheet No.___ of___

Customer__________ Date__________ Project No.__________
Address__________ Air Temp.__________ Rel. Humidity__________
Job Site__________ Date Last Inspection__________
Address__________ Last Inspection Report No.__________

Equipment Location__________

Owner Identification__________

Manufacturer__________ Type__________ Bus and Cubicle__________

Serial No.: AØ__________ BØ__________ CØ__________ __________

Name Plate Ratio__________ BIL__________

Thermal Rating__________ Dry ☐ Oil ☐ Oil Spec.__________

Fuse Type__________ Size__________ __________

Accy/Burden__________/__________ __________

____

TURNS RATIO TEST

Primary Excitation Voltage	Secondary Volts/TTR A	B	C		RATIO

INSULATION RESISTANCE (MEGGER) TESTS

		A	B	C		
Primary Windings to Ground @	______ KVDC					Megohms
Secondary Windings to Ground @	______ KVDC					Megohms
Primary to Secondary Windings @	______ KVDC					Megohms

COMMENTS

____

____

____

____

____

Equipment Used:__________ Tested By:__________

Figure 9-8. Potential Transformer (PT) Test Record

Nominal System Voltage	BIL (kV)	Test Dielectric Withstandability Field Test Voltage (kV)	
		ac	*dc*
2.4	45	11.3	15
4.8	60	14.3	19
8.32	75	19.5	26
13.8	95	25.5	34

Table 9-10. Dielectric Withstandability—Field Test Voltage

Cables and busways should be tested and maintained on a three-year cycle as a minimum. Tests should be performed more frequently for those systems showing any deterioration.

The following sections provide checklists and tabulations with recommended inspections, tests and general maintenance procedures for:

- Low voltage cables - Less than 600 volts
- 5kV-69kV cables
- Metal enclosed busways

Checklists for Low Voltage Cables—Less than 600 Volts

1. *Visual and Mechanical Inspection*
    - Inspect exposed sections of cables for physical damage.
    - Test cable mechanical connections to manufacturer's recommended values with a calibrated torque wrench. In lieu of above torquing, perform an infrared survey.
2. *Electrical Tests*
    - Perform insulation resistance test on each conductor with respect to ground and adjacent conductors. Applied potential to be 1000 volts dc for 1 minute. For best results, cables should be isolated from external equipment as much as possible (PT's, transformers, lighting arresters, etc.).
3. *Test Values*
    - Minimum insulation resistance values shall be not less than two megohms.

Test Checklists for 5KV-69KV Cables

1. *Visual and Mechanical Inspection*
   - Inspect exposed sections for physical damage.
   - Inspect for shield grounding, cable support, and termination.
   - Inspect fireproofing in common cable areas.
   - If cables are terminated through window type CT's, make an inspection to verify that neutrals and grounds are properly terminated for proper operation of protective devices.

2. *Electrical Tests*
   - Perform a shield continuity test on each power cable. Ohmic value shall be recorded.
   - Perform a dc high potential test on all cables. Adhere to all precautions and limits as specified in the applicable NEMA/ICEA Standard for the specific cable. Perform tests in accordance with ANSI/IEEE Std. 400. Test procedure shall be as follows, and the results for each cable test shall be recorded as specified herein. Test voltages shall not exceed 60% of cable manufacturer's factory test value or the maximum test voltage in Tables 9-11 or 9-12. Figure 9-9 provides a dielectric test report.
     - Current sensing circuits in test equipment shall measure only the leakage current associated with the cable under test and shall not include internal leakage of the test equipment.
     - Record wet and dry bulb temperatures of relative humidity and temperature.
     - Test each section of cable individually.
     - Individually test each conductor with all other conductors grounded. Ground all shields.
     - Terminations shall be properly corona suppressed by guard ring, field reduction sphere, or other suitable methods.
     - Ensure that the maximum test voltage does not exceed the limits for terminators specified in IEEE Standard 48 or manufacturer's specifications.
   - Apply a dc high potential test in at least five equal increments until maximum test voltage is reached. No increment shall exceed the voltage rating of the cable. Record dc leakage current at each step after a constant stabilization time consistent with system charging current.
   - Raise the conductor to the specified maximum test voltage and hold for a minimum of five minutes. Record readings of leakage current at 30 seconds and one minute and at one minute intervals thereafter.

Insulation Type	System Grounding	Rated Cable Voltage	Test Voltage kV dc
Elastomeric			
(Butyl, Oil Base, EPR)	Grounded	5 kV	27
	Grounded	15 kV	47
	Ungrounded	15 kV	67
Polyethylene	Grounded	5 kV	22
	Grounded	15 kV	40
	Grounded	25 kV	67
	Grounded	35 kV	88
	Ungrounded	15 kV	52

(Derived from ANSI/IEEE Std. 141-1986 Table 81)

Note: Selection of test voltage for in-service cables depends on many factors. The owner should be consulted and/or informed of the intended test voltage prior to performing the test. The above tables are consistent with ICEA recommendations. AEIC C55 and C56, and ANSI/IEEE Std. 400 specify higher voltages. If the cable fails during the test it will require repair or replacement prior to re-energizing.

Table 9-11. Maximum Maintenance Test Voltages (kV dc)—Pre-1968 Cable

- Reduce the conductor test potential to zero and measure residual voltage at discrete intervals.
- Apply grounds for a time period adequate to drain all insulation stored charge. Proper notification must be made to all concerned parties if grounds are left in place.

3. *Test Values*
   - Shielding must exhibit continuity. Investigate resistance values in excess of 10 ohms per 1000 feet of cable.
   - A graphic plot may be made with leakage current (x=axis) versus voltage (y=axis) at each increment.
     - The step voltage slope should be reasonably linear.
     - Absorption slope should be flat or negative. In no case should slope exhibit a positive characteristic.
     - Compare test results to previously obtained results.

Insulation Type	Insulation Level	Rated Cable Voltage	Test Voltage kV, dc
Elastomeric:			
Bultyl and Oil Base	100%	5 kV	19
	100%	15 kV	41
	100%	25 kV	60
	133%	5 kV	19
	133%	15 kV	49
Elastomeric:			
EPR	100%	5 kV	19
	100%	15 kV	41
	100%	25 kV	60
	100%	35 kV	75
	133%	5 kV	19
	133%	15 kV	49
	133%	25 kV	75
Polyethylene	100%	5 kV	19
	100%	15 kV	41
	100%	25 kV	60
	100%	35 kV	75
	133%	5 kV	19
	133%	15 kV	49
	133%	25 kV	75

Derived from ANSI/IEEE Std. 141-1986 Table 82

Note: Selection of test voltage for in-service cables depends on many factors. The owner should be consulted and/or informed of the intended test voltage prior to performing the test. The above tables are consistent with ICEA recommendations. AEIC C55 and C56, and ANSI/IEEE Std. 400 specify higher voltages. If the cable fails during the test it will require repair or replacement prior to re-energizing.

Table 9-12. Maximum Maintenance Test Voltages (kV, dc)—
1968 and Later Cable

DIELECTRIC TEST SHEET

SHEET NO. ______ OF ______

CUSTOMER ______________________ DATE ____________ PROJECT #
ADDRESS ______________________ AIR TEMP. ____________ REL. HUMIDITY ____________
OWNER/USER ______________________ LAST DATE INSPECTION ______________________
ADDRESS ______________________ LAST INSPECTION REPORT # ______________________
EQUIPMENT LOCATION ______________________
OWNER IDENTIFICATION ______________________

EXTERNAL EQUIPMENT INCLUDED IN TEST ______________________

MFR.	
RATED kV	
OPER kV	
LENGTH	
AGE	
NO. COND.	
SIZE	
INSUL. THICKNESS	
INSUL. TYPE	
INSTALLED IN	
TYPE OF TEST	
MAX. TEST kV	
TEST SET	

INSULATION RESISTANCE TEST

One Minute	@	kV	Megohms

Stabilization Time ______________________

STEP VOLTAGE TEST						LEAKAGE CURRENT VS. TIME		
LEAKAGE CURRENT MICROAMP				TIME	VOLTS	LEAKAGE CURRENT MICROAMP		
kV	1	2	3	MIN.	kV	1 △	2 ○	3 □

SUBMITTED BY: ______________________

Figure 9-9. Dielectric Test Report

Inspecting Metal-Enclosed Busways

1. *Visual and Mechanical Inspection*
   - Inspect bus for physical damage.
   - Inspect for proper bracing, suspension alignment, and enclosure ground.
   - Check tightness of bolted joints by calibrated torque wrench method in accordance with manufacturer's published data. In lieu of above torquing perform an infrared survey.
   - Check outdoor busway for removal of "weep-hole" plugs, if applicable, and the proper installation of joint shield.
2. *Electrical Tests*

   Measure insulation resistance of each bus run phase-to-phase and phase-to-ground for one (1) minute.
3. *Test Values*

   Bus bolt torque values shall be in accordance with manufacturer's recommendations.

   Insulation resistance test voltages and resistance values shall be in accordance with manufacturer's specifications or the following:

Equipment Voltage	*Test Voltage*	*Insulation Resistance*
250 Volts	500 Volts	25
600 Volts	1000 Volts	100
5000 Volts	2500 Volts	1000
8000 Volts	2500 Volts	2000
15000 Volts	15000 Volts	5000

MAINTAINING GROUNDING SYSTEMS

This section addresses the design and maintenance of your facility's grounding system. There are two major areas of a plant's grounding system that are addressed in these maintenance procedures: the personnel protection grounding system and the electrical equipment neutral grounding system.

Although these systems use common grounding facilities, they should be maintained and tested as two separate entities. The personnel protection grounding system, or the "bonding" system, is connected to all electrical equipment to prevent hazards from electrical shock. The electrical neutral grounding system provides the path for ground fault currents during short circuits.

The National Electrical Code permits grounding systems to have a resistance up to 25.0 Ohms. The lower the ground resistance, the better the overall grounding system will be; both for personnel safety and for the protection of electrical equipment during ground faults.

The National Electrical Code requires that a 480 volt system, having a capacity greater than 1000 amps, should be provided with ground overcurrent protection. These systems should be tested and maintained during the ground system maintenance procedure. However, they should also be tested during the protective relay procedures that were previously discussed.

Test & Inspection Checklists for Personnel Protection Grounding Systems

1. *Visual and Mechanical Inspection*
   - Inspect ground system for compliance with drawings.
   - Grounding systems, since they are not normally current carrying, are subject to deterioration from corrosion. Presence of green-colored conductors at the equipment bonding connections, would be a sign of possible breakdown of the grounding system.
2. *Electrical Tests*
   - Perform three (3) point fall-of-potential test per IEEE Standard No. 81, Section 9.04 on the main grounding electrode or system.
   - Perform the two (2) point method test per IEEE No. 81, Section 9.03 to determine the ground resistance between the main grounding system and all major electrical equipment frames, system neutral, and/or derived neutral points.
   - An alternate method is to perform ground continuity test between main ground system and equipment frame, system neutral, and/or derived neutral point. This test shall be made by passing a minimum of one (1) ampere dc current between ground reference system and the ground point to be tested. Voltage drop shall be measured and resistance calculated by voltage drop method.
3. *Test Values*

   The main ground electrode system resistance to ground should be no greater than five (5) ohms for commercial or industrial systems and one (1) ohm or less for generating or transmission station grounds unless otherwise specified by the owner/user's electrical engineer.

Inspection Checklists for Electrical Equipment Neutral Grounding Systems

1. *Visual and Mechanical Inspection*
   - Inspect for physical damage and corrosive deterioration.
   - Inspect neutral main bonding connection to assure:
     - Protective relay sensing system is grounded.
     - Ground strap sensing systems are grounded through sensing device.

- Ground connection is made ahead of neutral disconnect link on protective relay sensing systems.
- Grounded conductor (neutral) is solidly grounded.

• Manually operate ground overcurrent monitor panels, and record proper operation and test sequence per manufacturer's instructions.

• Inspect zero sequence ground overcurrent systems for symmetrical alignment of core balance transformers about all current carrying conductors.

• Set pickup and time delay settings for ground overcurrent relays in accordance with the owner/user's electrical engineer's provided settings.

2. *Electrical Tests*

• Measure system neutral insulation resistance to ensure that no shunt ground paths exist. Remove neutral-ground disconnect link. Measure neutral insulation resistance and replace link.

• Determine the relay pickup current by current injection at the sensor and operate the circuit interrupting device. Test the relay timing by injecting three hundred percent (300%) of pickup current, or as specified by manufacturer.

• Test ground overcurrent zone interlock systems by simultaneous sensor current injection and monitoring zone blocking function.

• Perform ground continuity tests between main ground system and the protective devices neutral connections. Inject one (1) ampere dc current and measure voltage drop.

3. *Test Parameters*

• System neutral insulation shall be a minimum of one hundred (100) ohms, preferably one (1) megohm or greater.

• Relay timing shall be in accordance with manufacturer's published time-current characteristic curves but in no case longer than one (1) second for fault currents equal to or greater than 3,000 amperes.

MAINTAINING THE EMERGENCY GENERATOR

Emergency generators are normally used as standby emergency systems when the utility power supply system is out of service. Since it is a "standby" system, it should be ready to operate quickly and should be kept in a dependable, well-maintained condition.

Emergency generators usually operate through one of the following systems:

- Automatic Transfer Switches
- Uninterruptible Power Systems
- Low Voltage Switchgear

A generator maintenance plan is the only way to be certain of dependable performance from your standby emergency generating system. Factory-trained and -authorized technicians will provide proper year-round inspection service and maintenance of your equipment. Lack of maintenance is the major cause of generator failures.

Your manufacturer's maintenance manual should provide the procedures to follow and their respective frequencies. Refer to Table 9-13 for typical schedules for good generator maintenance.

Performing Engine-Generator Specific Maintenance

The following procedures pertain to the electric generator portion of this system.

1. *Visual and Mechanical Inspection*
   - Inspect for physical damage, proper anchorage, and grounding.
2. *Electrical and Mechanical Tests*
   - Perform a dielectric absorption test on generator winding with respect to ground. Determine polarization index.

$$\text{Polarization Index} = \frac{\text{Ohms @ 10 Minutes}}{\text{Ohms @ 1 Minute}}$$

   - Test protective relay devices.
   - Functionally test engine shutdown for low oil pressure, over-temperature, over-speed, and other features as applicable.
   - Perform vibration base line test. Plot amplitude versus frequency for each main bearing cap.
   - Perform resistive load bank test at one hundred percent (100%) nameplate rating. Loading shall be:
     - 25% rated for 30 minutes
     - 50% rated for 30 minutes
     - 75% rated for 30 minutes
     - 100% rated for 3 hours
   - Record voltage, frequency, load current, oil pressure and coolant temperature at periodic intervals during test.
3. *Test Values*
   - Perform dielectric absorption at test voltage listed in below. Polarization index values shall be in accordance with IEEE Standard 43.
   - Load test results shall demonstrate the ability of the unit to deliver rated load for the test period.

SYSTEM	COMPONENT	FREQUENCY
FUEL	Main tank supply level	Weekly
	Day tank supply level	Weekly
	Fuel level switch	Weekly
	Transfer pump operation	Weekly
	Solenoid valve operation	Weekly
	Filter(s)	Quarterly
	Water in fuel system?	Weekly
	Flexible lines, connectors	Weekly
	Vent and return lines open?	Yearly
	Fuel piping	Yearly
	Gasoline supply	Six-Months
LUBRICATION	Oil level	Weekly
	Change oil	/Mfgr.
	Change filter(s)	/Mfgr.
	Crankcase breather	Quarterly
COOLING	Coolant level	Weekly
	Antifreeze protection	Six-Months
	Change coolant	Yearly
	Water supply to heat exchanger	Weekly
	Clean heat exchanger	Yearly
	Air inlets	Weekly
	Clean radiator exterior	Yearly
	Fan, alternator belts	Monthly
	Inspect water pump	Weekly
	Flexible hoses, connectors	Weekly
	Block heater operation	Weekly
	Air ducts, louvers	Yearly
	Louver motors, controls	Yearly
EXHAUST	Leakage	Weekly
	Drain condensate trap	Weekly
	Insulation, fire hazards	Quarterly
	Back pressure	Yearly
	Hangers and supports	Yearly
	Flexible connector (s)	Six-Months

Table 9-13. Emergency Generator Preventive Maintenance Service Checklist

SYSTEM	COMPONENT	FREQUENCY
BATTERY SYSTEM	Electrolyte level	Monthly
	Terminals clean, tight	Quarterly
	Remove corrosion, clean	Monthly
	Specific gravity, charge state	Monthly
	Charger operation, charge rate	Monthly
	Recharge after engine start	Monthly
ELECTRICAL SYSTEM	Inspection	Weekly
	Tighten wiring connections	Yearly
	Wire damage	Quarterly
	Safety and alarm operation	Six-Months
	Controller lamp test	Weekly
	Circuit breaker, fuse service	Monthly
	Transfer switch main contacts	Yearly
	Voltage sensing adjustment	Yearly
	Wire-cable insulation	3 yrs. or 500 hrs.
ENGINE	Inspection	Weekly
	Air cleaner	Six-Months
	Governor operation, lubrication	Monthly
	Governor oil, change	Yearly
	Ignition components	Yearly
	Choke, carburetor adjustment	Six-Months
	Fuel system inspection	Yearly
	Valve clearance	3 yrs. or 500 hrs.
GENERATOR	General inspection	Weekly
	Rotor and stator	Yearly
	Bearing condition, lubrication	Yearly
	Exciter	Yearly
	Voltage regulator	Yearly
	Test winding insulation	Yearly
GENERAL EQUIPMENT	Clean interior of room, housing	Weekly
SET SYSTEM	For automatic operation	Weekly

Table 9-13. Emergency Generator Preventive Maintenance Service Checklist (*continued*)

Equipment Rating	*Test Voltage (Min.)*	*Insulation Resistance Megohms*
250 Volts	500 Volts	25
600 Volts	1000 Volts	100
5000 Volts	2500 Volts	1000
8000 Volts	2500 Volts	2000
15000 Volts	2500 Volts	5000

MAINTENANCE PROGRAMS FOR BATTERIES AND D.C. CONTROL SYSTEMS

Batteries and their D.C. control systems are an integral part of a facility's power supply system. Their principal applications include some of the following systems:

- Switchgear D.C. control system
- Emergency generator starting & control system
- Uninterruptible power supply system
- Motor static control systems

Battery Types

The following represent some basic types of batteries you might find in your facility:

- Lead acid
- Lead calcium
- Lead Antimony
- Lead (pure)
- Lead acid sealed
- Nickel cadmium (nicad)
- Lead calcium gel cells

Refer to Table 9-14 for some general application comments and related life expectancies.

Typical Battery Maintenance Programs

1. *Sealed Cell Unit Construction*
    - Perform overall visual inspection of cells, connectors, racks and room.
    - Ambient temperature.
    - Perform visual check of jars, covers, leaks, and so on.

BATTERY TYPE	COMMENTS	LIFE EXPECTANCY
LEAD ACID		
A. LEAD CALCIUM	• Best trade cost *vs* reliability. • Generates less hydrogen gas and less maintenance than Lead Antimony.	20 Years
B. LEAD ANTIMONY	• Better suited for applications where they are charged and discharged more often. • Found in use where the power frequently goes out.	15-20 Years
C. LEAD (PURE)	• Contains no impurities in their structure. • Higher cost than Lead Antimony and Lead Calcium (two to three times). • Durable. • Frequent cycling. • Long floating periods. • Tolerate high operating temperatures. • Less maintenance. • Contain nearly all their capacity at the end of warranty period.	20-25 Years
LEAD ACID SEALED	• Available in sealed maintenance free versions. • Used for stationary applications.	5-10 Years
NICKEL CADMIUM (NICAD)	• Higher cost than Lead Calcium (three times more expensive). • Smaller in size and weight than Lead Calcium. • Can tolerate temperature change.	25 Years
LEAD CALCIUM GEL	• Gelled electrolyte. • Sealed. • Does not lead even with cracked case • Virtually no gas vented. • Suitable for use in unventilated areas. • Short useful life.	2-5 Years

Table 9-14. Battery Types

- Check for corrosion build-up throughout the entire system, observing trouble spots.
- 100% voltage readings, all cells.
- Total battery voltage, measured at positive and negative terminals.
- Rectifier output current, all phases.
- Rectifier charge current per voltage.
- Clean and neutralize cells and racks.
- Document all findings during inspection and provide recommendations for any corrective action deemed necessary.

2. *Wet Cell Unit Construction*
   - Perform overall visual inspection of cells, connectors, racks and room.
   - Check ambient temperature and verify adequate room ventilation.
   - Check electrolyte levels and adjust as necessary.
   - Check individual cells for post seals, sediment, plate condition, flame arresters, jars, covers, leaks, and so on.
   - Check for corrosion build-up throughout the entire system, observing trouble spots.
   - 100% voltage readings, all cells.
   - Total battery voltage, measured at positive and negative terminals.
   - Rectifier Output Current, all phases.
   - Rectifier Charge Current per voltage.
   - Take 100% specific gravity readings corrected for temperature, all cells.
   - Check battery cell temperatures, 20% sampling.
   - Clean and neutralize cells and racks.
   - Document all findings during inspection and provide recommendations for any corrective action deemed necessary.

Electrical Tests

The following are typical generic tests that should be performed on batteries. Refer to ANSI/IEEE Standard 450 (lead) or ANSI/IEEE Standard 1106 (nickel-cadmium). Figure 9-10 provides a battery data sheet.

- Measure bank charging voltage and each individual cell voltage.
- Measure electrolyte specific gravity and visually check fill level.
- Perform contact integrity tests across all connections between adjacent terminals.
- Perform an integrity load test.
- Perform a capacity load test at five year intervals.

BATTERY DATA SHEET

Sheet No.___ of___

Customer________________________ Date____________ Project No.________

Address ________________________ Air Temp.______ Rel. Humidity______

Owner/User________________________ Date Last Inspection ____________

Address________________________ Last Inspection Report No.________

Equipment Location________________________ Battery Type____________

Owner Identification________________________

Charger Float Voltage________ Charger Equalize Voltage______ Battery Voltage________

INDIVIDUAL CELL VOLTAGES OUT OF SERVICE: Amp Hours______ Battery Temp.________

	VOLTAGE	SPEC.GR.		VOLTAGE	SPEC.GR.		VOLTAGE	SPEC.GR.		VOLTAGE	SPEC.GR.
1)			26)			51)			76)		
2)			27)			52)			77)		
3)			28)			53)			78)		
4)			29)			54)			79)		
5)			30)			55)			80)		
6)			31)			56)			81)		
7)			32)			57)			82)		
8)			33)			58)			83)		
9)			34)			59)			84)		
10)			35)			60)			85)		
11)			36)			61)			86)		
12)			37)			62)			87)		
13)			38)			63)			88)		
14)			39)			64)			89)		
15)			40)			65)			90)		
16)			41)			66)			91)		
17)			42)			67)			92)		
18)			43)			68)			93)		
19)			44)			69)			94)		
20)			45)			70)			95)		
21)			46)			71)			96)		
22)			47)			72)			97)		
23)			48)			73)			98)		
24)			49)			74)			99)		
25)			50)			75)			100)		

Comments: ____

____

Submitted By: Time:

Figure 9-10. Battery Data Sheet

- Verify proper charging rates from charger during recharge mode.
- Verify individual cell acceptance of charge during recharge mode.
- For the acceptance level of the tests, compare measured values to manufacturer's specifications.

MAINTAINING UNINTERRUPTIBLE POWER SUPPLIES (UPS)

Static uninterruptible power supplies (UPS) have no moving parts and are the most widely used form of backup power. On-line systems are the system of choice for most installations and are discussed in this section because they provide continuously conditioned power to the critical load.

An online UPS provides:

- Isolation from both line voltage and line frequency variations.
- No switching transient generated when power is switched to the batteries.
- Faster battery recharge time.
- Power is always delivered to the load by the use of an online invertor and static transfer switch.

A sophisticated UPS system consists of the following:

- Emergency Generator
- UPS - rectifier/charger
  -invertor/filter
  -battery bank
- Battery monitoring system.

UPS Self-Diagnostic Procedures

The UPS is automated to a great extent so that it is simple to operate and contains self-diagnostic circuitry. A documented periodic program of inspection and preventive maintenance should be implemented to assure optimum operation. Troubleshooting and repairing will require trained personnel with a detailed technical knowledge of the UPS system. An outside service firm should be employed for this purpose.

In general, personnel from an outside service firm will have both high voltage power and technical electronic experience in UPS Systems. Experienced UPS maintenance personnel with a good knowledge of their UPS, with proper test equipment and with factory spare parts, should maintain the UPS unit.

The equipment of the UPS is divided into sub-assemblies which can be removed for repair. The entire sub-assembly should be replaced in case of failure. No repairs are to be made on the sub-assembly.

Refer to Table 9-15 for a diagnostic guide for common failures that may be experienced in a UPS.

I — INSTRUMENTATION — Consists of Metering or Digital Readout

A. Input

- Input Voltage (AC)
- Input Current (AC)

B. Output/By-Pass

- Voltage (AC)
- Hertz
- Current (AC)

C. DC Battery

- Charge/Discharge
- Voltage (DC)

II — MIMIC PANEL — Mimic depicts the complete single line diagram of UPS Unit

A. Input Circuit Breaker	On/Off
B. Battery Circuit Breaker	On/Off
C. Output Circuit Breaker	On/Off
D. By-Pass Input Circuit Breaker	On/Off
E. Static By-Pass Switch	On/Off

III — CONTROL PANEL

A. Audio Reset
B. Emergency Off—Local or Remote
C. Lamp Test Reset
D. Transfer UPS/By-Pass

IV — SELF-DIAGNOSTIC CIRCUITRY AND ALARMS — Sophisticated UPS Systems contain built-in diagnostic circuitry for assisting outside service firms in troubleshooting. Many of the systems contain system level alarm lights, usually of the long life "LED" type. After an audible alarm has cone on, and after the manufacturer's manual has been followed, the outside service firm should be called in for service. The following are some typical alarms:

- AC Overvoltage/Undervoltage
- Static Switch Defective
- Fan Failure
- Fuse
- Overload
- Overload Shutdown
- Load on By-Pass
- Control Power Failure
- Input Failure
- Emergency—Off
- DC Overvoltage
- Battery Discharge
- Battery C.B. Opened
- Low Battery Voltage
- Overtemperature
- Ambient Overtemperature

Table 9-15. Diagnostic Guide for Common Failures of UPS Systems

Battery Monitoring System Maintenance

In the event of a power failure, (UPS) can be rendered totally useless without properly maintained batteries. Refer to the battery maintenance procedures described previously.

Battery monitoring systems provide dependability to monitor the conditions which influence battery life and performance and warn of possible problems or failing conditions by means of user programmable alarms.

These systems are large data acquisition systems that continually monitor all critical parameters of a battery system. The system alarms on any out-of-tolerance condition and stores all pertinent data associated with an alarm, a load test, an unscheduled discharge or historical data readings.

The following represents some display features that a battery monitoring system would incorporate to ensure battery reliability:

- Individual cell and overall battery voltage
- Battery location and number
- Alarm settings and reports
- Historical settings and reports
- Directory
- Capacity and integrity testing
- Pilot cell temperature
- Ambient temperature
- Current

Battery monitoring systems are designed for long life and should only require a preventative maintenance check once every six months.

The items in these systems that require periodic inspection are described in the following sections.

Sense Leads

The sense leads connected to the battery terminal posts should be checked for possible corrosion problems due to acid leakage around the post seals.

If corrosion is evident, the leads should be removed, cleaned, greased with non-oxide grease and reconnected.

UPS Batteries

Batteries used in battery monitoring systems are sealed, lead acid cells that should operate trouble free for seven or more years. Just as any other backup battery system, it should be tested and inspected periodically.

Remove primary power to the system once every six months and verify that the local UPS inverter operates correctly and that the battery voltage holds up as expected. Refer to the UPS manufacturer's instruction manual to determine the rundown time expected on the batteries.

Load Modules

The Load bank modules should be covered when not in use. The units may gather dust and dirt from the surrounding environment because of the expanded metal sides and tops. These foreign particles will normally be burned off during a load test, but do create hot spots that could lead to premature failure of some of the resistors and may interfere with the normal operation of the cooling fans. If the units have been allowed to gather dust and dirt, they should be cleaned using high pressure air.

The copper pads on the front of the load units, used as connection points to the battery load cables, should be cleaned before every test to make sure that the best possible connection is made and minimize any heat build-up.

Calibration

Typical battery monitoring systems utilize highly accurate and stable measuring circuits that should only require calibration once a year. Refer to the manufacturer's manual for specific calibration procedures.

MAINTENANCE AND TESTING PROCEDURES FOR AUTOMATIC TRANSFER SWITCHES

This section deals with the maintenance and testing procedures for automatic transfer switches. There are two basic ATS types:

- *Open Transition*—This type functions as a "break-before-make" operation. Usually, the normal supply voltage has failed, an emergency generator has been started, and when the generator's voltage and frequency are at proper levels, the switch transfers its load to the emergency source. Most ATSs are the open transition type.
- *Closed Transition*—This type functions as a "make-before-break" operation. The normal supply voltage and the alternate supply voltage are usually present. The switch will momentarily parallel both sources (100 milliseconds, 6 cycles) and then open its normal supply side contacts. This type of transfer permits the load not to lose voltage and to continue its operation without an interruption.

Automatic transfer switches should be maintained at the same frequency as their respective emergency generators and/or their alternate power supplies.

1. *Visual and Mechanical Inspection*
   - Inspect for physical damage
   - Check switch to ensure positive interlock between normal and alternate sources. mechanical and electrical
   - Check tightness of all control and power connections
   - Perform manual transfer operation with no connected load
   - Ensure manual transfer warnings are attached and visible to operator
2. *Electrical Tests*
   - Perform insulation resistance tests phase-to-phase and phase-to-ground with switch in both source positions, where possible.
   - Perform a contact resistance test or measure millivolt drop across all main contacts. As an alternate, an infrared scan may be performed for both the normal and emergency loads.
   - Verify settings and operation of control devices in accordance with the owner/user's electrical engineer's specifications for, voltage and frequency sensing relays, all time delay relays, and engine start and shutdown relays.
   - Perform the following automatic transfer tests:
     - Simulate loss of normal power
     - Return to normal power
     - Simulate loss of emergency power
     - Simulate all forms of single-phase conditions
   - Monitor and verify correct operation and timing of the following:
     - normal voltage sensing relays
     - engine start sequence
     - time delay upon transfer
     - alternate voltage sensing relays
     - automatic transfer operation
     - interlocks and limit switch function
     - time delay and retransfer upon normal power restoration
     - engine cooldown and shutdown feature
3. *Test Values*
   - Insulation resistance test voltages and minimum values to be as follows:

Equipment Rating	*D.C. Test Voltage (Min.)*	*Minimum Insulation Resistance-Megohms*
250 Volts	500 Volts	25
600 Volts	1000 Volts	100

- Determine contact resistance in microhms. Investigate any value exceeding 500 microhms or any values which deviate from adjacent poles by more than fifty percent (50%).

CHECKLISTS FOR MAINTAINING MISCELLANEOUS ELECTRICAL EQUIPMENT

This section covers the maintenance and testing procedures associated with various types of electrical equipment not covered in other sections. The following items provide recommended inspections, tests and general maintenance procedures for:

- Capacitors
- Lightning/surge arresters
- 600 volt or less - surge protection devices
- 5kV-69kV lightning arresters

Maintaining Capacitors

The principal use of capacitors is to help maintain the service voltage level and to provide Vars for power factor correction.

1. *Visual and Mechanical Inspection*
   - Inspect capacitors for physical damage, proper mounting, and required clearances.
   - Compare nameplate information with single line diagram and report discrepancies.
   - Verify that capacitors are electrically connected in the proper configuration.
   - Inspect all bus or cable connections for tightness by calibrated torque wrench method. Refer to manufacturer's instructions.
2. *Electrical Tests*
   - Perform insulation resistance tests from pole(s) to case for one (1) minute. Test voltage and minimum resistance shall be in accordance with manufacturer's instructions.
   - Measure the capacitance of all pole-to-pole combinations and compare with manufacturer's published data.
   - Verify that internal discharge resistors are operating properly.
3. *Test Values*
   - Check that bolt torque levels are as specified by manufacturer.
   - Investigate any insulation resistance values that are less than the following table:

Equipment Rating	*D.C. Test Voltage*	*Resistance Megohms*
250	500	25
600	1000	100
5000	2500	1000
8000	2500	2000
15000	2500	5000
25000	5000	20000

- Investigate capacitance values that differ more than ten percent (10%) from manufacturer's published data.
- Reduce residual voltage of a capacitor to 50 volts in the following time intervals after being disconnected from the source of supply:

Rated Voltage	*Discharge Time*
≤ 600V	1 minute
> 600V	5 minutes

Maintaining Lightning/Surge Arresters

The principal usage of lightning and/or surge arresters is to protect the various electrical systems against lightning strokes and sharp voltage spikes created by utility company switching or the operation of line capacitors.

600-Volt or Less—Surge Protection Devices

1. *Visual and Mechanical Inspection*
   - Inspect for physical damage.
   - Inspect for proper mounting and adequate clearances.
   - Check tightness of connections by calibrated torque wrench method. Refer to manufacturer's instructions for proper torque levels.
   - Check ground lead on each device for individual attachment to ground bus or ground electrode.
2. *Electrical Tests*

   Air gap surge protection Devices (excluding valve and expulsion type devices). See ANSI/IEEE C62.32 for design test criteria.
   - Perform dc breakdown voltage tests.
   - Perform insulation resistance tests.

   Expulsion Type Arresters
   - Perform insulation resistance tests on each pole-to-ground. Maximum applied voltage must not exceed the rating of the device.
   - Ensure that exhaust ports are clear of any obstruction.

Gas Tube Surge Protection Devices. See ANSI/IEEE C62.31 for design test criteria.

- Perform dc breakdown voltage tests.
- Perform insulation resistance tests.

Valve Type Arresters - Silicone Carbide

- Perform insulation resistance tests on each pole-to-ground. Maximum applied voltage must not exceed the rating of the device.
- Perform dc voltage rise test to flashover value; terminate test immediately upon flashover. Flashover voltage should approximate the calculated 60 hertz RMS sparkover equivalent.

Varistor Surge Protective Devices

- Perform clamping voltage test in accordance with ANSI/IEEE Standard C62.33, Section 4.4.
- Perform rated RMS voltage test and rated dc voltage test in accordance with ANSI/IEEE Standard C62.33, Section 4.7.

Testing 5KV to 69KV Lightning Arresters

1. *Visual and Mechanical Inspection*
   - Inspect for physical damage.
   - Compare nameplate information with latest single line diagram and report discrepancies.
   - Inspect for proper mounting and adequate clearances.
   - Check tightness of connections by calibrated torque wrench method. Refer to manufacturer's instructions for proper torque levels.
   - Check ground lead on each device for individual attachment to ground bus or ground electrode.
   - Verify that stroke counter is properly mounted and electrically connected.
2. *General - Electrical Tests*
   - Test grounding electrode in accordance with IEEE Standard No. 81, Section 9.04 using the 3-Point Fall-of-Potential method.
   - Perform an insulation power factor test.
   - Perform an insulation resistance test. Use manufacturer's values.
3. *Silicone Carbide Arresters - Electrical Tests*
   - Perform RF noise test using a radio interference voltage test set. RIV must be not more than 10 microvolts above background, with an applied voltage of 1.18 times maximum continuous operating voltage (MCOV).
   - Perform dc voltage rise test to flashover value - terminate test immediately upon flashover. Flashover voltage should approximate the calculated 60 hertz RMS sparkover equivalent.

4. *Metal Oxide Arresters - Electrical Tests*
   - Perform RF noise test using a radio interference voltage test set. RIV must be not more than 10 microvolts above background with an applied voltage of 1.18 times maximum continuous operating voltage (MCOV).
   - Do not perform dc voltage rise to flashover tests on metal oxide arresters - damage will result due to the linearity of metal oxide elements.
   - Perform leakage current and watts loss tests.

10

Electric Motor Maintenance

W. Dumper, P.E.

Electric motors are found everywhere in our society; in homes, commercial facilities, and industrial plants. They consume more energy than any other utilization device in the United States. Since motors are so pervasive and important to all phases of our lives it is important that they be maintained properly. This chapter is devoted to the care and maintenance of electric motors. Since the types of electric motors are so numerous, a brief description of the types normally encountered is included in the beginning. Proper maintenance depends on an understanding of the motors entrusted to your care.

TYPES OF ELECTRIC MOTORS

The types of electric motors can be grouped into a few simple categories. The major categories are a.c. and d.c. motors. A.C. motors include induction and synchronous motors with further subdivision into single-phase and polyphase motors. D.C. motors are divided into permanent magnet and wound field categories. The following descriptions and associated maintenance tips should help you recognize and maintain the various types of electric motors you will encounter in the field.

Single-Phase Motors

These motors come in a large variety of designs. They tend to be more complicated than their three-phase counterparts because of the need for additional windings and mechanical devices required for starting. As a result, single-phase motors are somewhat less reliable and require more maintenance than three-phase motors. (See Table 10-1.)

Symptom	Possible Cause	Action/Correction
Motor won't start (dead motor)	Tripped or defective overload device in motor	If equipped with a manual device press reset button. Check that internal overload is not open. Replace device if defective.
	Power supply open	Check voltage at motor terminals. Correct as required.
Motor won't start (just hums)	Capacitor wrong size or defective (permanent split & capacitor start)	Check capacitor for short or open. Replace if defective or wrong size.
	Centrifugal starting switch bad (split-phase & capacitor-start)	Check centrifugal switch rotating and stationary members for proper operation. Check switch contacts. Replace either part if defective.
	Starting or running windings open, shorted or grounded	Check windings for burns, opens, grounds. Replace motor if windings are defective.
	Brush/commutator problems (repulsion motor)	Check brushes and commutator. See section on dc motors and text on commutator maintenance.
Motor trips after starting, draws excessive current	Centrifugal switch fails to open	Check switch as above. Possible that switch contacts welded closed.
	Defective windings	Check windings as above. If windings are defective replace motor.
	Defective capacitor (permanent-split cap. motor)	Replace capacitor—check to see if rating is proper.
	Defective running capacitor (capacitor-run motor)	Same as above

Table 10-1. Troubleshooting Guide for Single-phase Motors*

Symptom	Possible Cause	Action/Correction
Motor trips after starting	Defective over-load protector	If motor is not drawing current in excess of nameplate rating replace overload protector.
Motor comes to speed and cyclically slows and accelerates (centrifugal switch operates)	Motor heavily overloaded	Reduce load—increase motor hp rating.
	Motor connected for wrong voltage	Check nameplate for proper connections and correct.
	Running winding defective	Replace motor.

*Courtesy of the Lincoln Electric Company

Table 10-1. Troubleshooting Guide for Single-phase Motors* (*continued*)

Shaded-Pole Motors

These are the simplest and least expensive of the single-phase motors. They are used mainly for small blowers and fans ¼ hp and less. The stator consists of distinct pole pieces encircled by windings as shown in Figure 10-1. Starting is provided by a shading coil, generally a single turn of heavy wire, embracing a small portion of the pole face. These motors rotate in one direction only. The direction can be determined visually by looking at the stator and remembering that the rotor sweeps past the major portion of the pole face towards the minor portion encircled by the shading coil. It is the only type of motor whose direction of rotation can be determined by inspection of the stator.

This motor has very low starting torque, power factor, and efficiency. Its speed can be varied by reducing the terminal voltage. This is generally accomplished by tapping the winding with multiple leads brought out for speeds such as high, medium high, and so on. When installing or reconnecting such motors it is imperative that line potential be applied only between the "common" lead and any of the various speed winding leads. Never apply line voltage between any of the speed windings. Failure to observe this precaution will result in your maintenance effort becoming a replacement procedure, since the winding will fail very rapidly. Other than the universal criteria of keeping a motor clean, dry and properly lubricated, there is little required in the way of maintenance for these motors. It is generally uneconomical to repair these motors if a winding should fail.

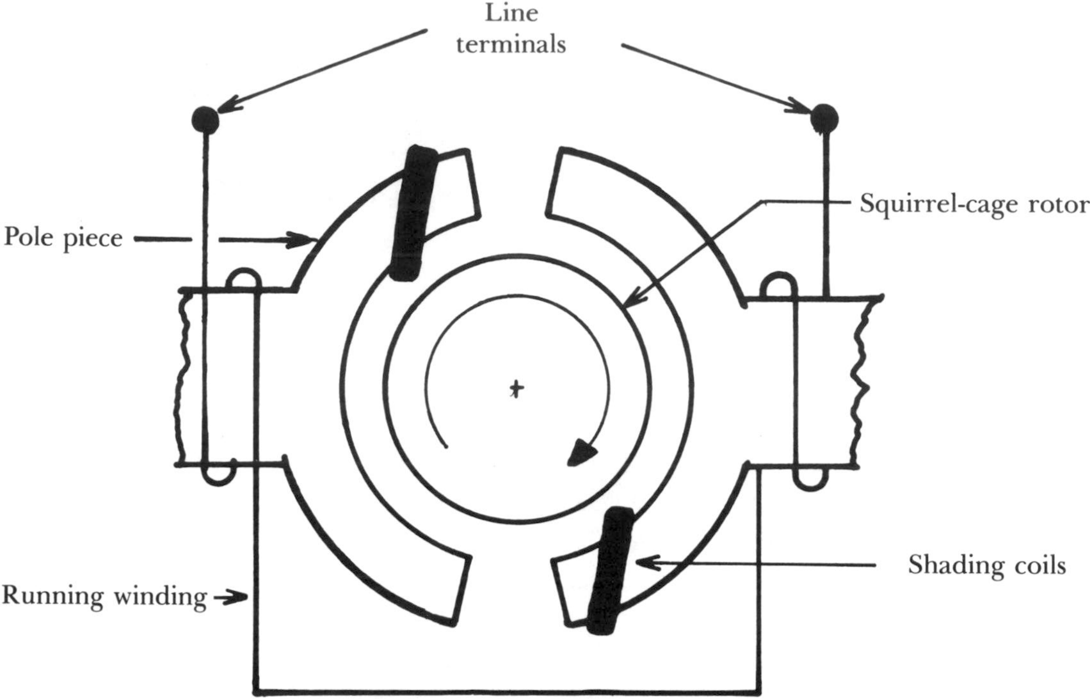

Figure 10-1. Shaded-Pole Motor

Permanent-Split-Capacitor Motors

These motors are a step more complicated, as you can see from Figure 10–2. An additional winding and a capacitor have been introduced. Like the shaded-pole motor it is used mainly on fans and blowers and sometimes on pumps up to about 1.5 hp. The stator consists of two separate windings placed in slots in the stator bore surface. These windings are called the running and auxiliary windings. The auxiliary winding is connected in series with a capacitor and provides the starting function. Both windings are also energized during normal operation and contribute to the output of the motor.

This motor has relatively low starting torque in the order of 80-100% of full load torque. It has high efficiency and power factor and the speed can be reduced by voltage reduction or by a tapped winding. An approximate reduction in speed of 25% can be achieved under load. If the motor is tested at no-load little change in speed will be noticed between the various windings. If a motor is replaced with a larger rating you may find that the range of speed reduction will be reduced. The direction of rotation can be reversed by reversing the connection relationship between the main and auxiliary windings.

If a motor fails to start, assuming that it is not locked by frozen bearings or a problem with the driven equipment, you should look first at the easiest item to check, the capacitor. The capacitors used in these motors are generally metal encased and failure may not be visually noticeable. Sometimes a slight bulging may be

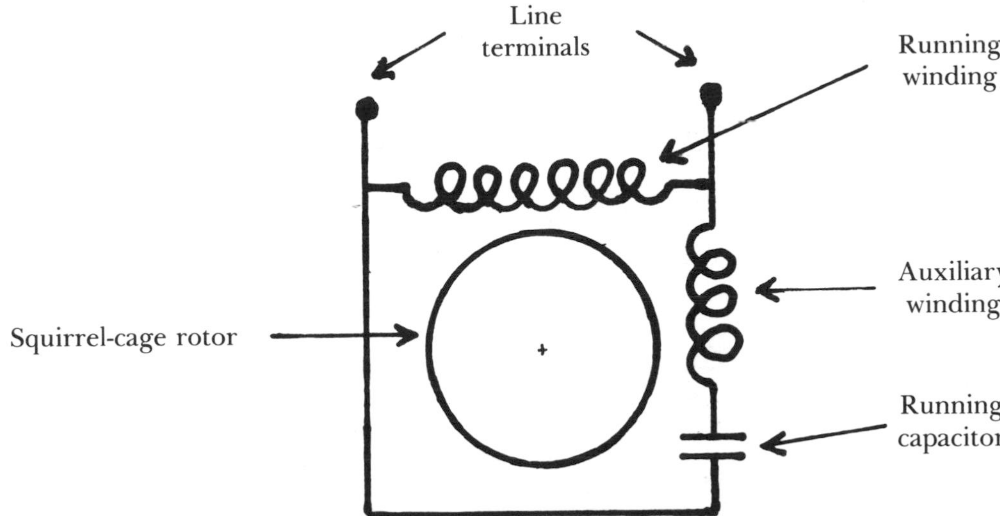

Figure 10-2. Permanent-Split-Capacitor Motor

evident at the capacitor ends. A capacitor checker or the substitution of a known good capacitor can be used to make a proper determination. The proper size capacitor for the motor is generally shown on the motor nameplate and a range of 4 to 20 mfd with a voltage rating of 370 volts is typical. A capacitor which is either open or shorted will prevent starting. If the capacitor proves to be healthy then the problem is usually a failed winding and the economical solution is replacement. Motors equipped with ball bearings may be worth repairing in the larger sizes if they become noisy.

Split-Phase Motors

These are generally used on belted fan drives, grinders, pumps, and other loads requiring less than 200% starting torque up to about ¾ hp. As shown in Figure 10-3 this motor contains a running winding, a starting winding and a centrifugal switch mechanism which disconnects the starting winding when the motor reaches about 80-85% of rated speed. As you can see, another component has been introduced which requires maintenance: the centrifugal switch mechanism with its rotating and stationary parts. Although there are many styles of centrifugal switches, the basic operating principle utilizes rotating spring restrained weights to activate a spoolpiece at a predetermined speed. Movement of this revolving spoolpiece permits the spring-loaded contacts on the stationary switch to open and disconnect the starting winding. The starting winding is not energized during normal operation and provides no help in driving the load.

Unlike the other motors discussed, this motor is a constant speed device. Its speed should not be varied by reducing the terminal voltage since the centrifugal switch mechanism would tend to close and reconnect the starting winding. This winding is designed for use only during the brief starting interval. Prolonged energization will cause overheating and failure.

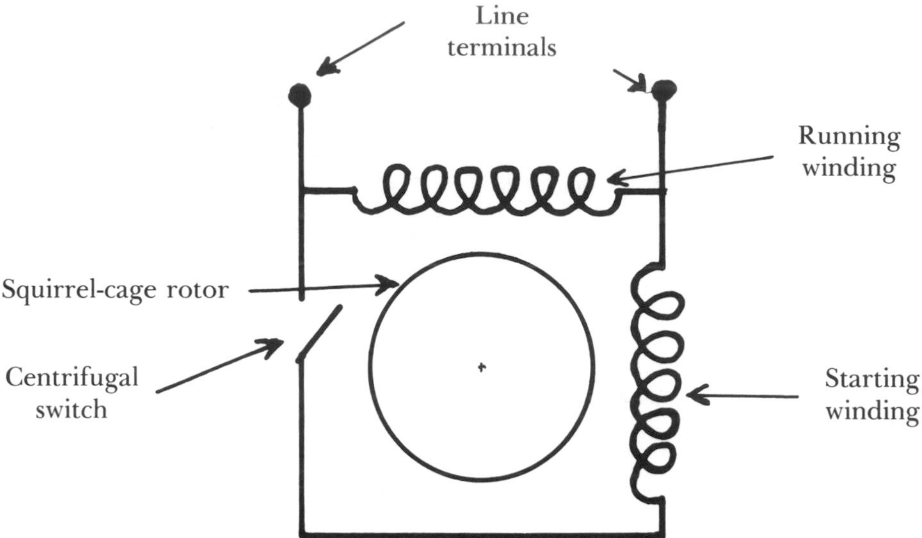

Figure 10-3. Split-Phase Motor

You will find that maintenance problems on this type of motor tend to center about the centrifugal switch mechanism. Failure to start is generally caused by the stationary switch contacts not closing when the motor was last shut down. The starting winding is not energized and the motor just hums. If the motor is not disconnected quickly, the heavy starting current drawn by the running winding will cause it to overheat and fail. Similarly, a switch with contacts unable to open will cause the starting winding to fail. Some motors are equipped with automatic thermal overload protection which will sense these elevated currents and disconnect the motor. However, if the motor is located in a remote area such as on a roof, it will continue to cycle on and off unnoticed and eventually fail due to repeated overheating. As you can see, the centrifugal switch mechanism is a very critical component. The stationary switch contacts must be kept clean and free from oil and grease and the centrifugal portion must be kept clean to permit mechanical operation at the proper speed. Adjustment or repair of this mechanism should be left to those experienced in this area due to the small tolerances involved and the possibility of destroying the motor if done incorrectly. Unless the motor is of non-standard design with special mounting or shaft configurations, it is not economical to repair split phase motors which have experienced winding failures.

Capacitor-Start Motors

These motors are used on applications requiring high starting torques such as compressors, hoists, and conveyors up to a rating of about 10 hp. Starting torque is in the order of 350% of rated full load torque. This motor is essentially a split-phase motor with a capacitor in series with the starting winding. Refer to Figure 10-4.

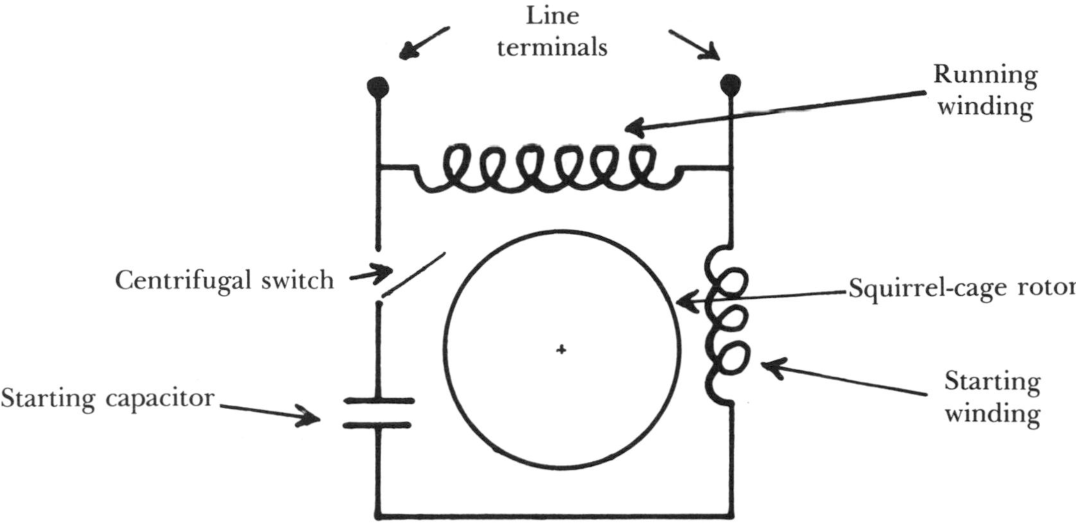

Figure 10-4. Capacitor-Start Motor

Reasons for failure to start include the centrifugal switch mechanism misoperations described above plus the possibility of capacitor failure. The capacitor is generally mounted on the motor exterior and should be checked first. A failed capacitor which is leaking or distorted will be immediately obvious. These capacitors are not usually metal enclosed. A failed capacitor may not be the initial cause of the problem, however. Starting capacitors, unlike running capacitors, are meant to be energized only during the starting interval. A centrifugal switch which fails to open and deenergize the starting circuit can cause capacitor failure. A capacitor checker is useful in this situation. If you substitute a known good capacitor as a test, be sure to measure the input current. If the motor is drawing more than rated current, possibly a multiple of such current, then a problem still exists internally and the motor should be disconnected before the capacitor fails.

Starting capacitor rated values are not indicated on the motor nameplate. The usual range is between 50 and 1200 mfd with larger motors utilizing the higher values. Voltage ratings vary from 125 Volts to 440 Volts. If at all possible, the capacitor should be replaced with one of the same capacity marked on the old capacitor. The capacitance is generally shown as a range, such as 161–193 mfd for example. The exact value is not critical and a replacement within 10% will usually work satisfactorily. The voltage rating should be duplicated. This rating is an internal design parameter and it may be different from the motor voltage rating.

Capacitor-Start, Capacitor-Run Motors

These combine the advantages of the capacitor-start motor with the high efficiency and power factor of the permanent-split-capacitor motor. (See Figure 10-5.) In this design, the centrifugal switch removes the starting capacitor from the circuit

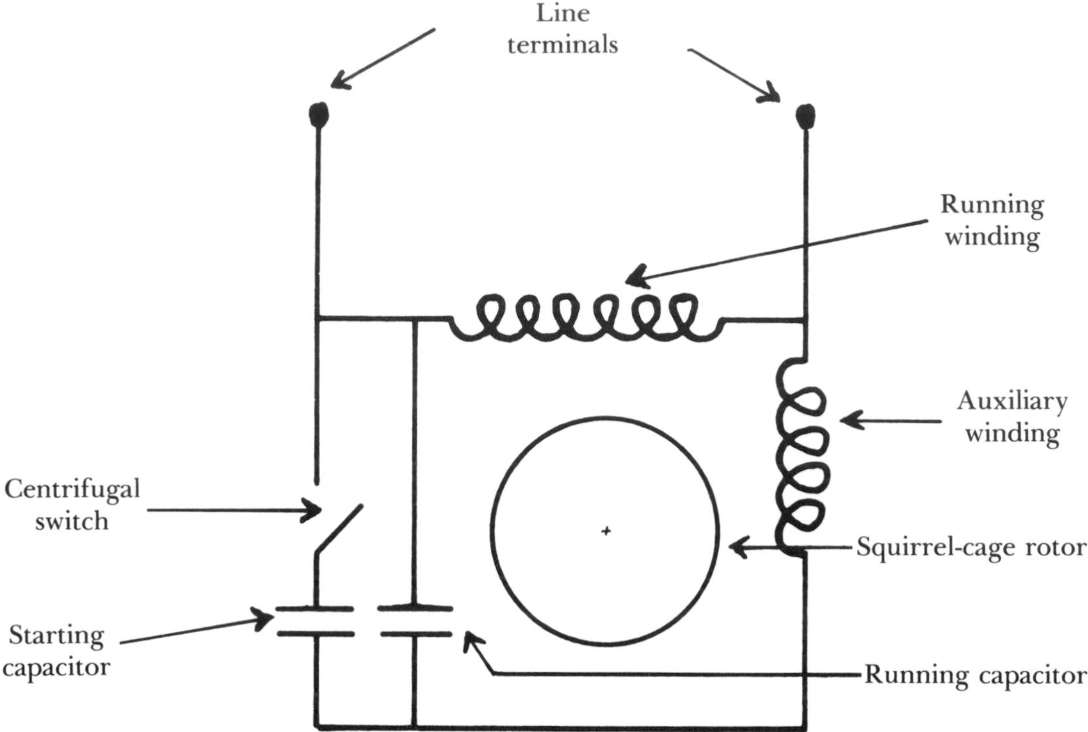

Figure 10-5. Capacitor-Start, Capacitor-Run Motor

when the desired speed is reached. The auxiliary winding and the running capacitor are energized during both starting and running.

These motors are built as large as 15 to 20 hp. More efficient use of materials is made in this motor since the auxiliary winding provides useful motor output in normal operation rather than taking up space as a deenergized winding. The price to pay for this is an additional capacitor to maintain.

Repulsion-Single-Phase Motors

These motors are not common today, as is true with all repulsion type motors. This motor contains a wound armature with a commutator and brushes which are shorted together. Maintenance is similar to that for a d.c. motor. The brushes are mounted on a moveable common support and by changing the brush axis with relation to the field windings, the speed can be varied. Starting torques of 400 to 600% can be obtained. This motor has been supplanted by the capacitor-start motor which is simpler in design, less expensive, and easier to maintain while still producing large starting torques.

Repulsion-Start Induction-Run Motors

These motors have an additional complex switch which shorts all the commutator bars together at running speed, thus forming a high resistance squirrel-cage winding. Provision may also be included to lift the brushes from the commutator after starting to reduce wear. These mechanical devices are additional concerns. In the case of older motors, repair parts may no longer be available. When replacing repulsion type motors with capacitor-start motors, make sure there is sufficient starting torque capability in the replacement motor. Sometimes it is necessary to increase the horsepower rating to assure proper starting performance.

Repulsion-Induction Motors

These start as repulsion-single-phase motors and then run on a low resistance squirrel-cage winding also contained in the rotor. A switch removes the commutated winding from service at a determined speed. This motor has the high repulsion motor starting torques combined with the high breakdown torques of the squirrel-cage motor. Breakdown torque is the maximum torque that a motor will produce and it generally occurs at 60-80% of rated speed in single-phase motors depending on the type and design of the particular motor.

Three-Phase Motors

These motors are built in sizes ranging from fractional horsepower to 100,000 hp and larger. The construction of these motors is relatively uncomplicated because no switches, capacitors or starting windings or circuits are required. While the general term for these motors is polyphase motors, you probably will not encounter other than three-phase motors. In the past there were some two-phase systems and motors in use, but today the three-phase system is pretty much universal. Because of its design, the three-phase motor is self-starting and the direction of rotation is easily reversed by interchanging any two of the three line connections.

The voltage ratings of three-phase motors range from 200 V to 13.2 kV. You will find the most numerous ratings to be 200, 208, 230 and 460 Volts. These motors are used on voltage systems rated 208, 240 and 480 Volts respectively. In rare instances, you may come across motors rated 575 V for use on 600 V systems. All these motors are designated low voltage motors. Motors rated 2.3, 4.0, 6.6 and 13.2 kV are known as medium voltage motors and are found in large industrial plants and power production facilities.

The most confusion for maintenance personnel seems to center around the 200, 208 and 230V ratings when replacing motors. You must know the rating of the system supply at your location. If it is a 240 V system, commonly referred to as 230 V, you should use motors rated 230 V. If you have a 208 V system, replacement motors should be rated 200 or 208 V or at least state on the nameplate "suitable for operation on a 208 V system." Industry standards state that a motor will operate satisfactorily over a range of + 10% to − 10% of rated voltage. A 208 V motor replaced by a 230 V motor will have a minimum operating voltage of 230 V × 0.9 or 207 V. If the 208 V system voltage should drop below 207 V the motor would be operating outside of its allowable range. The reverse situation occurs when 208 V motors are used on 240 V systems.

Terminal connection leads on three-phase motors vary between three and twelve. The nine-lead connection is probably the most common and it permits motors to have a dual voltage rating of 230/460 V. Motors may also be designed for "wye" or "delta" connection. Figure 10-6 shows the standard terminal markings for a wye or star connection and the connections for low or high voltage. Figure 10-7 shows similar information for delta motors. Connection information is usually shown on the motor nameplate or on a connection plate on the motor. To reduce starting currents, motors are sometimes started in the wye connection and then transferred to a delta connection for running operation. This transfer takes place in the motor starter.

Squirrel-Cage Induction Motors

These are by far the most common type of motor you will encounter. (See Table 10-2 for a checklist of maintenance problems and solutions for this type of motor.) The rotor construction is very simple, consisting of uninsulated bars either driven or cast into slots in the rotor surface and connected by end rings at each end. This is a very rugged winding which requires no maintenance. The usual failure is a cracked bar or end ring and this is a rare event. The motor will exhibit starting difficulties and running problems in this situation. Special test equipment is needed to physically locate such a condition, and on small motors the most economical solution is to replace the motor. Squirrel-cage windings can be damaged by excessive and rapid starting and also by a locked rotor condition.

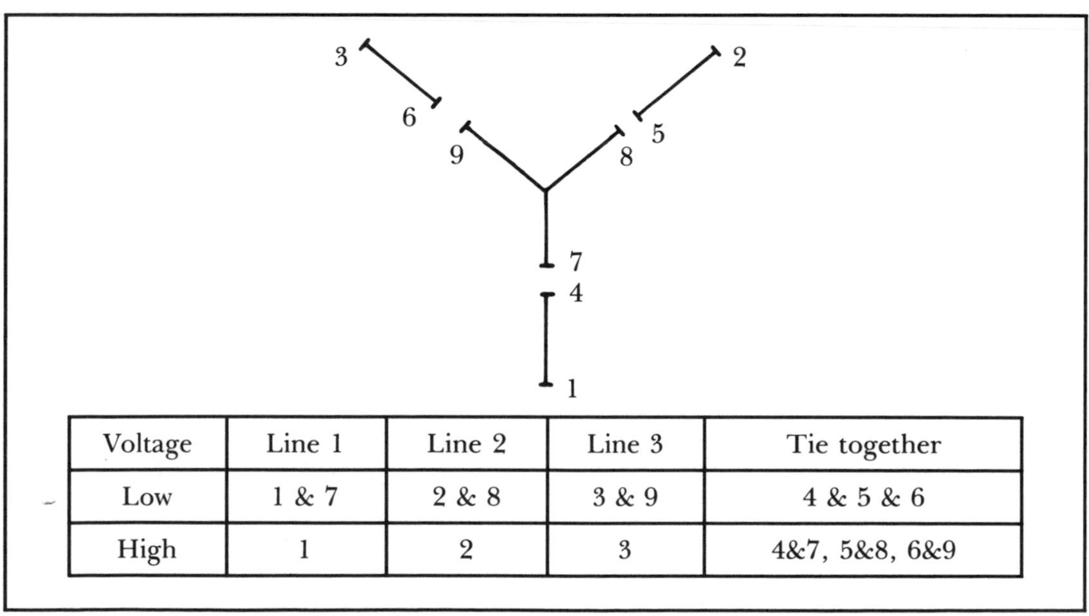

Voltage	Line 1	Line 2	Line 3	Tie together
Low	1 & 7	2 & 8	3 & 9	4 & 5 & 6
High	1	2	3	4&7, 5&8, 6&9

Figure 10-6. Three-Phase, Nine Lead, WYE Connections

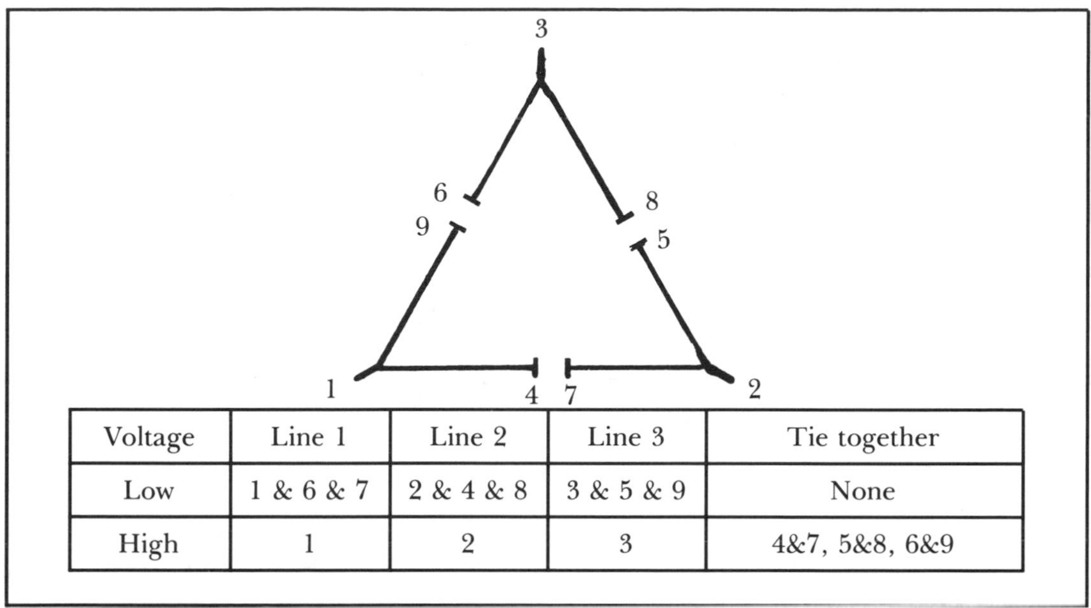

Voltage	Line 1	Line 2	Line 3	Tie together
Low	1 & 6 & 7	2 & 4 & 8	3 & 5 & 9	None
High	1	2	3	4&7, 5&8, 6&9

Figure 10-7. Three-Phase, Nine Lead, Delta Connections

The stator winding in low-voltage motors consists of enamel coated wires wound into coils and placed into insulated slots in the stator bore. Medium voltage motors employ formed and insulated coils placed directly into the slots. Low voltage stators are dipped in insulating varnish and baked to bind the wire conductors together and provide additional insulation covering. Stator windings require maintenance. They must be kept clean and dry since dirt and moisture are insulation's worst enemies. This becomes more important as the voltage rating increases.

Wound-Rotor Induction Motors

These motors are used in applications where speed control is required such as for hoists and cranes. Wound-rotor motors are capable of providing large starting and accelerating torques. The stator construction is the same as for squirrel-cage motors. The big difference lies in the design of the rotor which contains windings similar to those in the stator placed in slots in the rotor face. These windings are connected to three slip rings with associated stationary brushes. Speed and starting torque control are obtained by placing external resistance in the three-phase rotor circuits.

Maintenance considerations for the rotor windings are similar to those for the stator windings. (See Table 10-3.) However, these windings are subject to rotational forces that the stator windings do not experience. They should be inspected for signs of movement in the slots, loose slot wedges and movement in the end turn areas. The end turns are usually banded to restrain the centrifugal forces.

Symptom	Possible Cause	Action/Correction
Motor will not Start	Overload protection tripped	Check reset on motor or starter. Allow to cool, reset and try starting again.
	Motor not energized	Check power supply to starter. Check control circuits and fuses. Check voltage at motor terminals. Does contactor operate properly?
Motor trips starter on overload when attempting to start	Wrong voltage	Check motor nameplate for voltage rating. For dual-voltage motors, check diagram for proper lead connection.
	Voltage low	Check voltage at starter and at motor terminals if at a distance from starter. Motor terminal voltage should be within plus or minus 10% of nameplate rating.
	Starter overloads too small or defective	Check overload sizing per starter manufacturer's recommendations for hp and enclosure. Check overloads for defects.
	Voltage unbalanced or single-phase condition	Check for voltage unbalance between line terminals. See Figure 10-11 and associated text. Measure line currents. See section on unbalanced input current. Check for high resistance connections and broken leads.
	Load inertia too large	See section on slow starting.
	Load stalls motor	Check that load is free to rotate, not defective and properly lubricated. May require disconnection from motor to do.
	Motor bearings seized	Disconnect motor from load and turn by hand. If bearings are bad replace them. Check journals and housings for wear.

Table 10-2. Troubleshooting Checklist for Three-Phase Squirrel-Cage Induction Motors & General Motors

Symptom	Possible Cause	Action/Correction
	Worn bearings or housings	Bearings free to turn by hand but magnetic pull causes rotor/stator contact. Generally magnetic growling noise. Replace bearings and rebuild end frames, etc. to proper dimensions. Check for possible stator iron and winding damage. Also check rotor for damage.
	Motor stator windings defective	Measure line currents when starting. They should be balanced. Disconnect motor from supply. Measure line-to-line winding resistances as accurately as possible. Should be balanced. Test may be inconclusive on larger motors since values are very low. Micro-ohmmeter may be required.
	Defective rotor	Look for broken or cracked bars and rings. Replace small motors if defective. Larger rotors may be replaced, rebarred or recast.
Motor trips supply breaker or fuses on start	Supply breaker or fuses defective or improperly applied	Check breaker for proper operation. Check that fuses are suitable for motor starting currents, install time delay fuses if required. Fuse or breaker rating could be increased to code limit.
	Motor or supply circuit has ground or phase-to-phase fault	Disconnect motor from supply leads. Megger windings to ground and between phases if leads permit. Does motor smell? If megger values are low disassemble motor for inspection. If motor is good, check supply circuits.

Table 10-2. Troubleshooting Checklist for Three-Phase Squirrel-Cage Induction Motors & General Motors (*continued*)

Symptom	Possible Cause	Action/Correction
Slow starting (Starter may trip)	Low voltage	Check source voltage and motor terminal voltage. Check for high line drop. Starting torque decreases as square of voltage—80% V produces 64% torque. Correct system to produce proper voltage.
	Motor connected for incorrect voltage	Check motor nameplate for proper lead connections. Reconnect as required.
	Load inertia too high	Increase motor size if inertia can not be reduced.
Load speed appreciably below rated speed	Overload	Measure line current, probably a multiple of rated current. Reduce load or increase motor size.
	Excessively low voltage	Voltage would have to be very low. A 10-15% undervoltage would affect speed by only 1-2%. Check for correct voltage connection on dual voltage motors.
	Inaccurate speed measurement	Check speed with another device or method.
	Frequency low	This applies only to isolated systems. Speed proportional to frequency. Adjust prime mover speed to produce rated frequency.
Starter trips at load after acceleration	Overload	Reduce load or increase motor size. Check line currents and compare to nameplate. Overloads can be caused by increasing air flow, adding ducts, changing pulley ratio. Fan loads vary as cube of speed, twice speed requires eight times horsepower.

Table 10-2. Troubleshooting Checklist for Three-Phase Squirrel-Cage Induction Motors & General Motors (*continued*)

Symptom	Possible Cause	Action/Correction
Starter trips at load after acceleration (cont.)	Incorrect voltage	Supply voltage too high or low. Should be within 10% of rated. Check for proper connections for dual voltage motors.
	Voltage unbalanced or single-phased	Accurately check motor terminal voltages. See Figure 10-11 and associated text. See section on unbalanced input currents.
	Starter overloads improperly sized or defective	Refer to manufacturer's recommendations for proper sizing for hp and starter enclosure.
	Ambient temperature too high at starter	Check temperature at starter. Reduce temperature. Check with manufacturer for possible change in overload rating.
	Defective Rotor	Look for cracked or broken bars or end rings. Replace small motors. Large motor rotors may be repairable.
Unbalanced input current (5% or more deviation from average input current)	Unbalanced voltage due to: • Power supply • Unbalanced system loading • Poor connections • Undersized supply lines Note: A small unbalance in voltage will produce a large current unbalance which could exceed the current rating of the motor in one or more leads	Accurately check line-to-line voltages at motor terminals. To determine if trouble is in motor or power supply check as follows: Rotate all three input line leads to the motor by one position, ie., move line 1 to motor lead 2, line 2 to motor lead 3, and line 3 to motor lead 1. Measure input currents. If the current pattern follows the input lines the problem is in the power supply. The problem is in the motor if the pattern is unchanged with respect to the motor leads. Correct the voltage balance or system loading if a system problem.
	Defective motor	Replace or rewind motor if above test indicates motor problem.

Table 10-2. Troubleshooting Checklist for Three-Phase Squirrel-Cage Induction Motors & General Motors (*continued*)

Symptom	Possible Cause	Action/Correction
Incorrect rotation	Phase sequence to motor reversed	Interchange any two leads at motor or starter for three-phase motor.
Stator windings roasted or overheated, completely or in patterns	Excessive winding current flow for extended period due to:	Motor will have to be rewound or replaced.
	• Improper or nonexistent overcurrent protection	Install proper protection in accordance with industry codes and standards.
	• Overcurrent protection defective	Repair or replace defective device.
Motor windings faulted at a localized point	Insulation failure due to: • Moisture, dirt, chemicals (a) • Voltage surge (b) • Mechanical damage (c) • Age	Motor will have to be rewound or replaced. Causes a, b, c should be corrected to prevent recurrence
Motor noisy Electrical noise	Single-phased	Measure line currents for balance. Stop and try to restart motor. If it won't start it is single-phased. Correct condition causing single-phasing.
	Broken rotor bar	Measure line current, varying value with time may indicate broken bar. Inspect rotor for broken bars or rings. Replace small motor, repair of large rotor may be possible.
	Wrong voltage (excessively high)	Check nameplate for proper lead connections (dual voltage and wye-delta motors).
Motor noisy Mechanical noise	Bearings bad	Replace bearings, see section on bearings and text.

Table 10-2. Troubleshooting Checklist for Three-Phase Squirrel-Cage Induction Motors & General Motors (*continued*)

Symptom	Possible Cause	Action/Correction
Motor noisy Mechanical noise (cont.)	Rotor rubbing on stator	Generally result of severe bearing wear. Replace bearings and repair shaft and housing if required. Inspect windings, stator and rotor for damage.
	Loose parts	Look for loose bolts, fans, baffles, broken fan blades.
	Misalignment	Correct alignment.
	Driven equipment noise	Disconnect motor from load to determine source of noise.
Motor overheats	Overload or supply system problem	See section "Starter trips at load"
	External, internal fans broken or missing	Repair fans as required.
	Internal ventilation reduced by dirt	Blow out with compressed air. Disassemble and clean if required.
	Filters clogged	Replace filters.
Excessive vibration	Motor mounting	Check that motor is correctly and securely mounted.
	Misalignment	Check coupling or belt alignment, correct as required
	Unbalance due to: • Load	Disconnect coupling halves or belts. Start motor. If vibration stops problem was in belts or load.
	• Sheaves or coupling	Remove sheave or motor coupling half. Securely fasten ½ key in motor keyway. Start motor. If problem stops, problem was in sheave or coupling half.
	• Motor	If problem continues the unbalance is in the motor. Try balancing motor rotor. Look for missing rotor parts. On small motors replace entire motor.

Table 10-2. Troubleshooting Checklist for Three-Phase Squirrel-Cage Induction Motors & General Motors (*continued*)

Symptom	Possible Cause	Action/Correction
Hot bearings	Misalignment	Realign motor. See text.
	Excessive radial belt pull	Reduce belt tension.
	Excessive end thrust	Reduce end thrust from driven load. Furnish other means for carrying thrust. Thrust may be result of temperature expansions of shafts.
	Heat from hot motor or ambient	Correct source of high temperature
	Improper lubrication	Check for proper quantity and quality of lubricant. Antifriction bearing may have excess lubricant.
	Sleeve bearings:	
	• Oil grooves dirty	Clean oil grooves and bearing housing and refill with new oil.
	• Incorrect oil grade	Check manufacturer's recommendations and change oil to proper viscosity for application.
	• Oil rings not turning	Check for bent, worn or damaged rings. Replace as required.
	• Worn or scored bearings	Replace bearing, resurface journal if worn or grooved.
	Ball bearings:	
	• Broken balls, damaged raceway	Replace bearing, check for worn shaft or end frame.
	• Over lubrication	Reduce grease level to proper level. See text on bearing maintenance.
	• Bearings rotating on shaft or in housing	Restore worn surfaces to proper dimensions and concentricity and replace bearings.

Table 10-2. Troubleshooting Checklist for Three-Phase Squirrel-Cage Induction Motors & General Motors (*continued*)

Symptom	Possible Cause	Action/Correction
Motor won't Start	Open phase in rotor	Check rotor connections and external circuits, resistors for continuity.
	External resistance too high for load torque	Check resistance value for starting and accelerating torque values.
Motor runs at low speed	External resistance not cut out	Remove resistance—check controls.
	Too much circuit resistance between motor and control	Provide larger cables—move control closer to motor. Possible high resistance connections need correction.
	Brush pressure incorrect, brushes hung up in boxes	Adjust to proper pressure per manufacturer's recommendation. Check that brushes are free to move in boxes.
	Rough or dirty rings	Clean and polish rings—remove source of contamination.
	Eccentric rings	Turn rings to proper concentricity.
Sparking at brushes	Rings rough, dirty, out of round, brushes hung up	All sparking should be considered destructive. Clean, polish rings, turn in lathe if out of round, free brushes.

Table 10-3. Troubleshooting Checklist for Wound-rotor Induction Motors

Failure of this banding has resulted in the end turns being flung outward and damaging the stator windings. In addition, the slip rings must be kept in a smooth and clean condition and the brushes maintained with the proper contact pressure and length and kept free to move in their brushholders.

Synchronous Motors

These motors are generally found in the larger sizes, say, 250 hp to 100,000 hp or larger. This motor has the unique property of rotating at a constant speed regardless of load within its rating. This differs from the induction or asynchronous motor which runs at less than synchronous speed. For example, a 4-pole induction motor might run at 1760 rpm but its synchronous motor counterpart will always run at exactly 1800 rpm on a frequency of 60 Hertz. The synchronous motor has another advantage that it can be used to correct system power factor in a manner similar to the use of shunt capacitors.

The construction of the stator is similar to that for the induction motor. The major difference is the construction of the rotor which contains coils wound on pole pieces projecting from the rotor or possibly in rotor slots for two and four-pole motors. These windings are generally brought to slip rings and brushes for connection to an external d.c. excitation supply. Brushless designs incorporate rotating rectifiers inductively fed from a stator to provide excitation. Synchronous motors must be provided with a starting winding in the pole face which resembles a squirrel-cage winding. This winding accelerates the motor to near synchronous speed at which time it is pulled into step by application of field excitation. Control circuits assure proper timing and polarity of the excitation.

Synchronous motors are more expensive than induction motors and have rings, brushes or other equipment, a d.c. excitation supply and additional controls which require maintenance. (See Table 10-4.)

Direct-Current Motors

These motors were very common in the early days of electric power. The triumph of the a.c. distribution system over the d.c. system caused the decline of the d.c. motor for general application. The advent of solid-state controls has initiated a rebirth of the d.c. motor, permitting its use in areas where speed control or precise positioning and accurate indexing are required such as in machine tool applications. (See Table 10-5 for maintenance considerations.)

Permanent Magnet Motors

These motors are the simplest type of d.c. construction. High-strength permanent magnets provide the magnetic field, thus eliminating the need for field windings and the power they consume. Reversing the polarity of the input leads reverses the direction of rotation. When servicing permanent magnet motors, take care not to crack or chip the brittle magnets or cause them to separate from the frame. The magnets are usually cemented to the frame. When reassembling and inserting the armature, be careful to maintain a good grip on it since this configuration resembles a solenoid and the armature may be forcefully pulled into the pole structure along the motor axis.

The rotor, or armature, consists of coils wound into insulated slots and connected to individual commutator bars. Brushes riding on the commutator connect the coils to the power supply terminals. Most of the maintenance on d.c. motors centers around the brushes and the commutator.

Symptom	Possible Cause	Action/Correction
Motor will not reach speed	Field excited	Field application contactor must be open. Field must be connected across discharge resistor.
Motor fails to pull into step	No excitation voltage	Check source of excitation for proper voltage output.
	Field voltage not applied	Check that field application control is operating correctly with proper timing and polarity.
	Field circuit open, rheostat set too high, inoperative	Check circuit continuity, rheostat setting. Rated field current should be applied.
	Load inertia too high	Check application and NEMA Standard MG 1-21.42 for allowable inertia. Consult manufacturer for the application.
Motor pulls out of step	Excitation voltage low	Increase voltage to provide rated field current.
	Open in field or exciter circuit	Repair break.
	Short circuit in field coil	Check voltage drops across coils at reduced voltage. Repair failed coil.
Excessive pulsating line current	Possible misapplication (reciprocating load)	Pulsation magnitude—max.—min. current measured by oscilloscope—should not exceed 66% of rated current. See NEMA Standard MG 1-21.84. Consult motor manufacturer.
	Motor out of step field excitation not removed	Check field application control for proper operation.
Excessive line current	Over or under excitation	Adjust field current to rated value.

Table 10-4. Troubleshooting Checklist for Synchronous Motors

Symptom	Possible Cause	Action/Correction
Motor does not start	Motor not energized	Check supply voltage—supply circuit continuity, fuses, starter—check voltage at motor terminals—correct as required.
	Brushes not contacting commutator	Brushes worn too short, springs broken or weak. Replace as required.
	Brushes sticking holders	Clean brushes and holders.
	Field not excited	On shunt or separately excited motors, check that field is connected and energized.
Motor does not come up to speed	Voltage low	Apply proper voltage to motor terminals.
	Shorted or open armature windings, shorted commutator bars	Test armature with growler for shorts or opens—repair armature or replace. On small motors, replace motors.
	Field voltage too low for load (weak field)	Check field voltage—field rheostat may be set at too high resistance setting. Reduce resistance.
	Starting resistance not cut out	Check to assure controls are operating properly to remove resistance.
	Brush axis not on neutral	Set brushes to proper position.
	Overload	Reduce load to increase motor rating.
Excessive speed	Voltage above rated	Reduce to proper terminal voltage level.
	Motor under-excited (low load)	
	• Shunt field shorted	Repair or replace defective field coils.

Table 10-5. Troubleshooting Checklist for DC Motors

Symptom	Possible Cause	Action/Correction
Excessive speed (cont.)	• External field resistance too high	Decrease field rheostat setting.
	Series field coil reversed or shorted	Reconnect series field—replace field coils if faulty.
Commutator or brushes too hot	Overload	Reduce load or increase motor size.
	Brush pressure too high	Adjust brush pressure to proper amount with spring scale.
	Shorted bars	Check for signs of overheated bars—check armature windings with growler for shorts or opens.
	Brushes not set on neutral	Set brushes to proper position.
	Brushes too abrasive	Substitute brushes which are less abrasive.
	Ventilation low	Check for clogged ventilation openings, broken fans, ambient temperature too high.
Sparking at brushes	Shorted armature coils or bars	Test armature with growler—repair armature or replace.
	Commutator eccentric, high bars, rough surface, high mica, dirt	Turn and undercut commutator, reseat brushes.
	Brush pressure too light	Measure pressure with spring scale. Increase pressure to manufacturer's recommendation.
Note: Blue pinpoint sparking frequently harmless. White streamer sparking is always harmful	Brushes not on neutral	Set brushes to neutral position.
	Overload	If overload is temporary action is not necessary. If overload is permanent then install larger motor.
	Interpole air gap too large	Decrease air gap.

Table 10-5. Troubleshooting Checklist for DC Motors (*continued*)

Symptom	Possible Cause	Action/Correction
Sparking away from brushes between bars	Oil, dirt, carbon dirt shorting bars	Clean commutator and eliminate source of contamination.
Brush wear high	Brushes too soft	Generally excessive dust in motor—replace with proper grade as recommended by manufacturer.
	Brush pressure too high	Set to proper value per manufacturer's recommendation with spring scale.
	Oil, grease or abrasive dirt	Resurface commutator, reseat brushes or replace and seat to proper contour. Correct contamination problem.
	Commutator eccentric, high bars, rough surface, high mica, grooved.	Turn and undercut commutator, replace brushes.

Table 10-5. Troubleshooting Checklist for DC Motors (*continued*)

Brushes must be free to slide in their brushholders and be maintained under the proper spring pressure. They must be replaced periodically due to wear. If a brush wears to the point where shunt wires or rivets contact the commutator, damage will be done to the latter. The carbon brush material tends to coat the motor interior and windings with a conducting film due to brush wear. This material should be removed periodically by the use of clean, dry compressed air. The motor manufacturer selected a grade of brush material suitable for the motor design and application. When replacing brushes, the same grade material should be used. New brushes must be contoured to the cylindrical surface of the commutator for proper contact and current carrying ability.

The commutator must be kept clean, smooth, and free from oil and grease. The establishment of the proper surface film is very important. A healthy commutator generally has a coloration ranging from copper to a deep brown with a glossy appearance. The surface should be cylindrical with no grooving or flat spots. Burnt or blackened segments may indicate winding problems. There should be little or no sparking in operation. Additional coverage on commutator maintenance is included later in this chapter.

Wound-Field D.C. Motors

These motors have coils wound on pole pieces attached to the outer frame of the motor which provide the magnetic field. Wound fields are common in the larger sizes of d.c. motors, say 5 hp and up, since the cost and size of permanent magnets tends to prohibit their use. On large motors, you may find an additional set of poles and windings called interpoles which function to improve commutation.

There are various connection arrangements for the field windings. The most common is the shunt winding, where the winding is connected in parallel with the armature. (See Figure 10-8.) By placing a rheostat in series with the field winding, the motor can be made to operate at a speed higher than its base speed. The shunt machine exhibits excellent speed stability as the motor is loaded. Sometimes the field is connected to a separate source of excitation and the motor is called a separately excited motor. Shunt field windings consist of many turns of wire and draw very little current compared to the motor rated current.

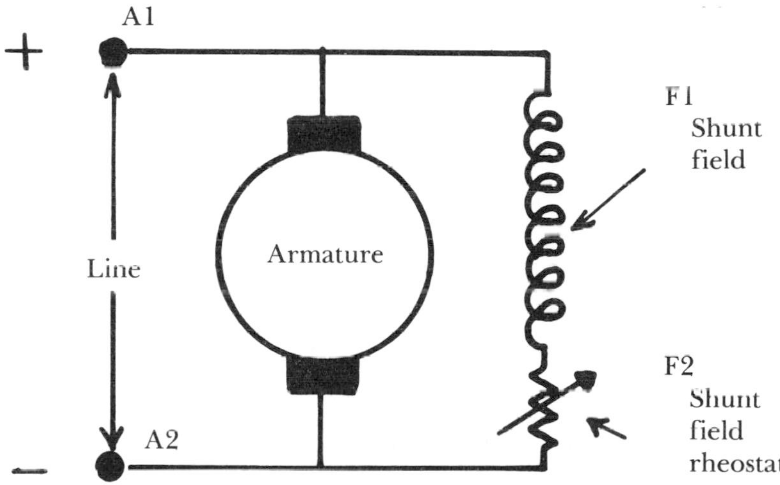

Figure 10-8. DC Shunt Motor

Another connection is to place the field winding in series with the armature. This is the series connected motor. Since the field winding must carry the full current drawn by the motor it is wound with heavy wire and consists of a relatively few turns. (See Figure 10-9.) It produces very high starting torques and is used in applications such as traction motors. The ability to produce high torques permits this connection to be used in small appliances and hand tools where it operates on a.c. This motor is known a "universal" motor because of its ability to operate on both a.c. and d.c. It also operates at very high speeds and produces more horsepower per unit of weight or volume than any other type used in the industrial market. If you need convincing, consider the weight of a portable circular saw with a 1.5 or 2 hp motor compared to an induction motor. Universal motors are not intended for continuous

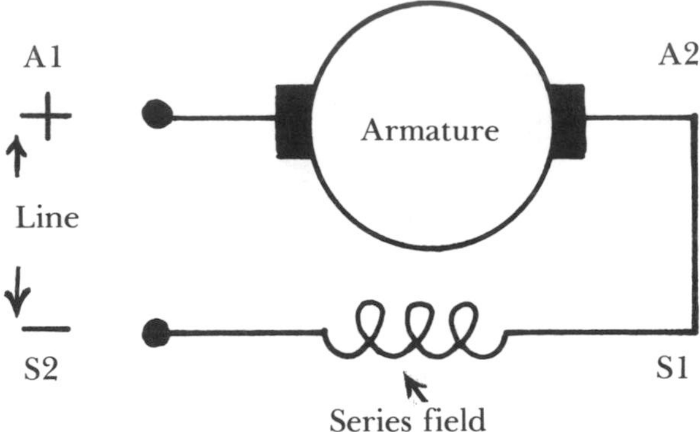

Figure 10-9. DC Series Motor

operation because the elevated speed and concentrated design leads to increased maintenance, especially in regard to the commutator and brushes.

Enclosures

The type of enclosure is an important factor in the application of an electric motor. Hopefully the individual who originally selected the motor chose the proper type for the environment. Sometimes plant conditions change and what was initially a benign environment becomes more severe. Knowledge of the types of enclosures available is important. You may find that an improper enclosure is adding to your maintenance problems.

Open Enclosure

These are the simplest ones. Some motors such as the skeleton type found in small fans or ventilators have no enclosure at all and rely on the fan unit to provide the necessary mounting and physical protection. The typical open enclosure provides physical protection, mounting features and consists of a frame around the motor with end plates or brackets which contain the bearings. Ventilating openings are placed in these parts to dissipate the heat generated by the motor internals. An internal fan or blade-like projections on the rotor circulates external air through the motor for proper cooling. Sometimes the internal fan is eliminated and the inlet air to the fan or blower driven by the motor provides the necessary cooling air flow. This arrangement is common in many air moving applications such as condenser fans and heating and air conditioner fans and blowers. This type of motor is called an "air-over" motor. Never use an air-over for any other application since it will not be able to cool itself properly without the air flow provided by the driven load.

Open Drip-proof Enclosures

These will be encountered very frequently. You will see this enclosure more than any other type. It is a variation of the open enclosure where the ventilating openings are placed so as to exclude dripping water at an angle of no more than 15 degrees from the vertical from entering the motor. This enclosure is generally used in locations that are clean and dry. A motor rated as drip-proof for horizontal mounting will not be drip-proof if it is mounted vertically and other drip shields will have to be added if dripping water is a possibility.

Totally-Enclosed Motors

These motors come in two classifications. The more common type is the totally-enclosed, fan-cooled motor which has a covered external fan on one end of the motor which blows air over the motor exterior. Heat generated in the motor is conducted through the frame and removed by the air flow. There is no direct interchange of the surrounding air with the air inside the motor other than natural breathing through the frame joints, terminal box and shaft openings. Sometimes a drain opening is provided in the bottom of the motor to eliminate possible condensation accumulation. In larger horsepower motors, external fins on the frame are used to aid in the dissipation of heat.

The second type of enclosure is totally-enclosed, non-ventilated. This is found in fractional horsepower motors where the frame is able to radiate the heat without the need for forced air flow over the motor surface.

Totally enclosed motors are used in dirty, damp, or wet locations and outdoors. Some manufacturers can furnish a severe duty motor which may have better shaft sealing, gasketed terminal boxes, seals around the wires entering the motor from the terminal box, and other features which might not be present in the usual totally-enclosed motor. Another variation is the "wash-down" enclosure which is more tightly sealed to withstand hose streams. It is found in dairies and food processing plant where equipment wash down is required for sanitary reasons.

Explosion-Proof Motors

These motors are a special category. These motors are used in hazardous locations and are defined by the National Electric Code with a Class, Group and Division designation. A typical designation might be Class I, Group D, Division I. These designations define the environment for which the motor is approved such as vapors, gases, dusts, and flying fibers. The classification is shown on a nameplate attached to the motor. Never replace such a motor with one that does not have the same classification.

The types of hazardous locations are defined as follows:

- Class I locations are those where flammable gases or vapors are or may be present in the air in quantities sufficient to produce explosive or ignitable mixtures.
  - Group A—Atmospheres containing acetylene
  - Group B—Atmospheres containing hydrogen or gases or vapors of equivalent hazards such as manufactured gases

- Group C—Atmospheres containing ethyl ether vapors, ethylene or cyclopropane
- Group D—Atmospheres containing gasoline, hexane, naphtha, benzene, butane, propane, alcohol, acetone, benzol, lacquer solvent vapors and natural gas

• Class II locations are those which are hazardous because of the presence of combustible dust.

- Group E—Atmospheres containing dusts of aluminum, magnesium or their commercial alloys.
- Group F—Atmospheres containing carbon black, coal or coke dust
- Group G—Atmospheres containing flour, starch or grain dust

• Class III locations are those which are hazardous because of the presence of easily ignitable fibers or flyings but in which such fibers or flyings are not likely to be in suspension in the air in quantities sufficient to produce ignitable mixtures.

Division I category is a location where the hazard is present in normal or expected operating conditions. Division II category is a location where the hazard will only be present due to equipment failure or abnormal operating conditions.

An explosion-proof motor may appear to be quite similar to a totally enclosed motor. However, the designs are considerably different. The surface temperatures may be kept lower than normal for atmospheres with low ignition temperatures. The design theory is not to exclude the hazardous atmosphere from the motor interior but to prevent any internal explosion from causing a similar event externally. Bearing fits and frame joints are elongated or of a labyrinth design to provide for cooling of exploded gases to a safe level before reaching the surrounding atmosphere. The frame and terminal box are designed for suitable strength, and the entrance from the terminal box to the motor interior is sealed. When installing the motor the conduit connection must also be done in conformance with the code requirements.

Repair of explosion-proof motors legally can only be done by certified facilities. If the motor is even so much as opened for inspection and reassembled by an uncertified facility the motor is no longer a qualified explosion-proof motor. You should be very careful to follow these requirements. Serious liability, legal and insurance problems could occur should an explosion be traced to an uncertified motor.

Weather-Protected Motors

These are open type motors generally found in the larger sizes in outdoor locations at industrial facilities and power plants. Design of the enclosure incorporates the principle that sudden changes in inlet air direction and velocity will cause dirt, water and snow to drop out before reaching the motor interior. These motors are not suitable for use in atmospheres where the dirt or dust is continuously suspended in the air since the dirt will eventually accumulate inside the motor.

MAJOR PROBLEM AREAS IN MAINTAINING ELECTRIC MOTORS

This section covers the major problem areas that affect the level of maintenance for electric motors. Understanding these problem areas should help you reduce your maintenance effort by not allowing problems to develop.

Dust, Dirt, and Moisture

These three villains have accounted for the demise of untold numbers of electric motors. The saying that cleanliness is a virtue certainly applies to motors. The general procedure for removing internally generated heat is to force cooling air through the motor. The motor becomes a filter of sorts since small amounts of contaminants are being continually trapped inside the motor. One way to keep the motor clean is to install filters on the air inlets. Be resolved that you will have to periodically clean or change the filters at appropriate intervals or you will have made the problem worse by cutting off the cooling air to the motor.

A growing build-up of foreign material will reduce the heat transfer from windings and other parts, resulting in increased operating temperatures. Carried even further, dirt can block air cooling passages and diminish air flow similar to a clogged external filter. Now add moisture from condensation during shutdown or directly from the cooling air and maintenance problems are sure to occur. Pure water is an excellent insulator, but water mixed with most contaminants probably has conductive properties which, combined with minute cracks in the insulation, can cause failures to ground or between coils or turns. Depending on the type of contaminant, the addition of water may form chemical solutions which could damage even today's improved insulation systems.

If you want healthy motors, keep them clean by blowing them out periodically with clean dry compressed air. If dirt is a problem, you should take steps to improve the surrounding environment, if possible. Perhaps a change in the type of motor enclosure is the ultimate solution.

Operating Temperatures

One of the major design considerations which determines the output of a motor is the maximum allowable operating temperatures of the insulation system and other parts of the motor. The Institute of Electrical and Electronics Engineers (IEEE) has established classes of insulation systems and their maximum operating temperatures. Since the maximum temperatures occur in the interior portions of windings the industry has agreed on limiting observable temperatures which can be checked by a measurement of the winding resistance when hot and compared to its resistance at a known temperature.

The standard ambient temperature for which motors are designed is 40°C (104°F). Motors which will operate in higher ambient temperatures must be designed for such service. The effect of elevated ambient temperatures must also be considered for bearing lubricants. When motors are run in ambient temperatures above 40°C they must be derated in horsepower output. If the plant design is changed by the addition of high temperature piping or heat sources, the change in ambient temperature could be to the detriment of nearby motors.

The life of an insulation system is heavily dependent on the temperature at which it operates. A rough rule is that the insulation life is halved for each 10 degrees C (18°F) of temperature increase. You can see why it is necessary to keep motors clean and operating at the temperatures for which they were designed. The limiting hot-spot temperatures for the most common classes of insulation are shown in Table 10–6.

Class	*Maximum temperature*
A	105 C
B	130 C
F	155 C
H	180 C

Table 10-6. Insulation Class Limiting Temperatures

Another point to keep in mind is that the useful lives of all the classes of insulation are roughly the same when operated at their allowable temperatures. The advantage of the higher temperature insulation system is that it permits a physically smaller motor for a given output, which uses less material and is less costly to build. Sometimes motors are specified with, say, Class B temperatures and Class F insulation. This would give approximately four times the insulation life or permit the motor to operate in a higher ambient temperature.

Because of the use of insulation with higher temperature ratings, the old touch test to determine if a motor is overheating is no longer valid. Winding and core iron temperatures have been increased to take advantage of the higher temperature capabilities. Table 10-7 shows typical measurable winding rises above ambient for the various insulation classes. For specific rises for the various types of enclosures, you are referred to National Electric Manufacturers Association (NEMA) Standard MG 1, Section 12.42.

Class of insulation	*Winding rise above ambient*
A	60 C to 70 C
B	80 C to 90 C
F	105 C to 115 C
H	125 C to 135 C

Table 10-7. Approximate Motor Winding Temperature Rises

The temperature of the external frame will not be at these temperatures but it will reflect to some degree these internal temperatures. A person can maintain hand contact with an object in the neighborhood of 66°C (150°F) temperature for a short period of time with varying degrees of individual discomfort. It is possible that some modern motors may approach or exceed this temperature on external surfaces during normal operation.

The sense of touch is, at best, a comparative test. A contact thermometer is a better indicator. Increases in surface temperature over the usual level experienced for a particular motor may indicate a developing problem. Sometimes the smell of insulation operating at elevated temperatures can be detected. Once you smell the acrid odor of burnt insulation or see signs of smoke, the insulation system probably has been damaged beyond repair. A burnt motor winding has an aroma that once smelled will not be forgotten.

Lubrication

Proper lubrication is essential for a long operating life for motors and all mechanical equipment. Lubrication must be done periodically and consistently. If you start lubricating when the motor audibly communicates its needs, you are too late. Many times service personnel try to quiet a noisy motor by pumping lubricant into the bearing. This may work for a short period of time but it is not a solution.

The other end of the spectrum is the person who thinks that if the recommended lubricant quantity is good then a larger quantity or a more frequent application is better. While the bearing may not complain, it is not in the best interest of the motor. Excess quantities of oil or grease tend to accumulate inside the motor. Windings become coated and this film collects even more dirt, moisture and carbon dust if brushes are involved. Lubricants which reach the centrifugal switch mechanism of single-phase motors will collect dirt and cause eventual misoperation. Oil and grease on the stationary switch contacts may cause them to overheat, arc or burn and in some cases to weld themselves closed. Lubricants will have a detrimental effect on the surface film on commutators and slip rings. Lubricants mixed with dirt and carbon dust can cause brushes to stick in their brushholders.

In general, the bearings used in electric motors utilize common industrial oils and greases. Oil is generally used in small motors and again in motors above about 500 hp. Grease is generally used in the integral horsepower NEMA frame sizes. If the motor manufacturer has lubricant recommendations they should be followed, especially in severe duty applications or environments.

Turbine oils are commonly used in electric motors. They contain oxidation inhibitors to reduce sludging. Light turbine oil, 150 SUS at 100°F, is satisfactory for all normal fractional horsepower applications and larger units rated 1800 and 3600 rpm. A medium turbine oil, 300 SUS at 100°F, may be used for lower speeds or higher temperature operation. Automotive oils are sometimes used because of their ready availability. These oils contain additives tailored to combustion engines and may cause operating difficulties such as foaming. A SAE 10W oil is roughly the same viscosity as light turbine oil and SAE 20W is roughly equivalent to medium turbine oil. The normal temperature limit for oil is about 200 F.

Grease is generally used in anti-friction bearings up to about 500 hp. Grease consists of oil held in a soap or sponge-like structure. Release of this oil over extended periods of time furnishes the necessary lubrication. Lithium greases are widely used today and are suitable for use over a temperature range of −20° to 250°F as are most petroleum greases. For temperatures outside this range synthetic greases are available. The reaction of grease with air generally limits the life of the grease. Oxidation of the oil within the grease structure results in drying, hardening and caking of the grease and loss of lubrication.

Alignment

In many motor applications alignment is not a problem. This is especially true for small motors where the motor is flange mounted to the driven equipment. Many fan applications have the fan blade mounted directly on the motor shaft. However, larger size motors are usually coupled to their loads. Proper shaft alignment is important to assure adequate bearing life. In belt-driven applications it is necessary that the belts be perpendicular to the shaft axes. If not, the belts will wear and bearing life will be adversely affected. The following discussion is not meant to be a treatise on alignment but it does show the general approach to the subject. Also, remember that alignments made under one set of circumstances can change due to temperature, loading, foundation movement and rotational speeds.

In coupled drives, proper alignment exists when a straight line through the motor shaft axis extends through the axis of the driven equipment shaft. Flexible couplings are not meant to make up for improper alignment practices. In other words, there should be no radial or angular displacement between the driving and driven shafts.

An exaggeration of a typical misalignment situation is shown in Figure 10-10 which contains both radial and angular errors. The object is to make distances A1 and A2 between the coupling faces equal and to reduce distance R between the coupling rims to zero.

A close approximation can be made initially by placing a straight edge across both coupling halves at several locations around the periphery and moving the motor in the desired direction to bring the coupling halves into concentricity. Generally, it is easier to move the motor since the driven equipment may be fixed in location by piping, ductwork or mechanical connections. Final alignment consists of a coupling rim check and a coupling face check. These surfaces are generally machined true to the centerline of the coupling. The object, however, is to align the shafts and not the coupling surfaces. The following method eliminates errors due to possible coupling inaccuracies.

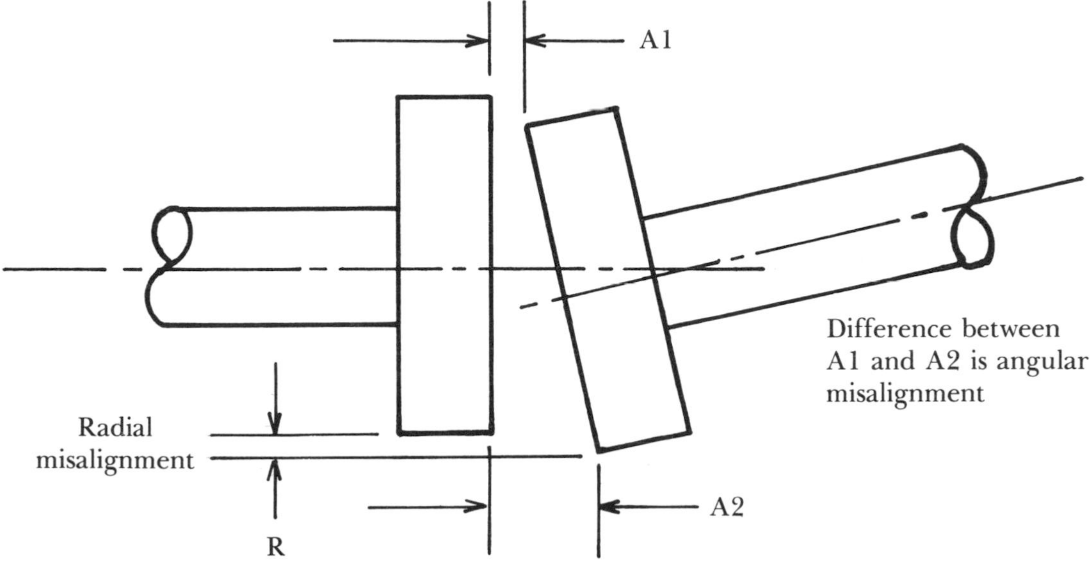

Figure 10-10. Typical Misalignment with Radial and Angular Errors

To perform a rim check, a dial gauge is attached to the driven equipment coupling half with the gauge button resting on the rim of the other coupling half such that the indicator plunger is radial to the axis of that shaft. The point of contact should be marked since all measurements will be made with the gauge button on that spot. Both shafts are rotated together and dial gauge readings are obtained at 90 degree intervals. The difference between the 0 and 180 degree readings divided by 2 indicates the vertical correction required. The difference between the 90 and 270 degree readings divided by 2 indicates the horizontal correction required. The dial gauge should read the same after 360 degrees of rotation as it did at the initial zero degree position. If there is a difference, the procedure was done incorrectly and should be repeated.

The face alignment can be accomplished by using feeler gauges for couplings with small gaps or with dial gauges on larger coupling spacings. Measurements of the A dimensions are made at the 0, 90, 180 and 270 degree positions. Both shafts are then rotated 180 degrees and the same set of readings taken. The readings at each position are averaged. If the averaged reading at the zero degree position is larger than the averaged reading for the 180 degree position then the coupling gap is open at the top by the difference in the two readings. The same procedure is followed in determining the right and left openings from the 90 and 270 degree readings. The motor is moved to cancel these coupling openings. It is a good idea to record the final coupling settings for future reference. Sometimes, machines with a long history of successful operation will indicate misalignment at standstill. This may be peculiar to the application and probably should be left unchanged.

System-Imposed Problems

Electric motors could be called man's best friend; they will go beyond the call of duty to be of service to the point where they will destroy themselves in the process. The typical electric motor can produce 2 to 3 times its rated horsepower; however, it will rapidly overheat and fail. Protection against overloads is provided by the starter or by thermal protection built into the motor. Electric motors are equally cooperative even when supplied with the improper voltages and frequency. Again, it is the job of the external protection to prevent damage. Many times it is thought that circuit fuses will protect the motor. Such fuses and circuit breakers generally are sized to protect the circuit wiring. A 15 amp fuse or breaker will provide no overload protection for a motor rated at 10 amps but it will protect the circuit from overloads and faults. For long life and proper operation, it is essential that the electric supply system provide the motor with power at acceptable voltage and frequency.

Single-Phasing

This is the situation where one leg of the three-phase supply system opens. This could be caused by a utility failure, a broken circuit in the plant, a bad contact in the motor starter or a single blown fuse.

Three-phase motors subjected to single-phasing while carrying a substantial load will draw elevated currents on the remaining two lines and continue the attempt to drive their loads. Unless removed from service, they will experience winding damage. Inspection of a motor which has been so damaged will show a repeatable pattern of burning in the end-turn region. Those windings that were connected to the phase that was deenergized will appear undamaged.

Strange things can occur when a system containing large, lightly loaded motors is single-phased. Since the motors are lightly loaded, it is possible that they could remain in service without drawing more than their rated currents and not be damaged. These motors will act as phase balancers and maintain some voltage level on the disconnected phase. An unbalanced three-phase system results, and small motors and other devices connected to the system might continue to operate. The length of time they will continue to operate until removed by their protection or damaged depends on the degree of balance maintained and the loading on each individual motor. Small motors which attempt to start on this degraded system may or may not be successful. Should the large motors stop, they will not be able to restart and will experience damage if not disconnected by their protective devices. Any three-phase motor is unable to start on a system that is completely single-phased.

Unbalanced Voltage Operation

This is a condition that is recognized in the industry standards. All polyphase motors can operate successfully when the unbalance at the motor terminals does not exceed one percent. Voltages preferably should be evenly balanced as closely as can be read on a voltmeter. In actual practice, there is always some degree of unbalance. For example, you might obtain three line-to-line voltage readings of 233, 230 and 228 Volts. The formula for calculating percent unbalance is:

$$\% \text{ unbalance} = \frac{100 \times \text{maximum deviation from average voltage}}{\text{average voltage}}$$

For the example shown, the average voltage is 230.3 V and the percent unbalance is 100 × 2.7/230.3 or 1.2%. Per NEMA Standard MG 1 14.34 any voltage unbalance larger than 1% requires derating. See Figure 10-11 for recommended derating values as a function of percent unbalance. It is not recommended that motors be operated with an unbalance condition of more than 5%. A 5% unbalanced condition carries a recommended motor derating to 75% of rated hp. This NEMA section also offers advice on the selection of overload protection for unbalanced voltage operation.

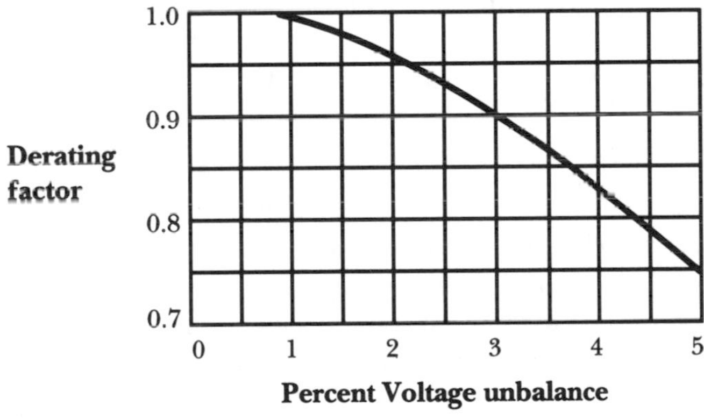

Figure 10-11. Unbalanced Voltage Derating Factors

Unbalanced voltages can be caused by conditions within the facility or they can be the result of the incoming supply from the power company. Measurement of the three line-to-line voltages at the service entrance will indicate where the trouble lies. Generally the power company will provide a fairly balanced three-phase supply. If it is determined that the power company is the cause of the problem then they will have to be contacted for a resolution. Usually, however, the problem is internal and caused by an unequal distribution of single-phase loads between the phases. Correction requires a redistribution of the single-phase loads between phases to produce a balanced supply to all motors affected.

Voltage and Frequency

Permissible limits for these are also defined by NEMA. A.c. and d.c. motors will operate satisfactorily over a range of plus or minus 10% of rated voltage and universal motors over a range of plus or minus 6% of rated voltage. The standards also permit a frequency variation of plus or minus 5% of rated frequency. A combined variation in both voltage and frequency of 10% (sum of absolute values) is permitted provided the frequency variation is not more than 5% and the voltage variation for universal motors does not exceed 6% (except for fan motors). For example, a plus 6% voltage variation combined with a minus 5% frequency variation totals 11% and is not acceptable. Likewise, a minus 9% voltage variation combined with a plus 2% frequency variation also is not acceptable since the sum of the absolute variations exceeds 10%. It should be noted that motor performance may not be in accordance with the standards or manufacturer's guarantees under these conditions.

Normally, frequency considerations are not a concern since power systems are closely regulated to maintain rated frequency, which is 60 Hertz in the United States. However, isolated systems for emergency service or portable units for field use are not as well regulated. A generator producing 50 Hertz or below at rated voltage could severely damage connected 60 Hertz motors due to overheating. When using such generators, make sure that the prime mover speed governor is correctly adjusted to provide the correct frequency and is operating correctly. The voltage regulator should be checked to assure proper voltage output.

Mechanical Overloads

These are a serious threat to the longevity of any motor. Unlike an internal combustion engine which lays down and dies when overloaded, the electric motor will rise to the occasion and continue operating. This level of effort will result in increased input current and overheating of the windings. Generally, the stator windings fare worse under overload conditions. A winding inspection of a motor which has been overloaded will show a uniform pattern of heating ranging from insulation discoloration to broken winding ties to a complete roasting of the coils. This is in contrast to localized failures which are due to dirt, water or insulation breakdown.

Many times motors will have a service factor rating on the nameplate. The term Service Factor (SF) is a measure of the overload capacity designed into a motor. A service factor of 1.0 means there is no overload capacity while a service factor of 1.15 indicates an overload capability of 15% above the nameplate rating. Motors will operate successfully in the service factor range but will have a shorter life since the additional output is obtained at the expense of higher winding temperatures. It is best not to utilize the motor service factor when replacing motors and keep it as a cushion for unexpected overloads or degraded supply system operating conditions.

Locked rotor conditions—where the motor is prevented from turning—will affect both the rotor and stator. The rotor may experience cracked or melted bars or end rings and the stator may have roasted windings. Motors should be able to accelerate their loads quickly. Prolonged starting intervals or repetitive frequent starting are detrimental to a motor with the rotor more likely to show signs of distress. This problem is more severe in large motors. Sometimes the motor manufacturer will limit both the number of starts per day and the time between successive starts. One of the toughest duties on a motor is the acceleration of its load to rated speed, especially if it is a high inertia load. NEMA Standard MG 1-12.50 specifies the maximum allowable inertia that a motor can accelerate.

The key to protection of a motor from the ills of overloads and supply system irregularities is the use of adequate overcurrent protection, generally in the form of a starter with overload elements. The best protection for a three-phase motor utilizes an element in each phase. In general, the protection should be set to trip at 115% and 125% of motor rated current for motors with service factors of 1.0 and 1.15 respectively. The proper size of the element should be obtained from the starter manufacturer. Some vendor designs incorporate three-phase overloads which are adjustable over a range of currents rather than being rated for one specific amperage. At least one manufacturer makes such a device which is also sensitive to single-phasing operation. These adjustable overloads permit you to replace a motor with one having somewhat different full load currents and still retain proper protection.

GENERAL MOTOR MAINTENANCE

General motor maintenance consists of preventative maintenance and repairs to failures which occur. A good preventative maintenance program can do much to prevent failures which generally choose to occur at inopportune times. This section contains typical preventative maintenance schedules, trouble shooting checklists and information on windings, bearing and commutator maintenance.

Preventative Maintenance

Schedules for preventative maintenance on motors will depend on the environment, access, size, use, redundancy, application, history of problems, and, of course, whether the motor is essential to plant or life safety. Economic considerations will dictate the necessity of spare motors for some applications.

An inventory record and history file of all motors is essential. A tickler system should be developed to schedule each motor for inspection and/or test. General inspection should be done at least annually, or as often as monthly, where large motors operate continually in dirty, wet environments. Figure 10-12 depicts some typical inspection and checking system record forms. The record card provides the historical data and evidence of scheduled maintenance and testing. An important aspect of such data is the profile of insulation resistance and load values which may be analyzed as to sudden change and potential trouble.

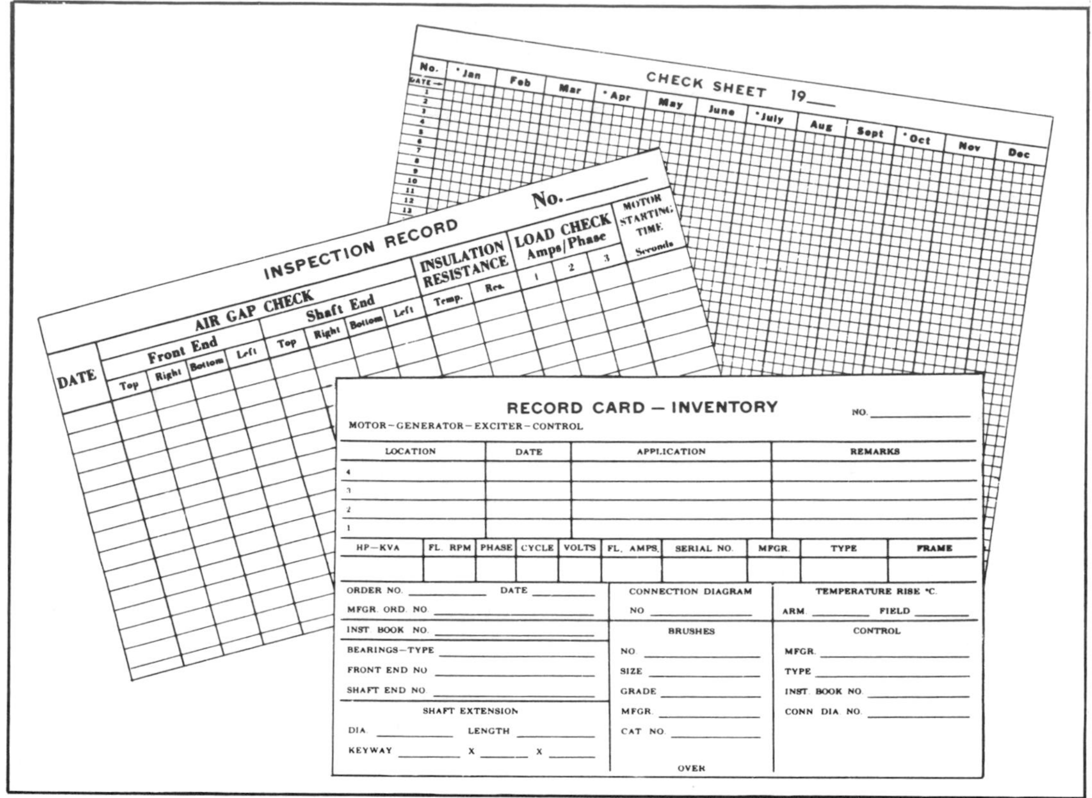
CHECK SHEET 19___

No. / DATE — | *Jan | Feb | Mar | *Apr | May | June | *July | Aug | Sept | *Oct | Nov | Dec

INSPECTION RECORD No.___

DATE | AIR GAP CHECK: Front End (Top, Right, Bottom, Left); Shaft End (Top, Right, Bottom, Left) | INSULATION RESISTANCE (Temp., Res.) | LOAD CHECK Amps/Phase (1, 2, 3) | MOTOR STARTING TIME Seconds

RECORD CARD — INVENTORY NO.___

MOTOR–GENERATOR–EXCITER–CONTROL

LOCATION	DATE	APPLICATION	REMARKS
4			
3			
2			
1			

HP—KVA	FL. RPM	PHASE	CYCLE	VOLTS	FL. AMPS.	SERIAL NO.	MFGR.	TYPE	FRAME

ORDER NO. ___ DATE ___
MFGR. ORD. NO. ___
INST. BOOK NO. ___
BEARINGS—TYPE ___
FRONT END NO. ___
SHAFT END NO. ___
SHAFT EXTENSION
DIA. ___ LENGTH ___
KEYWAY ___ X ___ X ___

CONNECTION DIAGRAM
NO. ___
BRUSHES
NO. ___
SIZE ___
GRADE ___
MFGR. ___
CAT. NO. ___
OVER

TEMPERATURE RISE °C
ARM. ___ FIELD ___
CONTROL
MFGR. ___
TYPE ___
INST. BOOK NO. ___
CONN. DIA. NO. ___

Figure 10-12. Typical Record Cards for Motor Data and Profiles

A summary for a typical motor preventative maintenance program is shown in Table 10-8. You should adjust or supplement this to suit your particular application.

Winding Testing and Maintenance

Motor windings consist of conductors and an insulation system. Insulation is required to support the voltage differences existing between the windings and the grounded frame, between conductors and turns and between phases or other windings. Motor windings are subjected to electrical, mechanical and thermal stresses and the effects of external contamination and moisture. The service life of a motor will largely depend on the level of care given to the insulation system. The type of maintenance program to be followed depends on the operator's experience and philosophy. It should take into account the importance of the motor to the operation of the facility. The accumulated records of tests performed on a motor can be extremely useful in evaluating its present insulation condition.

The motor manufacturer has designed an insulation system using materials with proper temperature capabilities and voltage-withstand levels to provide a normal life when operated in accordance with the nameplate rating. Motor windings are tested

Activity	Environment		
	Clean Dry	*Moderate*	*Dirty Wet*
General Inspection • Environment • Motor Cleanliness • Lubrication • Loose bolts, covers, fittings • Noise level • Belts, couplings, mountings	6–12 months	3–6 months	1–3 months
Test • Insulation Resistance • Voltage and current at load and no load • Bearing or frame temperatures • Vibration	12–18 months	8–12 months	3–8 months
Lubrication • Oil • Grease Refer to manufacturers recommendations	6 mos. annually	3 mos. 6–12 months	monthly 3–6 months
Bench Inspection • Disassemble, clean, check for moisture, condition of insulation, winding supports, leads, signs of overheating, bearing wear, replace if anti-friction, shaft journals, bearing housing fits, stator iron • Insulation resistance and winding continuity checks • Inspect rotor bars and end rings for signs of cracking or overheating. • Reassemble, megger windings, lubricate, operate, check for proper current draw, sound level, vibration.	3–5 years	2–3 years	1–2 years
Motors with brushes, slip rings or commutators. • Check brushholders for wear, binding, discoloration, burning, weak springs • Check rings for wear, discoloration, signs of sparking • Check commutators for good color, wear, high mica, etching or signs of sparking, loose bars, broken riser connections, burning • Check brushes for signs of unusual wear, replace brushes.	6–12 months	3–6 months	1–3 months

Table 10-8. Preventive Maintenance Program

at elevated voltage levels to search out defects in workmanship and material in the factory. In spite of this testing, motors will exhibit a higher failure rate immediately after initial service. After the infant mortality phase has passed, a lower failure rate will exist until the effects of service and aging begin to take hold. Proper maintenance is essential in postponing the onset of service aging problems.

High Voltage Tests

These tests applied to new motors at the factory are based on the rated voltage of the motor. All motors rated 1 hp and larger are tested with an alternating voltage of twice rated plus 1,000 V for a period of 60 seconds. Motors rated less than 600 V are tested at twice 600 V plus 1,000 V.

Factory test voltages are applied to a winding only once in its life. Field acceptance test voltages are usually at 85% of the factory test value. Routine field test voltages should be at no more than 65% of the factory value. An equivalent direct voltage test level is considered to be 1.7 times the above voltage values.

The philosophy of testing motor windings at elevated voltages above rating is controversial. One school of thought says that such testing may unduly stress the insulation and cause a failure that would require rewinding of a motor that may have operated for an extended period of time. The other school of thought maintains that this test searches out weak insulation which will probably fail sometime in the future. If such testing is done during plant shutdowns or slack seasons, correction of a deteriorated insulation problem is done at the operator's convenience rather than risking a failure at a future inopportune time. The decision as to the proper course to follow must be made on a case by case basis for each facility.

Low Voltage Tests

These tests are generally done in the field to determine insulation condition. Such testing may represent the complete program or it may be the necessary initial test prior to high voltage testing. It would be foolish to subject a winding with a correctable deficiency, such as dirt or absorbed moisture, to potentially damaging high voltage testing. It is unwise to expose windings which are suspected to be damp or very dirty to even line voltage which could cause an unnecessary destructive failure.

The standard low voltage test is done for one minute at a direct voltage of 500 or 1,000 V. The use of a megohmmeter will measure the insulation resistance directly while applying the required voltage level. You will hear this test referred to as a megger test. The term "megger" is a registered trademark of the G. Biddle Co. but it is used almost universally to describe the process and the instrument. The typical megger is hand cranked but motor driven and solid-state battery operated models are available. These are useful when making 10 minute tests on windings.

The advantage of the 500 V megger test is that it is non-destructive while searching out low insulation resistance due to dirt, moisture and some forms of degraded insulation. The applied current of about one milliampere is not sufficient to harm the insulation. Acceptable values of insulation resistance obtained by megger testing should not be less than 1 megohm per kV of motor rating plus 1 megohm. The minimum value for a motor rated 600 V or less is 2 megohms. These values are based on an insulation temperature of 40°C (104°F). Insulation resistance doubles for each 10°C reduction in temperature (above the dew point) and this approximate correction should be applied to tests made at other than 40°C winding temperatures.

The acceptable values of insulation resistance are for a complete three-phase winding. If a portion of the winding is measured, such as a single phase, the acceptable value should be adjusted upward to reflect the percentage tested. If portions of the winding can be isolated, the testing of each winding should be done with the others grounded. This permits location of the low insulation resistance by subdivision and also permits testing of non-ground insulation such as phase insulation.

The larger the insulation surface the lower will be the insulation resistance. Large motors have greater insulation surfaces and lower insulation resistances can be expected than on smaller motors. Motors as large as the 680 frame size should have insulation resistances above 50 megohms if clean and dry with insulation that is not deteriorated. Small motors with healthy insulation may register close to infinity on the megger scale.

The following example will underscore the value of keeping records of past insulation resistance measurements, temperature, and date. Consider two motors which are thought to have been subjected to moisture and are to be placed in service. One motor has a historical insulation resistance level of 120 megohms and the other a value of 20 megohms. At present, both motors test 18 megohms. These values should lead you to the conclusion that the winding of the first motor has absorbed a considerable amount of moisture. A drying period before placing the motor in service would be a prudent course of action even though the insulation resistance is well above the kV level plus one megohm criterion. The second motor appears to have absorbed little moisture and could be placed in service with minimal cause for concern. You can see that a record of periodic insulation resistance measurements is much more valuable than a reading at a single point in time.

Dielectric Absorption Tests

This provides additional valuable information, especially for higher voltage motors with form wound coils. This method requires the determination of the insulation resistance after one minute and 10 minutes of megger voltage application. The 10 minute value divided by the one minute value is called the polarization index. The recommended minimum value for a.c. and d.c. motors is 1.5 for Class A insulation and 2.0 for Classes B and F insulation.

Dielectric absorption tests are considered more meaningful than just a one-minute megger test in defining the condition of the insulation. Low values of this index generally indicate moist or contaminated windings which require drying out or cleaning. Windings with low values of insulation resistance and polarization index are definitely in need of maintenance before being energized or placed in service.

For more information on insulation testing and maintenance, refer to IEEE Standards 43 and 432.

Correction of Low Insulation Resistance

This is generally accomplished by cleaning or drying of windings where no permanent damage has occurred. Insulation within a coil which has been carbonized, burnt or lost its life due to age generally cannot be rehabilitated. The motor should be disassembled and the windings cleaned with an approved solvent compatible with the insulation system and not harmful to personnel. Adequate safety precautions should be followed. Water can be used to flush dirt from windings and cooling passages but the insulation will require careful drying. The newer insulating materials may be more amenable to this treatment than older winding insulation systems or aged windings.

Drying of windings is usually accomplished by the application of heat. Methods include baking in an oven at 80 °C-90 °C, blowing warm, dry air over the windings or by passing current through the windings to develop internal heat. Drying out a winding can take a considerable amount of time lasting as long as several days. Periodic megger and polarization index measurements will be required to monitor dryout progress.

Bearing Maintenance

All bearings require maintenance to perform properly and achieve their service lives. Bearings can be classified as sleeve or anti-friction types. There is considerable overlapping in the application of both types in electric motors. Proper maintenance consists simply of keeping the bearing clean, lubricated and loaded not in excess of its rating.

Sleeve Bearings

These depend upon a hydrodynamic oil film developed between the shaft and bearing for their separation. Theoretically, a properly applied sleeve bearing which is kept properly lubricated will last indefinitely. It can support heavy loads and is less susceptible to shock loads and vibration than anti-friction bearings. Sleeve bearings operate more quietly than anti-friction bearings, especially at higher speeds. In the larger sizes which contain an oil reservoir and rings the flushing action of the oil reduces its sensitivity to contamination and aids in cooling the bearing.

Bearings in fractional horsepower motors are generally bronze sleeve bearings. Some manufacturers utilize sintered bronze which is porous and is saturated with oil from a felt type reservoir.

Sleeve bearings should be lubricated periodically with the proper grade of oil as recommended by the manufacturer. When oiling a motor it is recommended that the motor be shutdown so that stray oil is not flung into the windings or onto commutators, slip rings or brushes. Emptying, flushing and refilling the oil reservoir will remove contamination and oxidized oil. Care should be taken to prevent overfilling. Replace all covers and caps.

Sleeve bearing wear can sometimes be determined without disassembly of the motor by taking air gap readings with feeler gauges if the end frames permit access to the gap. Readings should be taken at four equally spaced positions on each end of the motor.

Anti-Friction Bearings

They operate on the principle of rolling contact between elastic circular components. These parts are precision made and are sensitive to corrosion and abrasive materials. The life of a bearing depends on the fatigue resistance of the materials and has a defined life which decreases as speed and loading are increased. Anti-friction bearings have a life defined by a B-10 hour rating. This is the life at which 10% of a group of like bearings would have failed at a specified loading and speed.

Ball bearings come in three types of enclosures. The first type, having no enclosure, is the open type which has no protection from the entry of foreign particles and contamination. This type of bearing is usually greased from an external fitting and the bearing can receive too much grease. Excess grease can lead to churning by the balls and to overheating.

Another type of enclosure is the prelubricated sealed bearing incorporating close-fitting seals between the inner and outer races on each end. This bearing is not regreaseable and is usually replaced when a motor is reconditioned. A service life of about three years continuous operation at 3,000 rpm is typical. The seals exclude dirt and moisture and the bearing contains the proper amount of grease to prevent churning.

The third type of enclosure is the shielded design. Metal shields are provided on each side of the bearing with a small gap at the inner race. This bearing is also prelubricated but the shields permit regreasing in service while minimizing the entrance of foreign particles.

Regreasing of anti-friction bearings is best done with the motor at standstill so that excess grease which may be forced into the motor interior will not be spread throughout. Remove the drain plugs and add new grease under low pressure until the old grease is purged and new grease emerges from the drain openings. Run the motor for 10 minutes and shutdown. Wipe off excess grease which may have been expelled and replace all plugs which were removed. The manufacturers recommendations should be followed in regard to the type of grease used and the frequency of regreasing.

Commutator Maintenance

The commutator is a vital part of every d.c. motor. Fortunately, commutators tend to show signs of distress in advance of serious trouble. Periodic inspection is a valuable preventative maintenance tool if you recognize the difference between normal conditions and a developing problem.

Commutators provide two functions. The first is to provide a sliding electrical contact between the rotating armature and the stationary brushes connected to the external power supply. The circuit arrangement of armature coils and commutator bars between brush pairs provides a summation of individual coil voltages to correspond to the rated voltage of the motor.

The second function of the commutator is its performance as a reversing switch. Each armature coil experiences a reversal in the direction of current flow as the commutator bars to which it is connected pass from one side of the brush to the other. During the time interval that the brush spans two adjacent bars the associated coil is short-circuited by the brush. Successful commutation requires that this current reversal take place within the brush in a fraction of a millisecond. If the current reversal takes place external to the brush, the brush is not set on the neutral axis and damaging sparking will take place on the commutator surface. Large motors are provided with smaller interpoles midway between the main poles to accelerate the current reversal and keep the neutral axis under the brush. In a motor the polarity of an interpole is opposite to the polarity of the next main pole in the direction of rotation.

The brushes must maintain a good physical and electrical contact with the commutator surface to perform properly. The commutator surface must be smooth, polished and concentric with the axis of rotation. A commutator is a complex assembly of many pieces of copper and insulation which is subjected to centrifugal forces and temperature variations. As a result there will always be minor variations from the ideal in the surface. The pressure of the brush springs forces the brushes to follow these minor surface irregularities.

When speed or the surface irregularities become great enough the brushes are no longer able to maintain proper surface contact and the commutator surface is further deteriorated by sparking. The condition becomes progressively worse unless recognized and corrected. A slightly high bar can cause a brush to jump adjacent bars before regaining contact with the commutator thus causing arcing, burning and flat spots. Flat spots can also be caused by vibration or defective bearings. If the direction of rotation is reversed, the flat area developed behind a high bar will now cause the brush to fall into it and the high bar will shatter the leading edge of the brush.

In addition to mechanical surface considerations, it is necessary that the proper surface film be developed and maintained. This film is established by current flow and by the sliding action of the brush and is the result of electro-chemical deposition of carbon, graphite, copper oxide and water vapor on the commutator surface. Brush and commutator wear will be accelerated if this film is damaged mechanically, electrically, chemically or contaminated by dirt. Oil or oil vapors add an insulating layer and destroy the bonding ability of the surface film. This promotes brush wear and eventually commutator threading. Abrasive contamination will also cause brush wear and threading. The film condition can be judged by the appearance and color of the commutator. A smooth, even, glossy copper, brown, or chocolate color is proper. Sometimes a color pattern may be evident between bars and as long as the pattern is repeatable around the entire commutator there is no cause for concern.

A commutator which has flat spots, raised bars, burns or is threaded should be removed from the motor and turned by placing the armature in a lathe. The armature should also be tested for grounds and also for shorted and open windings on a growler. Burning between adjacent bars can indicate winding problems. The mica insulation between the bars should be undercut below the level of the commutator surface. When a commutator is worn or cut to the extent that the mica edges protrude, the mica flakes break off and become embedded in the brush surface. This disrupts commutation and causes accelerated wear on the commutator. Minor commutator surface restoration can be accomplished with the armature in the motor by using a commutator polishing stone. Never use emery cloth since this material is conductive and can short adjacent bars. The brushes should be replaced and contoured to the commutator surface. Brush springs which are weak should be replaced.

Deciding When to Repair Versus When to Replace a Motor

These decisions must be made on a case-by-case basis. Many factors enter into this decision but the usual considerations are repair time versus the availability of a new motor with cost as a common denominator.

It may be worth considering a rewind of an older T Frame or U Frame motor rather than a new replacement. Many motor rewind shops use Class F or Class H insulation and winding materials as a standard. A motor originally designed and wound as Class B would be upgraded with the use of Class F insulation and would have a longer insulation life, higher service factor rating or could be increased in horsepower rating with a change in winding design. Many older motors incorporated more iron and frame materials and a more robust design than their modern counterparts. A rewind and replacement of bearings may be a better overall value than a less expensive replacement. Most reputable motor repair shops can provide satisfactory rewinds complete with warranties that compare favorably or equal those for new motors. Check the type of warranty offered. The Electrical Apparatus Service Association (EASA) is a nationwide organization which provides its members with technical information on various manufacturers' designs, industry practices and trends, and a code for reliable service. Manufacturers also have organized networks of authorized service facilities and will advise of one in your area if asked.

Another consideration in deciding on motor repair or replacement is the matter of efficiency. The escalating cost of energy and the recent trend to reduce power consumption through conservation has spurred motor manufacturers to offer high efficiency motor designs. Although these motors are initially more expensive than standard motors, the savings of a few percent in losses can be recovered in a period of a few years or less, depending on the hours of use and the local cost of power. Some power companies have offered rebates on the purchase of a replacement motor which exceeds a defined efficiency level. This rebate generally approximates the difference in initial cost between a high efficiency motor and a standard motor. The important thing to remember is that the lower energy cost of a higher efficiency motor provides a continuing benefit year after year and becomes even more attractive as energy costs increase. You should evaluate the possible advantages of high efficiency motors for your particular application when replacing or repairing motors.

Motors with special shafts, frames, ratings, and mountings may not leave you with much choice other than to repair the motor in regard to time frame or even cost. Many times special or obsolete motors may not be available at any price. In this case repair is the only choice. Motors made for an "original equipment manufacturer" (OEM) by a motor manufacturer may not be available through that motor manufacturer's distributor network since it was made exclusively for the OEM. Often, an essentially equivalent design requiring little or no modification can be found in some manufacturer's standard line. Simple modifications such as cutting a shaft, utilizing a three speed motor in place of a two speed motor, using standard frame adapting bases or shaft adaptors can save time and cost compared to obtaining the exact replacement from an OEM.

Manufacturers' Warranties

Most manufacturers warrant their motors to be free from defects in factory workmanship and material. The usual manufacturer's warranty is for a period of one year from the date of service. Generally, there is a limit also on the time from date of manufacture. It may be as short as 18 months or as long as three years. It is essential that you understand the warranty offered by the motor manufacturer before trying to make a warranty claim for a motor failure.

Warranties do not include coverage for things that happen to a motor in the field which are not attributable to poor factory workmanship or material; for example, windings which are roasted uniformly are the result of excessive current flow for an extended period of time. Proper overload protection would prevent this damage and all motor manufacturers hold to this point of view. Proper protection not only includes correct rating and application, it also requires maintenance to assure operation when an overload develops. Often a user with a roasted motor will protest that he has the proper protection and that it tests satisfactorily. The proof of the pudding is in the eating and there is no contesting the fact that, on at least that occasion, injurious current was allowed to flow. Uniformly roasted windings may be caused by overloads, locked rotor conditions, low voltage or lack of ventilation. Locked rotor conditions may also be accompanied by severe damage to the rotor.

Another type of damage which is not warrantable is due to single-phasing. In this instance the winding is roasted in repeatable patterns around the winding according to the phases which were left connected and those which were disconnected. Again, proper protection will prevent this damage. If the single-phasing is plant-wide or affects a portion, you should look for damage to other motors which may have been connected at the same time.

Other types of failures which are not warrantable are caused by the environment around the motor. A motor which is misapplied, such as an open drip-proof motor located outdoors, which fails due to water or snow entry is not the responsibility of the manufacturer. Motors which have been submerged (assuming it is not a submersible design) are not warrantable. Motor bearings which fail because an adjacent pump seal fails and sprays them with water are not included in the motor manufacturer's warranty. Motors which fail due to misalignment with the load, are dropped or otherwise damaged in the field are not warrantable.

Winding failures which are usually warrantable are those of a localized nature providing they are not the result of contamination, voltage or lightning surges. These failures include ground faults in windings and leads, turn-to-turn failures within a coil, phase-to-phase and failures between windings and winding open circuits. Mechanical damage done to the motor by failure of an internal fan or baffle is covered. Capacitors which are furnished as part of the motor, centrifugal switches, internal overload devices and bearings which have not been abused are usually warrantable if they fail or cause damage to the motor.

When claiming a warranty it is required that the motor be sent to an authorized service facility for disassembly and inspection. The warranty is between you and the manufacturer, not the service facility. The service facility will contact the manufacturer in cases of doubt and the final decision is with the manufacturer. He may elect to have the service facility repair the motor or replace it in accordance with his standard policy. If you take the motor to an unauthorized service facility for repair or inspection the warranty will be voided. The same applies to repairs you may attempt to make yourself. The motor manufacturer is not liable for transportation costs, removal costs or production costs due to downtime.

Recommended Service Equipment

The complement of service equipment and instrumentation depends on the type and size of the facility. The following are recommendations for equipment to support minimal preventative maintenance and motor service for all but the smallest or simplest facilities. This list does not include tools to accomplish the repairs which may be determined necessary after inspection or diagnosis. All test equipment must be properly rated and used in accordance with the manufacturers' recommendations for safety. Equipment of good quality and rugged design will be found to be the best investment in the long run.

- Multimeter with voltage, amperage and ohmic scales—Overload protection is desirable to prevent damage to the meter. Digital meters are accurate and foolproof.
- D.C. megger, 500 V or possibly 1,000 V if higher voltage motors are involved—If absorption tests are to be made a motor driven or battery operated megger may be useful.
- Tachometer or speed indicator—A revolution counter can be used with a watch for timing. Stroboscopes are accurate but the basic speed must be known. Electronic tachometers which count interruptions in a reflected light beam are available.
- A.C. clamp around ammeter which permits measuring currents without breaking into the circuit being measured—Very useful for measuring motor line currents.
- Temperature measuring device, preferably a dial device with a quick reading sensitive tip, 0-150°C—Devices with metallic stems should be used with care around energized equipment. Glass thermometers are slow reading and make poor contact with the surface being measured.

- Vibration measuring instrument of the portable type.
- Dial and feeler gauges for checking coupling alignment and setting.
- Micrometers and bore measuring devices for measuring sleeve bearing and journal wear and anti-friction bearing fits in housing and on shafts.
- Micro-ohmeter for measuring low resistances such as windings and resistances across connections.
- Growler for detecting shorted windings in stators, rotors and d.c. armatures—Hand-held models are available for use on outside and inside diameters.
- Spring scale, 0-16 oz. for measuring brush spring pressures.

REFERENCES

The following references provide additional background on the subject of motor design, application, maintenance, industry practices, standards and codes.

- *Motor Application and Maintenance Handbook,* McGraw-Hill Book Company (New York, N.Y.)
- National Electrical Manufacturers Association, 2101 L Street N.W., Washington, D.C. 20037
  - MG 1 Motors and Generators
  - MG 2 Safety Standard for Construction and Guide for Selection, Installation and Use of Electric Motors and Generators
- *National Electrical Code* Published by National Fire Protection Association, 470 Atlantic Avenue, Boston, MA 02210
- Institute of Electrical and Electronic Engineers, 345 East 47th Street, New York, NY 10017
  - Std 1 Rating of Electrical Equipment, General Principles Upon Which Temperature Limits Are Based
  - Std 43 Recommended Practice for Testing Insulation Resistance of Rotating Machinery
  - Std 62 Guide for Making Dielectric Measurements in the Field
  - Std 112A Test Procedure for Polyphase Induction Motors and Generators.
  - Std 113 Test Code for Direct Current Machines
  - Std 114 Single-Phase Induction Motor Tests
  - Std 115 Test Procedure for Synchronous Machines
  - Std 288 Guide for Induction Motor Protection
  - Std 329 Synchronous Motor Protection Guide
  - Std 432 Guide for Insulation Maintenance for Rotating Electrical Machinery

11

Planning and Executing Lubrication Procedures

Robert Mayer

This chapter is designed as an aid to individuals who are responsible for preventive maintenance, including lubrication of all types of industrial equipment, heating, and ventilation found in industrial plants, and commercial and residential buildings. The necessary information is provided to set up a successful lubrication program including storing, handling and application of lubricants as part of plant maintenance in a modern facility where the maintenance manager does not have a large engineering staff to develop lubrication requirements.

BENEFITS OF PROPER LUBRICATION

The goal of any planned maintenance program is to reduce costs. A proper lubrication program will increase the life of equipment, reduce failure and repair costs, minimize downtime and reduce overall efforts by reducing the unplanned repairs of major equipment, and help to lower energy costs. Proper lubrication depends upon proper planning. Planning must take into account the characteristics of the equipment to be maintained. Accessibility and location of equipment, and lubricant storage, resources available in terms of staff and their skills, and the probable time interval between failures must all be considered in order to develop an effective plan.

Preventive maintenance has traditionally included the elements of inspection, lubrication, adjustment, and cleaning at regular intervals. Obviously, to plan such a program effectively, you must have on hand accurate and reliable data. It is not enough to tell a mechanic to *inspect* a certain piece of equipment. The instruction must advise him as to whether the inspection should be performed under running conditions or when the equipment is down. The instruction also should be specific in terms of whether the inspection should be visual only, or should include instrumentation such as, vibration testing, or the use of a listening device as aids in evaluating the condition of possible wear, misalignment, or lack of lubrication.

When designing the lubrication program, a decision must be made as to whether the inspection program will be limited to inspection only, or if the technician will also perform lubrication and adjustments at the time of inspection.

Inspection and adjustment should be an integral part of the lubrication program where possible. This means that maintenance instructions must tell the person who will perform the maintenance work to inspect at certain intervals when lubrication may not be normally required. Manufacturers' guidelines for inspection and lubrication frequencies and the actual needs of each individual device should be integrated in accordance with recommendations set forth in this chapter.

DETERMINING EQUIPMENT LUBRICATION REQUIREMENTS

Most reference works classify lubrication by types of bearing surfaces such as high and low friction, high rotating speed and low rotating speed, journal or antifriction bearings (roller, ball, etc) and drive mechanisms. Lubricants are classified by characteristics such as viscosity, or type, such as waxes, greases, petroleum, synthetic, or solid lubricants. While such information is absolutely essential for proper integration of good lubrication with machine design, it is less important when you are establishing a lubrication program for your plant. To classify all the operating equipment in the plant by rotation speed, and/or lubricant base (petroleum or synthetic), would not be very beneficial in setting up a program—unless lubrication frequency and effective routes for the maintenance mechanic to follow were also specified.

An effective starting point for the lubrication program is an inventory of the equipment in the plant. Classify or group together items in the same department or functional requirement.

Obviously, first priority is given to the most important equipment. Major production equipment would probably have priority in most plants. Environmental control items would also be high on the list. For instance, you might begin with the central power plant and group together auxiliary equipment for the steam generation system and the equipment generating chilled water, or compressors generating compressed air.

Next, record all available information that relates to lubrication of the equipment items listed. Information sources might include master equipment record cards, manufacturers' operating manuals, and the present lubrication procedure for each item. Usually this inventory will uncover many items that have little, if any, recorded information.

There is a high degree of probability that the lubrication procedures for unrecorded equipment are wasteful, inefficient, or nonexistent and are leading to early breakdown or retirement of the equipment.

Before proceeding further into the specifics of a lubrication program such as establishing proper lube cycles, evaluating labor requirements and operating the entire program, become familiar with some of the basic lubricant definitions, basic theory, and practical tables/charts, to aid in understanding this important subject.

TYPES OF LUBRICANTS

Lubricants can be broken down into four generic types: gaseous, liquid, plastic, and solids.

Gaseous Lubricants

These will not be given consideration here because they are used in very restricted applications such as very light bearings, extremely high speed spindles, inertial gyroscopes, etc.

Liquid Lubricants

The major group of liquid lubricants are made up of the mineral oils produced from petroleum crudes by various refining processes. These crudes are complex compounds of carbon and hydrogen, referred to as hydrocarbons.

A secondary group of liquid lubricants are called fixed oils. These are fatty substances extracted from animals, vegetable matter, and fish. They are called fixed oils because they will not volatilize without decomposing. This chapter concentrates on nondrying fixed oils, which are usually blended with mineral oils for specific requirements.

A third and rapidly growing classification of liquid lubricants is synthetic fluids. These fluids have been chemically developed to meet the tremendously growing demands placed on lubricants caused by technology developments in equipment of all types, such as increased speeds, pressure, and temperatures. In some applications, the use of synthetic lubricants is mandatory. For example, mineral petroleum oils cannot be used in lubricant systems exposed to ignitable environments, extreme heat, or extreme cold.

Synthetic oils must be used where high temperature may carbonize mineral oils. Such a breakdown will cause maintenance problems or failure in equipment such as ovens, draft dampers, ammonia and air compressors, or other extreme temperature-oriented equipment.

Plastic Lubricants

This is a general classification for greases, petrolatum and semi-fluids. The primary interest for this chapter is in greases. Plastic lubricants are indicated where operating conditions make it difficult to keep a liquid lubricant in place: for example, in a bearing located in such a position that any oil would leak right out.

Greases are generally made by saponifying a metal base with a fatty substance to form a soap to which any of the liquid lubricants discussed above, are added. This process, which includes agitation, is performed under controlled temperature conditions either in a pressure cooker or an open kettle. Some soaps are formulated to make greases withstand high temperatures; others give a waterproof (or steamproof) characteristic to the finished grease. A variety of application possibilities can be designed into greases by varying base oils and/or saponifying agents. In most instances, the grease selected should be recommended by the grease manufacturer under the guidance of the equipment manufacturer.

Solid Lubricants

These include such products as graphite, molybdenum disulfide, polytetrafluorotheylene (PTFE) talc, mica, borax, and wax. Solid lubricants are sometimes used in their dry states such as molybdenum disulfide, graphite, PTFE, but more often they are used as additives in oils and greases or are combined with a vehicle to place them on a specific surface and keep them there. Dry solid lubricants should be applied on surfaces subjected to high temperatures (such as oven chains) or in dusty environments where an oil or grease would attract dirt and hold it there.

THE IMPORTANCE OF VISCOSITY AND VISCOSITY INDEX

The measurement of viscosity and viscosity index is fundamental to the science of lubrication. Viscosity is the property of a fluid that resists the force, tending to cause the fluid to flow. It is also a measurement of the rate at which a fluid is capable of flowing under specified conditions. Fluidity decreases with higher viscosity; for example, flow rate varies inversely with viscosity, and vice versa.

Viscosity and viscosity indices are very important to the lubrication engineer. Viscosity plays a major role in developing the film thickness of a hydrodynamic wedge. It determines load-carrying capacity, the ability of an oil to flow, operating temperature levels, and often wear rate. Generally, high viscosity or heavy oils are used on parts moving at slow speeds under high loads, since the heavy oil resists being squeezed out from between the rubbing parts. A high viscosity index is required when a wide range of operating temperatures is indicated. Low viscosity or light oils are used when higher speeds and lower loads are encountered; they do not impose as much drag on high speed parts and the high speed permits a good oil wedge to form, even though the oil is less viscous.

Low temperatures call for lighter oils and high temperatures call for heavier oils. Oils increase in viscosity as their temperatures become lower and decrease in viscosity as their temperatures go up.

Kinematic viscosity is the time required for a fixed amount of an oil to flow through a capillary tube under the force of gravity. The unit of kinematic viscosity is the stoke or centistoke (1/100 of a stoke). Kinematic viscosity may be defined as the quotient of the absolute viscosity in centipoises divided by the specific gravity of a fluid while both are at the same temperature.

As stated above, viscosity is a measure of the "flowability" at a definite temperature. The unit of measure is time in seconds required for 60 milliliters of oil to flow through a standard orifice under a standard falling head and at a given temperature. (100°F and 210°F are common temperatures for reporting viscosity.) Saybolt Universal viscosimeters heat a test sample to the desired temperature and then pass the sample through the orifice while being timed. Saybolt Furol viscosity is obtained with the same instrument but with a larger orifice, producing results approximately 1/10 those of the Universal orifice readings. There are other methods of rating viscosity; see Figure 11-1.

Lubricating oils decrease in viscosity as their temperatures are raised. For example, an oil having a viscosity of 3,000 seconds at 100°F (SUS) might show a viscosity of 175 sec. at 210°F. This is said to be the rate of viscosity change for that oil. Another oil of 3,000 seconds at 100°F may have a viscosity of 160 seconds (SUS) at 210°F. This oil then would be considered to have a higher rate of viscosity change per degree temperature rise.

The rate at which an oil changes viscosity with a rise or drop in temperature is designated by a comparative number called viscosity index (VI). A VI of 100 indi-

SAE	SUS @ 100°F	SUS @ 210°F	CPS (Centipoise)*
10	200	48	44
20	325	57	70
30	550	68	119
40	850	84	184
50	1200	100	260
90	1500	110	326
140	2500	150	543
250	5000	200	1085

*Equivalent to SUS @ 100°F

Figure 11-1. Viscosity Conversion Chart

cates that an oil with this value thins out less rapidly than oils having a VI of zero. Values in between indicated intermediate viscosity change according to the VI number. (See Figure 11-2.)

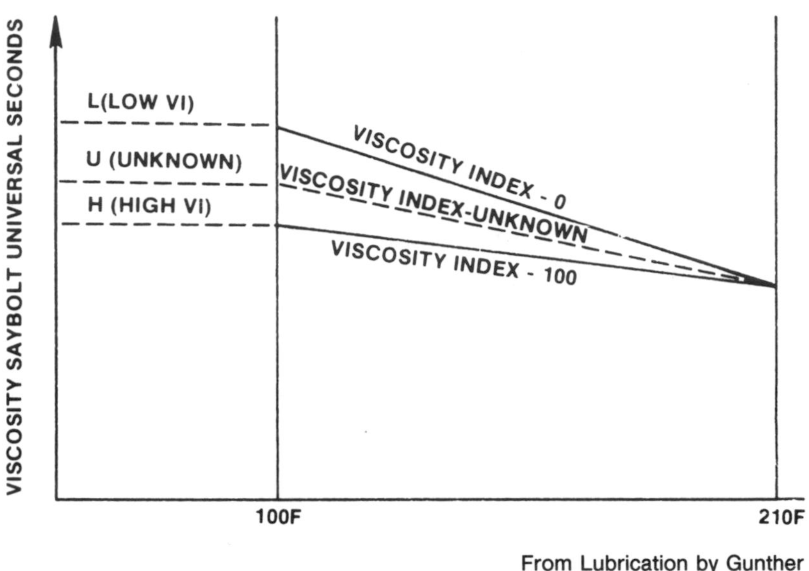

From Lubrication by Gunther

Figure 11-2. Viscosity Index

Figure 11-3 shows viscosity versus temperature relationship for SAE (Society of Automotive Engineers) multigrade motor oils. Note an SAE 30 oil compared to an SAE 10W40 motor oil. The 10W40 oil has more body (higher viscosity) at 210°F to provide more protection at high temperatures and a lower viscosity at 0°F to provide faster starts, and less wear in the winter cold because of better "pumpability." An SAE 10W40 oil has a higher VI than an SAE 30 oil.

MECHANISM PARTS REQUIRING LUBRICATION

All surfaces in any mechanism that move relative to other surfaces require lubrication. This is to separate these surfaces with a film of lubricant to reduce friction and wear. This occurs in bearings that support rotating shafts, in gears that have meshing teeth, and between pistons and the cylinders in which they move. All mechanisms, regardless of size or complexity, will employ one or more of the following basic moving parts: bearings, gears, and pistons.

Bearings

These can be divided into two basic types, plain bearings and anti-friction bearings.

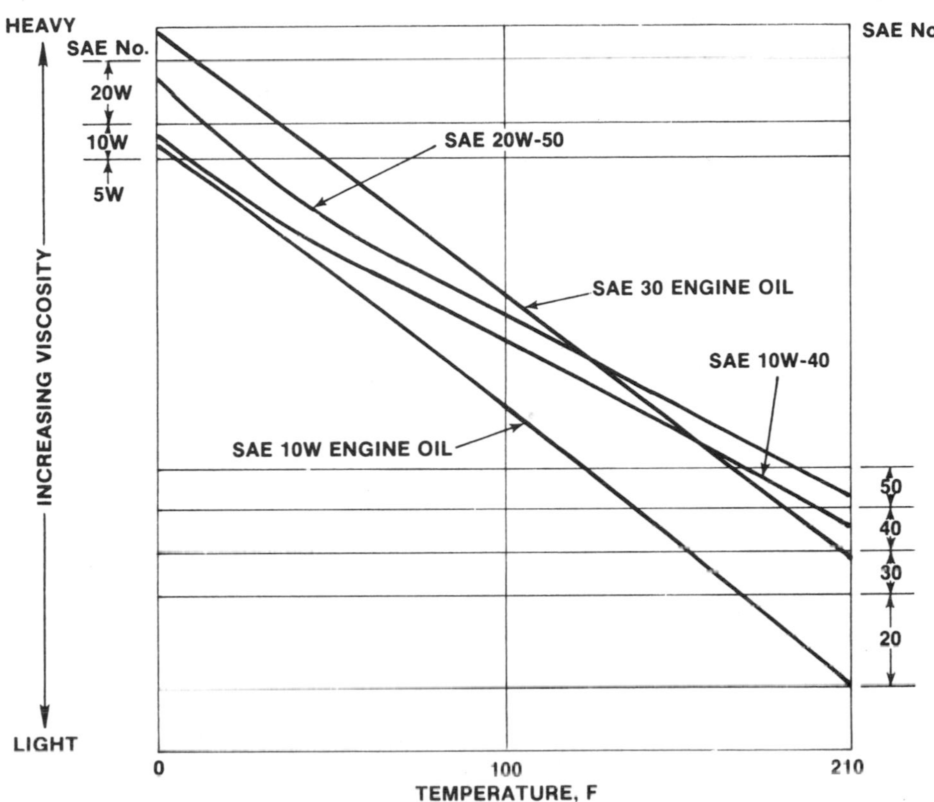

From Classification Systems for Auto, Truck, and Heavy Duty Equipment, Power Train Lubricants by Sun Petroleum Products Company, A Division of Sun Oil Company of Pennsylvania.

Figure 11-3. Viscosity—Temperature Relationship

Plain bearings can be further classified according to function. Journal, guide, and thrust bearings are usually made from bronze, babbit, plastic (nylon, teflon), or some material softer than steel so that any wear will be on the bearing instead of the steel part.

Journal bearings are so named because they support or operate against a rotating shaft. The portion of the shaft within the bearing is known as the journal. In a machine, when a journal starts to rotate, a wedge-shaped film of oil develops under the journal. This wedge lifts the journal away from the bearing, reducing friction, and thus guarding the journal and bearing against wear. Journal bearings are of various designs:

- The *solid bearing* as the name implies, is in one piece, and is sometimes called a *sleeve bearing* or a *bushing*.
- The *split bearing* is divided into two pieces lengthwise.

- The *half bearing* encircles only one-half of the journal, leaving the other half-exposed. It is used when the load is applied vertically between the bearing and the shaft.

- The *multipart bearing* normally consists of four separate parts or quarters. They are found on crankshafts on such machines as air compressors and steam engines. The four-piece construction allows for readjustment of the bearing to the crankpin to take up slack when wear takes place.

Guide bearings guide or hold in proper position reciprocating parts of a machine.

Thrust bearings are bearings that prevent a shaft from moving endwise.

Anti-friction bearings are bearings in which a series of balls or rollers are interposed between the moving parts. These balls or rollers are usually (but not always) mounted in a cage and enclosed between rings known as races. Normally the outer race is fixed in the equipment while the inner race is tightly fixed on the shaft with which it rotates. When the shaft rotates, the balls or rollers are rotating between the inner and outer races with a minimum of friction. Antifriction bearings are classified as follows.

- *Straight roller bearings* are where the axis of the rollers are parallel to the axis of the bearing.

- *Tapered roller bearings* are used not only to support a rotating shaft but also to prevent the shaft from moving endwise.

- *Ball bearings* are probably the most commonly known and used of all anti-friction bearings. They come in single- and double-row and in self-aligning types.

- *Ball thrust bearings* are usually flat bearings where the loads are applied to the flat of the race rather than the inner and outer diameters, and as the name suggests, are designed to prevent linear movement of the shaft rather than for radial movement.

- *Needle bearings* differ from the others in that they have no inner race, cage, or separator, and their small rollers or needles are just slightly separated by the lubricant. The name comes from the fact that the rollers are much longer than their diameter. They are primarily used for oscillating and intermittent motion.

Gears

Gears serve various functions, such as transmission of motion from one shaft to another; change of direction and orientation from one shaft to another; or change of speed from one shaft to another. Properly matched gear tooth surfaces are designed to provide a rolling motion for minimum tooth friction.

Pistons

The third fundamental moving part found in machinery is the piston, which operates in a cylinder. In an automotive engine, for example, fuel and air mixtures are pumped in above the piston and ignited. Combustion forces the piston to move downward to turn a crankshaft. Rings are located in grooves in top of the piston pushing outward against an oil film to seal the space between the cylinder wall and the piston and prevent loss of power escaping downward. The lubricating film of oil used to separate the rings from the wall of the cylinder along which it must slide also helps the rings seal the combustion chamber described above. The oil may be splashed onto the rubbing surfaces from the crankcase; however, in large pistons it is often pumped in small quantities through holes in the piston wall. Similar piston action may be found in other applications such as compressors, pumps, and even in a grease gun.

LUBRICATION REQUIREMENTS FOR MECHANISM PARTS

The lubricating instructions of the machine manufacturer should always be sought and followed. Lubricant suppliers can be helpful with suggestions for equivalencies, and sometimes even improvements through new lubrication products.

Grease Lubrication of Anti-Friction Bearings

There are several rules associated with grease lubrication of anti-friction bearings:

1. Use the correct grease according to the type of rolling element and the service for which it is used. Normally a grease should contain an oil having the same viscosity that would be used if the oil alone were providing the lubrication. Never use a grease containing solids in any needle bearing or in a roller bearing turning more than 100-150 rpm.
2. Use a grease of the lowest possible consistency commensurate with the temperature, speed, leakage rate, and position of the bearing. Lower consistency greases generally contain more oil, and oil is the principal lubricant upon which bearing life depends.
3. Never mix greases without the approval of the grease manufacturer.
4. Depending on the presence of water and other contaminants in the environment of the bearing during operation, plan the greasing interval to allow one complete change of grease scheduled on a basis of shaft rotation. In an extremely dirty environment this could mean full purging at every application, whereas purging at every eighth application might be sufficient in a clean, dry environment. The purging can be performed as described, i.e., feeding the grease into the bearing until all the old grease is removed or the new grease appears at the drain.

5. Before attaching a grease gun to a grease fitting, wipe the nozzle and the fitting clean with a lintless cloth, preferably dampened with an industrial cleaning solvent. This is a precaution against pumping foreign matter into the bearing along with the grease.
6. Never fill an anti-friction bearing to full capacity. Otherwise, it will run hot and subsequent expansion of the grease will eventually break the seals. The proper volume of grease in a bearing should be approximately equal to one-third to one-half of its space capacity. The exception to this are low-speed bearings where the cavity may be nearly filled to capacity. There is no one simple answer to knowing when to stop filling an anti-friction bearing except perhaps in a newly hand-packed bearing. One way of assuring against overpacking a bearing is by providing a pressure relief opening at the bottom of the housing; a threaded plug is temporarily removed from this opening during the greasing process. After greasing, the relief opening is left open for a short time to allow any excess grease to be forced out. The flow of grease will stop when the pressure in the housing has found its own comfortable level. The plug should then be replaced in the relief opening. Another method involves the use of an automatic lubing device, several types of which are on the market. These devices replace grease fittings and have their own reservoir and metering devices to feed grease to the bearing. Some have springs, some have pressure equalizing chemicals, and others have a diaphragm into which the grease is pumped. When inadequate grease is in the bearing, it creates an imbalance in pressure, causing the atmospheric pressure on the diaphragm to push the right amount *of* grease into the bearing *to reestablish and maintain equilibrium.*
7. Beyond these small individual lubricating devices there are, on many pieces of industrial machinery and available to industry, various types of automatic lubricating equipment which can be added to existing machinery to dispense a wide range of grease or oil lubricants. These systems, called Central Lubrication Systems, offer a way to improve production equipment performance.

Central Lubrication Systems*

A broad understanding of these systems will help you keep your equipment and these systems in top shape for extended trouble-free operation. You and your staff must recognize the importance of delivering the right quantities of the right lubricant to the right place at the right time. Selecting, designing and maintaining the proper centralized lubricating equipment is the best way to fulfill these four basic

*The section on Central Lube Systems was written by Peter M. Sweeney, Market Communications Manager, Bijur Lubricating Corporation Bennington, Vermont. Used with permission.

goals. It is part of your responsibility to see that the four rights of lubrication are met to assure ongoing peak performance of your equipment.

Sophisticated modern centralized lubrication systems are available that deliver precise amounts of oil or grease lubricants to as many as one thousand lubrication points. It is possible for these systems to deliver the lubricant to the application points in various forms,—droplets, spray etc. A wide choice of lubricators are available, pneumatic operation and self-contained electric motor-driven, being the most popular models. A wide variety of solid state electronic controls are also available to program and monitor fully automatic systems.

Basic Lubricating System Types Defined

Before delving into the characteristics and details of centralized lubricating systems to dispense grease and oils, it is helpful to have a clear understanding of the make-up of a basic system, as well as the various system types your staff is likely to encounter on a daily basis.

Centralized Lubrication System: This description is most commonly used to describe ALL lubricating systems in which the lubricant is fed from a central point or points from a common lubricant source or reservoir. A centralized lubrication system may be installed on a single machine or on a group of machines throughout the plant. Operation may be either manually activated, semi-automatic or fully automatic.

Manual (Hand Operated) Lubricating System: In this type of system, the lubricant is fed manually from a central point or points in the system. Manual systems may be classified as simple, zerk fed points or those receiving lubricant delivery through a hand fed distribution block or manifold. (See Zerk Fed Systems).

Semi-Automatic Lubricating System: Systems which utilize a manual pump to deliver lubricant from a central point or points are described as semi-automatic. The actual delivery of the lubricant to the point is completed automatically through a metering block mechanism after the pump is hand actuated.

Automatic Lubricating System: This type of system is automatically actuated by either the machine or an internal or external timing mechanism. Lubricant is fed from a central point or points and does not require manual operation.

Total Loss System: In this system, the lubricant is not recovered after it has been used at the bearing or lube point. While it is common practice for many oil systems to fall into this category, all grease systems are included in this group.

Re-circulating System: These systems recover the lubricant after it has been delivered to the bearing point. Oil is returned to a sump or reservoir for re-use within the lubrication system. A system is classified as either a "Total Loss" or "Recirculating" type from the way in which it is applied to the machinery or plant. It is possible to have a combination of the two types in which the lubricant is recovered from some points for re-use, but is allowed to run to waste from the remaining points in the system.

Description of Centralized Lubrication Systems

It is important for you and your staff to have a clear understanding of the individual type of system being serviced. Being able to identify the major elements of the system and accurately follow the sinuous paths of the system distribution lines will facilitate a quick and easy understanding of the particular system. System types fall under these general categories: Group fed (zerk), Single line, Dual line and Multiline systems.

Zerk Fed Systems:

Grouped Nipples —The simplest type of lubricating system on a machine is a hand operated system in which all individual lubrication points are fed at a centrally located battery of nipples. This simple and convenient arrangement permits lubrication of the machine during operation without danger to the operator. Further, the central grouping of nipples minimizes the chance of overlooked lube points.

The grouped nipple method can be used for oil or grease with different types of lubricant nipples being used to differentiate between lubricant types. Refer to Figure 11-4.

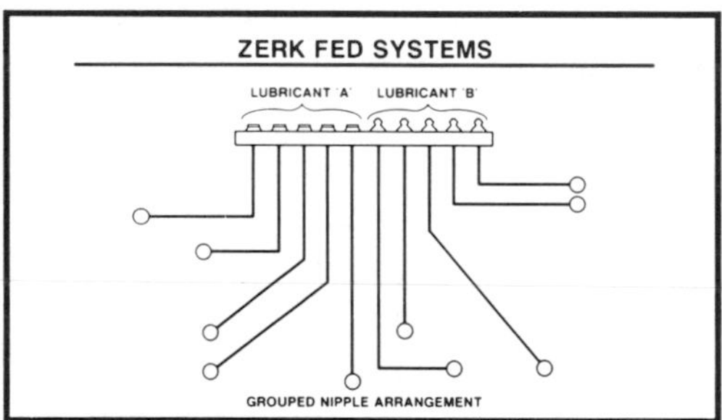

Figure 11-4. Zerk Fed Systems

Single Zerk Fed Progressive Movement Block —The grouped nipple method can be further simplified by combining the various feed tubes servicing a common lubricant into a progressive series movement manifolded block having a single, zerk fed delivery point. Refer to Figure 11-5.

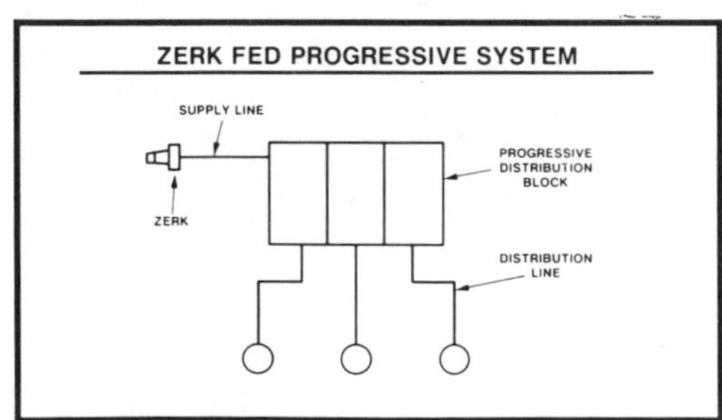

Figure 11-5. Zerk Fed Progressive System

Important Note: When zerk fed manual lubrication methods are in use in which no central pumping source is incorporated, ie. groups of lube points are being fed individually by hand, you should take extra steps to minimize contamination and subsequent ingress of foreign product to the bearing surfaces. This is best accomplished by keeping all dispensing equipment thoroughly free from dirt. Be sure also to clean off oil cups, grease fittings and the surrounding areas to prevent contamination when replenishing lubricants. Careful attention to this detail is vital to assure clean lubricant reaches the friction surfaces.

Single Line Centralized Systems

Resistance or restrictor type systems for oil —A semi-automatic or automatic pump forces a pre-determined amount of oil into the restrictor type distribution system. While under pressure, the oil is apportioned to each bearing or lube point to meet its individual requirement. A fixed orifice resistance fitting (flow unit) assures the specified lubricant amount to each point. Flow units are available with a wide range of rates to permit various discharge amounts. This compact system lends itself ideally to machinery with closely grouped bearing points. Refer to Figure 11-6.

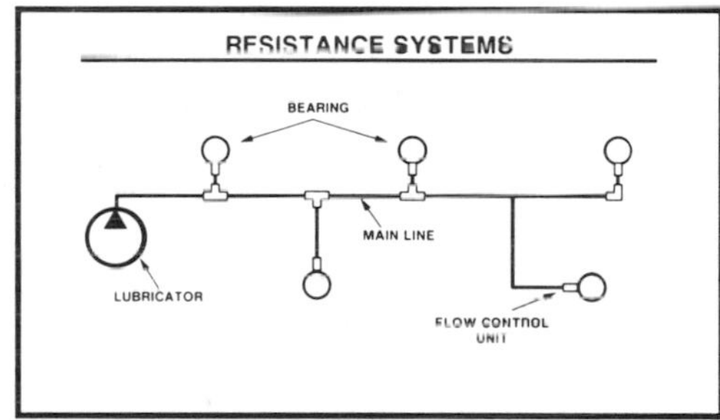

Figure 11-6. Resistance Systems

Oil Mist Resistance type systems (air operated) —While the principle of distribution is similar to restrictor type metering systems, oil is atomized at the pump and mixed with the operating air supply. The air/oil mix is carried along the distribution lines in mist form to the lube point. Generally, the mist is reclassified by a condenser type apportioning device at the bearing so that it can be efficiently delivered. The major advantage of the mist system is its ability to transmit small amounts of lubricant to the delivery points on a continuous basis at low pressure. Refer to Figure 11-7.

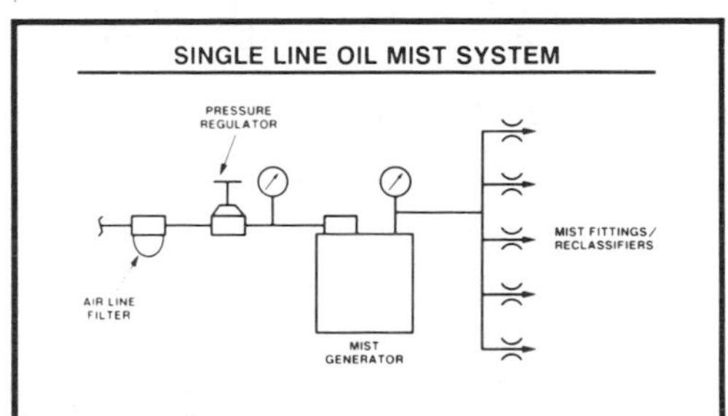

Figure 11-7. Single Line Oil Mist System

Both the Single Line Resistance and Oil Mist systems are for use with oil only and cannot be used with grease or semi-fluid greases. They are designed for a specific type of oil viscosity for the particular machinery application. It is critical that you use the machinery manufacturer's recommended lubricant in these systems.

Positive Displacement Systems —These systems are available with individual or manifolded injectors. Operation principles are identical in both injector types. . . the metering devices (injectors) displace a fixed volume of oil or grease by moving a piston to complete the discharge. A wide discharge range is possible through selection of appropriately rated injectors. These systems are relatively easy to adapt to machinery. Refer to Figure 11-8.

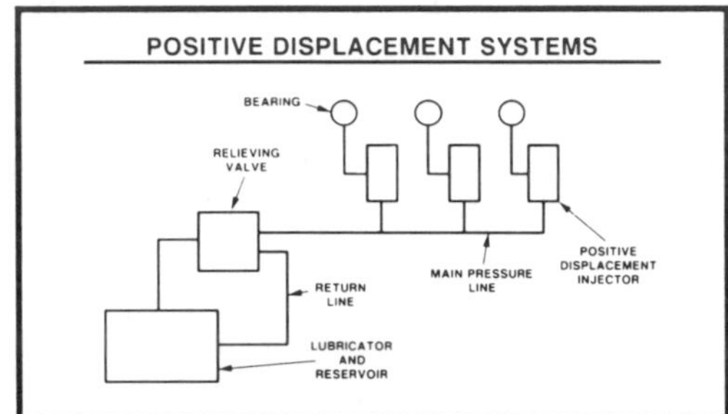

Figure 11-8. Positive Displacement Systems

An important feature in the design of these systems is their ability to be easily modified, ie. additional lube points can be added or removed from the system without affecting the balance of the system discharge.

Positive Displacement Systems with Progressive Series movement pistons —As lubricant is discharged from a non-reversing pump, pressure is directly linked to a manifolded distribution block. In sequence, a number of pistons are moved in an arranged sequence or series to complete the discharge cycle. The progressive movement of the pistons takes place due to the porting and valving to the pistons in the manifolded block. This movement or cycling will continue as long as lubricant is delivered from the pump to the blocks in the system.

To meet a wide range of lube requirements, this system typically utilizes a "master" distribution network to feed "secondary" distributors in the system. The overall functioning of this system is dependent on the successive movement of all other distribution components in the system. Refer to Figure 11-9.

Dual Line Systems (Non-progressive type)

These systems consist of twin delivery lines from the pump to all metering blocks in the system. Each metering block or dual line block contains a pilot piston and a discharge piston. The pilot piston or "slide valve" directs the pressurized lubricant to move the discharge piston to complete a half cycle. Output from the pump, typically grease, is directed to each metering block by means of a change-over valve in the system to complete the remaining half-cycle.

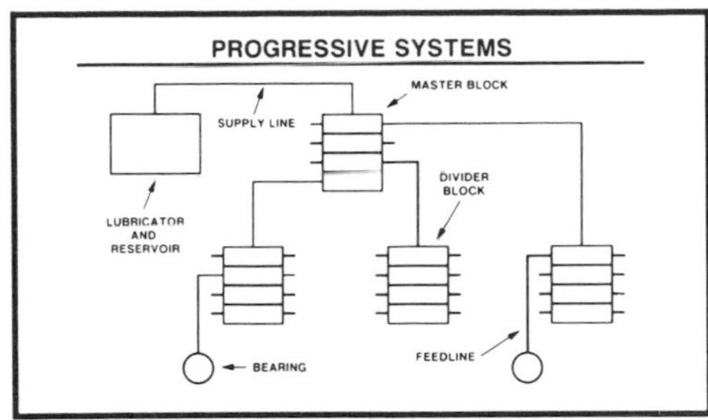

Figure 11-9. Progressive Systems

While this system is suitable for either oil or grease, it is most commonly used with grease. These systems are ideally suited for large machinery installations. Refer to Figure 11-10.

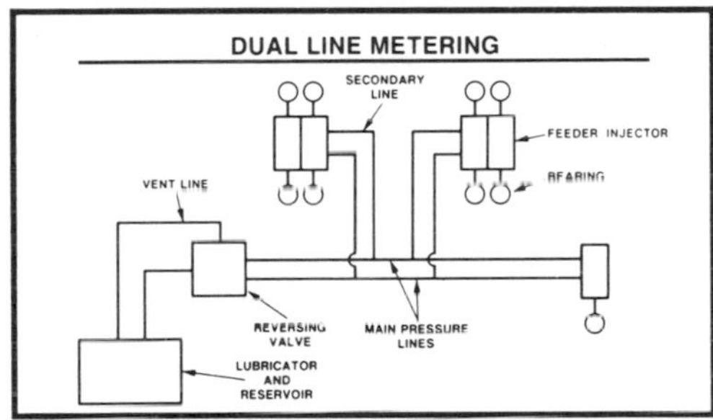

Figure 11-10. Dual Line Metering

Multi-Line Systems

Commonly referred to as "pump to point", these basically simple systems consist of a central pump source and a dedicated pumping unit or outlet for each application point. The lubricant from each pumping unit is delivered directly to the lube point by a direct line. This system utilizes considerable more tubing in the distribution network than other types; however, the system is relatively easy to design and troubleshoot. Refer to Figure 11-11.

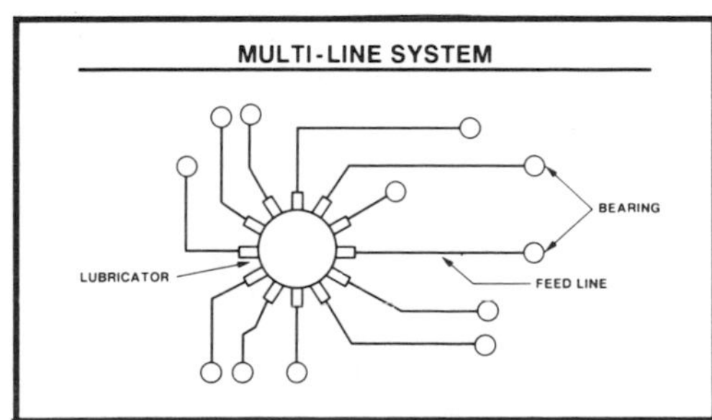

Figure 11-11. Multi-Line System

General Maintenance Guidelines for Centralized Lubrication Systems

Since most centralized lubrication systems today are fully automatic, there is a tendency to take them for granted and forget about them. Certain steps should be followed to keep them running at top efficiency:

ON A DAILY BASIS: Check the system for any changes in normal operating condition—Be on the look-out for abnormal drops or increases in system pressures. Check flexible hoses and look for leaky connections and repair or replace as necessary. Check lubricant levels in system reservoirs—replenish as necessary.

FREQUENTLY: Be on the look out for any changes in automatic system monitors. Also be sure to set up a regular routine timetable to inspect clean and replace as necessary all filters and screens.

BE ON THE LOOK OUT FOR: Any increase or decrease in the amounts of lubricant used. A broken or leaking flexible hose assembly or a crimped or crushed feed line can alter the consumption of lubricant.

BE ON GUARD AGAINST: Wrong lubricator setting. Always try to follow the equipment manufacturer's recommendations. While small quantities of lubricant can starve a bearing with resultant damage. . .keep in mind that too much lubricant delivery can cause severe problems—too much lube delivered to high speed bearings can cause increased internal fluid friction, overheating and premature failure.

Use of the wrong lubricant: using the wrong oil viscosity can cause a bearing to fail prematurely due to breakdown of protective film. Further use of an incorrect compound oil can damage the machine elements or clog the dispensing system devices.

In systems using grease lubricants, ensure that the proper NLGI (National Lubrication Grease Institute) grade is used. Normally, most systems can handle greases up to, and including, NLGI No. 1 grade. Some systems can also use Grade 2 greases. When in doubt, always use a lighter grade of grease.

Refer to Figure 11-12 Preventive Maintenance and Spare Parts Checklist. Following each of the procedures outlined forms the basis of a good preventive maintenance program for Centralized Lubrication Systems.

To keep downtime to a minimum and machines running at peak efficiency, many production oriented manufacturing plants today are increasingly adapting computerized scheduling of lubrication programs. The tabulated listings illustrated will go a long way towards establishing a basis for a program of this nature.

Lubrication of Gears

The American Gear Manufacturers Association (AGMA) has published the following standards for enclosed gear drives: *AGMA Standard Specification—Lubrication of Industrial Enclosed Gear Drives.* The following material has been extracted from AGMA Spec (AGMA 250.04) with the permission of the publisher,

	Grease	Oil
• Clean reservoir periodically (do not use cotton or fiber rags).	☐	☐
• Inspect suction filter and screens; replace filter at least annually and clean screens.		☐
• Remove strainer and clean regularly.	☐	
• Change line filter (pressure filter) annually.		☐
• Check flexible hoses for cracks and wear.	☐	☐
• Inspect tubing/pipe for breaks or flattening.	☐	☐
• Check all connections for leaking or "weeping"; check tightness of connection but avoid over-tightening.	☐	☐
• Monitor system pressure for unusual drops or increases in operating pressure.	☐	☐
• Use only recommended lubricants (be cautious about lubricants with additives that could clog filters or flow apportioning devices).	☐	☐
• Follow recommended lubricant storage and filling procedures to avoid introducing air and contaminants into the system.	☐	☐

Figure 11-12. Preventive Maintenance Checklist for Centralized Lubrication System

The American Gear Manufacturing Association, 1330 Massachusetts Ave., N.W., Washington, D.C. 20005. This standard covers the lubrication of industrial enclosed gear drives having the following types of gearing:

- Helical
- Herringbone
- Straight or Spiral Bevel
- Worm
- Spur

These lubrication recommendations apply only to enclosed gear drives that are designed and rated in accordance with current AGMA standards and are being operated up to 3600 RPM and/or at a pitch line velocity of up to 5000 ft. per min. or both. They are also applicable to enclosed worm gear drives which are being operated at or below 2400 RPM (worm speed). These recommendations may be useful in determining suitable lubricants at speeds greater than those stated above; however, the higher speeds may require additional consideration. For further guidance, refer to manufacturer's recommendations or the appropriate AGMA application standard.

The ambient temperature range is 15°F to +125°F, and is defined as the air temperature in the immediate vicinity of the gear drive. Gear drives exposed to the direct rays of the sun will run hotter and must therefore be given special consideration. Maximum lubricant sump temperatures are defined as follows:

Rust & oxidation inhibited (R & O) gear oils — +200°F (Max)
EP gear lubricants — +160°F to 200°F
Compounded gear lubricants — +160°F to 200°F

Figure 11-13.

NOTE: Extreme pressure lubricants formulated with lead naphthenate are no longer recommended because of limited availability and poor stability in comparison to the lubrication products being marketed at this time.

Gear drives operating outside of these temperature ranges, or those operating in extremely humid, chemical, or dust-laden atmospheres should be referred to the gear drive manufacturer.

Lubricant viscosity recommendations are specified by AGMA Lubricant Numbers. The corresponding viscosity ranges are shown in Figure 11-14. AGMA Lubricant Number recommendations for drives using all types of gearing except worm gearing are given in Figure 11-15. AGMA Lubricant Number recommendations for cylindrical and worm gear drives are given in Figure 11-16.

Rust and Oxidation Inhibited Gear Oils are commonly referred to as R & O gear oils. They are petroleum base liquids which have been formulated to include chemical additives which are more resistant to rust and oxidation than oils without these special features.

Extreme Pressure Lubricants are petroleum base liquids with chemical additives, such as, sulfur phosphorous or similar materials or soluble compounds which produce a protective film to withstand high pressures.

Note that extreme pressure lubricants formulated with lead naphthenate are no longer recommended because the EPA does not accept products containing lead.

Compounded Gear Oils: Compounded gear oils are a blend of a petroleum-based lubricant with 3 to 10 percent fatty or synthetic fatty oils. These lubricants are usually used for worm gear drives.

Synthetic Gear Lubricants: Diesters, polyglycols and synthetic hydrocarbons have been used successfully in enclosed gear drives for special operating conditions.

Rust and Oxidation Inhibited Gear Oils	Viscosity Range[a]	Equivalent ISO Grade[b]	Extreme Pressure Gear Lubricants[c]	Viscosities of Former AGMA System[d]
AGMA Lubricant No.	mm^2/s (cSt) at 40°C		AGMA Lubricant No.	SSU at 100°F
1	41.4 to 50.6	46		193 to 235
2	61.2 to 74.8	68	2 EP	284 to 347
3	90 to 110	100	3 EP	417 to 510
4	135 to 165	150	4 EP	626 to 765
5	198 to 242	220	5 EP	918 to 1122
6	288 to 352	320	6 EP	1335 to 1632
7 Comp[e]	414 to 506	460	7 EP	1919 to 2346
8 Comp[e]	612 to 748	680	8 EP	2837 to 3467
8A Comp[e]	900 to 1100	1000	8A EP	4171 to 5098

NOTE: Viscosity ranges for AGMA lubricant numbers will henceforth be identical to those of ASTM 2422.

[a] "Viscosity System for Industrial Fluid Lubricants, ASTM 2422. Also British Standards Institute, B.S. 4231.

[b] "Industrial Liquid Lubricants—ISO Viscosity Classification." International Standard, ISO 3448.

[c] Extreme pressure lubricants should be used *only* when recommended by the gear drive manufacturer.

[d] AGMA 250.03, May, 1972 and AGMA 251.02, November, 1974.

[e] Oils marked Comp are compounded with 3% to 10% fatty or synthetic fatty oils.

Extracted from AGMA Specifications-Lubrication of Industrial Enclosed Gear Drives. (AGMA 250.04) With the permission of the publisher, the American Gear Manufacturer's Association. Suite 1000, 1901 North Fort Meyer Drive, Arlington, VA 22209.

Figure 11-14. Viscosity Ranges for AGMA Lubricants

Type of Unit[a]	AGMA Lubricant Number[b,c]	
	Ambient Temperature[d,e]	
Low Speed Center Distance	−10°C to +10°C (15°F to 50°F)	10°C to 50°C (50°F to 125°F)
Parallel Shaft, (single reduction)		
Up to 200 mm (to 8 in.)	2-3	3-4
Over 200 mm, to 500 mm (8 to 20 in.)	2-3	4-5
Over 500 mm (over 20 in.)	3-4	4-5
Parallel Shaft, (double reduction)		
Up to 200 mm (to 8 in.)	2-3	3-4
Over 200 mm (over 8 in.)	3-4	4-5
Parallel Shaft, (triple reduction)		
Up to 200 mm (to 8 in.)	2-3	3-4
Over 200 mm, to 500 mm (8 to 20 in.)	3-4	4-5
Over 500 mm (over 20 in.)	4-5	5-6
Planetary Gear Units, (housing diameter)		
Up to 400 mm (to 16 in.) O.D.	2-3	3-4
Over 400 mm (over 16 in.) O.D.	3-4	4-5
Straight or Spiral Bevel Gear Units		
Cone distance to 300 mm (to 12 in.)	2-3	4-5
Cone distance over 300 mm (over 12 in.)	3-4	5-6
Gearmotors and Shaft Mounted Units	2-3	4-5
High Speed Units[f]	1	2

[a] Drives incorporating overrunning clutches as backstopping devices should be referred to the gear drive manufacturer as certain types of lubricants may adversely affect clutch performance.

[b] Ranges are provided to allow for variations in operating conditions such as surface finish, temperature rise, loading, speed, etc.

[c] AGMA viscosity number recommendations listed above refer to R&O gear oils. EP gear lubricants in the corresponding viscosity grades may be substituted where deemed necessary by the gear drive manufacturer.

[d] For ambient temperatures outside the ranges shown, consult the gear manufacturer. Some synthetic oils have been used successfully for high or low temperature applications.

[e] Pour point of lubricant selected should be at least 5°C (9°F) lower than the expected minimum ambient starting temperature. If the ambient starting temperature approaches lubricant pour point, oil sump heaters may be required to facilitate starting and insure proper lubrication.

[f] High speed units are those operating at speeds above 3600 rpm or pitch line velocities above 25 m/s (5000 fpm) or both. Refer to Standard AGMA 421, "Practice for High Speed Helical and Herringbone Gear Units," for detailed lubrication recommendations

Extracted from AGMA Specifications-Lubrication of Industrial Enclosed Gear Drives. (AGMA 250.04) With the permission of the publisher, the American Gear Manufacturer's Association. Suite 1000, 1901 North Fort Meyer Drive, Arlington, VA 22209.

Figure 11-15. AGMA Lubricant Number Recommendations for Enclosed Helical, Herringbone, Straight Bevel, Spiral Bevel, and Spur Gear Drives

Lubricant Maintenance

Lubricant change intervals. The lubricant in a new gear drive should be drained after four weeks' operation. The gear case should be thoroughly cleaned with a flushing oil. The original lubricant can be used for refilling if it has been filtered through a filter of 100 microns or less (50 microns or less for high speed gear units using sleeve bearings); otherwise, new lubricant must be used. Lubricants should not be filtered through Fullers Earth or other types of filters that remove lubricant additives.

Under normal operating conditions, the lubricant should be changed every 2,500 hours of operation or every six months, whichever comes first. Extended change periods may be established through periodic testing of oils.

A rapid rise and fall in temperature may produce condensation, resulting in the formation of sludge. Dust, dirt, and chemical fumes also react with the lubricant. Sump temperatures in excess of those listed in Figure 11-13 will result in accelerated degradation of the lubricant. Under these conditions, the lubricant should be changed every one to three months, depending on severity.

Cleaning and flushing. The lubricant should be drained while the gear drive is at operating temperature. The drive should be cleaned with a flushing oil. Used lubricant and flushing oil should be removed completely from the system to avoid contaminating the new charge.

Avoid the use of a solvent unless the gear drive contains deposits of oxidized or contaminated lubricant that cannot be removed with a flushing oil. When persistent deposits necessitate the use of a solvent, a flushing oil must then be used to remove all traces of solvent from the system.

The interior surface should be inspected where possible, and all traces of foreign material removed. The new charge of lubricant should be added and circulated to coat all internal parts.

*Lubricant Storage and Handling**

Lubrication storage and dispensing is almost as important as lubrication itself. Surveys conducted by a leading oil and grease manufacturer have shown some alarming statistics:

- For every gallon of lubricant used the average industrial user maintains an inventory of five gallons.
- The average industrial/institutional user spends $50.00 in material handling labor just getting the product from the receiving point to the point of use.
- One of the three most common lubrication related problems is the misuse of mis-identified lubricants.

---

*This section was prepared by Jim King, Jr, of Salk Valley, Rock Falls IL.

Type, Worm Gear Drive	Worm Speed[c] Up to (rpm)	AGMA Lubricant Number[a] Ambient Temperature[b]		Worm Speed[c] Above (rpm)	AGMA Lubricant Numbers[a] Ambient Temperature[b]	
		−10°C to +10°C (15° to 50°F)	10°C to 50°C (50° to 125°F)		−10°C to +10°C (15° to 50°F)	10°C to 50°C (50° to 125°F)
Cylindrical Worm[d]						
Up to 150 mm (to 6 in.)	700	7 Comp, 7 EP	8 Comp, 8 EP	700	7 Comp, 7 EP	8 Comp, 8 EP
Over 150 mm. to 300 mm (6 to 12 in.)	450	7 Comp, 7 EP	8 Comp, 8 EP	450	7 Comp, 7 EP	7 Comp, 7 EP
Over 300 mm. to 450 mm (12 to 18 in.)	300	7 Comp, 7 EP	8 Comp, 8 EP	300	7 Comp, 7 EP	7 Comp, 7 EP
Over 450 mm. to 600 mm (18 to 24 in.)	250	7 Comp, 7 EP	8 Comp, 8 EP	250	7 Comp, 7 EP	7 Comp, 7 EP
Over 600 mm (over 24 in.)	200	7 Comp, 7 EP	8 Comp, 8 EP	200	7 Comp, 7 EP	7 Comp, 7 EP
Double-Enveloping Worm[d]						
Up to 150 mm (to 6 in.)	700	8 Comp	8A Comp	700	8 Comp	8 Comp
Over 150 mm. to 300 mm (6 to 12 in.)	450	8 Comp	8A Comp	450	8 Comp	8 Comp

Over 300 mm. to 450 mm (12 to 18 in.)	300	8 Comp	8A Comp	300	8 Comp	8 Comp
Over 450 mm. to 600 mm (18 to 24 in.)	250	8 Comp	8A Comp	250	8 Comp	8 Comp
Over 600 mm (over 24 in.)	200	8 Comp	8A Comp	200	8 Comp	8 Comp

[a] Both EP and compounded oils are considered suitable for cylindrical worm gear service. Equivalent grades of both are listed in the table. Four double-enveloping worm gearing. EP oils in the corresponding viscosity grades may be substituted only where deemed necessary by the worm gear manufacturer.

[b] Pour point of the oil used should be less than the minimum ambient temperature expected. Consult gear manufacturer on lube recommendations for ambient temperatures below −10°C (14°F).

[c] Worm gears of either type operating at speeds above 2400 rpm or 10 m/s (200 fpm) rubbing speed may require force feed lubrication. In general, a lubricant of lower viscosity than recommended in the above table shall be used with a force feed system.

[d] Worm gear drives may also operate satisfactorily using other types of oils. Such oils should be used, however, only upon approval by the manufacturer.

Extracted from AGMA Specifications-Lubrication of Industrial Enclosed Gear Drives. (AGMA 250.04) With the permission of the publisher, the American Gear Manufacturer's Association. Suite 1000, 1901 North Fort Meyer Drive, Arlington, VA 22209.

Figure 11-16. AGMA Lubricant Number Recommendations for Enclosed Cylindrical and Double-Enveloping Worm Gear Drives

- The average automotive plant has per square foot output of $950.00. It also estimates that the equivalent of $400,000.00 worth of productive space is assigned lubricant container storage. When setting up movement of lubricants, consider these guidelines:
- How much product will be required per shift —This is particularly important when setting up refilling stations in your plant. You should also determine how large the containers and how many containers there should be for moving the products throughout your plant.
- Plant layout —How you will get the lubricants from the storage to the equipment. This will determine the type of transportation vehicle needed. It may be a two wheeled cart or a push cart with small containers on it. It may be a cart that is self contained or a large truck with lube equipment mounted on it.
- Will it be necessary for product to be transferred into areas that are hard to get to —Will the amount of product need to be measured in any way. Answers to these questions will help you determine if you need equipment such as hose reels, metered controls, special piping to other locations to transfer product through, or special pumping equipment for your product transfer.
- What type of power is most convenient in your plant, air, electricity, hydraulic? You must determine power supply to help determine transfer pumps if used.
- Who will do the lubrication and how often?
- Will the product need to be filtered, or strained?
- What other equipment will be needed to complete the lubrication process? Equipment such as tool boxes, storage cabinets, or absorbent material containers.
- When the project is complete, will the equipment grow with your needs, can it be modified as lubrication requirements change and have you met all the necessary safety controls both enforced by regulatory agencies and do they meet your companies safety standards.

Always enlist the help of experts. They will save you time, frustration and money. The job is not worth doing if it's not going to be done right.

Cold Temperature Starting

Low temperature gear oils. Gear drives operating in cold areas must be provided with oil that circulates freely and does not cause high starting torques. An acceptable low-temperature gear oil, in addition to meeting AGMA specifications, must have a low pour point and a viscosity low enough to allow the oil to flow freely at the startup temperature but high enough to carry the load at the operating temperature.

Sump heaters. If a suitable low-temperature gear oil is not available, the gear drive must be provided with a sump heater to bring the oil up to a temperature at which it will circulate freely for starting. The heater should be so designed as to avoid excessive localized heating, which could result in rapid degradation of the lubricant.

TYPES OF EQUIPMENT AND LUBRICANTS TO BE USED

Various pieces of equipment likely to be found in a typical operation should be considered. The parts to be lubricated and the lubricants to use are described in the following sections.

Lubricating Electric Motors

Proper lubrication is most important because most motors employ precision-built, anti-friction bearings and contain insulation on the windings, both of which can be damaged. Therefore, to keep the electric motor operating properly, the bearing must be lubricated properly. The lubricant must protect the bearing and, therefore, must stay in the bearing. If it leaks out onto the windings, it can interfere with cooling air flow, causing overheating and possible short circuiting or fire.

It is important that motor bearing housings should be sealed against dust, dirt, and grit. Because of this, the sealed-type, prelubricated ball bearing is frequently used, particularly in adverse environments, where speed control and low power consumption are important. Recommendations are typical but will vary for some applications according to temperature and speed range.

Bearings—Oil Lubricated (speeds up to 3,600 rpm)

Ambient Temperature (−°F)	Viscosity (SUS at 100°F)
Below 32	150 to 225 (pour −15°F)
32-200	300

Grease Lubricated (speeds up to 1800 rpm)

Ambient Temperature (−°F)	Consistency (N.L.G.I. No.)
−10 to 32	0
32 to 200	2

Lubricating Air Compressors

Compressors are divided into three basic types: reciprocating, screw, and rotary vane. In all instances, the oxidizing effect of air and high discharge temperatures tend to cause deterioration of the cylinder lubricating oil. Overapplication promotes the formation of carbonaceous deposits in the cylinder head, piston rings, and winds up, as well, out on the discharge valves, in the valve pockets, and on the valve springs.

Air compressor cylinder lubrication requires an oil of high chemical stability, high viscosity index, low carbon content, high flash point, and high thermal stability. To achieve the foregoing, a highly refined petroleum or synthetic oil containing corrosion and rust inhibitors is required.

Reciprocating compressors handling dry air, inert gases, nitrogen, CO_2, and so on., (not oxygen), up to 150 psi discharge pressure may be lubricated adequately with a nondetergent oil having a viscosity of 300 SUS at 100°F. For medium range pressures (up to 2,000 psi) use an oil of 600 SUS at 100°F; between 2,000 and 7,000 psi (cryogenic ranges) a 1,200 to 1,500 SUS at 100°F is required. Lubrication of crankcase components of low-pressure compressors (up to 150 psi) use the same oil as the cylinders. For higher pressure units, use a 600 SUS at 100°F oil regardless of application.

Screw compressors usually require a lower viscosity lubricant than a reciprocating compressor. A 150-200 SUS at 100°F oil is suggested here.

Rotary compressors require lubrication of both bearings and rotor cylinder. For most services, including refrigeration, a 300 SUS at 100°F nondetergent oil is recommended.

Lubricating Refrigeration Compressors

Proper lubrication for the moving parts of the refrigeration compressor is affected by a number of problems that relate to the thermal process of the refrigeration system. The piping and equipment beyond the compressor are involved, in addition to the internals of the compressor itself. Many problems are caused by improper oil selection. The interreaction of some refrigerants with oils, and/or incorrect piping designs may cause a number of problems, such as:

- *Loss of volumetric efficiency.* Vaporized oil mixed with the refrigerant reduces the rate of flow of refrigerant handled by the compressor.
- *Heat transfer losses.* A film of oil, or any of its constituents, including wax deposited on heat transfer surfaces, reduces the heat transfer efficiency of evaporators, condensers, and other heat exchangers.
- *Deterioration of dryers.* Desiccants lose their ability to absorb moisture when pores become oil-clogged.
- *Expansion valve clogging.* Poor quality oils containing waxes can cause trouble when exposed to solvent-type refrigerants. Wax deposits on expansion valves and other internal moving parts will cause them to malfunction.
- *Line clogging.* When refrigerant flows too slowly oil drops out in the refrigerant lines and drains into the lower parts of the system, interfering with refrigerant flow.
- *Foaming.* Foaming increases oil carryover to the discharge side of the system and reduces the lubricating quality of the oil.
- *Carbon deposits.* Carbonaceous deposits form on discharge valves, cylinder heads, pistons and in discharge lines, and are caused by oil deterioration under high temperature. This condition can cause serious damage.
- *Cylinder blow by.* Loss of compression results from the escape of gas around the piston back to the crankcase. This can be caused by insufficient or improper lubricant. In consideration of the above, oils selected for refrigeration compressor applications must possess the following properties:

Chemical stability	High dielectric strength
Low pour point	Proper viscosity
Low floc point	Low volatility

Lubricating Air Handling Units

Fans

Propeller and disk fans are sometimes mounted directly on the motor shaft, in which case only the electric motor bearings need be considered for lubrication. The motors of small fans are often factory lubricated for the life of the rolling bearings. An oil of 150 to 300 SUS at 100°F is recommended for oil-lubricated bearings applied by inverted wick, ring, or flooded periodically by oil can. A no. 2 NLGI lithium or calcium complex grease is recommended for temperatures between 40° and 200°F and speeds ranging from 600 to 3,600 rpm. For lower temperatures, use a No. 1 NLGI grease of similar soap bases and a high temperature grease for temperatures above 220°F.

Centrifugal fans, used primarily for ventilation, are considered light-duty types and are mounted on their own bearings which, are usually ring oil sleeve type. A 300 SUS at 100°F oil is recommended for ambient temperatures of 40° to 100°F, a 600 SUS at 100°F oil is recommended for ambient temperatures above 100°F. Centrifugal fans of the heavy-duty types, including forced and induced draft fans and large evaporative condenser fans, may be equipped with either sleeve or rolling bearings.

Basic lubrication requirements are as follows:

Bearing Type and Condition	Oil Viscosity (SUS) at 100°F
Sleeve bearings, ring-oiled, water-cooled temperatures 32-90°F	300-400
100 to 150°F ambient	400-600
Sleeve bearings, ring-oiled, not water-cooled*	600
Sleeve bearings, oil circulation with oil cooler	200-300
Roller bearings** oil bath:	
40 to 140°F ambient	300
140 to 220°F ambient	900

*Fans handling hot gases up to 150°F, such as induced-draft fans, may require an oil from 900 to 1,500 SUS at 100°F. This type of application invites mineral oil deterioration because of oxidation. Specially formulated synthetic oils should be considered for this type of duty.

**Fans handling hot gases or subject to temperature above 150°F require a heavier oil of 1,500 to 1,800 SUS at 100°F, preferably synthetic.

Blowers

The oil viscosity for blower bearings and geared transmissions depends on the horsepower rating of the blower. For 25 horsepower or less, use a 600 SUS at 100°F oil; for over 25 horsepower use a 900 SUS at 100°F oil. Turbo-blowers, single- or multi-stage, are constant pressure machine driven by an electric motor or steam turbine. Use a circulating oil of 150 SUS at 100°F for bearing speeds of 1,800 rpm and up. For speeds lower than 1,800 rpm, ring-oiled bearings use an oil of 300 to 400 SUS at 100°F.

Lubricating Pumps

Liquid pumps of the horizontal reciprocating types require lubrication as follows:

Crossband, crank, gears, and bearings	300 SUS 100°F
Vertical centrifugal pumps	
upper bearing	300 SUS 100°F
lower bearing	Grease #2 lime or calcium complex
Horizontal centrifugal pumps	Grease #2 calcium complex

Lubricating Couplings

Rigid and flexible couplings usually do not require lubrication. Flexible couplings with articulated joints attain flexibility by mechanical means and do require lubrication. Factors of temperature, angular velocity load transmitted, radial positions, and location of frictional parts, effectiveness of seals, etc., will affect the selection of a lubricant. The use of a lubricant specified by the coupling manufacturer is recommended. If the application involves unusual operating conditions (heavy shock loads, frequent axial movement, large speed variations, or extreme temperatures), such data should be submitted to the manufacturer when requesting lubricant recommendations.

The following are general suggestions for coupling lubrication. Cold operation in low temperature climates, cold storage rooms, high altitudes, etc., requires an oil with a pour point below 0°F and viscosity less than 220 SUS at 100°F. Intermediate temperature operation between 0 and 40°F requires a 225 to 250 SUS at 100°F lubricant. Low pour point leaded compounds of appropriate viscosity are also recommended. Temperatures between 40 and 120°F require an oil of 600 to 2,500 SUS at 100°F or a grease. A lead lime grease No. 0 or No. 1 NLGI consistency or a No. 2 aluminum soap-base grease is preferred. A leaded compound having a viscosity of about 400 SUS at 210°F is also highly satisfactory.

Lubricating Chain Drives

Chain drives should be lubricated such that the rollers are free-turning at all times. When the roller reaches the sprocket, it should not turn relative to the sprocket; all motion should be inside the roller. Frequently, a grease or oil applied to a chain will attract and hold airborne foreign matter that will gum up, so that these rollers do not turn freely. This causes excessive wear to the sprockets and the chain. A penetrating-type oil with a dry lubricant such as molybdenum disulfide, graphite, or PTFE suspended in it will lubricate the chain internally, thus keeping the rollers free-turning without attracting foreign matter to the chain. The vehicle used should contain rust and oxidation inhibitors to leave a protective coating on all metal surfaces.

Lubricating Hydraulic Systems

The majority of industrial hydraulic systems, including elevators, perform satisfactorily with oils having viscosities in the range of 150 to 325 SUS at 100°F for operating temperatures up to 165°F. In most cases a 10W40 grade motor oil (API service classification SE/CC) will operate in the wide temperature range that elevator hydraulic systems require for all seasons. Follow the equipment manufacturer's recommendations.

Other systems require the following:

Vane pumps	150-300 SUS at 100°F
Angle and radial piston pumps	150-900 SUS at 100°F
Axial piston pumps	150-300 SUS at 100°F
High pressure, high output pumps	300 SUS at 100°F
Gear pumps	300-600 SUS at 100°F

Lubricating Pneumatic Tools

Oils that are especially compounded to lubricate pneumatic tools and machine power cylinders are called air-line oils. They are similar to turbine-type oils, containing a small amount of emulsifying agent to combat the washing effect of condensate that precipitates from compressed air upon expansion. Viscosity ranges from about 200 SUS at 100°F for small tools to about 800 SUS at 100°F for larger cylinder wall areas.

Lubricating Laundry Equipment

Laundries contain various types of equipment that require special lubrication. Washing machine and water extractors, where lubricated, should use a waterproof grease, preferably made with an aluminum complex soap base. Mangles and/or large industrial ironing machines require lubrication frequently. They have rolls at each end with pillow block bearings over which steam pours. These bearings should also use a waterproof grease with good heat resistance.

Lubricating Kitchen Equipment

Here the primary concern is to use U.S.D.A. H-1 approved lubricants where possible contact with food might occur. Commercial kitchens, such as in hospitals or schools, frequently have conveyors that have the usual complement of motors, gear boxes, bearings, plus chains of various designs and configurations. Care should be taken to use waterproof lubricants in areas where steam or water exist, such as commercial dishwashers.

PLANNING THE PREVENTATIVE MAINTENANCE PROGRAM

You should have an accurate inventory and record of all plant equipment. All equipment to be lubricated should be listed. Equipment location should be identified by department and/or function: i.e., power house, machine shop, material handling equipment, HVAC, etc. Work center numbers can be assigned to these areas. Numbers can be assigned to the individual pieces of equipment within a given work center. Depending on the number of work centers and machines, these can be combined to make one number where the first or first two digits are the work center and the last digits are the individual piece of equipment in that work center. If convenient, a dash can be placed between the work center number and the machine number to improve legibility.

Starting with the power source (usually a motor) and working outward, each component of the equipment can be listed and the parts to be lubricated noted. It is desirable, but not always necessary, to indicate the number of points, such as oil holes, grease fittings, pillow blocks, and couplings. The correct lubricant to be used and the preliminary cycle for lubrication should also be noted. Provision should be made to record the results of oil analysis by a qualified laboratory.

Schedule a period of time in which to evaluate and update the survey. Card systems (see Figure 11-17) or computer printouts should be implemented so that the person performing the actual lubrication can periodically be handed a series of cards or a printout sheet listing his performance requirements for that day/week. With a card system, there can be a form on the back of the instruction card for date and initialing to verify the actual lubrication or inspection performed.

Space could be provided on the card to note service performed because of failures or overheating or during periodic servicing of the machine or component. This becomes a handy log on that component for trend analysis and program update.

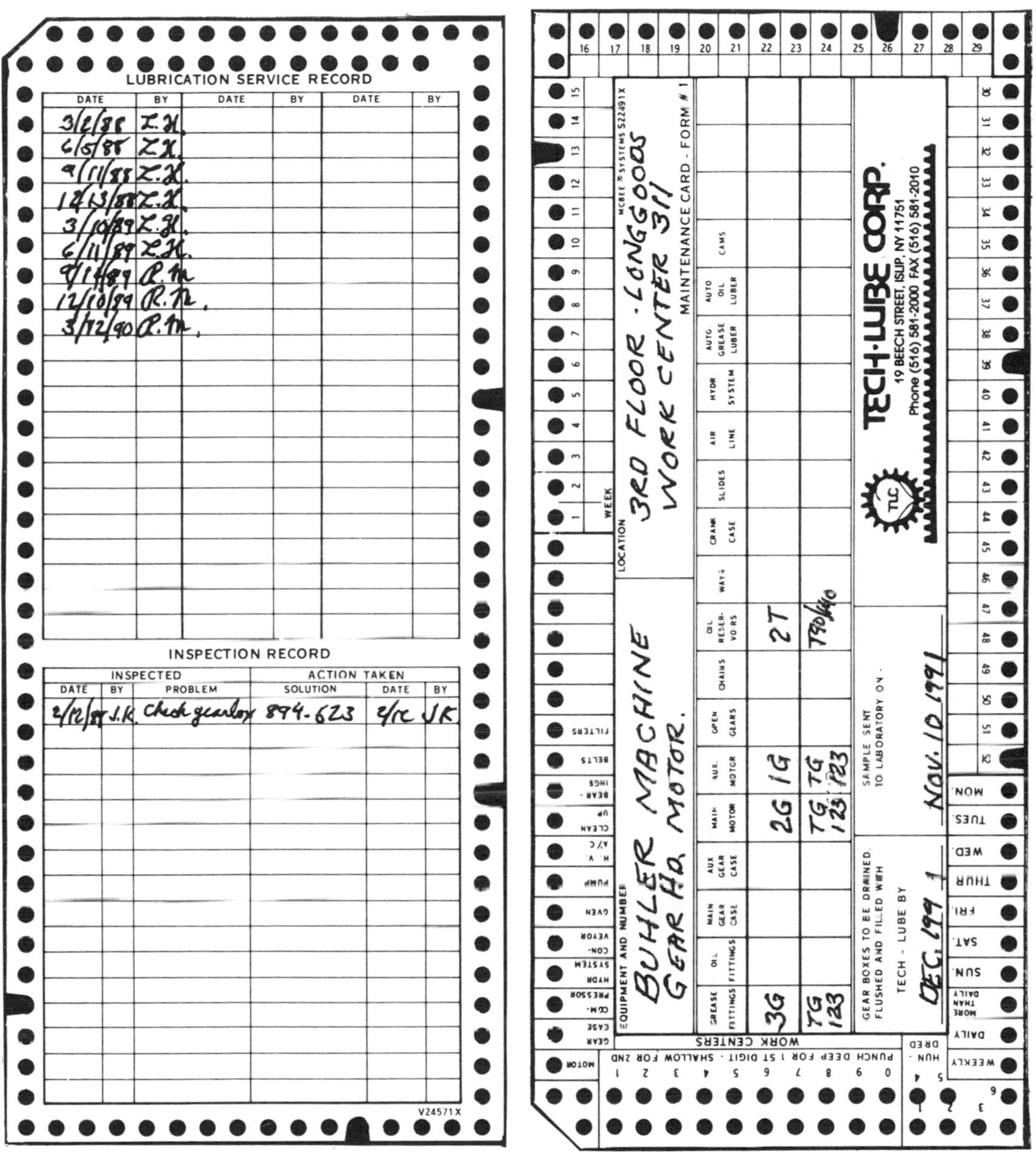

Figure 11-17. Sample Keysort Lubrication Card

Once the effectiveness of the program has been established, the lubricants in the storeroom, the transfer equipment to transport the lubricant from the storeroom to the machinery, and the actual lubrication point on the equipment should be color-coded to simplify the oiler's job and to reduce the possibility of using an incorrect lubricant.

Data sheets, and MSDS on all lubricants in the plant should be maintained in a central location.

TRAINING MAINTENANCE PERSONNEL IN LUBRICATION PROCEDURES

Mechanics should be aware of techniques of lubricant application; operating principles and maintenance of application devices; the importance of maintaining proper oil levels and lubricant feed rates; and the limitation of grease application to rolling bearings. They should also have a basic knowledge of lubricants, and the uses and necessity of applying each lubricant to its proper frictional component.

An initial period of on-the-job training with an experienced mechanic, followed by attendance at outside clinics or seminars, is an effective method of training new lubricating mechanics. Clinics and seminars are held by representatives of lubricant suppliers and equipment manufacturers, such as Tech-Lube Corp., Islip New York and Imperial Oil and Grease Company, A Division of ICI TRIBOL, Woodland Hill, CAL., and others. Such clinics are usually divided into a number of weekly sessions of approximately one hour each to provide a progressive training program and can be augmented by associated reading matter.

SCHEDULING LUBRICATION PROCEDURES

Lubricants are consumed or deteriorate from use. Consumption is caused by leakage (which can also have its own serious harmful side effects), evaporation, blow by and burning. Deterioration is caused primarily by heat, which with the catalysts of water and air, promotes oxidation and forms acids.

All-loss lubrication systems and defective piping, joints, and seals of circulation systems represent the areas of highest lubricant consumption. Be sure equipment is properly vented and sufficiently cooled. In the case of all-loss oil and grease application systems, the rate of lubricant consumption under normal operating conditions is fairly regular. On the basis of probability, such equipment will require lubrication at predictable intervals, whether in terms of hours, days, or months, depending on the equipment and its application.

Similarly, the life of a lubricant in a *reuse system* is a function of operating conditions. Where such conditions remain relatively constant, the lubricant cycles of that system are predictable within safe limits. Again, it requires experience, guided by periodic laboratory analysis, to determine the probable life span of a given system.

The time between introduction of an oil to a given system and the point when it reaches a specific level of deterioration, as determined by laboratory analysis, establishes the frequency of drain periods for that system and for that particular oil. Oil change can be scheduled correctly only after the interval is determined by competent chemical analysis. Ball and roller bearings of electric motors and machine spindles are often overlubricated. The trend, as previously noted, is to prepack and adequately seal such bearings, but they should be evaluated periodically by monitoring their operating temperatures. Mechanics should become familiar with all the lubricants used and where each is applied. Such information is found on the lubrication charts drawn up as a result of the plant-wide analysis previously mentioned.

These charts should be available to the oiler mechanic and updated periodically by his findings and recommendations.

HOW TO AVOID 7 COMMON MISTAKES IN LUBRICATION PROGRAMS

1. *Use of the wrong lubricant*—Do not use a common motor oil in a compressor; a high-temperature grease in a wet environment (such as a water pump); or a lightweight oil in a heavy-duty, slow-moving gear box. Care must be taken to follow proper directions.

2. *Improper lube frequencies*—The lube frequencies previously suggested are for guidance only. The final schedule must be determined from actual operational experience and chemical analysis. Adhere to the final schedule as rigidly as possible.

3. *Overlubing and underlubing*—It is very important to understand that overlubrication can be as bad and sometimes even worse than underlubrication. If a certain amount of lubrication is good, it doesn't necessarily follow that more is better.

4. *Proper inspection*—The system oil levels in gear boxes should be checked regularly, bearings should be checked for overheating, and lines should be checked for breakage. It is important that the program be reviewed and monitored by a qualified individual in order to ensure success.

5. *Mixing lubricants*—It is inadvisable to mix lubricants. Lubricants should be checked thoroughly before mixing to confirm compatibility, viscosity, performance, etc. Lubricant suppliers can frequently be helpful in this connection, but an independent chemical oil analysis should be performed when in doubt.

6. *Looking at price, not cost*—Real cost cannot be measured by price alone. Bottom-line cost at the end of the year is the criterion by which the purchase price of a lubricant should be evaluated.

7. *Using a solvent to clean out a gear box*—This should only be done when no other means works successfully, and if a solvent is used, it must be flushed out thoroughly with a good flushing oil.

SUMMARY

A PM program using adequate inspection, proper lubricants in the correct place, and quantities at the right time can save the organization many dollars over a time span. It can make the oiler's job more routine and reduce the possibilities of human error. This, with better performance, will reduce energy consumption, downtime, replacement parts cost, and prolong the smooth operation of the equipment.

12

Effective Maintenance of Filtration Systems

Clyde H. Gordon

The filter is the first line of defense in protecting heat transfer and building surfaces. This chapter describes the types of air filters commonly encountered in heating, ventilating, and air conditioning systems; how they are classified; what their purpose is intended to be; the limitations of each; and reasons for selection of each type. Instructions are provided for servicing each type of filter. How to maximize economy and utility in filter systems is discussed in detail for each element of cost, with common-sense rules for maximizing overall system life and economy in filtration systems.

Case histories are presented for upgrading filter systems to show the economics of using modern technology. Charts of standard times for testing pressure resistance across filter banks and for changing various types of filter media are included. Finally, this chapter will explain what to do when filtration has not been maintained effectively, including remedial cleaning of hot deck and cold deck coils, plenum chambers, duct work, and diffusers.

COMMON FILTER TYPE CLASSIFICATIONS

Filters in common use may be classified in several ways: by efficiency, by method, by use, or by construction.

Filtration Efficiency

- Low = 20 percent or less dust spot efficiency (The American Society of Heating, Refrigerating, and Air Conditioning Engineers Standard 52-76)
- Medium = More than 20 percent, less than 90 percent dust spot efficiency
- High = More than 90 percent dust spot efficiency, or more than or equal to 95 percent DOP Test (0.3 Micron Smoke)

Filtration Method

- Inertial Impingement: Particles are trapped on media fibers by being impinged by the force created by their weight and high velocity. An adhesive coating holds the accumulated dust in place.
- Interception: Particles too small and too light in weight to be impinged are removed most economically by interception, which occurs when their path is altered by air molecules after their velocity is slowed by passing through media. Filter media used in interception is pleated and usually of finer fiber than inertial impingement media.
- Electronic Agglomeration or Electrostatic Precipitation: This filter works by electronically charging dust particles so that they will collect on oppositely charged plates. When sufficient small particles bind together or agglomerate, they break away from the plates and are collected downstream, usually on inertial impingement or impingement-interception type media. Older designs require washing in place of collector cells to remove dust and oiling of plates for reconditioning.

Usage

- Prefilters: Protect both coils and final filtration systems by removing larger dust particles and contaminants upstream, usually right near outside air dampers. Prefilters are usually a panel-type using inertial impingement method.
- General Ventilation and Air Conditioning: Includes low, medium, and high efficiency ranges, dry, viscous-treated, washable, and disposable media types, in a variety of frames, construction, and arrangements. Economy in first cost and life cycle is the usual criterion for selection.
- Downstream High Efficiency Particulate Air Filter Beds: Designed to meet requirements of Department of Health, Education, and Welfare Resources Publication No. 76-4000 for Hospital and Medical Facilities. Efficiency of 90 percent minimum is required for most installations, to protect the environment of patient care, treatment, diagnostic and related areas, and sensitive

areas such as operating rooms, delivery rooms, recovery rooms, and intensive care units.

- Industrial (nonatmospheric dust): Includes special media and filter arrangements to remove lint, press ink mist, and other nonatmospheric contaminants.

Construction

- Panel Media with Coarse Fibers: Usually called throwaway filters. This construction is the common type found in prefilters. The medium most common in disposable panel filters is glass fiber, although polyester fiber is also used. The fiber is treated with an oil or adhesive spray to create a viscous impingement medium. This is usually found in an efficiency range of 20 percent or less (ASHRAE average with atmospheric dust). A variation of this type of construction is the renewable pad of glass fiber in a permanent metal and wire frame. When the pad collects its full dust load, it is discarded and replaced with another inexpensive pad.
- Pleated Media with Less Coarse Fiber, Few Pleats: There are typically two distinct layers of media material having different fiber sizes and packing densities. The less dense, coarse fibered upstream layer removes large, heavy particles, while the finer fibered, more dense downstream layer may also be treated with a special adhesive on the air-leaving side to prevent blowoff of collected dust. The pleats are generally supported and held in place by wire retainers or welded-wire fabric. Efficiency will range from 25 to 40 percent, according to various manufacturers.
- Pleated Media with Fine Fibers and Precisely Spaced Pleats: Efficiencies from 50 to 95 percent are available in this type of construction, depending upon the media fiber size, thickness, and dispersion. Capacities for operation in cfm vary with filter media area (net effective) provided. Although generally found in supported media construction, there are nonsupported pleated media filters in this range of efficiencies. Nonsupported media should never be used in variable volume air systems nor in areas of critical care where media blowout could cause contamination.
- HEPA-Type Finely Fibered Media: High efficiency particulate air filters were developed originally for clean rooms and gauge and metrology labs used in military and aerospace hardware production, which explains why the military standard testing procedure (M16-STD-282) is used for this type of filter. HEPA filters are now used in pharmaceutical manufacturing, food and beverage processing, surgical and other hospital applications, electronics and aerospace assembly, photosensitive film production, nuclear uses, and many other areas. The distinguishing construction features are high ratio of filter area to face area, and fine fibers with tightly controlled spacing, closely pleated. Performance characteristics typically include an efficiency of 99.97 percent on

0.3 micron size particles at an initial resistance of 1.0″/w.g. and final resistance of 2.0″ to 3.0″, depending upon construction.

- Roll Media: Although typically furnished as a prefilter or upstream filter for a higher efficiency filter system, the roll media filter may be used where 25 percent average efficiency will suffice. The usual roll media installation is an automatically advanced roll of adhesive-coated glass fiber filter media that is fed into the air stream and rerolled after it has collected its dust load. The roll may be advanced by a signal from a timer (at a predetermined rate), or better by an inclined draft gauge to advance the roll only when the design final pressure drop is reached. Manual crank models are also available.
- Side Access Housing: Packaged, side access housings are available for combinations of almost all types of prefilter and final filter construction. The advantages include ease of access for servicing, installation where limited headroom is a factor, and ability to service from outside the duct, usually from either side.
- Activated Carbon Filtration: Activated carbon will absorb up to 50 percent of its weight with odors and retain them in the network of tiny pores within the body of the carbon. One pound of activated carbon (approximately 50 cubic inches) contains an estimated six million square feet of surface area. Activated carbon filtration use is limited to odor control in air conditioning systems, although its widest usage is in the chemical processing industry where activated carbon is used to purify gases, to separate gases, to recovery solvents, and to carry catalysts. Activated carbon filters in environmental control systems are usually constructed of trays or panels of activated carbon and placed in the final filtration air stream in a housing designed either for front or side access to the trays or panels.
- Electronic Filters: Two types are generally found in HVAC systems: (1) the agglomerator-type with disposable collection media, or (2) the precipitator/collector-type, which requires washing and renewal of collector plates, automatically or manually. The principle of electrostatic precipitation is the same—dust and smoke particles are given positive or negative static charges by the electrostatic field set up by the charged ionizing wires and the grounded struts. Charged particles then enter the collecting section, which is made up of alternately positive charged and grounded plates. The charged particles are attracted to and held by the oppositely charged plates. In the agglomerator-type with disposable collection media, particles build up on the plates until they break off in larger chunks called agglomerates and flow downstream to be captured by the final collecting (disposable) filter. In the precipitator/collector-type, which requires washing and renewal, an automatic or manual wash cycle is required to remove the collected particles to prevent their unloading and passing downstream.
- Automatic Vacuum Renewable: This system is not designed for ordinary HVAC installations but for specific industrial situations, primarily laundries

and textile mills, where a stable interior environment is essential and large quantities of conditioned air must be recirculated. The system consists of a lint filter that traps airborne particles. An automatic vacuum or pneumatic sweep removes the particles through a manifold to a secondary collection point for recovery or disposal.

UNDERSTANDING EFFICIENCY TERMINOLOGY

Three performance test procedures for general HVAC air cleaning devices are in current use:

- Atmosphere Dust Spot Efficiency
- Synthetic Dust Weight Arrestance
- DOP Smoke Penetration Method

The first two procedures are covered by ASHRAE standard 52-76 (which supersedes 52-68). The third is prescribed by a U.S. Government standard, MIL-STD-282. Much past confusion has been eliminated by standardizing terminology. In the ASHRAE standard, the term *efficiency* now applies only to tests made by the dust spot procedure on atmospheric air and its contaminants. Results of tests measuring *weight* of an injected synthetic dust removed by the air cleaning device are now reported as *arrestance.* The DOP Smoke Penetration method, which reports results in *count percent,* is designed to distinguish between filters whose air cleaning efficiency exceeds 98 percent.

Of primary importance in building maintenance is the ability to reduce staining by trapping small particles found in atmospheric dust and contaminants. This is the performance measure meant by *efficiency* in the dust spot procedure and is the rating commonly spoken of in classifying filters. High Efficiency Particulate Air Filters (HEPA) are found in industrial "clean rooms," operating theaters, and other areas of hospitals, as final filter beds to meet critical requirements.

The relationships among particle size, efficiency requirements, filter design, materials, and construction are illustrated in Figure 12-1.

SERVICE PROCEDURES FOR AIR FILTRATION SYSTEMS

The general rule for servicing environmental air filters among too many maintenance people has been:

Look at it occasionally.

It if looks too dirty, clean it or change it.

Unfortunately, this approach ignores all principles of engineering economy. Let's look first at some guidelines for maintenance that apply to *all* filter systems, then to specific procedures for each type.

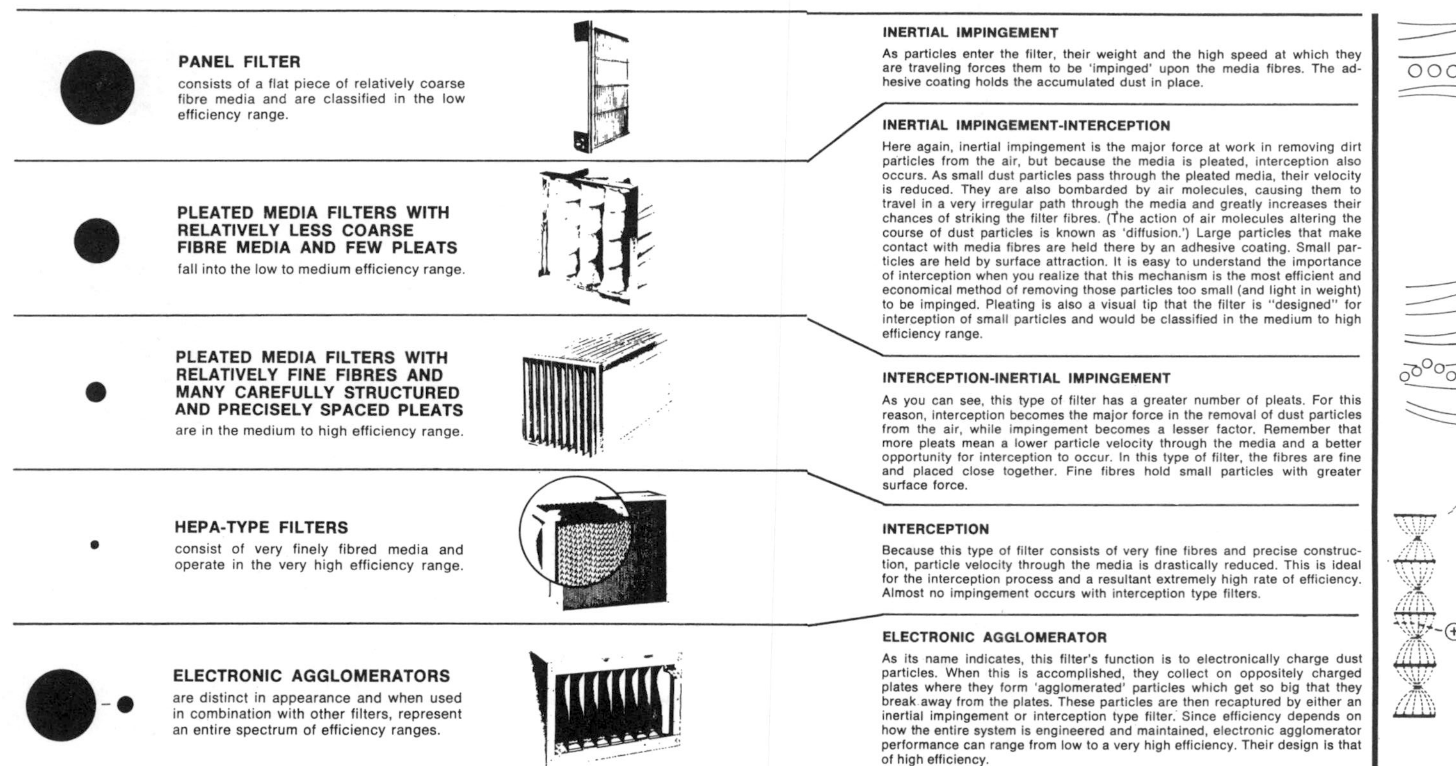

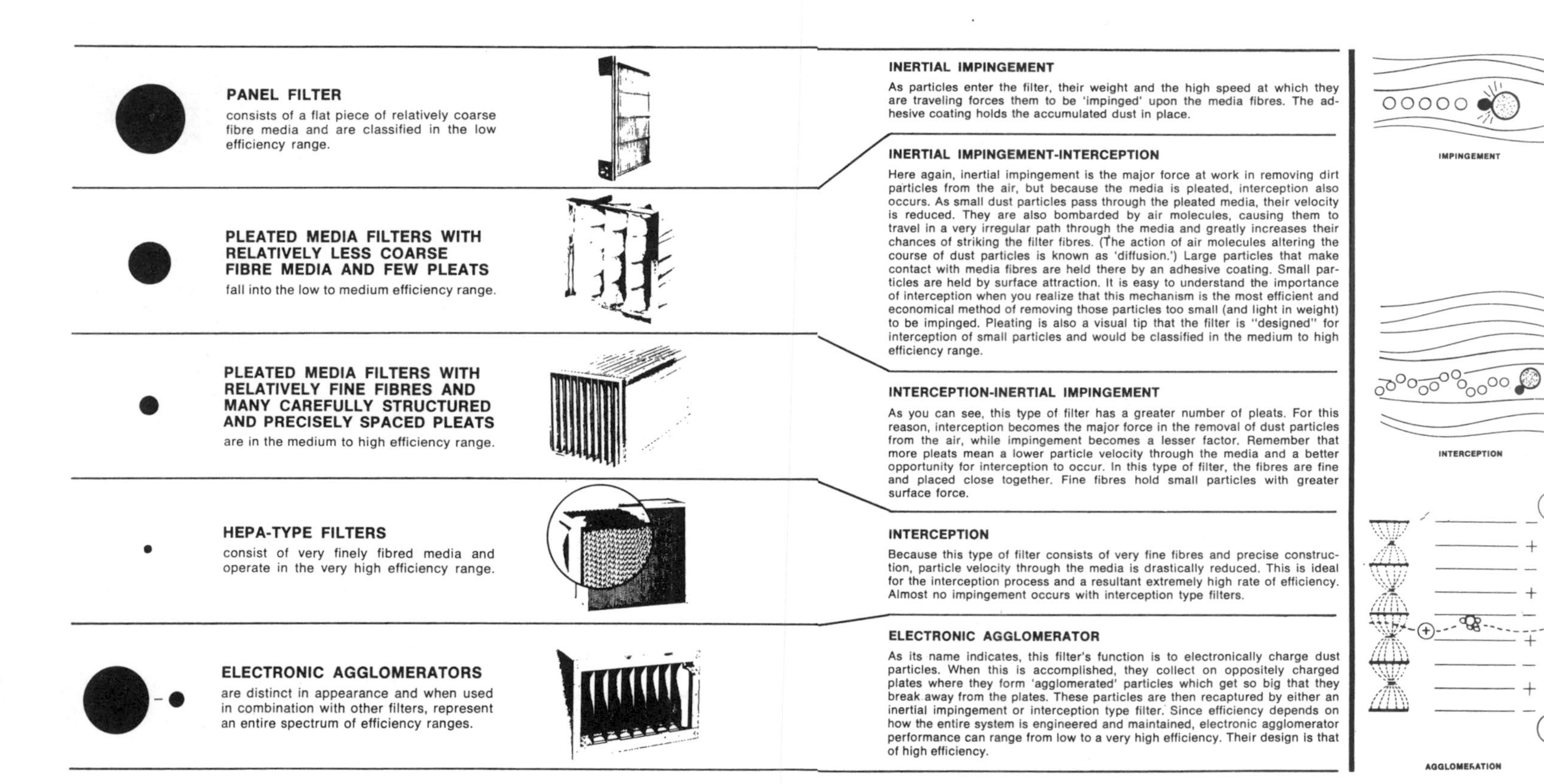

Courtesy of Cambridge Filter Corporation, Syracuse, New York

Figure 12-1. Various Types of Filters and How They Work

Filtration system economy is based on trade-offs among three major cost components: filter media, labor, and energy. It is easy to see that *filter media cost* would be minimized by using the cheapest media for the longest possible time. The labor cost component, however, has three elements: labor to service filters; labor to clean coils, ductwork, and registers in the distribution system and labor to clean building ceilings, wall surfaces, and furnishings. While the strategy that minimizes filter media cost will also minimize service labor cost, the likely effect will be to increase the cost of labor to clean air distribution system internals and building surfaces. Many filters have a tendency to *unload* excess contaminants once their capacity has been reached. They discharge downstream, fouling heat exchange surfaces, coating the inside of distribution ducts, and spilling into conditioned spaces, showing up as sooty soil on ceilings, diffusers, drapes, and wall surfaces.

Energy cost is affected by two elements related to filtration maintenance—heat exchange at coil surfaces and static pressure on the fan caused by resistance to the air stream. Optimum energy costs demand the best possible heat transfer between the coils and the airstream. The work required of the fan is minimized when the least resistance to the airstream is presented. Since the filter's resistance to air flow increases as it does its work of removing contaminants, allowing the filter to "load up" to its design capacity may create extra energy cost in excess of the cost of additional service labor and filter media. A good case can be made for changing filters at approximately 80 percent of design final resistance, since the energy cost rises sharply with resistance and the first 80 percent of resistance (and filter life) takes a much longer time period than would be proportional to resistance.

In order to optimize total system economy, the engineer would need to evaluate the fan characteristics (volume, static pressure, velocity pressure, and horsepower required) and the filter characteristics in order to determine design, initial and final resistances, efficiency, and downstream requirements.

How to Inspect Filter Installation

There are four important items in any filter installation to check:

- *Support of the filter frames.* If the bank of filter frames is not rigid and well supported, it can collapse as the filters load up.
- *Leaks around frames.* If light is showing anywhere between filter frames and/or between the frames and duct walls, caulk these cracks to prevent unfiltered air from leaking by.
- *Fit of filters in frames.* Incorrectly installed filters will also allow air to bypass. Set them in plumb and square.
- *Condition of the media.* Extended surface (bag) filters should be fully open and extended, not pinched shut. If you see any tears, rips, or holes in any media, replace the filter at once.

Tools Required for Servicing Filtration Systems

Any HVAC mechanic servicing filtration systems needs, at a minimum, these tools:

- A good flashlight, preferably with magnetic handle
- Screwdrivers and nut drivers
- Locking pliers
- A good quality manometer (portable)
- Caulking gun
- Duct tape

How to Change Filters

The following steps should be followed in changing filters in most environmental air filter systems:

1. Remove the old filters and set them to one side in the dust or filter housing.
2. Vacuum-clean or brush the holding frames to remove settled dust.
3. Inspect the holding frame gaskets, and replace any that are damaged.
4. Remove the new filters from their cartons and install according to manufacturers' instructions. It pays to read instructions.
5. Check the filter to make sure the media is not caught or damaged. If a filter was damaged during shipment, contact the delivering carrier and file your claim for loss. If you are using bag filters, check to make sure the pleats are free to open fully.
6. Check all fasteners to make sure they are holding filters in place securely.
7. Remove the dirty filters in the cartons that held the new ones. Vacuum the duct floor, close the duct door, and zero the manometer.

Use of the Manometer

A simple inclined gauge device is inexpensive and easy to use. Manometers may be portable or installed permanently at major filter bank locations. In either case, the purpose is the same—to remove guesswork from servicing filters. All filters have design specifications giving both initial and recommended final resistance to air flow. These specifications are usually stated in inches of water. The manometer is installed to measure pressure drop across the filters in inches of water.

Since oil can evaporate from the gauge, every three months or so the manometer should be *zeroed.* This requires the system to be off if the manometer is permanently installed, or at least the tubing disconnected, taking care not to lose oil from the gauge.

When pressure drop reaches the predetermined point, filters should be changed. Since panel-type prefilters have a relatively short life, measure their pressure drop weekly to prevent having them start to "unload" contaminants. For other types of filters, monthly readings should be adequate to tell the rate at which they are loading. Remember, the last 20 percent of resistance comes more quickly.

SPECIFIC INSTRUCTIONS FOR SERVICING FILTERS

Servicing Panel Filters

There are four main types of panel filters: permanent metal, permanent foam, disposable, and replaceable media. The general service procedures apply for each with the following modifications:

Servicing Permanent Metal Panel Filters

Unit filter cells of this type use an impingement fluid that may or may not be coated on the cell at the factory, but must be applied to "charge" the filter before it is placed in operation. The cleaning cycle will depend upon the amount of dust in the air, fan characteristics, and the number of hours in operation. Cells are usually cleaned when the resistance reaches 0.25″ to 0.50″ w.g., but can be operated at 0.75″ pressure drop or higher, depending upon the capacity of the fan. By cleaning one-sixth to one-fourth of the cells during each cleaning period, a constant operating resistance can be more nearly maintained, which will give a more uniform air distribution for the system.

Remove the dirty cells and replace immediately with clean sprayed cells. Clean the cell by washing with a water spray, soaking in a detergent solution, or immersing in a steam-heated cleaning tank. Water temperature is not critical, but a heated detergent solution only should be used in the solution. *Do not* use caustic (high pH) solutions for cleaning filter cells. After draining off dirty water, the cells should be recharged. It is not necessary for cells to be completely dry prior to recharging. Never try to clean filters by brushing off the dirt. This mats the dirt over the front of the cell, making it harder to wash.

For recharging, a pump-up pressure spray (such as the Hudson sprayer) is ideal. Alternate methods are aerosol spray, or shaking and draining dry. The impingement fluid (VIA Viscosine or equal) must coat both faces, penetrating into the depth of the cell. For a 20″ × 20″ size filter, approximately one ounce fluid per half-inch thickness of filter is required.

Servicing Permanent Foam Panel Filters

Open-cell foam or synthetic material (polyurethane or polystyrene) is used for impingement-type media in residential and small commercial equipment. This material generally will not have the capacity of permanent metal and will require more frequent cleaning. The foam is cleaned by washing with a hose or tap water. Impingement fluid is not generally used.

Servicing Disposable Panel Filters

Fiberglass or polyester fiber woven media are used in paperboard frames to construct a variety of types and thicknesses of panel filters. Typically the media are coated with an impingement adhesive similar to permanent metal, although with a different viscosity than that used with metal. Servicing is simple—follow the general rules for exchanging clean for dirty filters, and discard the dirty filter units.

Servicing Replaceable Media

Permanent frames with disposable pads of filter media are popular replacements for disposable panel filters. The media are treated with the same impingement adhesive as the disposable filters. The procedure for servicing is the same as other panel filters except that the frames snap open for exchanging clean pads in place of dirty pads.

Servicing Extended Surface Filters

These filters will operate until they are completely plugged without unloading. Actually, their filtration efficiency increases as more and more dust is collected. However, their resistance also increases, eventually reducing air flow. This makes monitoring of the pressure drop across the filter bank most critical. Also, the consequences of a torn or ruptured bag, releasing contaminants downstream, make it imperative to examine carefully the pleats to assure free opening to full extent of the surface. Supported pleat design filters may be treated essentially as disposable panel filters except for the capacity and pressure drop.

Servicing Roll Renewable Media

Roll media filters come in a variety of media constructions and resulting efficiency characteristics that allow their use as a prefilter, used alone as the principal filtration device, or to retain agglomerated dust from electrostatic precipitators. The distinguishing feature roll media filters have in common is the use of two spools on either side of the plenum, advancing clean filter media as the used media reaches its final resistance, just as film is advanced in a camera after taking a picture.

The media usually is marked for the last few feet of the roll to show that a clean roll will be needed soon. Some models have a media runout switch actuated through an arm resting on the clean media roll. When the supply is exhausted, this switch opens the circuit, stopping the drive motor and turning on a warning light.

To change media rolls, first the old roll must be advanced to run off the remaining media onto the rewind spool. Then the drive switch must be moved to the *off* position. Remove the used roll and store in carton for removal and final disposal. Take the empty spool from the supply side and put it into the rewind side. The new roll will be covered with a paper wrapper. *Leave this on* until the roll is completely in place, as it keeps the media from unrolling as the roll is on a spool under tension.

After placing the new roll in the metal media, cover; and assuring that all pins, slots, latches, etc., are in proper position, remove the paper wrapper and unwind the media to the rewind spool. The media must be spread evenly across the spool and face the duct or plenum with all air seals and retainers in proper place to assure good fit and prevent air leakage. Engage the rewind spool into the drive socket. Install the keeper.

Engage momentary contact switch and test operation by allowing rewind spool to make two revolutions. The filter should now be ready.

Servicing Activated Carbon Filters*

The primary task in servicing activated carbon filters is to replace the odor-saturated carbon with new carbon. The trick is to know when to replace the carbon, since pressure drop and visual methods are meaningless. Waiting until objectionable odors become noticeable is not recommended. The ability of an activated carbon filter to adsorb is not unlimited, and all gases and vapors are not adsorbed in the same amounts. The easiest way to determine if a carbon filter is loaded to capacity is to smell its outlet. If any odor comes through, it is loaded and the activated carbon should be changed.

Most activated carbon units are constructed of a number of metal trays filled with activated carbon and stacked like building blocks with alternate ends sealed together by a gasket so that the entire surface of all the trays is presented to the air stream. Some carbon filter units are front-opening and some are side-opening. In either case, the trays slide out and are released by gasketing.

The number of trays required is a function of the amount of carbon per tray and the velocity of the air flow. Typically, as few as six trays containing 2.5 pounds each would be required for 500 cfm at a velocity of 250 feet per minute. A large bank for 36,000 cfm at 500 feet per minute would require 432 trays of 3.75 pounds of carbon each.

Filter manufacturers can provide reference tables listing adsorption ability of activate carbon for various materials. You can thereby determine if activated carbon will be effective for the materials in your case. The reference tables will also give you an idea of how much activated carbon will be needed based on total cubic feet of space, and approximately how long it will remain effective before needing replacement.

One obvious solution is to have extra trays on hand at all times. This allows exchanging a fresh tray for one in service so that it may be laboratory tested to predict remaining useful life of the carbon. Some manufacturers will perform the test at no charge. If the useful life remaining is sufficient, new trays may be exchanged for old, a few at a time so that not all trays must be renewed at once. Carbon, available in bulk or trays, may be returned to the manufacturer for renewing.

There is also a relatively simple field test to determine the remaining life of activated carbon filters. Basically, the test is as follows:

Put a one-cup sample of carbon from the filter into a mason jar, add a half-teaspoon of carbon tetrachloride to the sample, and shake it for about three minutes.

---

*All information on Activated Carbon Filtration provided courtesy of the Farr Company, Inc., El Segundo CA.

Then cautiously sniff the contents. The test is completed when you sniff and there is an odor, ANY ODOR. If there is no odor, repeat the test by adding one more half-teaspoon of carbon tetrachloride and repeat the shaking process. Keep a record of how many half-teaspoons of carbon tetrachloride were added to the sample until the test is complete. The manufacturer's reference tables can guide as to how much remaining life the carbon activated filters have, based on the number of half-teaspoons adsorbed by the sample.

Activated carbon can also be rated in terms of retentivity. This is a direct measure of how much gas or vapor the activated carbon will release after it has adsorbed as much as it can, all at specified test conditions. The activated carbon filter has a carbon tetrachloride (CC_{14}) retentivity of 30% minimum—a good rating for an air conditioning grade of activated carbon.

Another rating applied to activated carbon is its break-through time or how much time it takes a gas or vapor to get through an activated carbon bed under specified test conditions. Farr rates its activated carbon in terms of carbon tetrachloride break-through time at 25 minutes. The other break-through time rating commonly used is with chloropicrin, a war gas. This test is not used because of the hazardous chemicals involved and, since chloropicrin break-through times are always twice carbon tetrachloride break-through times, the safer method is adequate.

All of these tests for activated carbon were originated by the Army Chemical Warfare Service (CWS).

Servicing Electronic Air Cleaners

Agglomeration with Disposable Collector Media: The more modern electronic air cleaners for large installations employ the agglomeration principle. This means that dust particles build upon the collection plates until they break off in larger chunks and are carried downstream to the collection media, which may be bag filters or roll media. The pressure drop across the filter bank gives the cue for removal of soiled media. Standard servicing procedures are followed for the particular media type.

Household Size Electronic Air Cleaners: The typical electronic cleaner for household use fits into an air plenum register or duct. The entire collector unit slides out and may be washed in a conventional dishwasher.

Electrical / Mechanical Servicing: Besides the agglomeration collection media, all electronic air cleaners have two components in common that require periodic maintenance: the power pack, and the ionizer section.

The power pack is a rectifier that supplies high voltage DC current to ionizer wires and plates. It should be checked regularly for proper operation.

The ionizer section should also be checked frequently to confirm proper operation. When operating properly, the wires will be surrounded by a corona visible in the dark as a pale blue glow extending the full length of the wires. The corona is evidence of ionization and the absence of the corona may indicate low voltage or dirty wires. Short circuits will show up as arcs and tripping of the circuit breaker. Broken ionizer wires should be replaced immediately. Occasionally it will be necessary to wipe down or brush dust from ionizer wires, struts, and plates.

TIME STANDARDS FOR SERVICING MAJOR FILTER TYPES

Time allowances or standard times, which can be applied with confidence to maintenance operations, are very valuable in planning and analyzing work requirements. The following time allowances are a combination of observed or experienced times for performing filter service work under average conditions. They should be attainable by any reasonably well-trained mechanic with proper tools and instructions.

Filter Type and Operation	Unit Service Time
Permanent Metal Panel Filters	
Remove and replace	0.75 minutes each
Wash, dry, and recoat	4 minutes each
Permanent Foam Panel Filters	
Remove and replace	0.75 minutes each
Wash and dry	2 minutes each
Disposable Panel filters	
Remove and replace	0.75 minutes each
Replaceable Media/Metal Frames	
Open frame, remove, replace, and close	1.5 minutes each
Extended Surface Media Filters up to 24″ × 24″ × 36″	
Remove and replace	1-2 minutes each
Roll Renewable Media up to 8 ft. width	
Remove and replace	45-120 minutes per roll
Activated Carbon Filter	
Remove and replace trays	1.5 minutes each tray
Electronic Air Cleaners	
Agglomerator type: See Renewable Roll Media	
Removable Collection Unit:	
Remove and Replace	
Wash in place collection chambers;	4 minutes each
Wash, rinse and spray on adhesive	30 minutes each

Access Time Allowances

An allowance must be added to the total filter servicing time to provide time for removing fasteners, removing or opening service or access panels, wiping or vacuum-cleaning the chamber, and reinstalling panels and fasteners after servicing the filter.

Some typical access time allowances are shown below.

Window Air Conditioning Unit		
Remove, reinstall screws	2 × 5 × 0.7 min. =	7.0 min.
Remove, reinstall panel	2 × 1 × 1.8 min. =	3.6 min.
Vacuum interior of unit	1 × 5.5 min. =	5.5 min.
		16.1 min.
Package Air Conditioning Unit, 3-9 Tons		
Remove, reinstall screws	2 × 12 × 0.7 min. =	16.8 min.
Remove, reinstall panels	2 × 2 × 1.8 min. =	7.2 min.
		24.0 min.
Package Air Conditioning Unit, 10-49 Tons		
Remove, reinstall screws	2 × 12 × 0.7 min. =	16.8 min.
Remove, reinstall panels	2 × 2 × 2.6 min. =	10.4 min.
		27.2 min.
Perimeter Baseboard Fan Coil Unit		
Lift, reclose panel	2 × 1 × 1 min. =	2.0 min.
Overhead Plenum Chamber (from ladder)		
Remove, reinstall screws	2 × 6 × 0.7 min. =	8.4 min.
Remove, reinstall panels	2 × 1 × 8.9 min. =	17.8 min.
Vacuum chamber	1 × 6.0 min. =	6.0 min.
		32.2 min.

Other items should be calculated to add to filter changing and access time allowances to get a total picture of service time requirements for filter installations. These items include allowances for reading and zeroing manometers, travel time to and from filter banks, material handling time, etc.

MODIFYING AND UPGRADING FILTER SYSTEMS

Many existing filter systems were designed for different purposes than those for which the conditioned spaces are now being used. Codes and requirements have changed. Certainly awareness of the costs associated with energy and maintenance practices affecting consumption and conservation has been heightened. For any number of valid reasons, engineers today are taking a second look at their filtration systems to determine whether a modification is feasible to lower costs, upgrade air quality, or both.

A look at some recent case histories will illustrate the possibilities.

Case 1—Changing Filter Type for Greater Efficiency

The typical air handler plenum had a filter section measuring a nominal 4′ high × 8′ wide, and was covered with Roll Renewable media. The upgrading consisted of replacing the filter section with holding frames for 24″ × 24″ × 4″ extended area pleated filters. Although both the original and replacement media are rated as medium efficiency (20 to 30 percent) by ASHRAE 52-76 standards, the replacement filters are 300 percent as efficient on small particles in the 5 to 10 micron range. Consequently, coil cleaning has been reduced from two times per year to once per year or longer. Labor to change filters has been decreased from 1.5 labor hours per change to 10 minutes. Filter media cost is a standoff. List price for eight filters 2′ × 2′ × 4″ comes to a few dollars less than the roll. Frequency of changes is slightly less also.

Total results: Improved efficiency, lower cost of labor in both filter servicing and coil cleaning, and lower resistance to the fan, which could lower energy costs.

Case 2—Improving Filter Efficiency Through Housekeeping Procedures

The main air handling units were provided with an oversize filter bank of non-supported bag filters, which had an average of 50 percent dust spot efficiency. The design velocity was less than 500 cpm at the filter bank. The engineering staff expected to experience two to three changes of filters per year at 1″ w.g. final resistance. Instead, the first year of occupancy saw most filters changed only once at 0.80″ w.g. The main factors in efficiency were a conservative design of the filter system and an excellent program of housekeeping in the building, preventing the expected accumulation of dust. The housekeeping program included daily vacuum cleaning of all the traffic areas of the carpets, which are very high quality, tight-woven construction. Very little of the usual “fuzz” from the top fibers of the carpet was seen during the initial wear-in period.

The engineering staff plan to add 2″ pleated disposable filters in front of the bag filters when changed. This will extend the life of the bag filters at very little expense in labor and materials since the prefilters will be very easily changed and are expected to cost $7.00 each, versus $28.00 each for the bag filters.

Case 3—Adjusting Filtration and Building Air Conditioning Settings

The building engineer needed to improve air quality in the most cost effective manner. The building provided 100,000 square feet of occupancy. The present filter system was low efficiency, throw away filters—60 each of 30/30 filters. The 300 ton air conditioning system was set for the lowest outside air to save energy, but the 667 building occupants complained about stuffiness, odors, etc.

Coil cleaning was accepted as a cost of doing business up to now, but the building engineer investigated the true cost of dirty coils, and discovered that cost of cleaning coils plus loss of efficiency which, added to energy cost totaled $50,000 per year, or $0.50 per leased square foot!

The solution:

Current ASHRAE Code (62-1989) recommends 5 cfm per person as lowest acceptable ventilation. However, this allows 4000 ppm of CO2, which is a cause of the stuffiness. To maintain CO2 at 1000 ppm a minimum 15 cfm/person is required; the new code actually recommends 20 cfm/person for offices. Using an IAQ rate method 15 cfm/person is acceptable, which also maintains CO at 1000 ppm. This also allows for dilution of bioaffluents (which cause odors), bacteria, fungi, and viruses (both odor producing and physically harmful, caused by coils and equipment in a dirty environment).

For irritations caused by VOCs (Volatile Organic Chemicals, found in outgassing of carpeting, synthetic materials used in carpeting and wall coverings, cleaning fluids, etc.), the new standard recommends activated carbon filtration. Carbon is now also tested on toluene, which is found in jet fuel (which finds its way into building air intakes near airports), resins and lacquer.

Our building engineer decided to increase filtration efficiency, increase outside air to 15 cfm per person, and add carbon filtration for odor control.

ASHRAE Standard 62-89 identifies particle size of 0.20 to 0.50 micron as respirable and lung damaging. The mean size of these particles is 0.30 micron; therefore replacement filters should be rated at 0.30 micron, which happens to be the least efficient point of any filter. Even though current filters are rated by ASHRAE using dust spot method at 25%, they are only 2% effective at 0.3 micron. His choice for replacement is a rigid filter, equivalent to Farr RF 200, which has 95% dust spot % rating and 75% efficiency at 0.30 micron. His selection for carbon filtration has average life efficiency on VOC's of 85%, 95% on ozone, 93% on SO2, and 82% on NO2.

The amortized costs per year come to only an incremental $0.36 per square foot, leaving a saving of $0.14 per sf ($14,000 per year) after solving the stuffy air and odor problems. His boss was highly pleased, to say the least!

WHAT TO DO WHEN THE FILTRATION SYSTEM HAS NOT BEEN MAINTAINED

The results of a poor service program are expensive and easily traced from maintenance records. They include dirty coils, poor heat transfer with resulting high energy bills, freezing of direct expansion coils, fire hazards in distribution system ducts and registers, dirt spills at registers and diffusers, etc.

The correction requirements are all many times as expensive as a good service program. Some of the obvious corrective measures are described in the following sections.

Cleaning Coils and Plenum Chambers

With a flashlight, determine the extent of residual dirt and fouling on the coils. Dust can be removed by brushing, vacuuming, or blowing with compressed air (and then vacuum-cleaning the settled dust). Oily residue, biological contaminants, and fungus will require chemical cleaning. A number of good chemical cleaning products are available for use with a pressure sprayer, which will remove fouling contamination and leave coils bright without damaging the metal surfaces. Rinsing is not generally required. Plenum surfaces should be brushed or vacuum-cleaned to remove all loose dust and debris.

Cleaning Ductwork, Registers, and Grilles

Heavily contaminated air distribution systems constitute a real challenge. Seldom is enough of the ductwork available for conventional vacuum cleaning to remove accumulated dust. The most successful method is to put throwaway filters over openings and to blow the dust *downstream* with high volume blowers, collecting it with the filters at diffuser and register openings. Dampers may be closed to section off parts of the system so that blowing and collecting will be limited to only a part of the system at a time. Heavily soiled registers and grilles should be cleaned with the same solution as the coil cleaner.

Cleaning Ceilings and Walls Near Conditioned Air Units

Try cleaning first with a portable vacuum cleaner, using a soft bristled brush tool on the end of the wand. If smears result or the area will not come clean, use a neutral detergent solution and a damp cloth. Heavily soiled walls should be washed from the bottom up to avoid streaking. Flat paint on walls may not be possible to clean satisfactorily and may require repainting, but a clean surface is required prior to painting.

"TIGHT BUILDING SYNDROME" AND "BUILDING-RELATED ILLNESSES"

This phenomenon is variously referred to as: "*Sick Building Syndrome,*" "*Tight Building Syndrome*" or "*Closed Building Syndrome.*" All three terms refer to indoor air quality problems that manifest themselves in the same way: tenant or occupant complaints that are difficult to trace. The syndrome should not be confused with "*Building-related illness,*" which refers to specific physical maladies that are traceable to causes in a facility. The most attention in recent years has been paid to "*Legionnaire's Disease*" or "*Humidifier Fever.*"

Public health officials and physicians are definitely needed to treat victims of these diseases. However, there are steps that building managers can undertake to prevent and correct the tight building syndrome and related problems.

How to Recognize Symptoms

- Sensory irritation of the skin and upper airways:
  * irritation of eye, nose and throat
  * dry mucous membranes and skin
  * erythema (superficial skin irritation)
  * headache
  * abnormal taste
- Building odors
- Lower airway and gastrointestinal symptoms:
  * airway infections
  * hoarseness of voice
  * wheezing
- General symptoms:
  * fatigue
  * dizziness
  * nausea
- Hyperactivity illness:
  * hypersensitivity pneumonitis
  * dermatitis

Causes

One study* was able to classify its findings by primary type of problem found:

- Inadequate ventilation—52%
- Contamination from inside the building—17%
- Contamination from outside the building—11%
- Microbiological contamination—5%
- Contamination from the building fabric—3%
- Unknown—12%

*National Institute for Occupational Safety and Health, *Guidance for Indoor Air Quality Investigations,* A Report Prepared by the Hazard Evaluations and Technical Assistance Branch, Division of Surveillance, Hazard Valuations and Field Studies (Cincinnati, Ohio 45266, 1987), pp. 4-6.

The primary offenders have been identified as:

- Noxious odors from formaldehyde containing finishes
- Volatile organic compounds (VOCs)
- Molds and mold spores
- Pesticides and their vehicles (oils)
- Carpets, fabrics and their cleaning agents
- Tobacco smoke
- Human metabolism (consumption of oxygen, production of carbon dioxide, heat, water vapor and body odors).

Since the causes for tight building syndrome and their subsequent effects are well documented now in the literature (one hygienist says it has achieved "seminar status"), we shall deal primarily with what to do.

How to Prevent Tight Building Syndrome

Good building design and good preventive maintenance generally prevent the occurrence of tight building syndrome.

Local exhaust systems are employed to capture air contaminants—dust, fumes, mists, vapors, hot air and even odors—at or near their point of generation or dispersion, to reduce contamination of the breathing zone of workers. Local ventilation is frequently used and is generally the preferred method for controlling atmospheric concentrations of airborne materials that present potential health hazards in the work environment.

General ventilation refers to the commonly encountered process of flushing a working environment with a constant supply of fresh air. General ventilation differs from local ventilation in that it is a dilution process rather than strictly an exhausting process.*

The emphasis on energy conservation in many facilities has led to reduced fresh air intake and infiltration. This drive for energy efficiency represents a tradeoff with indoor air quality. New air quality standards from ASHRAE attempt to overcome the deleterious effects by raising the fresh air circulation rate for occupants. In older buildings, the circulation may need to be increased to the new standards. (See discussion of ASHRAE 62-89 following).

Chemical levels can be determined by testing for formaldehyde, carbon monoxide, ozone, VOC's, etc.

---

*John E. Mutchler, "Local Exhaust Systems," *The Industrial Environment—its Evaluation & Control,* p. 597.

George D. Clayton & Associates, "Design of Ventilation Systems," *The Industrial Environment—its Evaluation & Control,* p. 609.

Temperature and humidity settings also definitely affect occupants and their comfort. When people are uncomfortable, they are more likely to react to other irritants. A person who is comfortable with temperature and humidity might well ignore the "new smell" of carpets and drapes, where an uncomfortable person might complain of headaches, nausea, etc. This is not to suggest that the complaints are invalid, but that the response threshold is affected by temperature/humidity discomfort.

Applicable standards, together with the nature of the contaminant, dictate the control levels. In an odor problem, the desired control level may be below the odor threshold or at some other level prescribed by the appropriate guidelines. For example, formaldehyde causes discomfort at 0.1 ppm or lower, but occupational regulations describe it as a health hazard at 3 ppm. The sulfur dioxide limit for workers is set at 5 ppm, but it causes electronic apparatus to corrode at 0.05 ppm. The engineer must determine whether the control motive is legal compliance, comfort, or protection. These factors determine the design criteria for the equipment and control devices. The standards of the American Conference of Governmental Industrial Hygienists, the American Industrial Hygiene Association, and the Occupational Safety and Health Administration are widely known and used.**

Microbial levels may also be established by sampling and culturing. The process is not quite as straightforward as sampling for chemical levels, but is still within reason.

Changing filters, cleaning coils, pans, ducts, and so on as described elsewhere in this chapter are very important in preventing occurrences and outbreaks of tight building syndrome complaints. Even a cursory inspection of conditioned building spaces will reveal problems that need attention and correction: visible dirt and soot around air intake vents or on return air grills; excessive dust accumulation on desktops and file cabinets; visible and lingering tobacco smoke or smoke from cooking equipment; uneven temperatures, indicating poor air distribution.

The use of office landscape furniture configurations with low movable partitions has worked against air distribution, sometimes causing short cycling of conditioned air from supply grilles and diffusers to return air grilles or registers.

Periodic cleaning of walls and ceilings, especially in areas known to be offenders, can include disinfection of these surfaces.

Dealing with Complaints About Tight Building Syndrome

It is important to respond promptly, professionally and sympathetically to complaints. The worst thing a building manager can do is to ignore complaints or even suggest that the complainers may have a sickness of the mind. One fact that has been

*American Society of Heating, Refrigerating and Air Conditioning Engineers, Inc., *1987 ASHRAE Handbook—Heating, Ventilating, and Air-Conditioning Systems and Applications,* p. 50.1.

established in investigations of such building complaints is that currently available instruments cannot measure to the level of sensitivity of the human nose. Hypersensitive persons with severe allergic reactions can pick up the presence of allergens long before their concentration allows measurement. Some of us can detect the presence of mold spores before they are highly visible on shower curtains or tile grout lines.

So the first step is always a site investigation. This normally includes:

- Discussion with complainants to isolate specifics of the problem
- Conducting an informal IAQ survey looking for odors, mold, etc.

Further investigative steps include:

Chemical measurements, sampling for airborne microbial agents, measurement of temperature and humidity on a continuing basis, and distribution of questionnaires.

Corrective Actions

The primary response involves the HVAC SYSTEM: proper operation, cleaning with biocides, replacement of system components (drip pans, filters, registers, etc.) if warranted.

Cleaning and disinfecting finishes and furnishings. Vacuuming carpets with a HEPA filtered vacuum. Occupants may be required (or at least urged) to clean books, files, etc.

Because of the role of plants in consuming carbon dioxide and producing oxygen, some building managers and occupants might be tempted to increase the use of plants. This is likely to be a big mistake. The increase of dirt, dust and mold spores are likely to more than offset any benefits.

Finally, building managers should not be reluctant to seek professional help—not a psychiatrist for being driven crazy, but an environmental hygienist from the local health department or a private consulting firm. These trained experts can aid not only the screening process, but psychologically in terms of visible action for the benefit of occupants and complainants.

Understanding and Complying with the ASHRAE Ventilation Standard

This section deals with the newest standard on "Ventilation for Acceptable Indoor Air Quality," ASHRAE 62-1981.

This most recent standard is called a "National Voluntary Consensus Standard," and as such becomes the standard of the industry for ventilation design and operations. The importance of the term "consensus" in the title is seen by the definition of "consensus" according to the American National Standards Institute (ANSI), of which ASHRAE is a member: ". . . Consensus is defined as 'substantial agreement reached by concerned interests . . . after a concerted attempt at resolving objections. Consensus implies much more than the concept of a simple majority, but not necessarily unanimity.' "

ASHRAE 62-89 is the newest revision of a Standard which began as ASHRAE 62-73 "Standard for Natural and Mechanical Ventilation," which provided a prescriptive approach to ventilation, i.e., it specified both minimum and recommended outside air flow to obtain acceptable indoor air quality (IAQ) for many types of spaces. The 1973 standard is still referenced by many building codes and ASHRAE's first energy standard, 90-75. The revised energy standard, 90A-1980, also took this approach.

This revised standard combines two procedures: (1) the ventilation rate procedure introduced in 1973, and (2) an alternative procedure, the innovative air quality procedure introduced in the 1981 revision.

The standard states in part, "the goal of achieving acceptable indoor air quality and of minimizing energy consumption appear to imply a compromise. An interdisciplinary committee of engineers, architects, chemists, physiologists, product manufacturers and industry representatives has endeavored to achieve the necessary balance between energy consumption and indoor air quality in this standard. It must be recognized, however, that the conditions specified by the Standard must be achieved during the operation of buildings as well as design of the buildings if acceptable IAQ is to be achieved."

The standard begins with Purpose, Scope and Definitions which apply to a well-drawn generic HVAC system schematic before presenting the classification section which specifies the alternative procedures.

Following a section which specifies standards for systems and equipment, the two procedures are outlined step by step, with several tables of values to be used by system designers. Much of the rationale is presented along with the procedures, but appendices contain the details of sources. The series of appendices (A through H) present the rationale for each of the various parts of the standard.

This standard is so very well developed and documented that it might well be used as a text for teaching ventilation design as well as a standard.

CONCLUSION—THE PAYOFF

The benefits of a good air filtration service program are definitely worth the investment of management time to get it established and followed through to successful implementation.

- Enhanced building appearance—this definitely affects marketability of tenant space and suitability for other types of occupancy.
- Energy savings—through reduced resistance to air flow, permitting lower fan speed, and resulting in lower horsepower requirements at design air volume.

- Improved air quality—occupants of the building always benefit from reduction in the carried over contaminants in the building's conditioned air.
- Labor savings—a good filtration service program will reduce the time to change filters, clean coils and ductwork, and reduce corrective work requirements. Servicing on a rational, planned basis always saves over breakdown or catch-up-type programs.

13

Proper Maintenance of Belts and Bearings

Robert Mayer

Belts and bearings are major work horses of industry. It is important, therefore, that they be given high priority in your maintenance program. Both preventative maintenance (Chapter 2) and predictive maintenance (Chapter 3) techniques are applicable in these two areas.

Two reasons may be given to justify this view: first, the performance of this equipment affects the efficiency of the machines of which they are components, and second, their failure to perform properly usually exerts a great influence on the quality and quantity of production, or service of the company utilizing them.

Factors to be considered in planning the maintenance of these critical pieces of equipment will be presented in general, where procedures and instruments will be discussed. Specific recommendations for each category are discussed under separate headings.

MAINTAINING BELTS*

Belt drives are one of the most popular and widely used methods of power transmission for industrial applications. There are four generic types: V-Belts, joined, linked and synchronous belts.

*The information in this section was provided Courtesy of the Gates Rubber Company, Denver Colorado

V-Belts can be divided into four classifications.

- Classical belts have a variety of cross sections and can be teamed up in multiples. Multiple belt drives can deliver up to several hundred HP continuously and absorb heavy shock loads.
- Narrow V-Belts usually provide substantial weight and space savings over classical belts.
- Variable-speed V-Belt drives offer an infinite selection of speed ratios.
- Light-duty V-Belts. These belts are often referred to as fractional HP belts and are typically designed for less than 1 HP

Joined V-Belts are recommended for drives that produce excessive vibration and heavy shock loads. These belts are two or more classical or narrow type belt strands connected across the top by a tie-band. If you are experiencing a belt turn over or belts jumping off the drive, use a joined V-Belt.

Linked belts consist of links that are fastened together to form an endless belt. Though they command a premium price they are more economical in the long run in cases of exposure to: oil, heat, chemical, water and abrasive conditions.

Synchronous belts also known as timing belts have moulded teeth meshing with grooves on the pulleys to produce a positive no-slip transmission of power. There are two basic types of synchronous belts - conventional trapezoidal and high performance curvilinear. A new type of belt has recently appeared on the market made of polyurethane with Dupont Kevlar tensile chords that can better withstand sudden power surges, shock loading and racheting.

Belts require only a minimum of maintenance, but certain procedures can help reduce equipment downtime, increase safety and improve efficiency.

Checklist for Installing or Replacing Belts and Pulleys

- Turn power off. Lock control box and tag with warning sign "Down For Maintenance - Do Not Turn Power On."
- Place all machine parts in safe position.
- Remove guard and inspect for damage, corrosion and cleanliness. Check for signs of wear or rubbing against drive components. Clean as needed. Make sure that guard has adequate air access.
- Inspect belt(s) for wear or damage.

  When installing a belt, either as a replacement or on a new drive, follow these steps:

  1. Loosen motor mounting bolts and move the motor until the belt is slack and it can be removed and installed without prying or rolling. Make sure guide rails are properly lubricated. Never pry or roll a belt off or onto a sheave! Prying the belt onto the sheave can shorten the belt's life even if no

visible damage is evident. And rolling the belts onto the sheaves can be very dangerous because if the sheaves turn, clothing and fingers can be caught and result in severe injuries.

2. Remove old belts. Check for unusual wear. Excessive wear may indicate problems with drive design or maintenance procedure.
3. Select proper replacement belt or belts being sure all belts are matched. Do not mix old and new belts.
4. Clean belts and sheaves with a rag dampened with a light, non-volatile solvent. Do not soak or brush solvent on belt, or sand or scrape belt with sharp object to remove grease or debris. Belts must be dry before using.
5. Inspect sheaves for wear and nicks. Check alignment. Use a groove guage to see if grooves are worn. If more than 1/32 of wear can be seen, replace sheave (see Figure 13-1). To properly install and align pulleys, check alignment of shaft and location of sheaves on the shaft. This can be accomplished by measuring the distance between the shafts at three or more locations. (see Figure 13-2) If the distances are equal, then the shafts are parallel. Use a straight edge or piece of string to check the location of the sheaves on the shafts. (See Figure 13-3.) Rotating each sheave a half revolution will determine whether the sheave is wobbly or the drive shaft is bent. Correct any misalignment.

Figure 13-1. Inspecting Sheaves

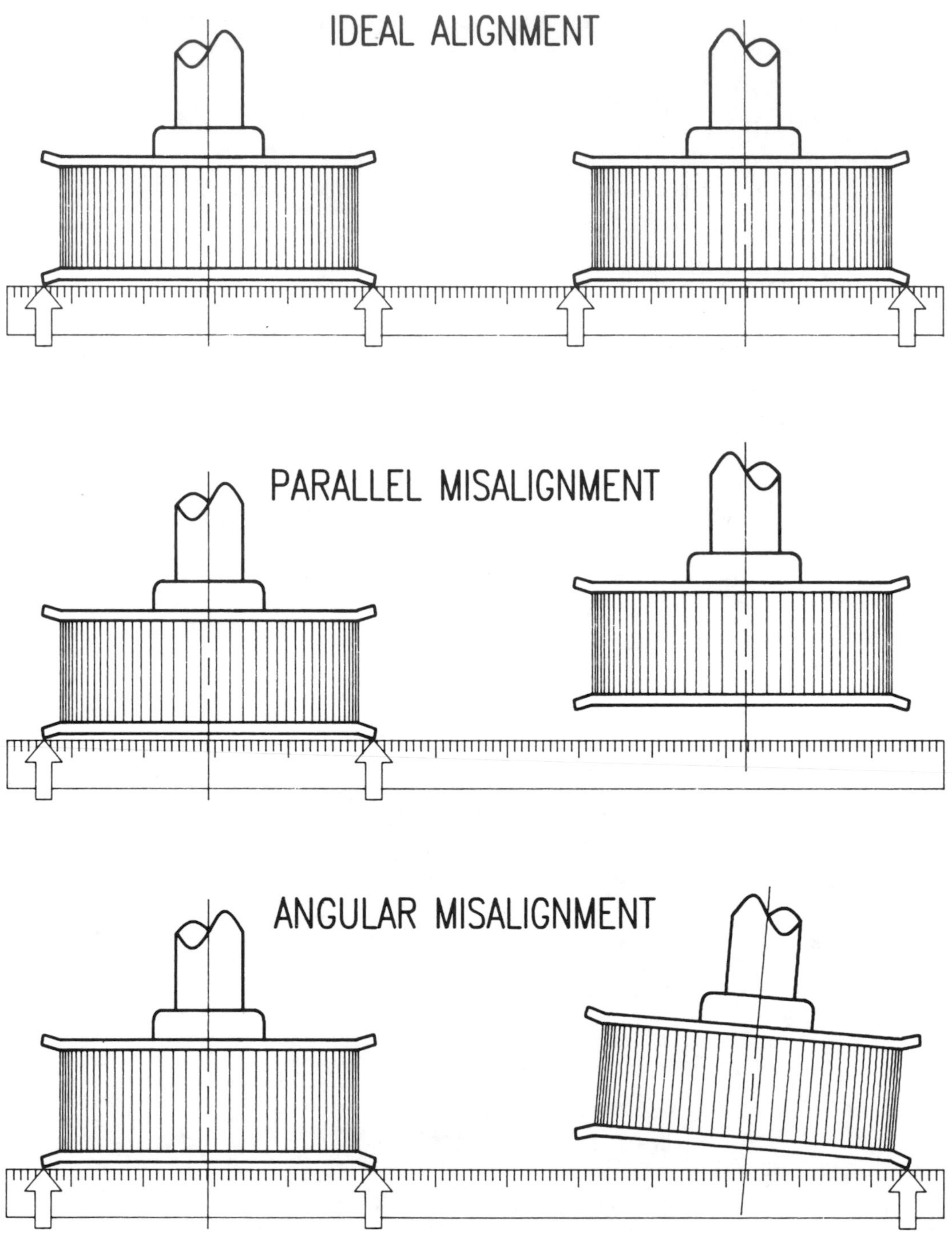

Courtesy of The Gates Rubber Company, Denver, CO

Figure 13-2. Checking Alignment of Sheaves

Figure 13-3. Checking Location of Sheaves on Shafts

6. Inspect other drive components such as bearings and shafts for alignment, wear, lubrication dirt, etc.
7. Install new belt or belt set. On multiple belt drives, use only matched sets from one supplier. Do not mix new belts with old belts.
8. Take up center distances on drive until proper tension is achieved. (See Figure 13-4 for static check.) Under dynamic conditions there are three ways to decide whether a drive is properly tensioned. First, adjusting by sight entails setting the center distance until the belts have only a slight bow on the slack side when they are operating at full load.
   Second, adjusting by sound is appropriate for machines, such as fans, that require peak torque on starting. If the belts squeal as the motor comes on, they are not tight enough. Belts will last longer if the squeal is eliminated.
   Third, adjusting by touch involves measuring belt slippage by the amount of frictional heat developed from it. Excessive slippage generates enough heat that with the drive stopped, you cannot hold a finger in one of the grooves. You must tighten the belts until the sheaves run cool.
   On multiple drives rotate the drives by hand for a few revolutions to help seat the belts into the pulley grooves and even out tension.

How to use Gates Tension Tester

1. Measure the span length (t).
2. Position the lower of the two O-rings using *either* of the following methods:
   a. On the scale reading "Deflection Inches," set the O-ring to show a deflection equal to 1/64" per inch of span length (t).
   b. On the scale reading "Inches of Span Length," set the O-ring to show a deflection equal to the inches of measured span length (t).

Read scales at the bottom edge of O-ring. Leave the upper O-ring in maximum up position.

3. At center of span (t) apply force, with Gates Tension Tester perpendicular to the span, large enough to deflect one belt on the drive until the bottom edge of the lower O-ring is even with the tops of the remaining belts.

   A straight edge across the belt tops will insure accuracy of positioning.
4. Find amount of deflection *force* on upper scale of Tension Tester. The Sliding Rubber "O-Ring" slides down scale as tool compresses—stays down for accurate reading of pounds pressure. Read at *top* edge of ring. (Slide ring up before re-using.)
5. Compare deflection *force* with range of forces in tables below. If *less* than *minimum* recommended deflection force, belts should be tightened.

   If *more* than *maximum* recommended deflection force, drive is tighter than necessary.

Note: There will normally be a rapid drop in tension during the "run-in" period. Tension new drives with a ½ greater deflection force than the maximum force recommended. Check tension frequently during the first day of operation.

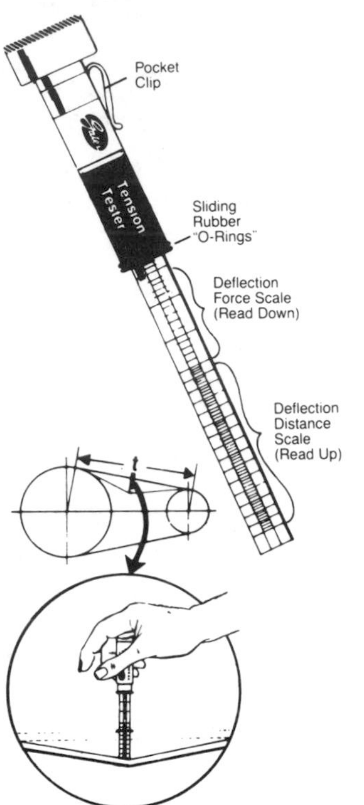

Recommended Deflection Force Per Belt For Super HC® V-Belts, Super HC PowerBand® Belts, Super HC Molded Notch V-Belts or Super HC Molded Notch PowerBand Belts

V-Belt Cross Section	Small Sheave Diameter Range (Inches)	Small Sheave RPM Range	Speed Ratio Range	Recommended Deflection Force (Pounds)	
				Minimum	Maximum
3VX	2.20	1200-3600	2.00 to 4.00	2.8	4.1
	2.35- 2.50	1200-3600		3.2	4.7
	2.65- 2.80	1200-3600		3.5	5.1
	3.00- 3.15	1200-3600		3.8	5.5
	3.35- 3.65	1200-3600		4.1	6.0
	4.12- 5.00	900-3600		4.8	7.1
	5.30- 6.90	900-3600		5.8	8.6
5VX	4.40- 4.65	1200-3600	2.00 to 4.00	9.0	13
	4.90- 5.50	1200-3600		10	15
	5.90- 6.70	1200-3600		11	17
	7.10- 8.00	600-1800		13	19
	8.50-10.90	600-1800		14	20
	11.80-16.00	400-1200		15	23
5V	7.10- 8.00	600-1800	2.00 to 4.00	11	16
	8.50-10.90	600-1800		13	18
	11.80-16.00	400-1200		14	21
8V	12.50-17.00	600-1200	2.00 to 4.00	28	41
	18.00-24.00	400- 900		32	48

*If replacing 3V or 5V belts with the same belt length and same number of belts in 3VX or 5VX cross section, belt tension does not need to be increased.

NOTE: New drives designed with 3VX or 5VX belts should be tensioned at the respective deflection force value shown in above table.

Recommended Deflection Force Per Belt For Hi-Power II™ V-Belts, Hi-Power II PowerBand Belts or Tri-Power® Molded Notch V-Belts

V-Belt Cross Section	Small Sheave Diameter Range (Inches)	Small Sheave RPM Range	Speed Ratio Range	Recommended Deflection Force (Pounds)			
				Hi-Power II		Tri-Power Molded Notch	
				Minimum	Maximum	Minimum	Maximum
A	3.0	1750 to 3600	2.0 to 4.0	2.7	3.8	3.8	5.4
	3.2			2.9	4.2	3.9	5.6
	3.4- 3.6			3.3	4.8	4.1	5.9
	3.8- 4.2			3.8	5.5	4.3	6.3
	4.6- 7.0			4.9	7.1	4.9	7.1
B	4.6	1160 to 1800	2.0 to 4.0	5.1	7.4	7.1	10
	5.0- 5.2			5.8	8.5	7.3	11
	5.4- 5.6			6.2	9.1	7.4	11
	6.0- 6.8			7.1	10	7.7	11
	7.4- 9.4			8.1	12	7.9	12
C	7.0	870 to 1800	2.0 to 4.0	9.1	13	12	18
	7.5			9.7	14	12	18
	8.0- 8.5			11	16	13	18
	9.0-10.5			12	18	13	19
	11.0-16.0			14	21	13	19
D	12.0-13.0	690 to 1200	2.0 to 4.0	19	27	19	28
	13.5-15.5			21	30	21	31
	16.0-22.0			24	36	25	36
E	21.6-24.0	435 to 900	2.0 to 4.0	32	47		

Figure 13-4. Testing Tension

9. Secure motor mounting bolts to correct torque.
10. Replace guard.
11. Provide adequatc run-in period, either over a lunch or even over night. Then re-tension as needed.
12. During start up, look and listen for unusual vibration or noise. Shut equipment down and check bearings and motor. If they feel hot, belt tension may be too tight, or bearings may be misaligned or improperly lubricated.

Possible Minor Changes to Improve Belt Performance

- Increase pulley diameters
- Increase the number of belts, or use wider belt
- Add vibration dampening to system
- Improve guard ventilation to reduce operating temperature
- Use correct, minimum pulley diameters on inside and backside idlers
- Use premium belts rather than general purpose types
- Replace pulleys when they are worn
- Keep pulleys properly aligned
- Place idler on span with lowest tension and as close to drives as possible
- Re-tension newly installed belts after a 4-24 hour run-in period
- Review proper belt installation and maintenance procedures

Troubleshooting Guide for Belts

The following guide indicates common problems encountered with belts, their probable cause and their solutions.

Premature Belt Failure

Symptoms	Probable Cause	Corrective Action
• Broken belt(s) 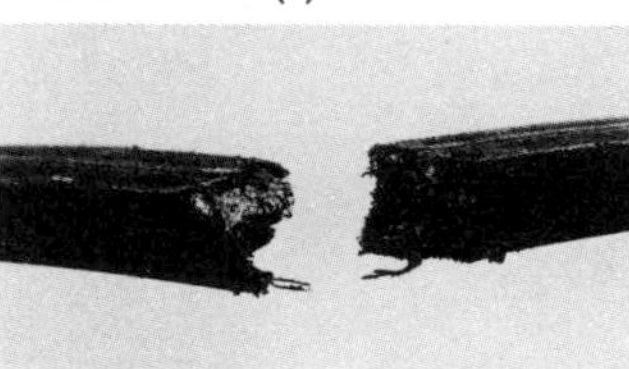	1. Under-designed drive 2. Belt rolled or pried onto sheave 3. Object falling into drive 4. Tampering 5. Severe shock load	1. Redesign 2. Use drive take-up when installing. 3. Provide adequate guard or drive protection. 5. Redesign to accommodate shock load.
• Belts fail to carry load, no visible reason 	1. Underdesigned drive 2. Damaged tensile member 3. Worn sheave grooves 4. Center distance movement	1. Redesign 2. Follow correct installation procedure. 3. Check for groove wear, replace as needed. 4. Check drive for center distance movement during operation.
• Edge cord failure	1. Pulley misalignment 2. Damaged tensile member	1. Check alignment and correct. 2. Follow correct installation procedure.
• Belt de-lamination or undercord separation	1. Too small sheaves 2. Use of too small backside idler	1. Check drive design, replace with larger pulleys. 2. Increase backside idler to acceptable diameter.

Severe or Abnormal V-belt Wear

Symptoms	Probable Cause	Corrective Action
• Wear on top surface of belt	1. Rubbing against guard 2. Idler malfunction	1. Replace or repair guard. 2. Replace idler.
• Wear on top corner of belt 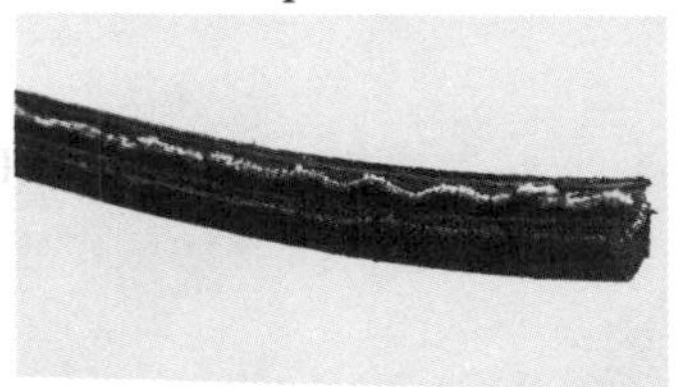	1. Belt-to-sheave fit incorrect (belt too small for groove)	1. Use correct belt-to-sheave combination.
• Wear on belt sidewalls 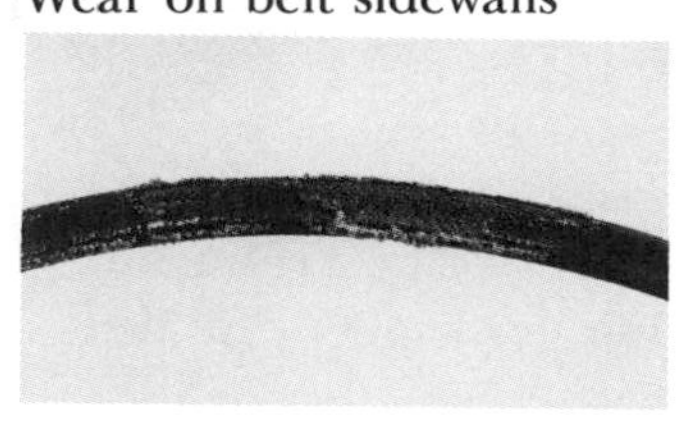	1. Belt slip 2. Misalignment 3. Worn sheaves 4. Incorrect belt	1. Retention until slipping stops. 2. Realign sheaves. 3. Replace sheaves. 4. Replace with correct belt size.
• Wear on bottom corner of belt	1. Belt-to-sheave fit incorrect 2. Worn sheaves	1. Use correct belt-to-sheave combination. 2. Replace sheaves.
• Wear on bottom surface of belt 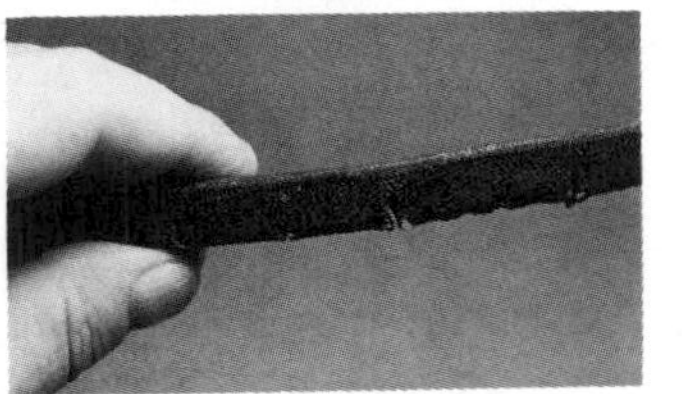	1. Belt bottoming on sheave groove 2. Worn sheaves 3. Debris in sheaves	1. Use correct belt/sheave match. 2. Replace sheaves. 3. Clean sheaves.

(*continued*)

Severe or Abnormal V-belt Wear (continued)

Symptoms	Probable Cause	Corrective Action
• Undercord cracking	1. Sheave diameter too small	1. Use larger diameter sheaves.
	2. Belt slip	2. Retention.
	3. Backside idler too small	3. Use larger diameter backside idler.
	4. Improper storage	4. Don't coil belt too tightly, kink or bend. Avoid heat and direct sunlight.
• Undercord or sidewall burn or hardening	1. Belt slipping	1. Retention until slipping stops.
	2. Worn sheaves	2. Replace sheaves.
	3. Underdesigned drive	3. Refer to drive manual.
	4. Shaft movement	4. Check for center distance changes.
• Belt surface hard or stiff	1. Hot drive environment	1. Improve ventilation to drive.
• Belt surface flaking, sticky or swollen	1. Oil or chemical contamination	1. Do not use belt dressing. Eliminate sources of oil, grease or chemical contamination.

V-Belts Turn Over Or Come Off Drive

Symptoms	Probable Cause	Corrective Action
• Involves single or multiple belts	1. Shock loading or vibration	1. Check drive design.
	2. Foreign material in grooves	2. Shield grooves and drive.
	3. Misaligned sheaves	3. Realign the sheaves.
	4. Worn sheave grooves	4. Replace sheaves.
	5. Damaged tensile member	5. Use correct installation and belt storage procedure.
	6. Incorrectly placed flat idler pulley	6. Carefully align flat idler on slack side of drive as close as possible to driveR sheaves.
	7. Mismatched belt set	7. Replace with new set of matched belts. Do not mix old and new belts.
	8. Poor drive design	8. Check for center distance stability and vibration dampening.

Belt Stretches Beyond Available Take-Up

Symptoms	Probable Cause	Corrective Action
• Multiple belts stretch unequally	1. Misaligned drive	1. Realign and retension drive.
	2. Debris in sheaves	2. Clean sheaves.
	3. Broken tensile member or cord damaged	3. Replace all belts, install properly.
	4. Mismatched belt set	4. Install matched belt set.
• Single belt, or where all belts stretch evenly	1. Insufficient take-up allowance	1. Check take-up. Use allowance specified in Gates design manuals.
	2. Grossly overloaded or under designed drive	2. Redesign drive.
	3. Broken tensile members	3. Replace belt, install properly.

Performance Problems

Symptoms	Probable Cause	Corrective Action
• Incorrect driven speed	1. Design error	1. Use correct driverR/driveN sheave size for desired speed ratio.
	2. Belt slip	2. Retention driveR.

Problems With Pulleys

Symptoms	Probable Cause	Corrective Action
• Broken or damaged pulley	1. Incorrect pulley installation	1. Do not tighten bushing bolts beyond recommended torque values.
	2. Foreign objects falling into drive	2. Use adequate drive guard.
	3. Excessive rim speeds	3. Keep pulley rim speeds below maximum recommended value.
	4. Incorrect belt installation	4. Do not pry belts onto pulleys.
• Severe Groove Wear	1. Excessive belt tension	1. Retension, check drive design.
	2. Sand, debris or contamination	2. Clean and shield drive as well as possible.

Problem With Other Drive Components

Symptoms	Probable Cause	Corrective Action
• Bent or broken shaft	1. Extreme belt overtension	1. Retension
	2. Overdesigned drive*	2. Check drive design, may need to use smaller or fewer belts.
	3. Accidental damage	3. Redesign drive guard.
	4. Machine design error	4. Check machine design.
	5. Accidental damage to guard or poor guard design	5. Repair, redesign for durability.

*Using too many belts, or belts that are too large, can severely stress motor or driveN shafts. This can happen when load requirements are reduced on a drive, but the belts are not redesigned accordingly. This can also happen when a drive is greatly overdesigned. Forces created from belt tensioning are too great for the shafts.

Hot Bearings

Symptoms	Probable Cause	Corrective Action
• Drive needs overtensioning	1. Worn grooves - belts bottoming and won't transmit power until overtensioned*	1. Replace sheaves. Tension drive properly.
	2. Improper tension	2. Retension.
• Sheaves too small	1. Motor manufacturer's sheave diameter recommendation not followed	1. Redesign using drive manual.
• Poor bearing condition	1. Bearing underdesigned	1. Check bearing design.
	2. Bearing not properly maintained	2. Align and lubricate bearing.
• Sheaves too far out on shaft	1. Error or obstruction problem	1. Place sheaves as close as possible to bearings. Remove obstructions.
• Belt slippage	1. Drive undertensioned	1. Retension.

*See footnote on page 489.

Belt Noise

Symptoms	Probable Cause	Corrective Action
• Belt squeals or chirps	1. Belt slip	1. Retension.
	2. Contamination	2. Clean belts and sheaves.
• Slapping Sound	1. Loose belts	1. Retension.
	2. Mismatched set	2. Install matched belt set.
	3. Misalignment	3. Realign pulleys so all belts share load equally.
• Rubbing Sound	1. Guard interference	1. Repair, replace or redesign guard.
• Grinding Sound	1. Damaged bearings	1. Replace, align & lubricate.
• Unusually loud drive	1. Incorrect belt	1. Use correct belt size. Use correct belt tooth profile for sprockets on synchronous drive.
	2. Worn sheaves	2. Replace
	3. Debris in sheaves	3. Clean sheaves, improve shielding, remove rust, paint, or remove dirt from grooves.

Unusual Vibration

Symptoms	Probable Cause	Corrective Action
• Belts flopping	1. Loose belts (under tensioned)	1. Retension.
	2. Mismatched belts	2. Install new matched set.
	3. Pulley misalignment	3. Align pulley
• Unusual or excessive vibration	1. Incorrect belt	1. Use correct belt cross section in pulley. Use correct tooth profile and pitch in sprocket.
	2. Poor machine or equipment design	2. Check structure and brackets for adequate strength.
	3. Pulley out of round	3. Replace with non-defective pulley.
	4. Loose drive components	4. Check machine components and guards, motor mounts, motor pads, bushings, brackets and framework for stability, adequate design strength, proper maintenance and proper installation.

Problems With Banded (Joined) Belts

Symptoms	Probable Cause	Corrective Action
• Tie band separation 	1. Worn sheaves 2. Improper groove spacing	1. Replace sheaves. 2. Use standard groove sheaves.
• Top of tie band frayed or worn 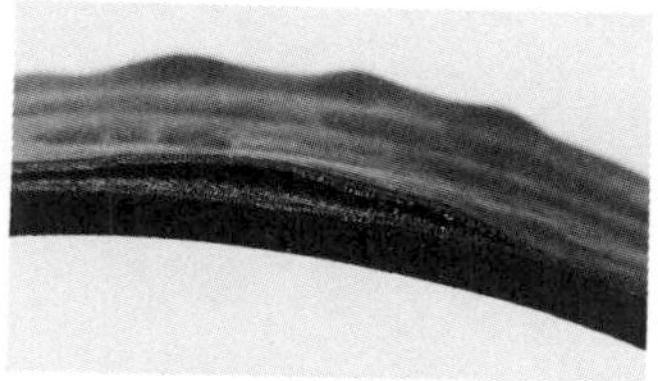	1. Interference with guard 2. Backside idler malfunction or damaged	1. Check guard. 2. Replace or repair backside idler.
• PowerBand® Belt comes off drive repeatedly	1. Debris in sheaves	1. Clean grooves. Use single belts to prevent debris from being trapped in grooves.
• One or more "ribs" runs out of pulley 	1. Misalignment 2. Undertensioned	1. Realign drive. 2. Retension.

Problems With Synchronous Belts

Symptoms	Probable Cause	Corrective Action
• Unusual Noise	1. Misaligned drive	1. Correct alignment.
	2. Too low or high tension	2. Adjust to recommended value
	3. Backside idler	3. Use inside idler.
	4. Worn sprocket	4. Replace.
	5. Bent guide flange	5. Replace.
	6. Belt speed too high	6. Redesign drive.
	7. Incorrect belt profile for sprocket (i.e. HTD, GT, etc.)	7. Use proper belt/sprocket combination.
	8. Subminimal diameter	8. Redesign drive using larger diameters.
	9. Excess load	9. Redesign drive for increased capacity.
• Tension Loss	1. Weak support structure	1. Reinforce structure.
	2. Excessive sprocket wear	2. Use alternate sprocket material.
	3. Fixed (non-adjustable) centers	3. Use inside idler for belt adjustment.
	4. Excessive debris	4. Remove debris, check guard.
	5. Excessive load	5. Redesign drive for increased capacity.
	6. Subminimal diameter	6. Redesign drive using larger diameters.
	7. Belt, sprocket or shafts running too hot	7. Check for conductive heat transfer from prime mover.
	8. Unusual belt degradation	8. Reduce ambient drive temperature to 185°F maximum.

Symptom	Probable Cause	Corrective Action
• Excessive Belt Edge Wear	1. Damage due to handling	1. Follow proper handling instructions.
	2. Flange damage	2. Repair flange or replace sprocket.
	3. Belt too wide	3. Use proper width sprocket.
	4. Belt tension too low	4. Adjust tension to recommended value.
	5. Rough flange surface finish	5. Replace or repair flange (to eliminate abrasive surface).
	6. Improper tracking	6. Correct alignment.
	7. Belt hitting drive guard or bracketry	7. Remove obstruction or use inside idler.
• Tensile Break 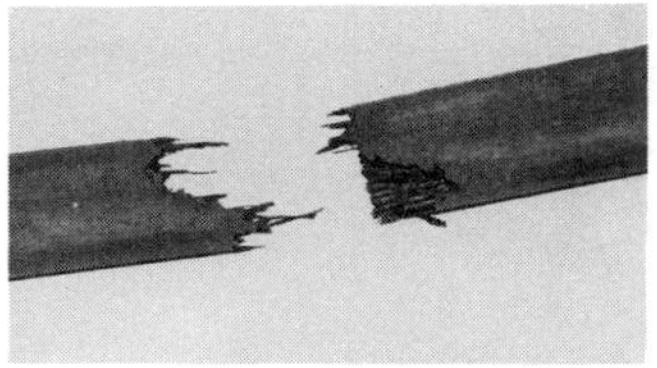	1. Excessive shock load	1. Redesign drive for increased capacity.
	2. Subminimal diameter	2. Redesign drive using larger diameters.
	3. Improper belt handling and storage prior to installation	3. Follow proper storage and handling procedures.
	4. Debris or foreign object in drive	4. Remove objects and check guard.
	5. Extreme sprocket run-out	5. Replace sprocket.
• Belt Cracking	1. Subminimal diameter	1. Redesign drive using larger diameter.
	2. Backside idler	2. Use inside idler or increase diameter of backside idler.
	3. Extreme low temperature at start-up.	3. Pre-heat drive environment.
	4. Extended exposure to harsh chemicals	4. Protect drive.
	5. Cocked bushing/sprocket assembly	5. Install bushing per instructions.

(continued)

Problem	Cause	Remedy
• Premature Tooth Wear  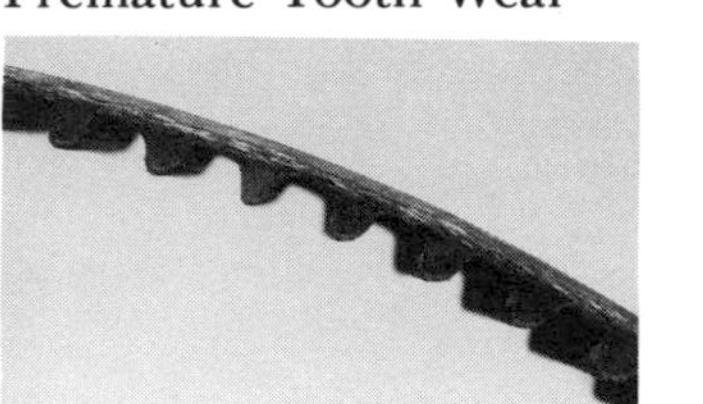	1. Too low or high belt tension	1. Adjust to recommended value.
	2. Belt running partly off unflanged sprocket	2. Correct alignment.
	3. Misaligned drive	3. Correct alignment.
	4. Incorrect belt profile for sprocket (i.e. HTD, GT, etc)	4. Use proper belt/sprocket combination.
	5. Worn sprocket	5. Replace.
	6. Rough sprocket teeth	6. Replace sprocket.
	7. Damaged sprocket	7. Replace.
	8. Sprocket not to dimensional specification	8. Replace.
	9. Belt hitting drive bracketry or other structure	9. Remove obstruction or use idler.
	10. Excessive load	10. Redesign drive for increased capacity.
	11. Insufficient hardness of sprocket material	11. Use a more wear-resistant sprocket
	12. Excessive debris	12. Remove debris, check guard.
	13. Cocked bushing/sprocket assembly	13. Install bushing per instructions.
• Tooth Shear	1. Excessive shock loads	1. Redesign drive for increased capacity.
	2. Less than 6 teeth-in-mesh	2. Redesign drive.
	3. Extreme sprocket run-out	3. Replace sprocket.
	4. Worn sprocket	4. Replace.
	5. Backside idler	5. Use inside idler.
	6. Incorrect belt profile for the sprocket (i.e. HTD, GT, etc.)	6. Use proper belt/sprocket combination.
	7. Misaligned drive	7. Realign.
	8. Belt undertensioned	8. Adjust tension to recommended value.

• Flange Failure	1. Belt forcing flange off	1. Correct alignment or properly secure flange to sprocket.
• Unusual Sprocket Wear 	1. Sprocket has too little wear resistance (i.e. plastic, aluminum, soft metals)	1. Use alternate sprocket material.
	2. Misaligned drive	2. Correct alignment.
	3. Excessive debris	3. Remove debris, check guard.
	4. Excessive load	4. Redesign drive for increased capacity.
	5. Too low or high belt tension	5. Adjust tension to recommended value.
	6. Incorrect belt profile (i.e. HTD, GT, etc.)	6. Use proper belt/sprocket combination.
• Belt Tracking	1. Belt running partly off unflanged sprocket	1. Correct alignment.
	2. Centers exceed 8 times small sprocket diameter and both sprockets are flanged.	2. Correct parallel alignment to set belt to track on both sprockets.
	3. Excessive belt edge wear	3. Correct alignment.
• Excessive Temperature (Belt, Bearing, Housing, Shafts, etc.)	1. Misaligned drive	1. Correct alignment.
	2. Too low or high belt tension	2. Adjust tension to recommended value.
	3. Incorrect belt profile (i.e. HTD, GT, etc.)	3. Use proper belt/sprocket combination.
• Shafts Out of Sync	1. Design error	1. Use correct sprocket sizes.
	2 Incorrect belt	2. Use correct belt with correct tooth profile for grooves.
• Vibration	1. Incorrect belt profile for the sprocket (i.e. HTD, GT, etc.)	1. Use proper belt/sprocket combination.
	2. Too low or high belt tension	2. Adjust tension to recommended value.
	3. Bushing or key loose	3. Check and reinstall per instructions.

MAINTAINING BEARINGS*

Bearings are the principal means of supporting shafts in all types of mechanisms. There are many types of bearings - plain bearings and anti-friction bearings. Anti-friction bearings include ball bearings, roller bearings and needle bearings. Each of these, in turn, have many variations - deep groove ball bearings, Y-bearings, self-aligning ball bearings, angular contact ball bearings, cylindrical roller bearings, spherical roller bearings, tapered roller bearings, angular contact thrust ball bearings, and spherical roller thrust bearings.

Bearings are extremely vulnerable to environmental conditions and require care in maintaining their performance. A bearing malfunction could cause complete stoppage to an entire system. This can be avoided by carefully utilizing the recommendations provided in the following sections of this chapter.

Mounting & Dismounting Bearings

Proper care begins in the stock room. Store all bearings in the original unopened packages, in a dry place. The bearing number is plainly shown on the box or wrapping. Before packaging, the manufacturer protected the bearing with a grease coating. An unopened package means continued protection. Do not open the carton until ready to use.

Don't work under the handicap of poor tools, dirt, a rough bench, or cluttered area. Clean tools and surroundings will help increase bearings performance. Handle bearings with clean, dry hands and with clean rags. Lay bearing on clean paper and keep it covered. Never expose a bearing on a dirty bench or floor. Never use a bearing as a guage to check either the housing bore or the shaft fit.

Don't wash a new bearing —it is already clean and the slushing oil should not be removed. Old grease can be washed from a used bearing with a solvent but fluid and container must be clean. After this cleaning, wash the bearing out thoroughly with light oil and then relubricate. Bearings should be washed only when necessary.

Before mounting, be sure the shaft size is within the specified tolerances recommended for the bearings. The bearing seat should be perfectly round and not tapered. It should be clean and free from nicks and burrs. Support shaft firmly in a clean place - if in a vise, protect it from vise jaws. Protectors can be soft metal, wood, cardboard or paper.

To press a bearing on a shaft, fit a tubular tool over the shaft and rest it on the inner ring. Before pressure is applied to the bearing, apply a coat of light oil, micronized graphite or molybdenum disulfide dry film to the bearing seat and the bearing bore. Be sure the bearing is square on the shaft, then apply pressure by tapping the end of the pipe with a hammer or by using an arbor press.

*All bearings related information, courtesy of SKF Bearing Services Company.

To shrink an open bearing on a shaft, expand the bearing either by:

- Boiling in an emulsion of 10% to 15% soluble oil in water for 15 to 30 minutes. Be sure to place supports under the bearing to isolate it from the bottom of the container, as contact will overheat the bearing. Or by:
- Heating in a clean temperature-controlled electric oven or on a hotplate to a maximum of 121°C (250°F) for about 15 minutes. Thoroughly heat the bearing but do not overheat. This will prevent seizing on the cold shaft. After the bearing is in place against the shaft shoulder, lock it immediately with a locknut. Otherwise in shrinking, the bearing may move away from its proper position against the shaft shoulder. Or by:
- Using an induction heater.

When mounting in a split housing, check the bore of the housing to see that it is within specified tolerances and is perfectly round. Bearing must not be pinched by a small bore or because of a cocked outer ring. Don't switch housing caps - they are not interchangeable. An undersized housing bore will pinch the bearing and cause early failure.

Some precautions must be exercised when mounting bearings in a solid housing; ie, the outer ring should be perfectly square with the housing bore before any pressure is applied. Here again, the housing bore should be within the specified tolerances for the bearing size and should be perfectly round. The housing bore and bearing outside diameter should be coated with light oil or micronized graphite or molybdenum disulphide dry film to facilitate assembly.

Cover an unfinished job, even if it is left for only a few hours. Rewrap each bearing with greaseproof paper to keep out dirt and moisture.

The type of lubricant usually depends on operating conditions; follow the machine builder's instructions. When oil is used, cover about half of bottom ball or roller. It is preferred that a sight oil gauge be used and marked so as to show static and operating oil levels. This helps to determine when additional oil is required. The operating level is different from the static level and can be determined only when the bearing is in operation.

Be sure the bearing is square with and held firmly against the shaft shoulder. Secure it with a locknut and lockwasher. Housing covers must be tight to keep lubricant in and dirt out. After held bearing has been positioned, the free bearing should be located centrally in its housing to permit expansion and contraction of the shaft.

Small and medium-sized bearings may be cold-dismounted using a conventional puller. If the bearing has been mounted with an interference fit on the shaft, the puller should preferably engage the inner ring. To avoid damage to the bearing seating, the puller must be accurately centered. The use of a self-centering puller eliminates the risk of damage and dismounting is simpler and quicker.

One of the easiest mounting and dismounting procedures is the SKF Oil Injection method which is frequently used for larger-sized bearings. (See Figure 13-5.)

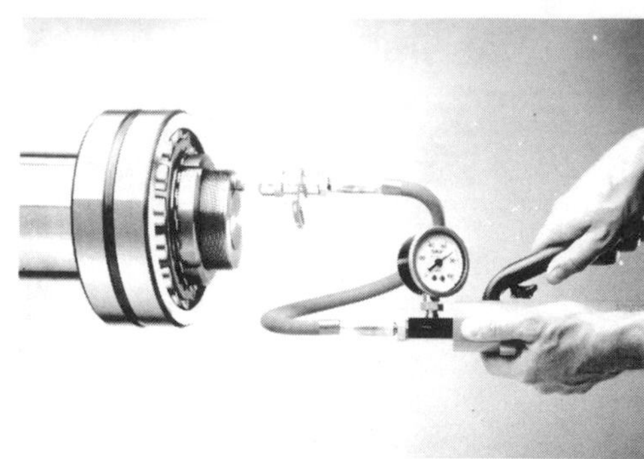

Figure 13-5. SKF Oil Injection Method of Mounting and Dismounting Bearings

An arbor press is equally good for both mounting and removing bearings. Bearing pullers may be used separately or in various combinations to pull or push complete bearings or individual rings.

Never pound directly on a bearing or a ring. This may damage both the shaft and the bearing. To drive a shaft out of a bearing, use a soft metal slug which will not mar the shaft.

Inspection and Assembly of Used Bearings

Don't try to judge the condition of a bearing until after it has been cleaned. Never spin dirty bearings, but rotate them slowly while washing. Don't spin any bearings with an air hose. Rotate one ring by hand when using air to expose all parts of the bearing.

Bearings with a shield or seal on the one side only should be washed, inspected and handled in the same manner as bearings without shields or seals. Bearings with shields or seals on both sides should not be washed. Wipe them off to keep dirt from working inside. Smooth turning bearings can be coated with protective lubricant, then wrapped and stored or used in their original application.

If a small tank with wire baskets for soaking and washing bearings is not available, a clean grease can or bucket filled with solvent can be used. Let the bearings soak long enough to loosen the grease and dirt (several hours or longer). Then slosh the bearing around near the top of the container, giving it a turn now and then until it is clean. Rinse it in a clean container of solvent. Petroleum solvents intended for bearing cleaning are preferred. Some solvents are highly flammable, and precautions should be taken to prevent fires. A short, clean bristle brush from which the bristles will not come out or break off is helpful in removing dirt, scale or chips.

After the bearings have been thoroughly cleaned, inspect them immediately. Inspected bearings that are considered good enough to be used again, but can't be reassembled in the equipment on the same day, should be dipped in slushing compound and stored overnight in a tightly covered pan.

If inspected bearings are to be stored more than a few days, dip them in a protective lubricant or coat all surfaces with a light grease, rotating them to work the grease thoroughly around the rolling members and on the raceways. Wrap the bearings in greaseproof paper and place in a clean box. Be sure to mark the outside of the package to identify the bearing. Do not leave bearings in partial assemblies exposed. Keep them covered with greaseproof paper to prevent damage by moisture, dirt or other foreign matter.

Lubricating Bearings

Adequate lubrication of rolling bearings is essential to achieving their calculated life expectancy.

Functions of Lubrication

- To lubricate the sliding contact which exists between the retainer and the other parts of the bearing.
- To lubricate that part of the contact between the raceways and rolling elements, which is not true rolling.
- To lubricate the sliding contact between the rollers and guiding elements in roller bearings.
- To protect all the elements from corrosion.
- To help seal the bearing against foreign matter.
- To provide a heat transfer medium. Both oil and grease are used for rolling bearings.

Oil Lubrication

The friction torque in a bearing is lowered with a very small quantity of oil, just sufficient to form a thin film over the contacting surfaces. However, that friction will increase with a greater quantity and with higher viscosity of the oil. With more oil than just enough to make a film, the friction torque will also increase with the speed.

When bearings have to operate in a wide range of temperatures, an oil that has the least changes with temperature variations, i.e., an oil with high viscosity index should be used. See Chapter 14.

For most applications, pure mineral oils are most satisfactory, but they should, of course, be free from contamination that may cause wear in the bearing. They should show high resistance to oxidation, gumming, and to deterioration by evaporation of light distillates. They must not cause corrosion of any parts of the bearing during standing or operation.

It is evident that for very low temperatures, an oil must be selected that has a sufficiently low pour point so that the bearing will not be locked by oil that is frozen solid.

In special applications, various compounded oils may be preferred and in such cases, the recommendation of the lubricant manufacturer should be obtained.

Relubrication Intervals

The frequency at which the oil must be changed depends mainly on the operating conditions and on the quantity of oil used. Where oil bath lubrication is employed, it is normally sufficient to change the oil once a year, providing the temperature of the oil does not exceed 50°C (120°F) and there is no contamination. Higher temperatures or more arduous running conditions necessitate more frequent changes, e.g.; at a temperature of 100°C (220°F) the oil should be changed approximately every three months. You should have the oil checked by analysis during a period of run-in and determine the change cycle from that.

For calculating oil systems, the period between complete oil changes depends on how often the oil is circulated over a given period of time and whether it is cooled, etc. The most suitable period can best be determined by trial runs and frequent examinations of the oil. The same practice also applies to oil jet lubrication. In oil mist lubrication, most of the oil is lost, as it is conveyed to the bearing only once.

Grease Lubrication

Rolling bearing greases are usually a suspension of oil in a soap or non-soap thickener, plus protection additives. When moving parts of the bearing come in contact with the grease, a small quantity of oil will adhere to the bearing surfaces. Oil is, therefore, removed from the grease near the rotating parts. The oil that is picked up by the bearing is gradually broken down by oxidation or lost by evaporation, centrifugal force, etc. Bleeding of the grease should, therefore, take place to continue to supply a small quantity of oil, which is usually sufficient for satisfactory operation. The bearing cannot operate properly unless the oil supply meets the demand and this cannot continue indefinitely. In time, the grease will oxidize or the oil in the grease near the bearing will be depleted. Relubrication is essential. You should consult with the bearing manufacturer and your grease supplier to determine the grease most suitable for the particular application, especially for temperatures below -29°C (-20°F) and above 121°C (250°F).

All grease shall be free from dirt, abrasive matter, fillers, excessive amounts of moisture, free acid or free alkali.

In greasing rolling bearings, the use of high pressure equipment is not only unnecessary, but is actually undesirable unless used with great care. High pressure may damage the bearing, damage the seals, create a danger of overheating by over greasing, cause a loss of grease and a messy piece of equipment around the bearing. A ball or roller bearing in most applications is assured of adequate lubrication if the level of grease is maintained at ⅓ to ½ the volume of the bearing or the bearing housing space.

There may be applications where it will be necessary to grease the bearings with either more or less than the recommended ⅓ to ½ of volume. In applications where low torque is a requirement, the bearings may be lubricated with a very small amount of grease. In applications where the speed is very low and the bearing is exposed to excessive amounts of dirt or moisture, the bearing may be packed nearly full.

Special synthetic oils for non-petroleum fluids are used to produce greases that can be used at extremely low or extremely high temperatures. Be careful to avoid mixing greases. In no case should petroleum oils be mixed with synthetic oils whether as oils themselves or as oils in greases, without consulting your lubricant manufacturer.

Operating Conditions

Temperature is the key here. Since a temperature rise of 8° to 11°C (14° to 20°F) can double the rate of oxidation, it is obvious that the higher the temperature, the more care in lubrication must be exercised.

Lubrication and Maintenance

Relubrication should always be undertaken at a time when the lubrication of a bearing is still satisfactory. Relubrication must be more frequent when water and/or solid contaminants can penetrate the bearing arrangement. In such cases, grease should frequently be renewed in order to remove contaminants from the bearing.

Troubleshooting Guide for Bearings Maintenance

The following Trouble Shooting Guide lists common problems encountered with bearings, their probable causes and their solutions.

Overheated Bearing

Probable Cause	Solution	
Wrong type of grease or oil causing break-down of lubricant.	Consult reliable lubricant manufacturer for proper type of lubricant. Check bearing manufacturer's instructions to determine if oil or grease should be used.	
Low oil level. Loss of lubricant through seal. Insufficient grease in housing.	Oil level should be just below center of lowest ball or roller in bearing. Using grease, lower half of pillow block should be ½ to ⅔ full.	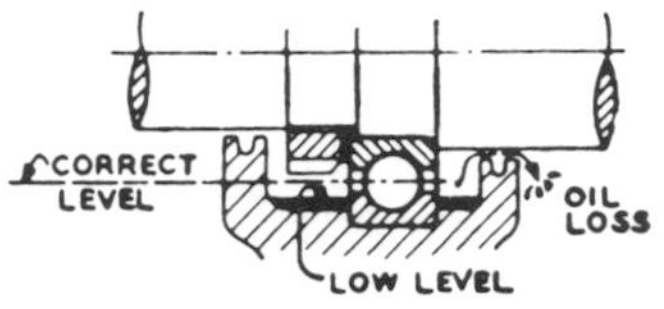
Housing packed with grease, or oil level too high . . . causing excessive churning of lubricant, high operating temperature, oil leakage.	Purge bearing until only lower half of housing is ½ to ⅔ full of grease. Using oil lubrication, reduce level to just below center of lowest ball or roller.	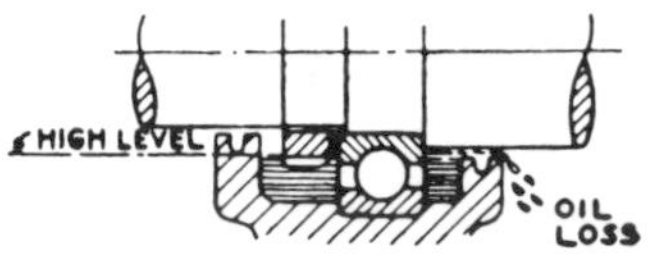
Bearings selected with inadequate internal clearance for conditions where external heat is conducted thru shaft, thereby expanding the inner ring excessively.	Replacement bearing should have identical marking as original bearing for proper internal clearance. Check with bearing manufacturer if bearing markings have become indistinct.	
Housing bore out of round. Housing warped. Excessive distortion of housing. Undersized housing bore.	Check and scrape housing bore to relieve pinching of bearing. Be sure pedestal surface is flat, and shims cover entire area of pillow block base.	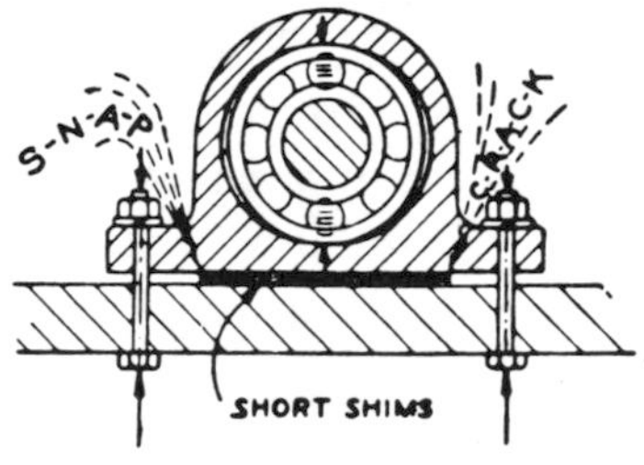
Leather or composition seals with excessive spring tension or dried out.	Replace leather or composition seals with ones having reduced spring tension. Lubricate seals.	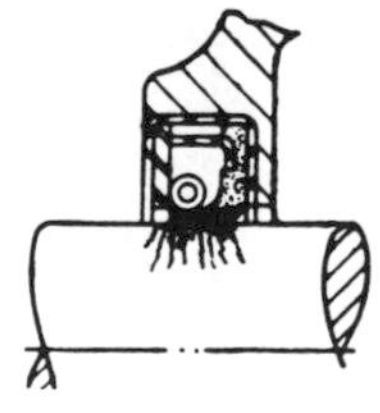

(*continued*)

Probable Cause	Solution
Rotating seals rubbing against stationary parts.	Check running clearance of rotating seal to eliminate rubbing. Correct alignment. 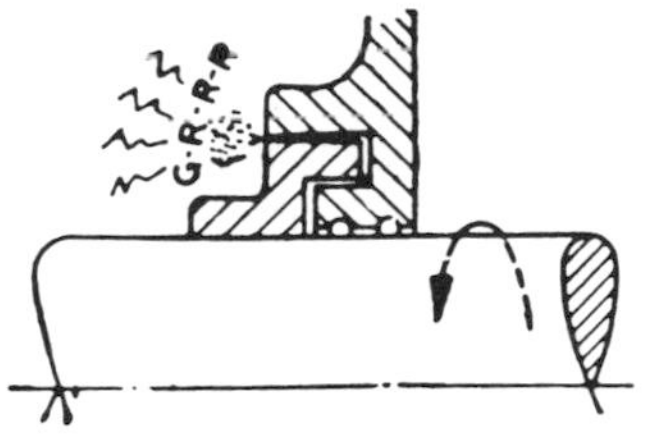
Oil return holes blocked—pumping action of seals cause oil leakage.	Clean holes. Drain out used oil — refilling to proper oil level with fresh lubricant. 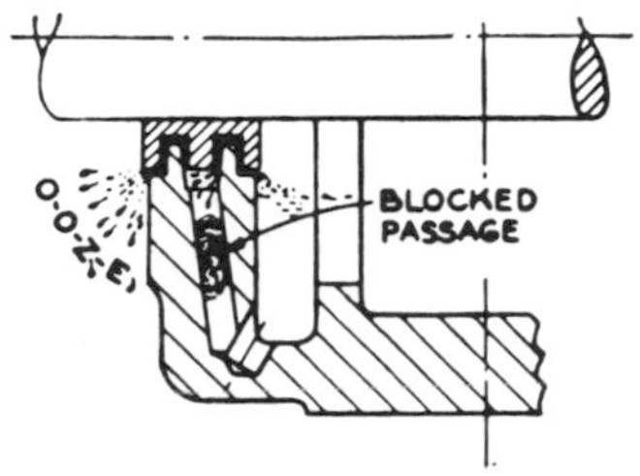
Opposed mounting.	Insert gasket between housing and cover flange to relieve axial preloading of bearing. 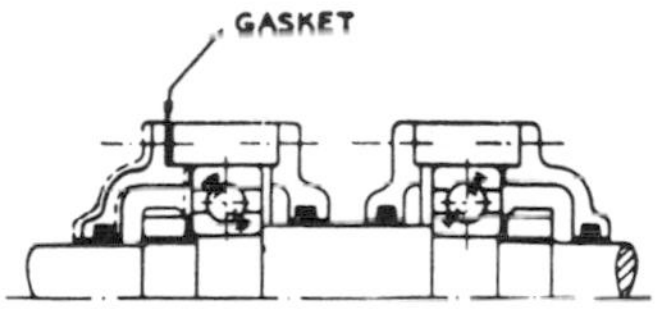
Two "held" bearings on one shaft. Excessive shaft expansion.	Back off covers in one of the housings, using shims to obtain adequate clearance of outer ring, to permit free axial bearing motion. 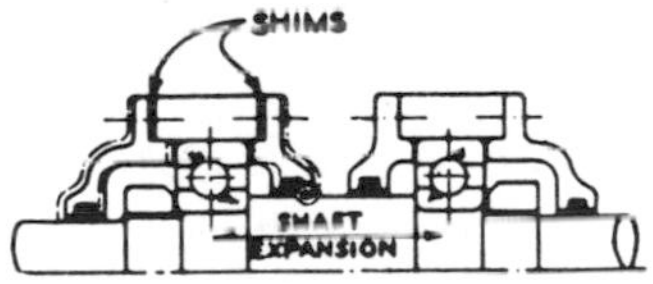
Adapter tightened excessively.	Loosen locknut and sleeve assembly. Retighten sufficiently to clamp sleeve on shaft but be sure bearing turns freely. 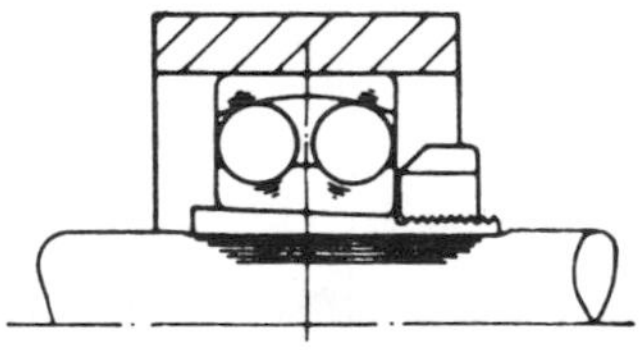
Unbalanced load. Housing bore too large.	Rebalance machine. Replace housing with one having proper bore. 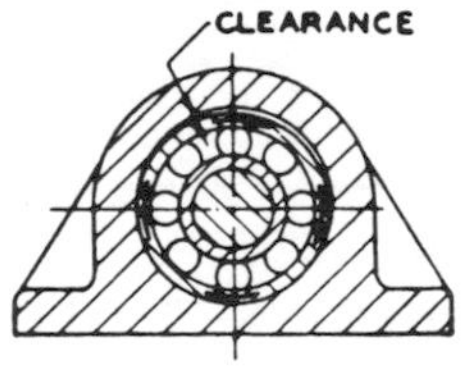

(continued)

Probable Cause	Solution
Rubbing of shaft shoulder against bearing seals.	Remachine shaft shoulder to clear seal. 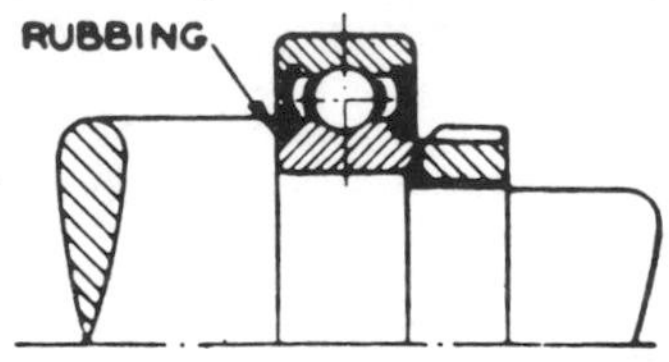
Incorrect oil level. Result: no lubricant in bearing.	Clean out clogged hole to vent oil gauge. 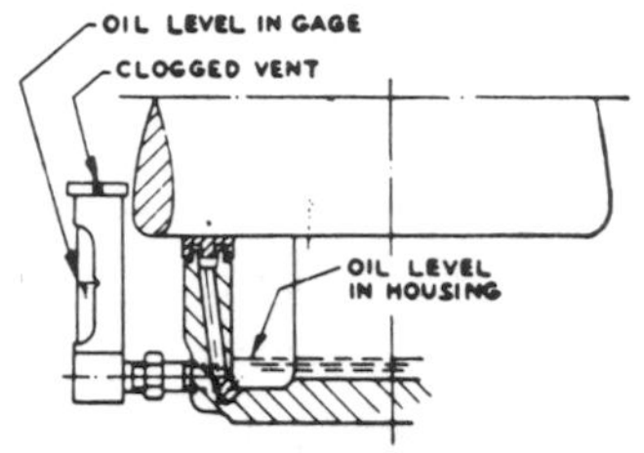
Incorrect linear or angular alignment of two or more coupled shafts with two or more bearings.	Correct alignment by shimming pillow blocks. Be sure shafts are coupled in straight line—especially when three or more bearings operate on one shaft. 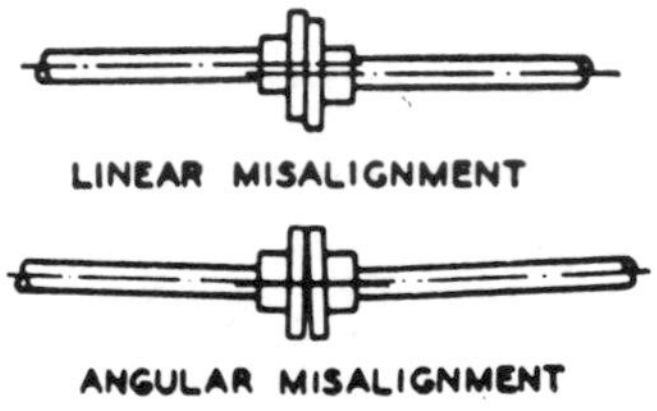
Incorrect mounting of constant oil level cup. (Too high or too low.) Cup located opposite rotation of bearing permitting excessive flow of oil, resulting in too high oil level.	The oil level at standstill must not exceed the center of the lowermost ball or roller. Sketch illustrates correct position of constant level oil cup with respect to rotation. Better replace constant level oiler with sight gage. 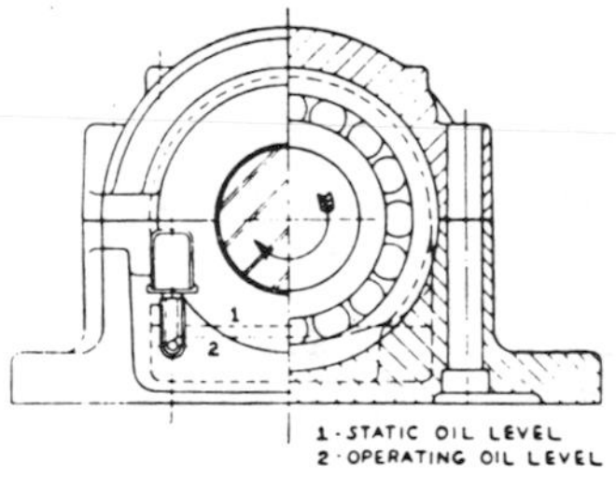
Prong rubbing against bearing.	Remove lockwasher — straighten prong or replace with new washer. 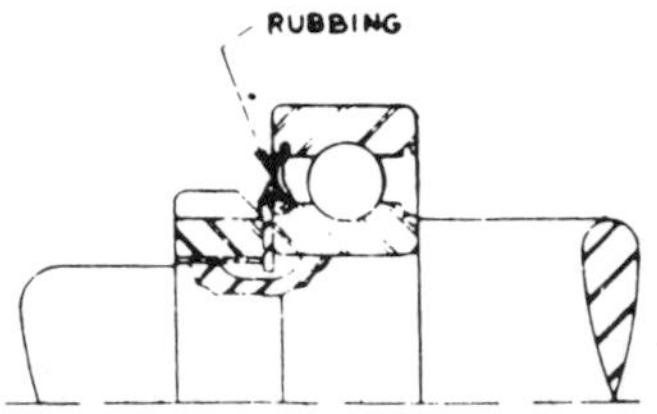
Knurling and center punching of bearing seat on shaft.	Unsatisfactory because high spots are flattened when load is applied, when fit is loose, metalize shaft and regrind to proper size.

(*continued*)

Probable Cause	Solution
Bearing seat diameter machined oversize, causing excessive expansion of bearing inner ring, thus reducing clearance in bearing.	Grind shaft to get proper fit between inner ring of bearing and shaft.
"Pounding-out" of housing bore due to soft metal. Result: enlarged bore . . . causing spinning of outer ring in housing.	Rebore housing and press steel bushing in bore. Machine bore of bushing to correct size.

Noisy Bearing

Probable Cause	Solution
Wrong type of grease or oil causing break-down of lubricant.	Consult reliable lubricant manufacturer for proper type of lubricant. Check bearing manufacturer's instructions to determine if oil or grease should be used.
Low oil level. Loss of lubricant through seal.	Oil level should be just below center of lowest ball or roller in bearing. 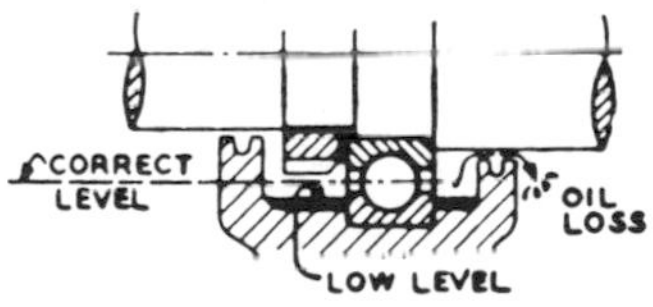
Insufficient grease in housing.	Using grease, lower half of pillow block should be ½ to ⅔ full.
Bearings selected with inadequate internal clearance for conditions where external heat is conducted thru shaft, thereby expanding the inner ring excessively.	Replacement bearing should have identical marking as original bearing for proper internal clearance. Check with bearing manufactuer if bearing markings have become indistinct.
Foreign matter (dirt, sand, carbon, etc.) entering bearing housing.	Clean out bearing housing. Replace worn-out seals or improve seal design to obtain adequate protection of bearing.
Corrosive agents (water, acids, paints, etc.) entering the bearing housing.	Addition of a shroud and (or) flinger to throw off foreign matter.

(*continued*)

Probable Cause	Solution	
Housing bore out of round. Housing warped. Excessive distortion of housing. Undersized housing bore.	Check and scrape housing bore to relieve pinching of bearing. Be sure pedestal surface is flat, and shims cover entire area of pillow block base.	
Failure to remove chips, dirt, etc. from bearing housing before assembling bearing unit.	Carefully clean housing, and use fresh lubricant.	
Rotating seals rubbing against stationary parts.	Check running clearance of rotating seal to eliminate rubbing. Correct alignment.	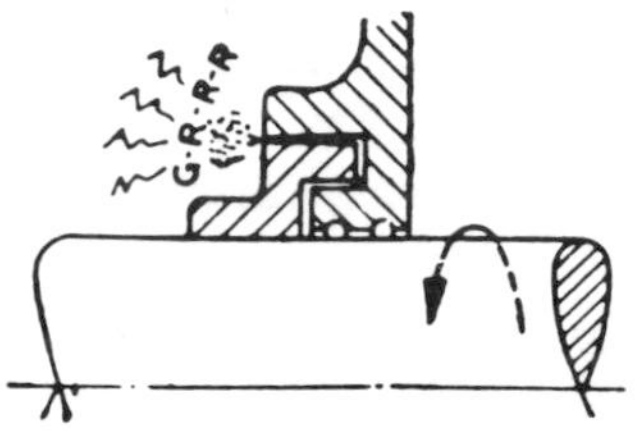
Opposed mounting.	Insert gasket between housing and cover flange to relieve axial preloading of bearing.	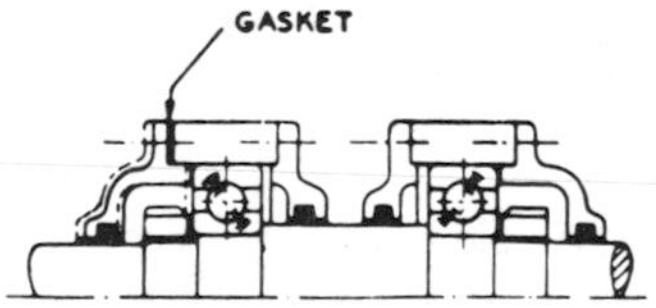
Two "held" bearings on one shaft. Excessive shaft expansion.	Back off covers in one of the housings, using shims to obtain adequate clearance of outer ring, to permit free axial bearing motion.	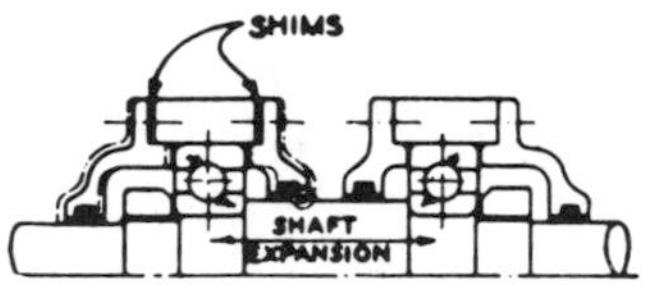
Shaft diameter too small. Adapter not tightened sufficiently.	Metallize shaft and regrind to obtain proper fit. Retighten adapter to get firm grip on shaft.	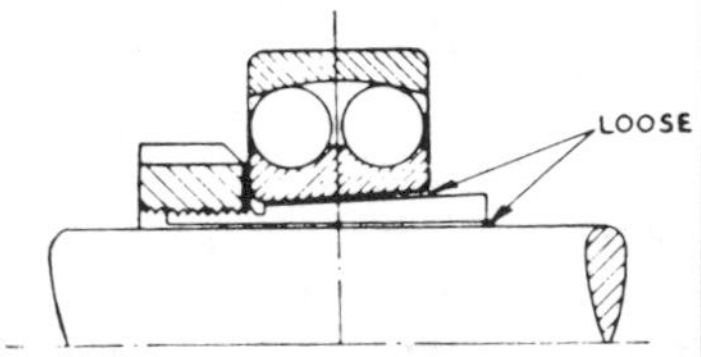

(*continued*)

Probable Cause	Solution
Adapter tightened excessively.	Loosen locknut and sleeve assembly. Retighten sufficiently to clamp sleeve on shaft but be sure bearing turns freely. 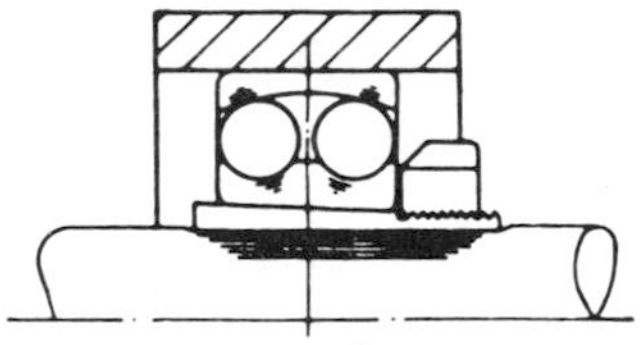
Flat on ball or roller due to skidding. (Result of fast starting.)	Carefully examine balls or rollers, looking for flat spots on the surface. Replace bearing.
Rubbing of shaft shoulder against bearing seals.	Remachine shaft shoulder to clear seal. (RUBBING) 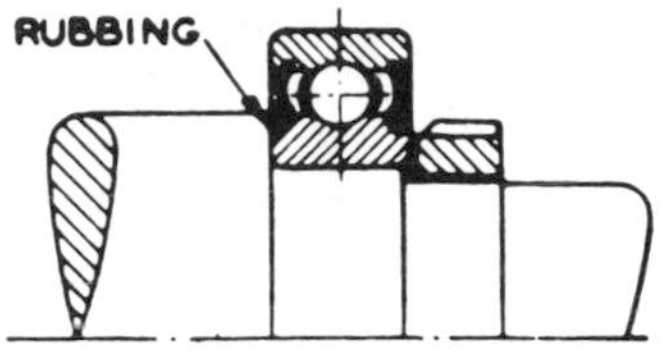
Distortion of bearing seals.	Remachine housing shoulder to clear seal. (INTERFERENCE) 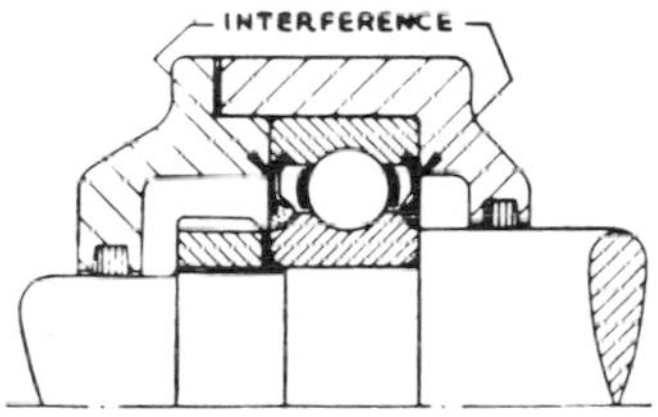
Prong rubbing against bearing.	Remove lockwasher — straighten prong or replace with new washer. (RUBBING) 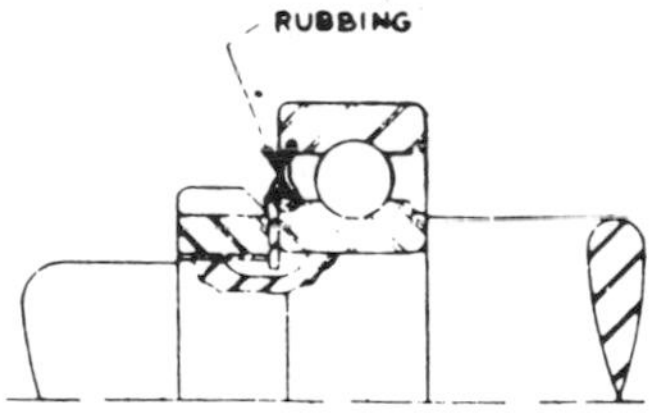
Incorrect method of mounting. Hammer blows on bearing.	Replace with new bearing. Don't hammer any part of bearing when mounting.
Interference of other movable parts of machine.	Carefully check every moving part for interference. Reset parts to provide necessary clearance.
Distorted shaft and other parts of bearing assembly.	Only in extreme cases should a torch be used to facilitate removal of a failed bearing. Care should be exercised to avoid high heat concentration at any one point so distortion is eliminated.

(*continued*)

Probable Cause	Solution
Bearing seat diameter machined oversize causing excessive expansion of bearing inner ring, thus reducing clearance in bearing.	Grind shaft to get proper fit between inner ring of bearing and shaft.
Unbalanced load. Housing bore too large.	Rebalance unit. Replace housing with one having proper bore.
"Pounding-out" of housing bore due to soft metal. Result: enlarged bore . . . causing spinning of outer ring in housing.	Rebore housing and press steel bushing in bore. Machine bore of bushing to correct size.
Bearing exposed to vibration while machine is idle.	Carefully examine bearing for wear spots separated by distance equal to the spacing of the balls. Replace bearing.

(*continued*)

Replacements Are Too Frequent

Probable Cause	Solution
Wrong type of grease or oil causing break-down of lubricant.	Consult reliable lubricant manufacturer for proper type of lubricant. Check bearing manufactuer's instructions to determine if oil or grease should be used.
Low oil level. Loss of lubricant through seal.	Oil level should be just below center of lowest ball or roller in bearing.
Insufficient grease in housing.	Using grease, lower half of pillow block should be ½ to ⅔ full.
Bearings selected with inadequate internal clearance for conditions where external heat is conducted thru shaft, thereby expanding the inner ring excessively.	Replacement bearing should have identical marking as original bearing for proper internal clearance. Check with bearing manufacturer if bearing markings have become indistinct.

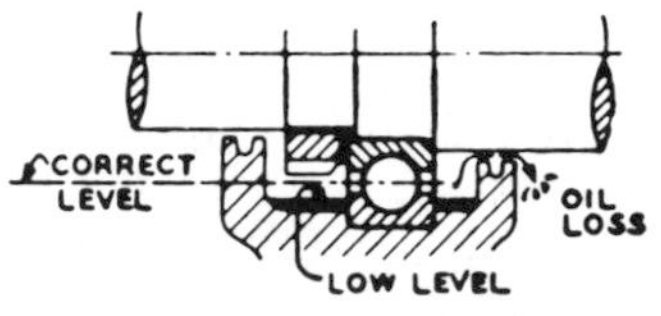

(*continued*)

Probable Cause	Solution	
Foreign matter (dirt, sand, carbon, etc.) entering into bearing housing.	Clean out bearing housing. Replace worn-out seals or improve seal design to obtain adequate protection of bearing.	
Corrosive agents (water, acids, paints, etc.) entering the bearing housing.	Addition of a shroud and (or) flinger to throw off the foreign matter.	
Housing bore out of round. Housing warped. Excessive distortion of housing. Undersized housing bore.	Check and scrape housing bore to relieve pinching of bearing. Be sure pedestal surface is flat, and shims cover entire area of pillow block base. 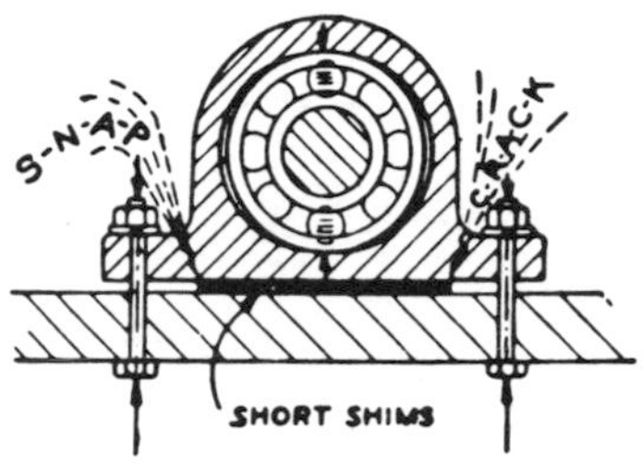	SNAP CRACK SHORT SHIMS
Failure to remove chips, dirt, etc. from bearing housing before assembling bearing unit.	Carefully clean housing, and use fresh lubricant.	
Oil leakage resulting from air flow over bearings. (Example: forced draft fan with air inlet over bearings.)	Provide proper baffles to divert direction of air flow.	
Opposed mounting.	Insert gasket between housing and cover flange to relieve axial preloading of bearing. 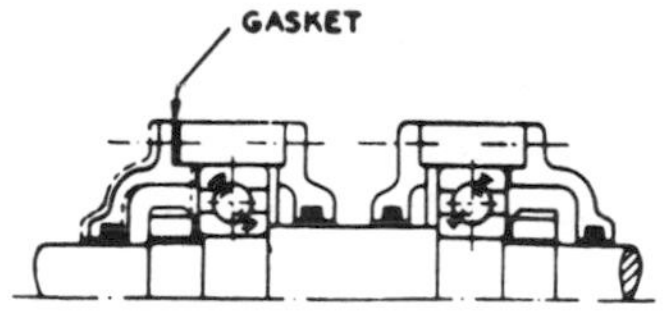	GASKET
Two "held" bearings on one shaft. Excessive shaft expansion.	Back off covers in one of the housings, using shims to obtain adequate clearance of outer ring, to permit free axial bearing motion. 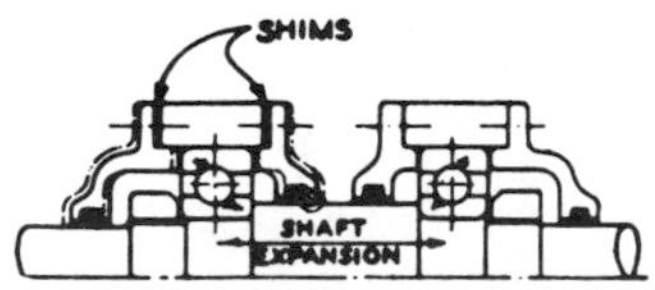	SHIMS SHAFT EXPANSION
Shaft diameter too small. Adapter not tightened sufficiently.	Metallize shaft and regrind to obtain proper fit. Retighten adapter to get firm grip on shaft. 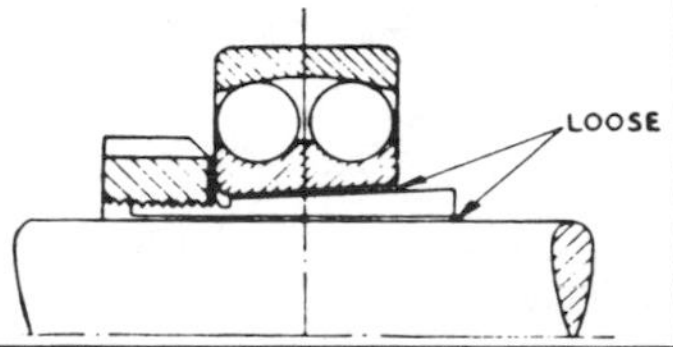	LOOSE

(*continued*)

Probable Cause	Solution	
Adapter tightened excessively.	Loosen locknut and sleeve assembly. Retighten sufficiently to clamp sleeve on shaft but be sure bearing turns freely.	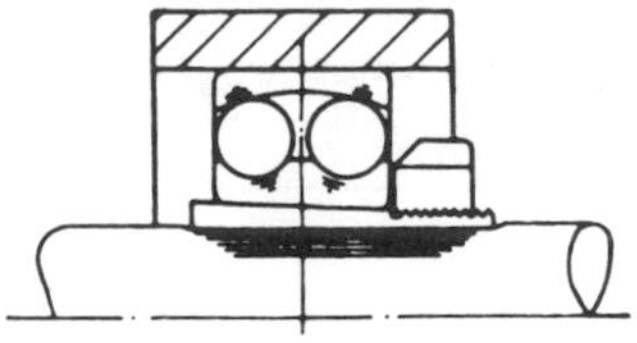
Oil leakage at housing split. Excessive loss of lubricant.	If not severe, use thin layer of gasket cement. Don't use shims. Replace housing if necessary.	 UNEVEN SURFACES DRIP DRIP
Unbalanced load. Housing bore too large.	Rebalance machine. Replace housing with one having proper bore.	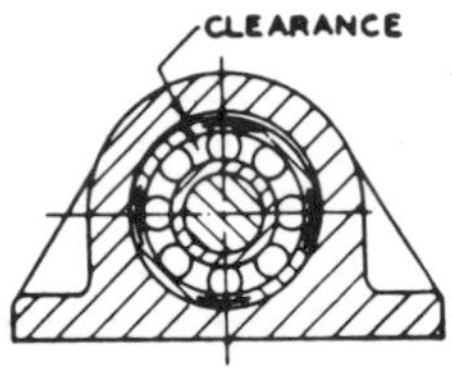 CLEARANCE
Unequal load distribution on bearing.	Rework shaft, housing, or both, to obtain proper fit. May require new shaft and housing.	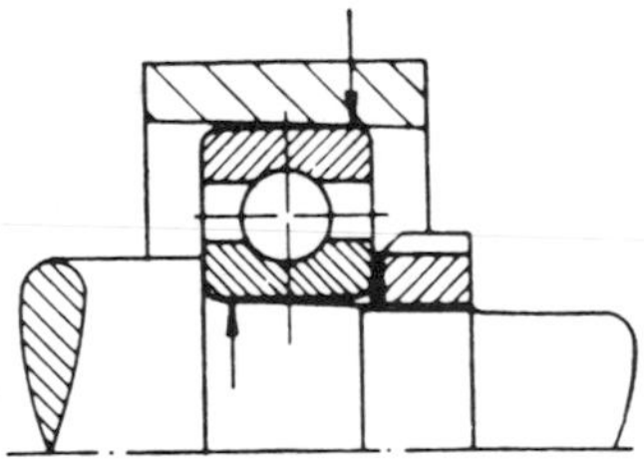
Inadequate shoulder support causing bending of shaft.	Remachine shaft fillet to relieve stress. May require shoulder collar.	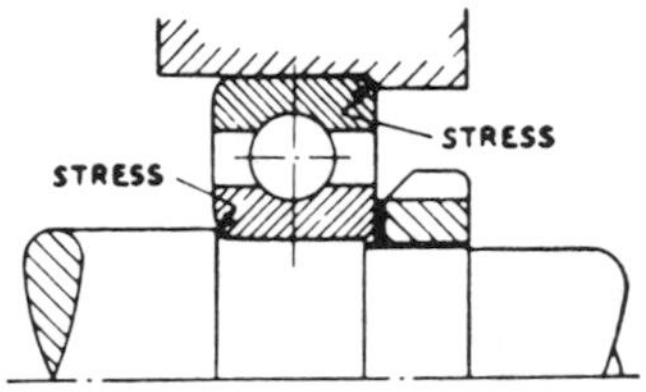 STRESS STRESS
Inadequate support in housing causing cacking of outer ring.	Remachine housing fillet to relieve stress. May require shoulder collar.	
Rubbing of shaft shoulder against bearing seals.	Remachine shaft shoulder to clear seal.	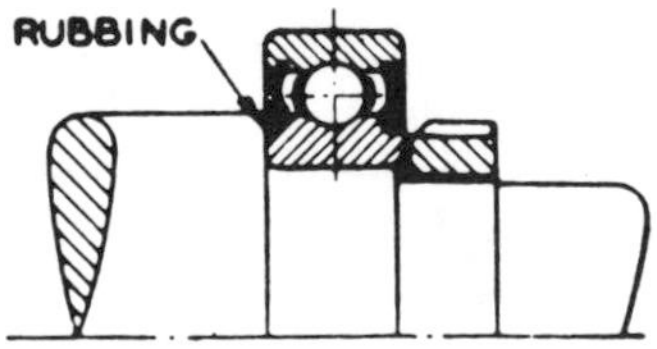 RUBBING

(*continued*)

Probable Cause	Solution	
Distortion of bearing seals.	Remachine housing shoulder to clear seal.	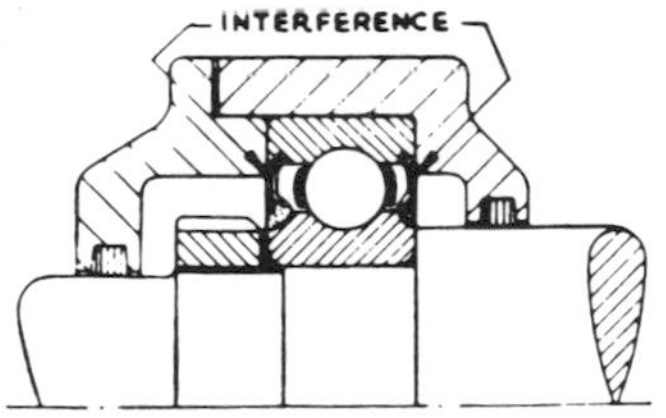 INTERFERENCE
Distortion of shaft and inner ring. Uneven expansion of bearing inner ring.	Remachine shaft fillet to obtain proper support.	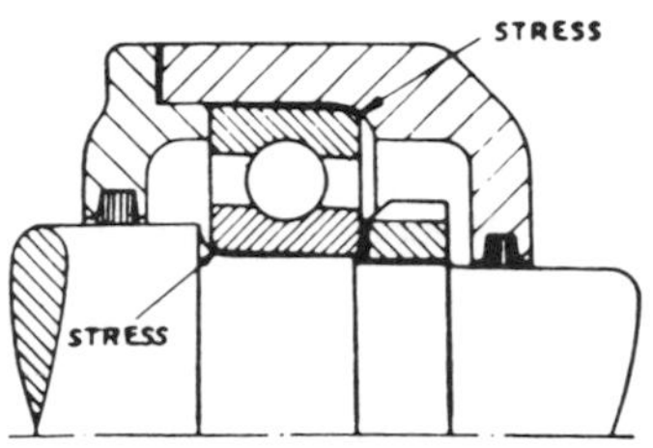 STRESS
Distortion of housing and outer ring. Pinching of bearing.	Remachine housing fillet to obtain proper support.	STRESS
Rotating seals rubbing against stationary parts.	Check running clearance of rotating seal to eliminate rubbing. Correct alignment.	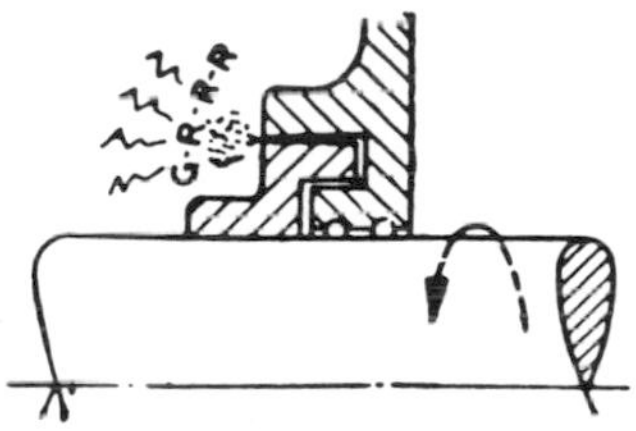 G-R-R-R
Incorrect oil level. Result: no lubricant in bearing.	Clean out clogged hole to vent oil gauge.	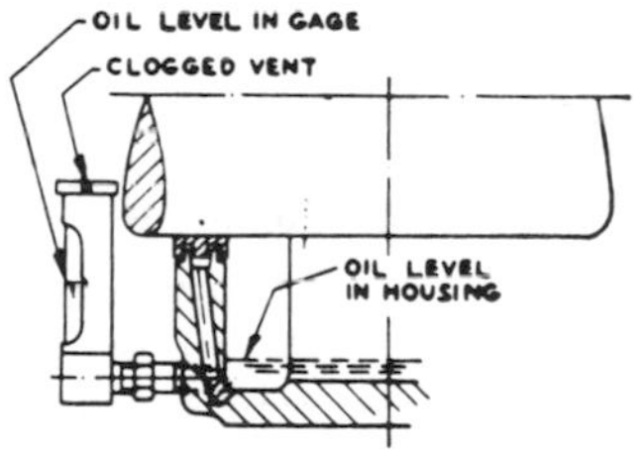 OIL LEVEL IN GAGE CLOGGED VENT OIL LEVEL IN HOUSING
Incorrect linear or angular alignment of two or more coupled shafts with two or more bearings.	Correct alignment by shimming pillow blocks. Be sure shafts are coupled in straight line—especially when three or more bearings operate on one shaft.	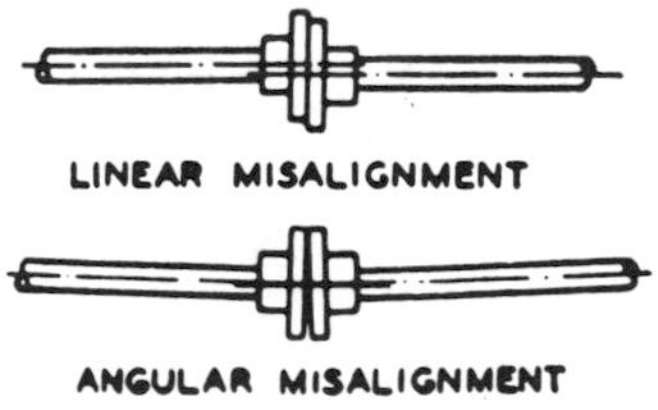 LINEAR MISALIGNMENT ANGULAR MISALIGNMENT

(continued)

<table>
<tr><th>Probable Cause</th><th colspan="2">Solution</th></tr>
<tr><td>Incorrect mounting of constant oil level cup. (Too high or too low.) Cup located opposite rotation of bearing permitting excessive flow of oil, resulting in too high oil level.</td><td>The oil level at standstill must not exceed the center of the lowermost ball or roller. Sketch illustrates correct position of constant level oil cup with respect to rotation. Better replace constant level oiler with sight gage.</td><td>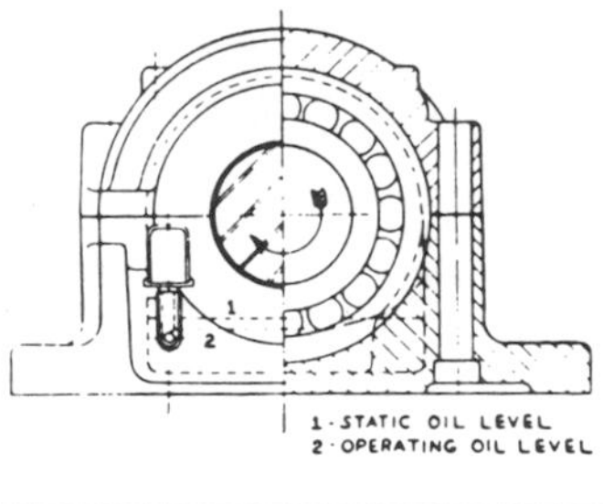
</td></tr>
<tr><td>Incorrect method of mounting. Hammer blows on bearing.</td><td colspan="2">Replace with new bearing. Don't hammer any part of bearing when mounting.</td></tr>
<tr><td>Excessively worn leather (or composition), or labyrinth seals. Result: lubricant loss; dirt getting into bearing.</td><td>Replace seals after thoroughly flushing bearing and refilling with fresh lubricant.</td><td>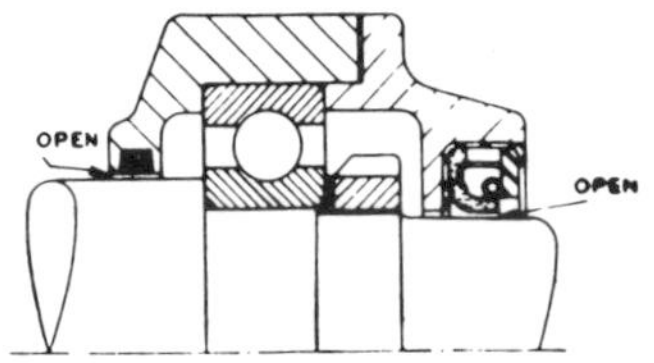
</td></tr>
<tr><td>Shaft and housing shoulders and face of locknut out-of-square with bearing seat.</td><td colspan="2">Remachine parts to obtain squareness.</td></tr>
<tr><td>Bearing seat diameter machined oversize, causing excessive expansion of bearing inner ring, thus reducing clearance in bearing.</td><td colspan="2">Grind shaft to get proper fit between inner ring of bearing and shaft.</td></tr>
<tr><td>"Pounding-out" of housing bore due to soft metal. Result: enlarged bore . . . causing spinning of outer ring in housing.</td><td colspan="2">Rebore housing and press steel bushing in bore. Machine bore of bushing to correct size.</td></tr>
</table>

Vibration

Probable Cause	Solution	
Foreign matter (dirt, sand, carbon, etc.) entering bearing housing.	Clean out bearing housing. Replace worn-out seals or improve seal design to obtain adequate protection of bearing.	
Corrosive agents (water, acids, paints, etc.) entering the bearing housing.	Addition of a shroud and (or) flinger to throw off foreign matter.	
Housing bore out of round. Housing warped. Excessive distortion of housing. Undersized housing bore.	Check and scrape housing bore to relieve pinching of bearing. Be sure pedestal surface is flat, and shims cover entire area of pillow block base.	
Failure to remove chips, dirt, etc. from bearing housing before assembling bearing unit.	Carefully clean housing, and use fresh lubricant.	
Shaft diameter too small. Adapter not tightened sufficiently.	Metallize shaft and regrind to obtain proper fit. Retighten adapter to get firm grip on shaft.	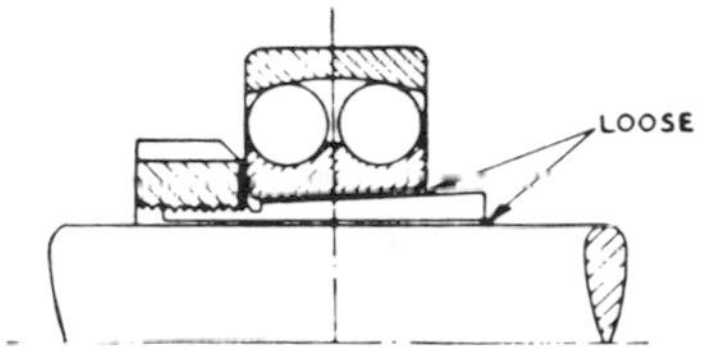
Unbalanced load. Housing bore too large.	Rebalance machine. Replace housing with one having proper bore.	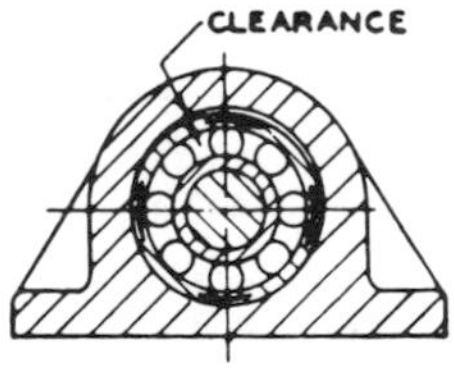
Flat on ball or roller due to skidding. (Result of fast starting.)	Carefully examine balls or rollers, looking for flat spots on the surface. Replace bearing.	

(*continued*)

Probable Cause	Solution	
Unequal load distribution on bearing.	Rework shaft, housing, or both, to obtain proper fit. May require new shaft and housing.	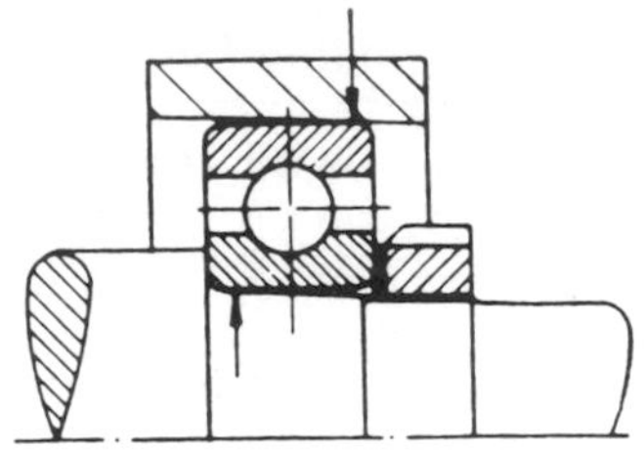
Inadequate shoulder support causing bending of shaft.	Remachine shaft fillet to relieve stress. May require shoulder collar.	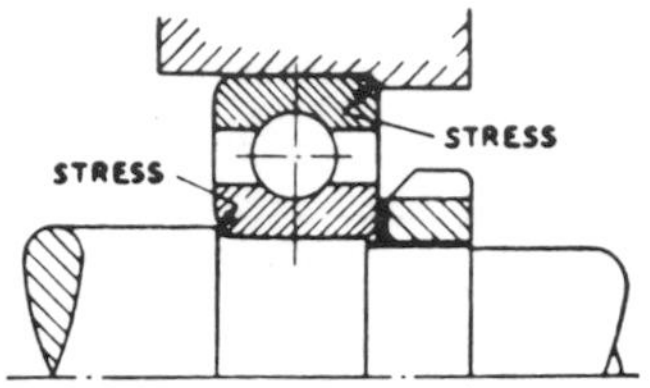 STRESS STRESS
Inadequate support in housing causing cocking of outer ring.	Remachine housing fillet to relieve stress. May require shoulder collar.	
Distortion of shaft and inner ring. Uneven expansion of bearing inner ring.	Remachine shaft fillet to obtain proper support.	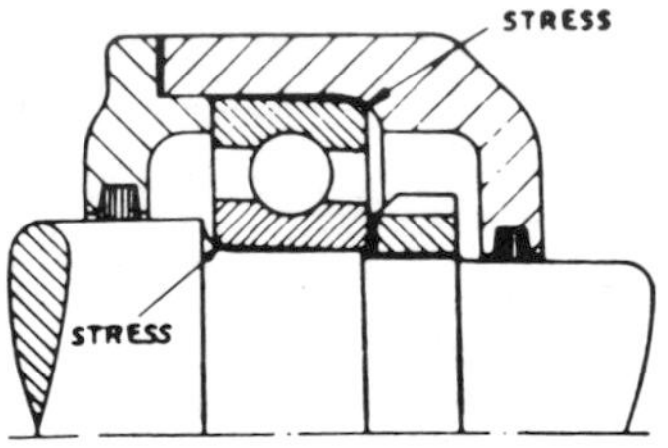 STRESS STRESS
Distortion of housing and outer ring. Pinching of bearing.	Remachine housing fillet to obtain proper support.	
Incorrect linear or angular alignment of two or more coupled shafts with two or more bearings.	Correct alignment by shimming pillow blocks. Be sure shafts are coupled in straight line—especially when three or more bearings operate on one shaft.	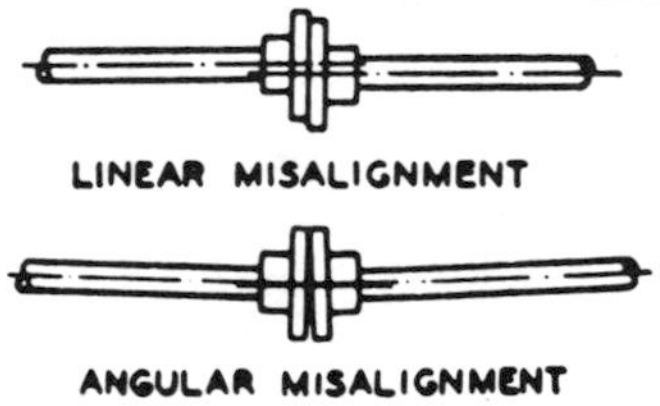 LINEAR MISALIGNMENT ANGULAR MISALIGNMENT
Incorrect method of mounting. Hammer blows on bearing.	Replace with new bearing. Don't hammer any part of bearing when mounting.	
Excessive clearance in bearing, resulting in vibration.	Use bearings with recommended internal clearances.	
Vibration of machine.	Check balance of rotating parts. Rebalance machine.	

(*continued*)

Probable Cause	Solution
"Pounding-out" of housing bore due to soft metal. Result: enlarged bore . . . causing spinning of outer ring in housing.	Rebore housing and press steel bushing in bore. Machine bore of bushing to correct size.

Unsatisfactory Equipment Performance

Probable Cause	Solution	
Bearings selected with inadequate internal clearance for conditions where external heat is conducted thru shaft, thereby expanding the inner ring excessively.	Replacement bearing should have identical marking as original bearing for proper internal clearance. Check with bearing manufacturer if bearing markings have become indistinct.	
Foreign matter (dirt, sand, carbon, etc.) entering bearing housing.	Clean out bearing housing. Replace worn-out seals or improve seal design to obtain adequate protection of bearing.	
Corrosive agents (water, acids, paints, etc.) entering the bearing housing.	Addition of a shroud and (or) flinger to throw off foreign matter.	
Housing bore out of round. Housing warped. Excessive distortion of housing. Undersized housing bore.	Check and scrape housing bore to relieve pinching of bearing. Be sure pedestal surface is flat, and shims cover entire area of pillow block base. 	

(*continued*)

Probable Cause	Solution	
Failure to remove chips, dirt, etc. from bearing housing before assembling bearing unit.	Carefully clean housing, and use fresh lubricant.	
Two "held" bearings on one shaft. Excessive shaft expansion.	Back off covers in one of the housings, using shims to obtain adequate clearance of outer ring, to permit free axial bearing motion.	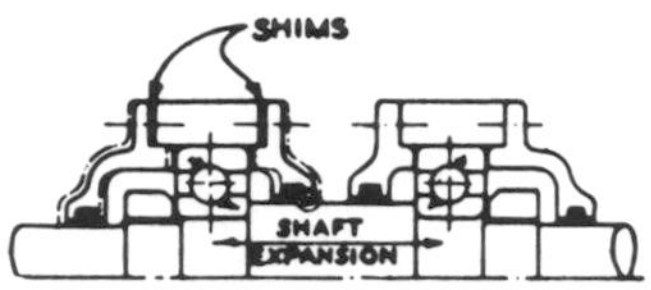
Shaft diameter too small. Adapter not tightened sufficiently.	Metallize shaft and regrind to obtain proper fit. Retighten adapter to get firm grip on shaft.	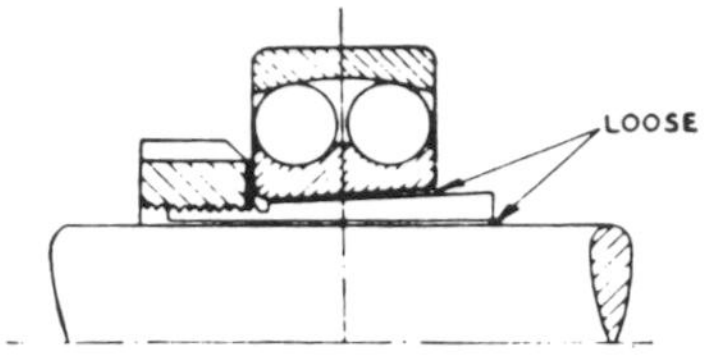
Adapter tightened excessively.	Loosen locknut and sleeve assembly. Retighten sufficiently to clamp sleeve on shaft but be sure bearing turns freely.	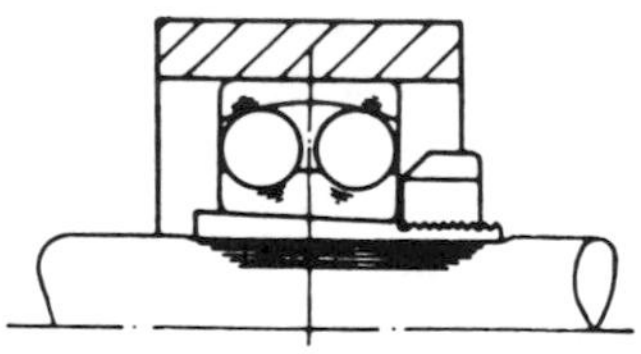
Unbalanced load. Housing bore too large.	Rebalance machine. Replace housing with one having proper bore.	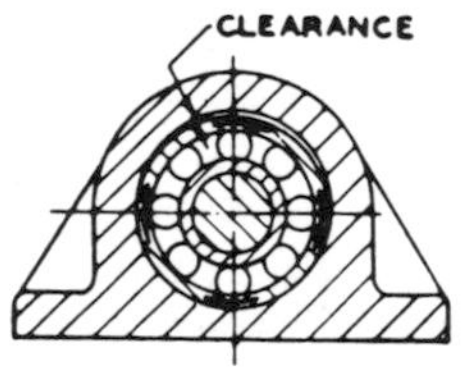
Flat on ball or roller due to skidding. (Result of fast starting.)	Carefully examine balls or rollers, looking for flat spots on the surface. Replace bearing.	
Unequal load distribution on bearing.	Rework shaft, housing, or both, to obtain proper fit. May require new shaft and housing.	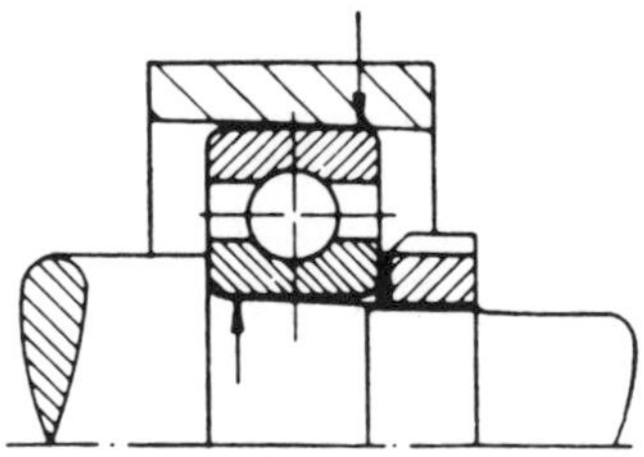

(continued)

Probable Cause	Solution
Inadequate shoulder support causing bending of shaft.	Remachine shaft fillet to relieve stress. May require shoulder collar. 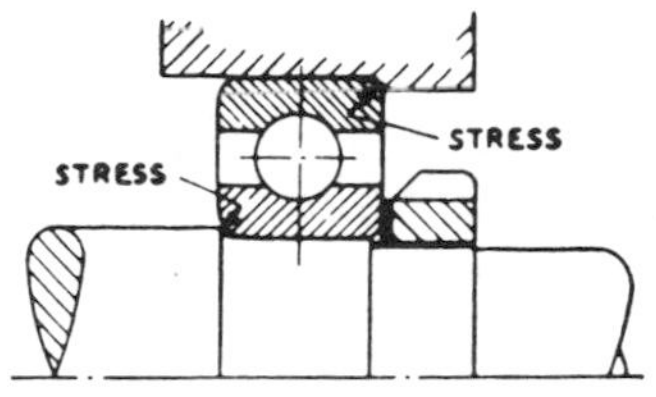
Inadequate support in housing causing cocking of outer ring.	Remachine housing fillet to relieve stress. May require shoulder collar.
Distortion of shaft and inner ring. Uneven expansion of bearing inner ring.	Remachine shaft fillet to obtain proper support. 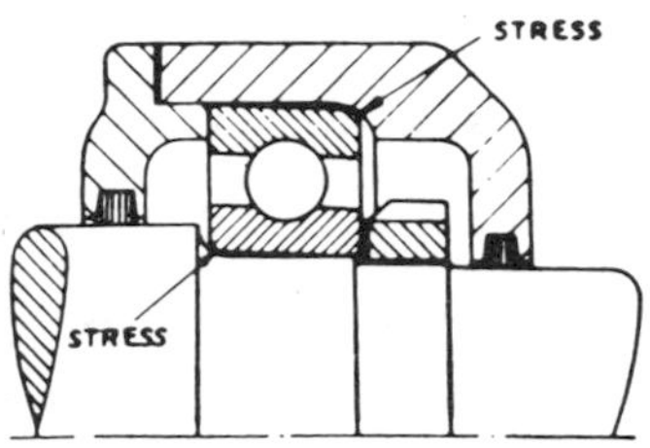
Distortion of housing and outer ring. Pinching of bearing.	Remachine housing fillet to obtain proper support.
Incorrect linear or angular alignment of two or more coupled shafts with two or more bearings.	Correct alignment by shimming pillow blocks. Be sure shafts are coupled in straight line—especially when three or more bearings operate on one shaft. 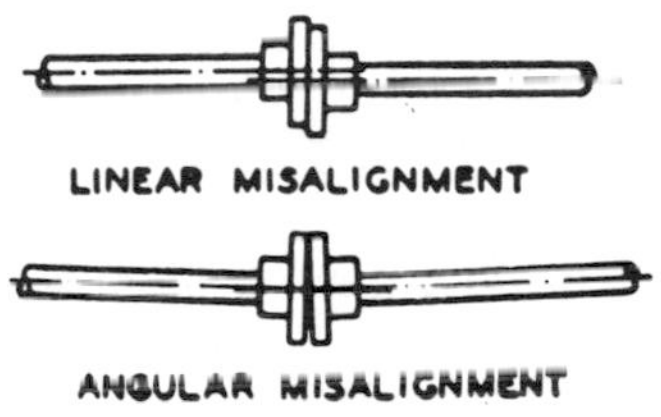
Prong rubbing against bearing.	Remove lockwasher — straighten prong or replace with new washer. 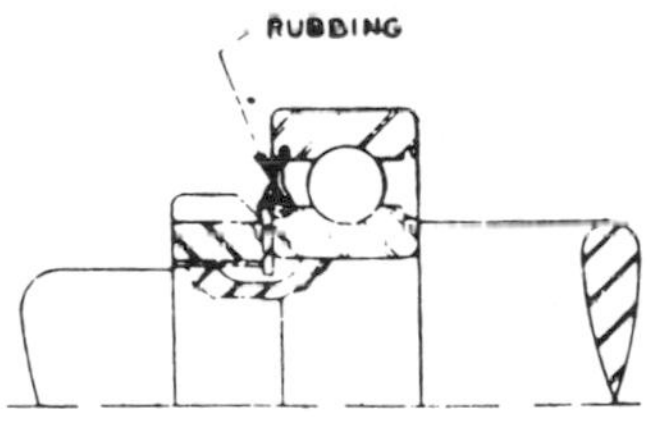
Incorrect method of mounting. Hammer blows on bearing.	Replace with new bearing. Don't hammer any part of bearing when mounting.
Excessive clearance in bearing, resulting in vibration.	Use bearings with recommended internal clearances.
Vibration of machine.	Check balance of rotating parts. Rebalance machine.
Shaft and housing shoulders, and face of locknut out of square with bearing seat.	Remachine parts to obtain squareness.

(*continued*)

Probable Cause	Solution
Bearing seat diameter machined oversize, causing excessive expansion of bearing inner ring thus reducing clearance in bearing.	Grind shaft to proper fit between inner ring of bearing and shaft.
"Pounding-out" of housing bore due to soft metal. Result: enlarged bore . . . causing spinning of outer ring in housing.	Rebore housing and press steel bushing in bore. Machine bore of bushing to correct size. If loads are not excessive, tighter fit in housing, without the use of the steel bushing, may correct the trouble.

Bearing Is Loose on Shaft

Probable Cause	Solution	
Shaft diameter too small.	Metallize shaft and regrind to obtain proper fit.	LOOSE
Adapter not tightened sufficiently.	Retighten adapter to get firm grip on shaft.	
Knurling and center punching of bearing seat on shaft.	Unsatisfactory because high spots are flattened when load is applied. When fit is loose, metallize shaft and regrind to proper size.	

Shaft Turns Hard

Probable Cause	Solution
Wrong type of grease or oil causing break-down of lubricant.	Consult reliable lubricant manufacturer for proper type of lubricant. Check bearing manufacturer's instructions to determine if oil or grease should be used.

(*continued*)

Probable Cause	Solution	
Housing packed with grease, or oil level too high . . . causing excessive churning of lubricant, high operating temperature, oil leakage.	Purge bearing until only lower half of housing is ½ to ⅔ full of grease. Using oil lubrication, reduce level to just below center of lowest ball or roller.	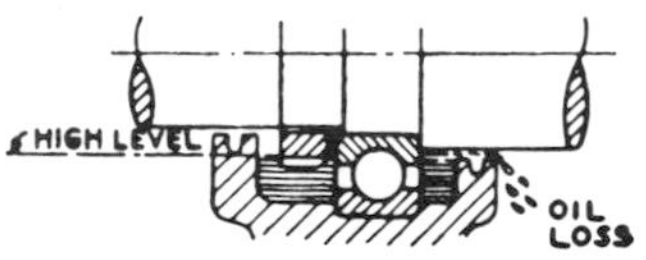 HIGH LEVEL; OIL LOSS
Bearings selected with inadequate internal clearance for conditions where external heat is conducted thru shaft, thereby expanding the inner ring excessively.	Replacement bearing should have identical marking as original bearing for proper internal clearance. Check with bearing manufacturer if bearing markings have become indistinct.	
Low oil level. Loss of lubricant through seal. Insufficient grease in housing.	Oil level should be just below center of lowest ball or roller in bearing. Using grease, lower half of pillow block should be ½ to ⅔ full.	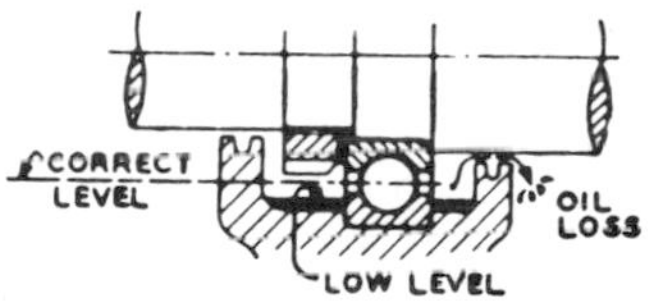 CORRECT LEVEL; OIL LOSS; LOW LEVEL
Foreign matter (dirt, sand, carbon, etc.) entering bearing housing.	Clean out bearing housing. Replace worn-out seals or improve seal design to obtain adequate protection of bearing.	
Corrosive agents (water, acids, paints, etc.) entering the bearing housing.	Addition of a shroud and (or) flinger to throw off foreign matter.	
Housing bore out of round. Housing warped. Excessive distortion of housing. Undersized housing bore.	Check and scrape housing bore to relieve pinching of bearing. Be sure pedestal surface is flat, and shims cover entire area of pillow block base.	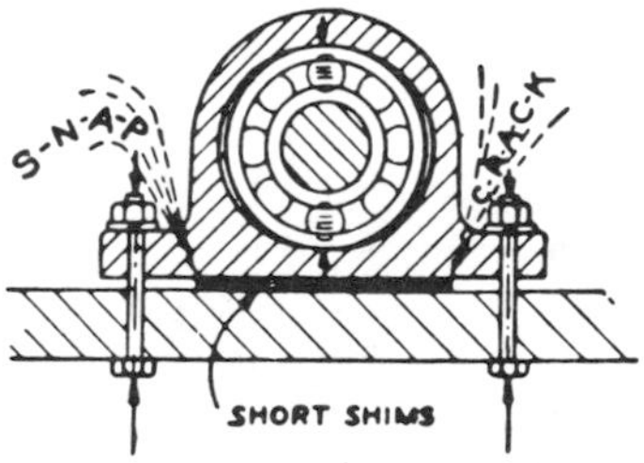 S-N-A-P; S-L-A-C-K; SHORT SHIMS
Failure to remove chips, dirt, etc. from bearing housing before assembling bearing unit.	Carefully clean housing, and use fresh lubricant.	

(*continued*)

Probable Cause	Solution	
Leather or composition seals with excessive spring tension or dried out.	Replace leather or composition seals with ones having reduced spring tension. Lubricate seals.	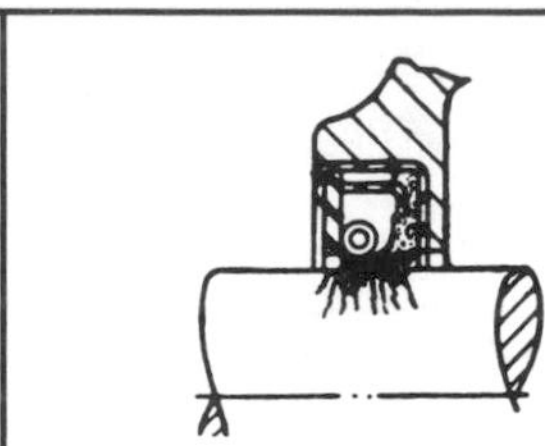
Rotating seals rubbing against stationary parts.	Check running clearance of rotating seal to eliminate rubbing. Correct alignment.	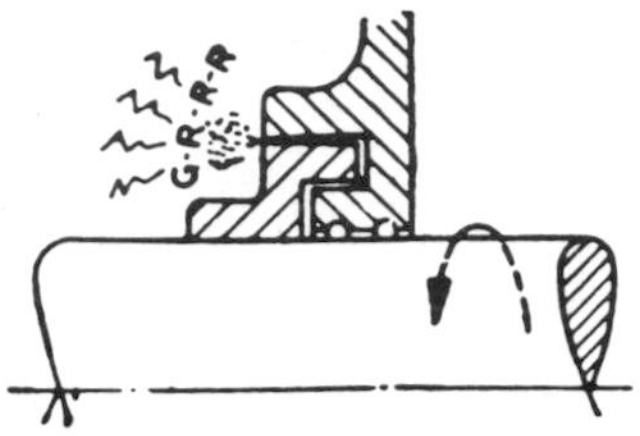
Opposed mounting.	Insert gasket between housing and cover flange to relieve axial preloading of bearing.	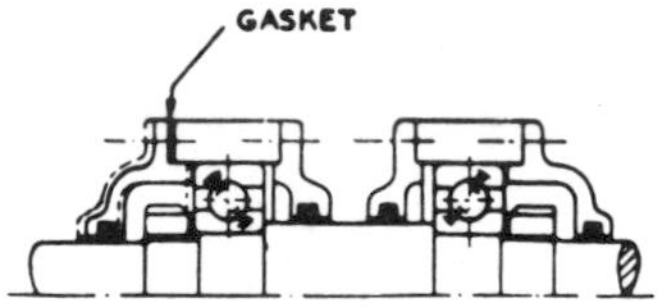 GASKET
Two "held" bearings on one shaft. Excessive shaft expansion.	Back off covers in one of the housings, using shims to obtain adequate clearance of outer ring, to permit free axial bearing motion.	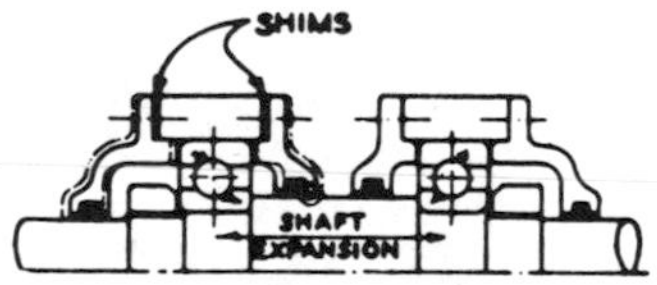 SHIMS SHAFT EXPANSION
Adapter tightened excessively.	Loosen locknut and sleeve assembly. Retighten sufficiently to clamp sleeve on shaft but be sure bearing turns freely.	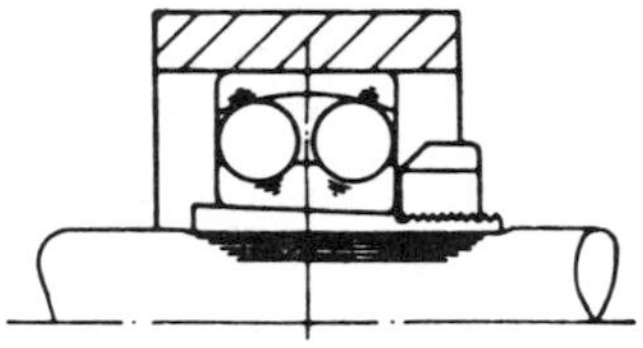
Prong rubbing against bearing.	Remove lockwasher — straighten prong or replace with new washer.	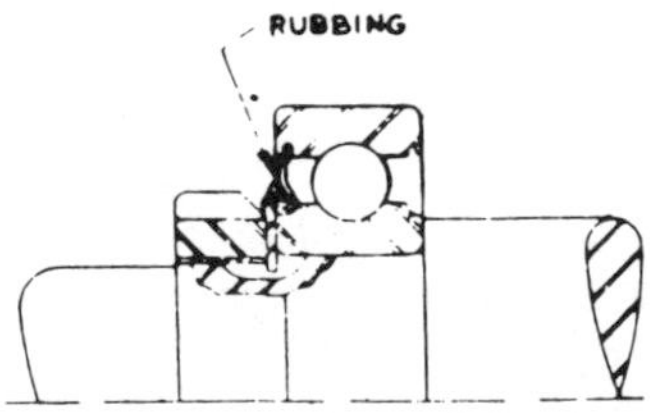 RUBBING

(continued)

Probable Cause	Solution	
Inadequate shoulder support causing bending of shaft.	Remachine shaft fillet to relieve stress. May require shoulder collar.	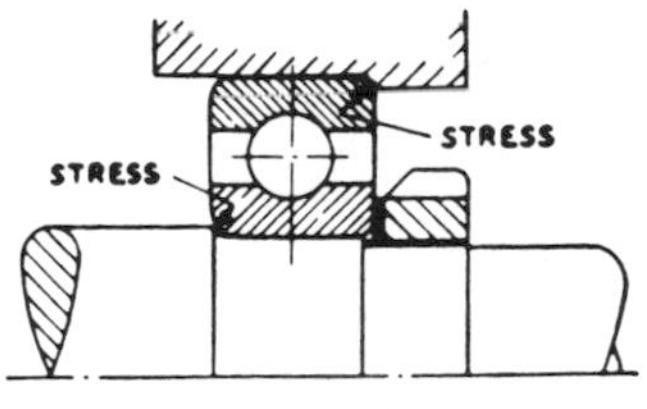 STRESS STRESS
Inadequate support in housing causing cocking of outer ring.	Remachine housing fillet to relieve stress. May require shoulder collar.	
Distortion of bearing seals.	Remachine housing shoulder to clear seal.	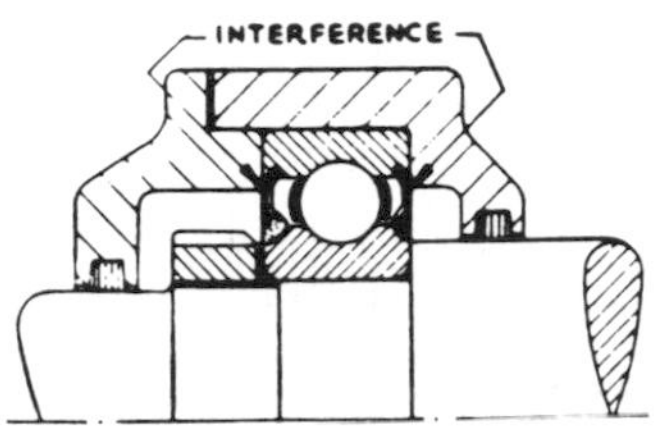 INTERFERENCE
Distortion of shaft and inner ring. Uneven expansion of bearing inner ring.	Remachine shaft fillet to obtain proper support.	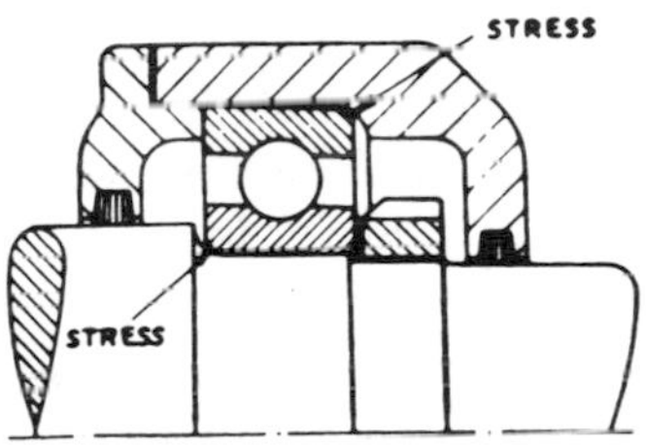 STRESS STRESS
Distortion of housing and outer ring. Pinching of bearing.	Remachine housing fillet to obtain proper support.	
Incorrect linear or angular alignment of two or more coupled shafts with two or more bearings.	Correct alignment by shimming pillow blocks. Be sure shafts are coupled in straight line—especially when three or more bearings operate on one shaft.	 LINEAR MISALIGNMENT ANGULAR MISALIGNMENT
Shaft and housing shoulders, and face of locknut out of square with bearing seat.	Remachine parts to obtain squareness.	
Bearing seat diameter machined oversize, causing excessive expansion of shaft and bearing inner ring, thus reducing clearance in bearing.	Grind shaft to get proper fit between inner ring of bearing and shaft.	

Causes of Bearing Failures

Since the bearings of a machine are among its most vital components, the ability to learn as much as possible from bearing failures is of utmost importance.

Unfortunately, too many ball and roller bearings installed in mechanisms never attain their calculated life expectancy because of something done or something left undone in handling installation and maintenance.

The calculated life expectancy of any bearing is based on four assumptions:

- Good lubrication in proper quantity will always be available.
- The bearing will be mounted without damage.
- Dimensions of parts related to the bearing will be correct. Eg. Shaft diameters and concentricity and housing bores and concentricity.
  There are no defects inherent in the bearing itself.

However, even when properly applied and maintained, the bearing will be subjected to one cause of failure—fatigue of the bearing materials. Fatigue is the result of shear stresses typically applied immediately below the load carrying surfaces, and is observed as spalling away of surface metal as seen in Figure 13-6 thru 13-8. Although, spalling can be easily observed, it is necessary to discern between spalling produced at the normal end of a bearing's useful life and that triggered by causes found in the three major classification of premature spalling - lubrication, mechanical damage, and material defects. The actual beginning of spalling is invisible because its origin is usually below the surface.

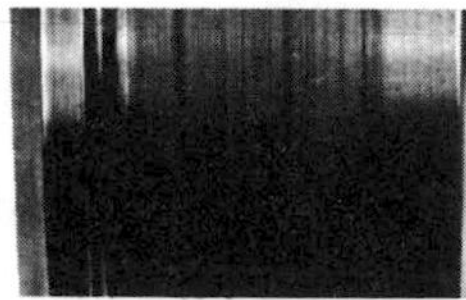

Figure 13-6. Incipient Fatigue Spalling

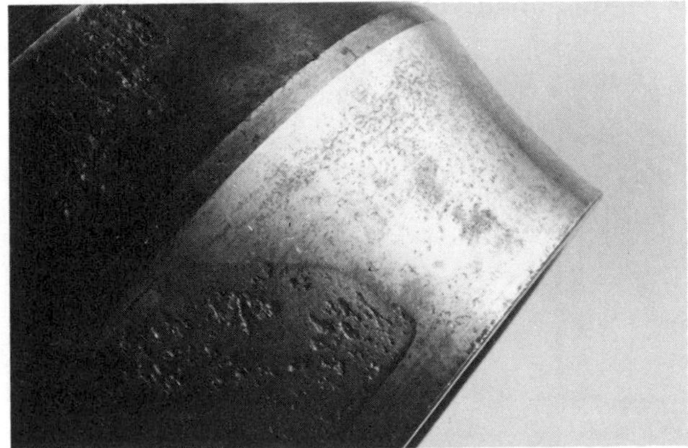

Figure 13-7. More Advanced Spalling

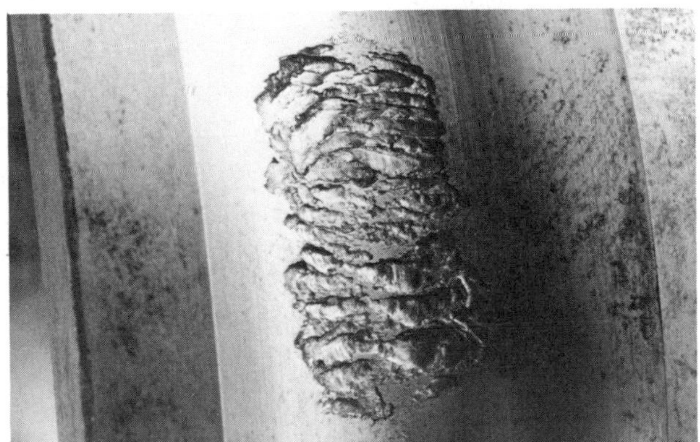

Figure 13-8. Greatly Advanced Spalling

The first visible sign is a small crack which can not be seen nor can its effects be heard while the machine is operating. Figures 13-6 thru 13-8 illustrate the progression of spalling. The spot on the inner ring on Figure 13-6 will gradually spread to the condition seen in the ring of Figure 13-8 where spalling extends around the ring. By the time spalling reaches proportions shown in Figure 13-7 the condition should make itself known by noise. If the surrounding noise level is too great, a bearing's condition should be periodically evaluated by some monitoring device such as a stethoscope or a shock pulse analyzer. The time between incipient and advanced spalling varies with speed and load, but in any event is not a sudden condition that will cause destructive failure within a matter of hours. Complete bearing failure and a consequent damage to machine parts is usually avoided because of the noise the bearing will produce and the eratic performance of the shaft carried by the bearing. Most bearing failures can be attributed to one of the following causes:

- Defective bearing seats on shafts and in housing
- Misalignment
- Faulty mounting practices
- Incorrect shaft and housing fits
- Inadequate lubrication
- Ineffective sealing
- Vibration while the bearing is not rotating
- Passage of electric current through the bearing.

14

Installation and Maintenance of Pumps and Air Handling Systems

William Krause

This chapter is concerned with the systems that transfer all types of fluids. It is divided into two major categories—pumps for liquids and semisolids and air handling systems for gases.

MAINTENANCE GUIDELINES FOR PUMPS*

The pumping of various materials such as fluids and semi solids frequently creates problems. These problems can be created not only by the pump itself, but by all other components of the pump system. The engineering of a pumping system must consider the proper power source, pump, piping, and system application. The next important consideration is its installation. If the system is to be trouble-free, maintaining and troubleshooting it is most important.

*All pump-related information was provided courtesy of WAUKESHAW PUMPS/Cherry Burrell Fluid Handling Division, 611 Sugar Creek Road, Delavan, Wisconsin, 53115.

Installing Pumps

Location

- Locate pump as close to liquid source as possible.
- Clean the area and allow adequate head room for the use of installation equipment.
- Make sure area is dry and has adequate space for air circulation, inspection, and maintenance.
- Illuminate the area sufficiently for inspection and maintenance.

Mounting

The unit can be mounted in four different ways.

- The most common method of mounting a pump is on a concrete foundation. Foundations should be solid and substantial enough to absorb vibration. Use concrete foundations with bolts imbedded for pump footings. Bolts should be long enough so that a minimum of four full threads protrude beyond the nut to insure proper tightening. These bolts should also be fitted with a pipe sleeve that is approximately 2-1/2 times the bolt diameter and twice as long as the I.D. of the sleeve.

  This applies to both standard bolts and "J" bolts. When using a standard bolt, use a washer to support the head of the bolt in the sleeve. After the concrete foundation has been poured, the pipe sleeve remains in place, which allows for proper alignment. If using "J" bolts, the washer is not required. If installing on an existing slab, you drill appropriate holes and use expansion bolts to anchor the pump and motor. (See Figure 14-1.)
- Another method is by using leveling and/or vibration isolation pads. Many commercial types of such pads are available. (See Figure 14-2.)
- For areas where it is necessary to wash-down under the base, an adjustable leg base can be used. This most commonly occurs with sanitary pumps. (See Figure 14-3.)
- Pumps can also be mounted on a portable hand truck for movement to different locations. (See Figure 14-4.)

When a pump arrives in the plant, uncrate it, and remove all instructions and literature. Maintain all paperwork in a safe and convenient place for future reference. No matter which of the above bases is used:

- Place leveling plates and shims at each foundation bolt.
- Level motor perpendicular and parallel to shaft. (See Figure 14-5) shimming as necessary, and tighten the nuts on the foundation evenly.
- Align motor and pump vertically and horizontally. (See Figure 14-6.)

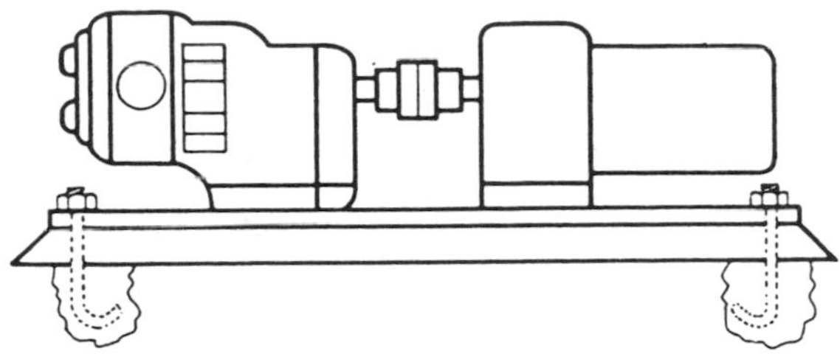

Figure 14-1. Permanent Installation on Foundation

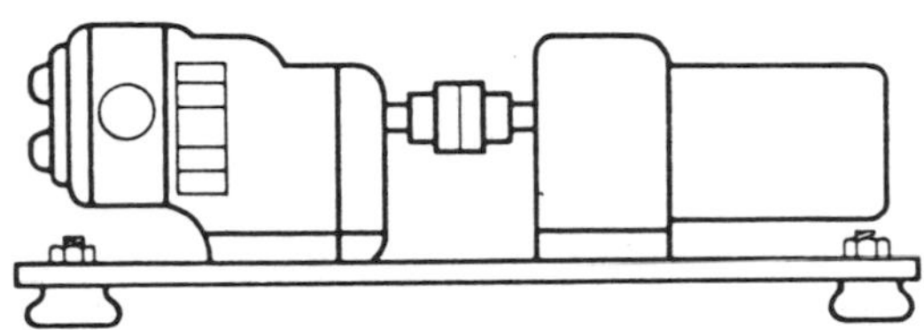

Figure 14-2. Leveling and/or Vibration Isolation Pads

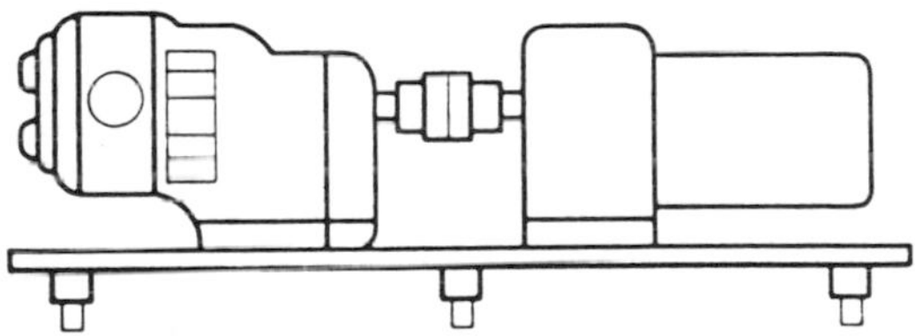

Figure 14-3. Adjustable Leg Base

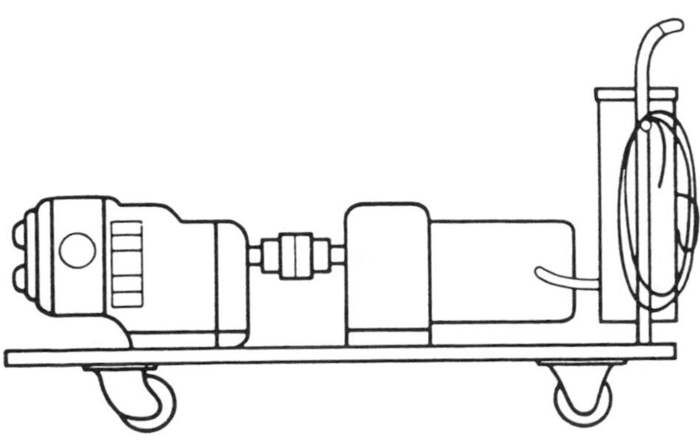

Figure 14-4. Portable Bases

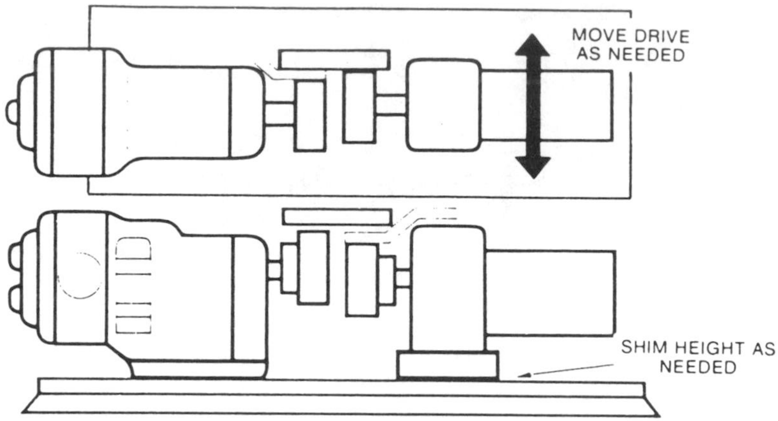

Figure 14-5. Checking Parallel Misalignment Using Straight Edges and Shims

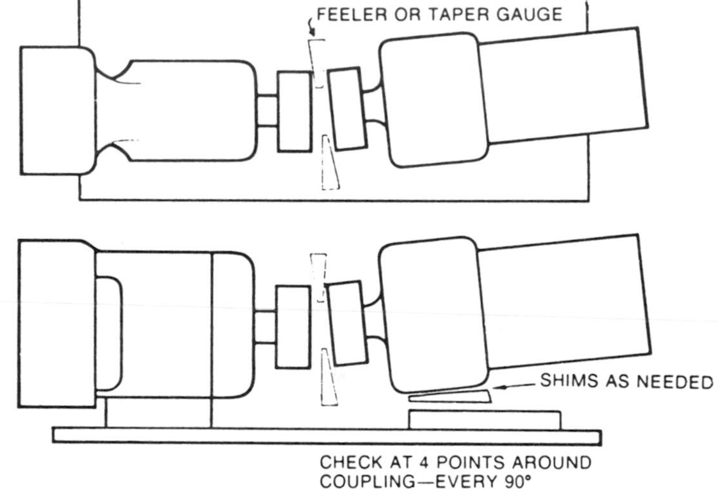

Adjust to get equal dimension at all points—at the same time set space between coupling halves to manufacturer's recommended distance.

Figure 14-6. Checking Angular Alignment Using feeler Gauges or Taper Gauges

Piping

- Make sure suction and discharge piping is as short and direct as possible. It should be supported by hangers and/or pedestals and not by the pump. (See Figure 14-7.) Carefully align it with the pump. Fittings and bends should be minimal. If bends are necessary, maximize the radius. Piping at the pump—both suction and discharge ends—should be as large as or larger than the openings in the pump and should be supported as near the pump as possible.
- Install gate valves and pressure gages in both suction and discharge lines to facilitate performance and maintenance.

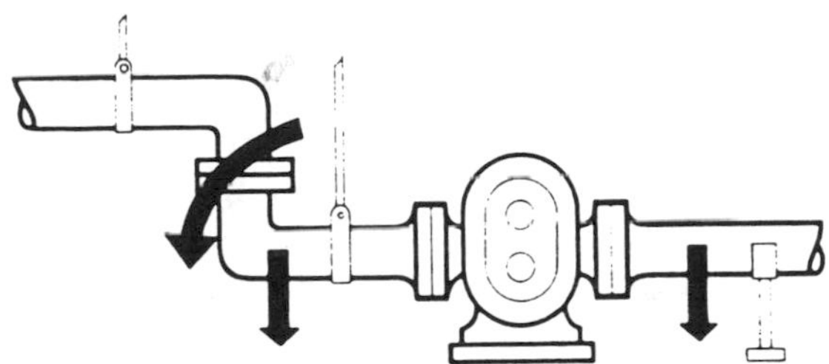

Figure 14-7. Piping Support

- Make sure suction piping contains no vapor traps. See Figure 14-8. (Use eccentric-reducer in horizontal inlets to pump.) This line must be free from air leaks. Allow for expansion of hot lines by usings thermal expansion joints. You can also use flexible joints to additionally limit the transmissions of mechanical vibration. Anchor free ends of any flexible hose in the system. (See Figure 14-9.)

  Install a strainer near the pump suction inlet to catch scale or other foreign matter. Install pressure gages on each side of the strainer to facilitate maintenance. Make sure the strainer has a free area of 2-1/2 to 4 times the area of the suction pipe.
- Locate discharge piping valves as near the pump as possible. In some installations, consider using a check valve to prevent backup of liquid in case of motor failure. You could use a gate valve to prime start and shut down the pump.
- Install isolation valves to permit pump maintenance and safe removal without emptying the entire system.

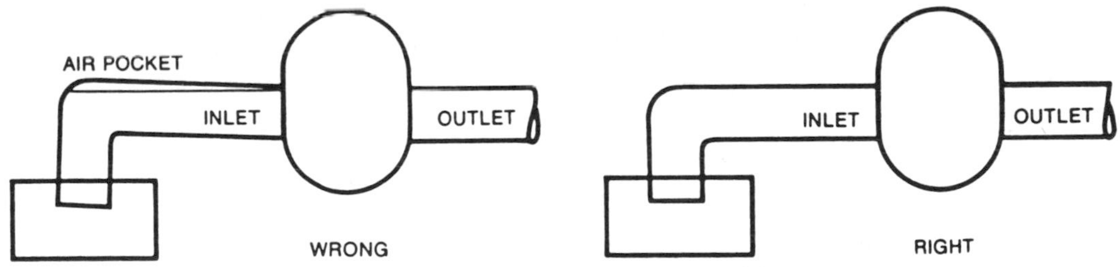

Figure 14-8. Piping Layout: Inlet Side—Slope Piping up to Inlet to Avoid Air Pocket

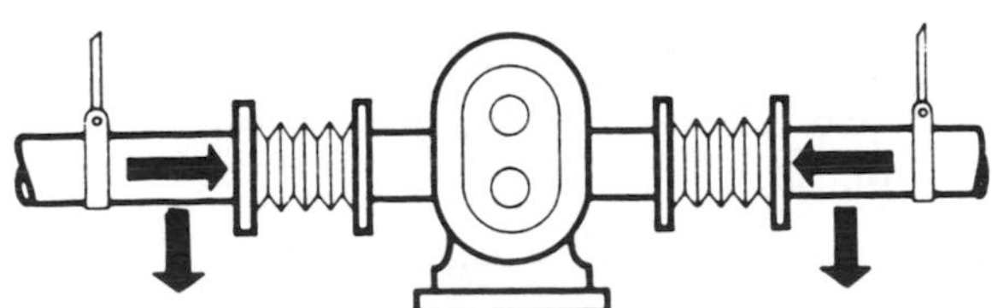

Figure 14-9. Flexible Joints

- To protect the pump and piping system against excessive pressure, install a relief valve. Do not use an integral relief valve (which is designed to bypass the fluid internally from the pump outlet to the inlet) on applications where the discharge must be closed for more than a few minutes. Prolonged operation of the pump with closed discharge will cause heating of the fluid circulating through the relief valve. When such operation is necessary, the relief valve, whether integral, attachable, or line-mounted, should discharge externally through piping connected to the fluid source, or if that is not practical, into the inlet piping near the source. (See Figure 14-10.)
- Wire motor properly to maintain equipment warranties.

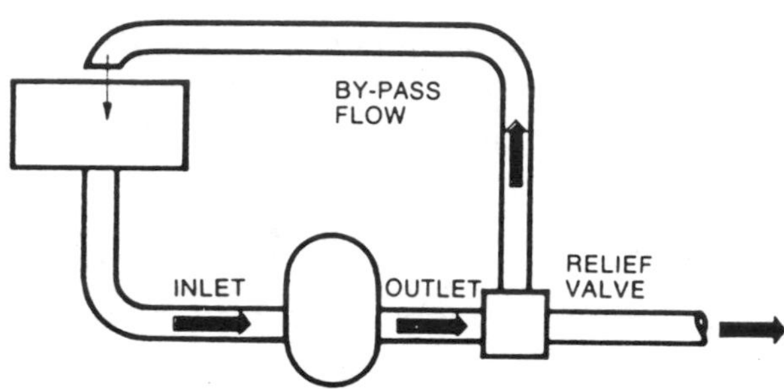

Figure 14-10. Relief Valve

Starting the Pump Operation

Use the following checklist when starting the new pump.

- Check that piping and pump are clean and free of foreign material, such as welding slag, gaskets, etc. **Do not use pump to flush system.**
- See that all piping connections are tight and leak-free. Where possible, check system with "nonhazardous" fluid.
- Check to see that pump and drive are lubricated. See manufacturers instructions. Install breather screw. Check drive lubrication instruction.
- Check that all guards are in place and secure.
- Check the packing on the seals: supply flushing fluid if needed. Leave packing gland loose for normal 'weepage'! Make adjustments as initial conditions stabilize, to maintain normal weepage. On mechanical seals with flushing, supply adequate flow of clean flushing fluids.
- See that all valves are open on discharge system and that there is a free flow path open to destination.
- See that all valves are open on inlet side and that fluid can reach pump.
- Check direction of pump and drive rotation (jogging is recommended). See Figure 14-11.

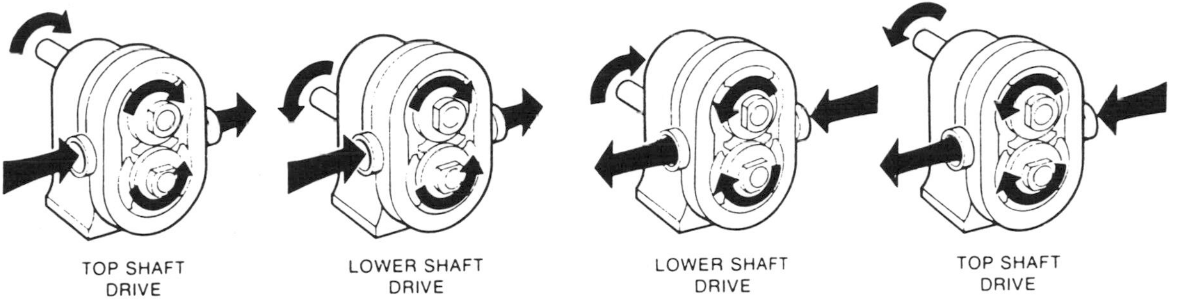

Figure 14-11. Check Rotation Direction of Drive to See That Pump Will Rotate in Proper Direction, Facing "Liquid End" of Pump

- Start pump drive. Where possible, start at slow speed, or jog. Check to see that liquid is reaching pump within several minutes. If pumping does not begin and stabilize, check items under "no flow" or "insufficient flow" in the troubleshooting guide, Table 14-1, on pages 534-538.

Pump Maintenance

Though there are numerous types of pumps, you will deal primarily with three major types—centrifugal, rotary, and reciprocating. The troubleshooting guide (Table 14-1) lists common problems encountered with these three types of pumps, their possible causes, and solutions.

MAINTAINING AIR HANDLING SYSTEMS

Never before has the ventilating system taken such an important role in the overall importance of the maintenance manager's list of priorities. Years ago, only a satisfactory comfort level was resultant of a properly operating ventilating system. The advent of the revised ASHRAE Ventilation standard 62 coupled with the increasing amount of building occupants' complaints and the lawsuits filed by employees have forced attention to the building ventilation maintenance practices.

Increasing concerns about indoor air quality (IAQ) have developed in the last decade. On one hand, the attempt to conserve energy has resulted in tighter and better insulated buildings and has minimized air movement in occupied spaces. On the other hand, these efforts have frequently resulted in elevated levels of contaminants which may directly affect many occupants' health.

You must take all steps necessary to make sure that your ventilating systems are receiving top-notch maintenance. Ventilating systems consist of the prime air moving force (fans) and the various inlet and delivery systems. This section is divided into these two components.

Problem	Probable Causes	Solutions
No Flow, pump not turning	Drive motor not running	Check resets, fuses, circuit breakers
	Keys sheared or missing	Replace
	Drive belts, power transmission components slipping or broken	Replace or adjust
	Pump shaft, keys, or gears sheared	Inspect; replace parts
No flow, pump turning	Wrong direction of rotation	Reverse
No flow, pump not priming	Valve closed in inlet line	Open valve
	Inlet line clogged or restricted	Clear line, clean filters, etc.
	Air leaks due to bad seals or pipe connections	Replace seals; check lines for leakage (can be done by air pressure, or by filling with liquid and pressurizing with air)
	Pump speed too slow	Increase speed. Filling inlet line with fluid may allow initial start-up. Foot valve may solve start-up problems permanently.
	Liquid drains or siphons from system during off periods	Use foot valve or check valves
	"Air" lock. Fluids which "gas off," or vaporize, or allow gas to come out of solution during off periods	Manual or automatic air bleed from pump or lines near pump
	Extra clearance rotors, worn pump	Increase pump speed. Use foot valve to improve priming

Table 14-1. Troubleshooting Guide for Pump Maintenance

Problem	Probable Causes	Solutions
No flow, pump not priming (cont.)	Net inlet pressure available too low	Check NIPA, NIPR*, recalculate system. Change inlet system as needed.
	On "Vacuum" inlet system: On initial start-up, atmospheric "blow back" prevents pump from developing enough differential pressure to start flow.	Install check valve in discharge line
No flow	Relief valve not properly adjusted, or held off seat by foreign material (flow is being recirculated to inlet)	Adjust or clear valve
Insufficient flow	Speed too low to obtain desired flow	Check flow-speed curve
	Air leak due to bad seals or pipe connections	Replace seals, check inlet fittings.
Fluid vaporization ("starved" pump inlet)	Strainers, foot valves, inlet fittings or lines clogged	Clear lines. If problem continues, inlet system may require change
	Inlet line size too small, inlet line length too long. Too many fittings or valves. Foot valves, strainers too small.	Increase inlet line size. Reduce length, minimize direction and size changes, reduce number of fittings.
	NIPA too low	Raise liquid level in source tank
	NIPA too low	Increase by raising or pressurizing source tank
	NIPA too Low	Select larger pump size with smaller NIPR

(*continued*)

*NIPA—Net Inlet Pressure Available at Pump
NIPR—Net Inlet Pressure Required by Pump

Problem	Probable Causes	Solutions
Fluid vaporization ("starved" pump inlet) (cont.)	Fluid viscosity greater than expected	Reduce pump speed and accept lower flow, or change system to reduce line losses.
	Fluid temperature higher than expected (vapor pressure higher)	Reduce temperature, reduce speed and accept lower flow or change system to increase NIPA
Insufficient flow, fluid being bypassed somewhere	Relief valve not adjusted or jammed	Adjust or clear
	Flow diverted in branch line, open valve, etc.	Check system and controls
Insufficient flow, high slip	Hot (HC) or extra clearance rotors on "cold" fluid, and/or low viscosity fluid	Replace with standard clearance rotors
	Worn pump	Increase pump speed (within limits). Replace rotors, recondition pump.
	High pressure	Reduce pressure by system changes
Noisy operation	• Cavitation	
	High fluid viscosity, High vapor pressure fluids, High temperature	Slow down pump, reduce temperature, change system
	NIPA less than NIPR	To increase NIPA or reduce NIPR, see Manufacturers instruction
	• Air or gas in fluid	
	Leaks in pump or piping	Correct leaks
	Dissolved gas or naturally aerated products	Minimize discharge pressure. Also see "Cavitation" above.

Problem	Probable Causes	Solutions
Noisy operation (cont.)	• Mechanical noises Rotor to body contact	
Improper assembly	Check clearance with shims. See Manufacturers instructions	
	• Rotor to body contact	
	Distortion of pump due to improper piping installation	Reassemble pump or reinstall piping to assure free running
	Pressure higher than rated	Reduce pressure if possible
	Worn bearing	Rebuild with new bearings, lubricate regularly
	Worn gears	Rebuild with new gears, lubricate regularly
	• Rotor-to-rotor contact	
	Loose or mis-timed gears, twisted shaft, sheared keys, worn splines	Rebuild with new parts
	• Relief valve chattering	Re-adjust, repair or replace
	• Drive component noise-gear trains, chains, couplings, bearings.	Repair or replace drive train
Pump requires excessive power (overheats, stalls, high current draw, breakers trip)	• Higher viscous losses than expected	If within pump rating, increase drive size
	• Higher pressure than expected	Reduce pump speed, increase line sizes

(*continued*)

Problem	Probable Causes	Solutions
Pump requires excessive power (overheats, stalls, high current draw, breakers trip) (cont.)	• Fluid characteristics	
	Fluid colder than expected, viscosity high	Heat fluid, insulate or heat trace lines. Use pump with more running clearances.
	Fluid sets up in line and pump during shut down	Insulate or heat trace lines. Install "soft start" drive. Install recirculating bypass system. Flush with other fluid.
	Fluid builds up on pump surfaces (example, latex, chocolate, fondants)	Use pump with more running clearance
"Short" pump service life	High corrosion rate	Upgrade material of pump
	Pumping abrasives	Larger pumps at slower speeds can help
	Speeds and pressures higher than rated	Reduce speeds and pressures by changes in system
	Worn bearings and gears due to lack of lubrication	Set up and follow regular lubrication schedule
	Misalignment of drive and piping. Excessive overhung load or misaligned couplings.	Check alignment of piping. Check drive alignment and loads.

Fan Usage and Maintenance

There are two major classifications of fans: axial and centrifugal. Axial fans, where the air flow through the wheel is parallel to the axis of rotation, consists of three basic categories: vane-axial, tube-axial and propeller. Like all equipment, each category has specific advantages for different tasks. *Tube-axial* is enclosed to improve efficiency and pressure capability. It has no inlet or outlet guide vanes and is typically used for ducted HVAC applications, paint spray booths, drying ovens, or food exhausts.

Vane-axial fans are the most efficient type of axial fans. Guide vanes improve efficiency and pressure capability because they straighten the pattern of air discharged thereby converting spin velocity pressure into static pressure which is more useful. Vane-axial fans have blades, and the pitch of the blades is fixed, adjustable (when the wheel is stopped), or controllable (adjustable while the wheel is rotating). These fans provide good air distribution down stream and are used for HVAC applications where compactness and straight-through flows are necessary.

Centrifugal fans develop more static pressure than axial fans, and they have very high discharge pressures. Upper limits for pressures from axial fans would range from one to four inches w.g. Centrigual fans tend to be more forgiving. If a centrifugal fan encounters slightly more static pressure than it was designed for, a little less air is moved and the pressure needed is achieved. An axial fan, under like circumstances, might become unstable.

For low-pressure general ventilation, the common choice might be a propeller fan generating pressure of 1/2 inch w.g. For higher pressure up to three inches w.g., a tube-axial style would be a better selection where short runs of ductwork to direct the air would be needed. If efficiency is critical, a vane-axial configuration would be most desirable because of its guide vanes.

To maintain maximum efficiency in operating your air handling system, implement the following checklist at start up and periodically thereafter.

Starting a Fan

Before actually starting a fan, all safety accessories must be properly installed and checked. At no time and under no circumstances should the maximum safe speed recommended by the fan manufacturer be exceeded. The startup procedures should include a complete check of the entire installation. On the next page are steps to follow before operating the equipment.

- Be sure the system is electrically isolated and all disconnected switches and other controls are locked in the "off" position. If these switches or controls are in locations remote from the fan, be sure to place prominent "DO NOT START" signs on them. Even when the fans are locked out electrically, if they are located outdoors or in a location where windmilling can take effect, the impeller should be secured physically to restrict rotational movement.
- Make sure the starting torque is adequate for the speed and inertia of the fan.
- Inspect the entire installation. Check both interior and exterior for loose items or debris that could be sucked into the wheel or dislodged by the discharge. Turn the wheel by hand to make sure it does not bind.
- Check the belt tension and the drive installations.
- Check to be sure all nuts, bolts, and setscrews are tight.
- Be sure all safety devices and guards are in place.
- Make sure the voltage supply is correct and the motor rotation direction is correct before coupling the motor to the fan.
- Carefully check the fan for unusual noises, excessive vibration, and correct fan speed. Perform these checks before connecting any ductwork. Be very careful, however, not to run the fan for any long period of time without the ductwork connected, as this can overload the motor. This operation should not exceed a few seconds.
- Recheck setscrews, etc. after several minutes of operation and again after a week or two, and tighten if necessary.
- Test centrifugal fans with a straight length of duct on both the inlet and outlet. A radial bladed paddlewheel fan tolerates poor inlet and outlet conditions better than an airfoil fan. A good rule of thumb is not to install an elbow within 2-1/2 duct diameters of the fan.
- Isolate the fan to allow for ductwork expansion and movement. Always use a flexible connection in a hot air application. It is also valuable to reduce vibration transmission from the ductwork to the fan, which can cause cracking and place stress on the fan.
- Be sure the foundation is adequate for the installation. A fan is properly supported by the structural steel of the building or on its own platform, which should be a concrete slab equal to three times the weight of the fan or five times the weight of the rotating element. It should extend at least six inches beyond the base of the fan.

Maintaining Fans After Installation

Steady, dependable service does not occur without routine maintenance. A regular maintenance program should include attention to all moving parts: drives, belts, motors, bearings, and the fan wheel. Two common mistakes are overgreasing the bearings and overtightening the drive belt. Too much grease reduces the bearing's ability to dissipate heat. Also, use the right lubricant; check with the fan manufacturer and your lubricant supplier. Improper tensioning of belt drives can be a problem. Overtensioning puts too much pressure on the bearings and can cause failure.

Some squealing on startup is normal. The noise does not always indicate that the belt is loose. Belts can cause vibration. Shut down fans immediately at any sudden increase in vibration. Vibration can be hard to detect and may require a meter to locate the problem.

Keep records so that variations of vibration are apparent and take steps to alleviate developing problems before they become serious. (See Chapter 3 on Predictive Maintenance.)

Use the troubleshooting guide in Table 14-2 on the next page. If, however, you are unsuccessful in eliminating all problems after exhausting all suggestions herewith, contact your fan manufacturer for suggestions, as the manufacturer will be more familiar with the characteristics of your particular fan.

Maintaining Ventilating System Components

Although the fan is obviously the major component of a ventilating system, its other component parts must be kept both clean and in proper operating condition. For systems that are not used during the entire year, both the start-up and shut-down procedures should include a thorough inspection of all components. For systems that are operational during the entire year, inspect the system at regular intervals.

Inspect all components as described in the following sections.

Inspecting Outdoor Air Louvers, Screens, Fresh Air Intakes

Note exterior conditions to reduce or eliminate pickup of roof dust, snow accumulations, and foreign materials, particularly if the materials are combustible. Note potential cross contamination from areas such as kitchens, toilets, or other areas that may contaminate the ventilation system. Make sure louver dampers and linkage that control air quantities, by-pass heat transfer equipment, or control and mix outdoor and return air are in perfect operating condition; check these on a regular schedule. Make sure they are free to operate, are not corroded, are not bent, are not restricted, and are well and properly lubricated.

Filtration is a critical component to the ventilating system. (See Chapter 15). Establish and maintain a filter cleaning or replacement schedule. Because schedules for cleaning or replacement is predicated on the resistance to air flow, conditions may change and, therefore, you should periodically check the resistance to see if schedules should be modified. Draft gauges are available that will monitor the operating pressure differential and produce an audible signal to indicate that the filter must be replaced or cleaned.

Check to make sure filters are installed properly, insuring that no air is bypassing around the cells or media. Do not clean or reuse disposable filters. If using liquid adhesive, be careful, because their low flash point could create substantial hazards.

SYMPTOMS	PROBABLE CAUSE	REMEDY
1. Fan will not start	1. Blown fuses. 2. Broken belts. 3. Loose pulleys. 4. Electricity turned off. 5. Impeller touching scroll. 6. Wrong voltage. 7. Open electrical contact.	1. Replace fuse. 2. Replace belts. 3. Tighten pulleys. 4. Turn electricity on. 5. Re-align impeller. 6. Check voltage rating. 7. Check open contact.
2. Fan too noisy to operate	1. Impeller hitting inlet cone. 2. Fan parts, motor & motor base not securely anchored. 3. Improper drive and belt selection. 4. Defective bearing. 5. Bent or undersized shaft. 6. High outlet velocity. 7. Wheel unbalanced. 8. Foreign material on housing. 9. Electrical.	1. Check clearance. 2. Tighten bolts. 3. Check or replace. 4. Replace bearing. 5. Replace shafting. 6. Check outlet duct size. 7. Balance impeller. 8. Remove foreign material. 9. Check relay or motor bearing, check phase current.
3. Insufficient Air Flow	1. Wrong fan rotation. 2. Fan speed too slow. 3. Dampers, registers in closed position. 4. Coil and filters dirty. 5. Restricted inlet/outlet airflow. 6. Fan too small for application.	1. Correct fan rotation. 2. Check sheave ratio. 3. Inspect dampers and registers and adjust. 4. Clean coil and replace filter. 5. Check for restriction & allowable pressure drop. 6. Check design or replace fan.
4. Too much air flow	1. Wrong fan rotation. Backward inclined impeller installed backwards, HP too high. 2. Fan speed too fast. 3. Registers or grilles not installed. 4. Access door open. 5. Fan too large for application.	1. For SWSI impeller change wiring connection for correct impeller rotation. Inspect. 2. Check design RPM. 3. Install registers or grilles and adjust for desired airflow. 4. Close access door. 5. Check fan selection.
5. Horsepower too high	1. Backward inclined impeller installed backwards. 2. Fan speed too high. 3. Fan size or type not best for application. 4. Gas density is heavy (cold start-up).	1. Check rotation. 2. Check design RPM. 3. Check design selection. 4. Check calculated HP based on light gas against fan starting cold at 70°F or below.

Courtesy of Ammerman Company, Incorporated.

Table 14-2. Fan Troubleshooting Guidelines

Clean out all liquid adhesive reservoir regularly to avoid unnecessary build ups. Check electrical and mechanical equipment for automatic filtration systems on a regular basis to make sure that it is functioning properly. Items to check are:

- Level mounting.
- Oil bath clean and oil level correct in gear boxes and filter baths.
- Lubrication.
- Alignment, clearance, and free movement.
- Drive tension.

Inspecting Fire Doors and Dampers

Check moving parts regularly to make sure that they are free to operate. Check control linkages for:

- Free movement;
- Freedom from excessive lost motion;
- Rigidity of mountings;
- Stiffness of members;
- Fit of pins;
- Bearing lubrication;
- Tightness of locking devices;
- Correct alignment.

Checking the Duct System

Check the duct system for air leaks in duct joints, missing access hole covers, and badly fitted or damaged flexible connections. Check for missing sections of duct, particularly when passing through builders voids. Check for dislodged duct lining, construction or maintenance debris, dust, waste material, etc. left in the duct systems. Inadequate duct stiffening at one branch can result in creating a natural bellowing effect that can be transmitted throughout the remaining system.

Inspect plenum chambers more frequently. Make sure they are not used for any storage. The control of microbial contaminants is essential. The detection and repairing of all areas where water collection or leakage has occurred is essential. Stagnant water in areas such as drain pans, cooling coils, etc. must be eliminated. Regular inspection is necessary to these areas, and direct access to these areas is required. If there are no access panels or doors or if they are inadequate, you should install an adequate means of access.

Where microbid growth has occurred, clean and disinfect the areas with detergents, chlorine-generating slimicides (bleach), and/or proprietary biocides. Remove these cleaners before reactivating the ventilating systems.

Cleaning Ductwork

The need to clean air-handling systems in existing buildings is widespread and acknowledged by many building owners. The problem of how to do it without putting the occupants or the integrity of the HVAC system at risk was explored in an article in the January/February 1991, issue of *Engineered Systems* magazine.* Following are excerpts, with permission, from that article.

Truck-mounted vacuum systems, developed for residential duct cleaning, have limited value in larger commercial structures. A new break-apart combination air handler and vaccum machine can be transported in a standard elevator. It requires 120-volt electricity at 70 amps. Its final HEPA filter and a biocide duct coating assure building owners that dirt and contaminants will not recontaminate the occupied spaces.

Duct and duct liner manufacturers are concerned about damage to ductwork from high vacuums and mechanically agitated brushes; therefore, the machine's vacuum and air flow can be adjusted to safe levels. Manual brushing by horsehair brushes protects against glass fiber duct or duct liner damage. Stiffer plastic bristles are used for bar galvanized ductwork.

In a typical job, a three- to five-person team works at night when the building is unoccupied. They clean only as much ductwork as can be treated with the biocide before the occupants arrive in the building the next day.

They start with the return-air system and work back to the the air handler. All but one of the openings—in this case, near the air handler—are sealed and each duct run is handled individually. The adjustable vacuum machine produced an airflow of approximately 4,500 cfm and a vacuum of 1 to 3 in. wg.

From existing or created access points, workers use pole-mounted brushes to move debris from the inside of the duct. The airflow and the brushes push the debris toward the removal point. Access points usually are about 20 ft apart but may be as far as 50 or 60 ft.

The agitation caused by the brushing removes most—but not all—of the debris. Air tools move the heavier debris to the collection point. This removes all of the visible debris. What is left is encapsulated in a clear latex spray coating that contains a biocide to prevent future contamination. Fogging doesn't work and can present a danger to the occupied spaces.

On the supply air side, the approach is reversed. Again, all but one outlet is sealed with plastic and the duct runs are handled individually.

Protection and Disposal

Designed to meet asbestos abatement regulations, knife-edge seals between components of the air handler-vacuum machine and a three-stage filtration system assure a filtration efficiency of 99.97% down to 0.3 microns (Federal Standard

*Courtesy of Gordon Duffy, Publishing Director of *Engineered Systems* magazine.

209B). An activated carbon prefilter removes any gases or odors resulting from the latex biocide spraying operation.

Much like the equipment used in asbestos removal, the cleaning crews wear special "Tyvek" clothing that is disposed of after each shift. They also wear special full-face PAPR respirators so they do not inhale any of the debris or pathogens they are working with. Cloth-covered vinyl gloves complete the ensemble.

Once collected at the vacuum machine, the debris is double-bagged and deposited in a dumpster. The debris is the property of the building owner, who is ultimately responsible for its disposal. Usually, it can be taken to a landfill with the proper certificate.

Tests have shown that clean glass fiber duct and duct liner is virtually inert and absorbs moisture at a rate well below the 10% safety threshold. Dirty ductboard, on the other hand, can produce moisture levels as high as 54%, according to a University of Florida, Gainesville, study.

The Thermal Insulation Manufacturers Association (TIMA) has strong reservations about both cleaning methods and the use of biocides. The Environmental Protection Agency (EPA) and the National Air Duct Cleaners Associations (NADCA) are considering the possibility of certifying duct cleaners and cleaning methods. TIMA would support this effort. TIMA also is concerned that the addition of any coating voids the Underwriter's Laboratories (UL) label that qualifies glass fiber liners and ducts as Class 1 air duct systems.

15

Water Treatment for Mechanical Equipment

J. Berkley

This chapter provides information essential to purchasing proper water treatment chemicals and using those chemicals correctly. This should allow you to have more control over your water treatment program and to run your building more efficiently. In addition, it will allow you to communicate more effectively with your local water treatment vendor.

To accomplish this you will need to:

- Understand and define your chemical water treatment requirements.
- Evaluate your current program and consider available options.
- Monitor the water treatment program to keep it on track.
- Evaluate the new program's effectiveness.
- Comply with the Occupational Safety Hazard Act (OSHA) and the Environmental Protection Agency (EPA) regulations.

The scope of this chapter is confined to the chemical water treatment of the waterside of heating, comfort air conditioning, and computer cooling equipment. External treatment, softeners, and filters are discussed only as they effect the chemical water treatment program. The subject of chemical water treatment for process heating and cooling is not discussed at all except to advise you that it is a very special area of consideration directly related to manufacturing standards. Any significant changes in the water treatment of process heating and cooling equipment being administered by a maintenance department should be coordinated in advance with the manufacturing department.

UNDERSTANDING AND DEFINING YOUR WATER TREATMENT REQUIREMENTS

Start by listing all your heating and cooling equipment that requires water treatment. Generate a formal written survey record that defines each system's:

- physical configuration.
- rated and actual operating capacity.
- water content when full.
- daily make-up water requirements.
- relationship to other heat exchange systems.
- existing external water treatment and equipment.
- existing chemical water treatment and equipment.

You should survey the following types of equipment:

- **Heating**
  - Steam Boilers
  - Hot Water Boilers
  - Steam Condensate Piping or Return Systems
  - Condensate Collection Receivers
  - Return Tanks
  - Boiler Feed Water Systems
  - Feed Water Pumps
  - Deairators
  - Hot Water/Steam Heat Exchangers
  - Closed Loop Systems, i.e.; Perimeter, Radiation, Secondaries, Reheats, Dual Temperatures, Make-Up Water Systems, Circulating Pumps, City Steam Lines, Softeners and Filters.
- **Cooling**
  - Condensers
  - Absorbers
  - Evaporative Coolers or Condensers
  - Freon Machines
  - Lithium Bromide Machines
  - Centrifugals
  - Reciprocating
  - Cooling Towers

- Chillers
- Glycol Systems
- Dual Temperature
- Air Cooled Condensers
- Circulating Pumps
- Mixed Systems
- Filters
- Strainers
- Softeners.

Some of the above names define the same system piece of equipment or overlap. Your staff's knowledge, especially if you are a new manager, is the best initial source of system information, and your facility may be using any of the above listed names. Start by asking your staff to sketch the systems and help you gather this survey information. Check your buildings drawings, maintenance records, water bills, fuel bills, electric bills service bills from boiler, burner, air conditioning, and computer cooling equipment suppliers. Do not overlook your current or previous water treatment vendors' bills, service reports, and bid proposals. The purpose of this formal survey is to help everyone understand the relationship of one water side heat exchange system to another and create a permanent written document to use as a basic reference for current and future water treatment programs. When you finish you should have most of the following information:

- For *Heating* Equipment:
  - Equipment manufacturer
  - Equipment type
  - Boiler horse power rating
  - Heating surface in square feet
  - Pounds of steam per hour
  - Boiler operating pressure
  - Fuel consumption rate
  - Days of operation/year
  - Hours of operation/day
  - Percentage of returns
  - Make-up water/year
  - Temperature of feed water
  - Pressure in feed water tank, deairator and lines
  - Water content in boiler and condensate and other closed loop systems (piping)
  - Existing chemical feed equipment and controllers

- For *Cooling* Equipment:
  - Equipment manufacturer
  - Equipment type
  - Tons rating
  - Pump recirculating rate in gallons/minutes.
  - Hours of operation/day
  - Days of operation/year
  - Chilled water return temperature
  - Chilled water leaving temperature
  - Condenser water inlet temperature
  - Condenser water outlet temperature
  - Water content in chilled and condenser systems and line pressures
  - Make-up water/year
  - Existing chemical feed equipment and controllers

The following charts and formulas will help you make estimates when no specifications are available. Keep in mind that some accuracy is lost.

Boiler Horse-Power	Pounds of Steam per Hour	Oil Nozzle Rating (GPH) (1)	Gas Consumption (Ft^3/H) (2)	Boiler Input (MBH) (3)	Boiler Output (MBH) (3)	Boilers Heating Surface (Sq. Ft.) (4)	Steam Boiler Capacity (Gallons of Water)
50	1700	17	2600	2100	1700	275	340
100	3500	35	5300	4200	3350	550	590
150	5200	52	7800	6300	5000	825	900
200	7000	70	10500	8400	6700	1100	1000
250	8700	87	13000	10500	8400	1350	1350
300	10500	105	15800	12600	10000	1650	1600
350	12000	120	18000	14700	11700	1900	1900
400	14000	140	21000	16800	13400	2150	2300
450	15500	155	23300	18900	15000	2450	2500
500	17300	173	26000	21000	16700	2700	2700

(1) Gallons per hour
(2) Cubic feet per hour
(3) Thousands of BTU/hour
(4) For boilers built since 1950

Table 15-1. Boiler Sizing Chart for Fire Tube Steam Boilers

Circulation rate in gals/min = Tons rating × 3

$$\text{Evaporation rate in gals/min} = \text{Circulation rate in gals/min} \times \text{inlet to outlet temperature differential in} \div 1000$$

$$\text{Bleed-off rate in gal/min} = \frac{\text{Evaporation Rate in gal/ min}}{\text{Cycles of Concentration} - 1}$$

Table 15-2. Cooling Tower Estimates

34.5 lbs. water evaporated per hr.	= rated boiler H.P.
1 U.S. gallon	= 231 cu. in.
1 U.S. gallon (water)	= 8.33 lbs.
1 cu. ft. (water)	= 7.48 gallons
1 cu. ft. (water)	= 62.3 pounds
1 ft. head (water)	= 0.434 psi
1 psi	= 2.3 ft. head (water)
1 pound	= 7,000 grains
pounds per 1000 × 120	= parts per million (ppm)
1 grain per gallon (GPG)	= 17.1 parts per million (ppm)
gallons per minute × 1440	= gallons per 24 hours
$\frac{\text{Gallons per 24 hours}}{1440}$	= Gallons per Minute (gpm)
Steam Generation	
Oil: lbs of steam	= gallons of oil × 100
Gas: lbs of steam	= $\frac{\text{cu. ft gas} \times 1000}{1500}$

Boiler Water to be Treated:
lbs. of steam generated = total lbs. of water (approx.) to be treated − lbs of condensate return

Table 15-3. Table of Equivalents

To approximate the amount water in a firetube-boiler's hot water system, add approximately 20% to its steam boiler capacity. Typical cast-iron sectional water tube steam boilers contain only 55% of a similarly rated cylindrical fire tube boiler. The most accurate way to determine the number of gallons of water in a closed loop piping system is to estimate the piping lengths and sizes and then use the following chart.

To estimate the number of gallons of water contained in the piping of a closed system (chilled or hydronic) multiply the length (in feet) of pipe, according to size, by the following factors:

1″ pipe by 0.044	8″ pipe by 2.65
2″ pipe by 0.174	10″ pipe by 4.23
4″ pipe by 0.661	12″ pipe by 5.87
6″ pipe by 1.5	16″ pipe by 10.44

Now that you have walked around your building to collect the survey information and scheduled repairs for any steam or water leaks, you have reduced your make-up water requirements already and also reduced your water treatment costs. The quantity of make-up water you use and its chemical characteristics are the key factors that determine the nature and cost of your water treatment program. The next step is to sample your make-up water and have it analyzed for those chemical characteristics that effect water treatment.

Make-up water is an often misused term and should be defined exactly. For our purposes, make-up water is any water that enters a heating or cooling system to replace water that is lost from the system. Let's further define it by keeping it separate from other water and external treatments. For example, the water fed to a steam boiler that contains a combination of water from your well and your condensate receiver is *feed water,* not make-up water. In this case, the well water alone is considered your make-up water. The city water that enters your cooling towers through your tower reservoirs water level control float, from pipes directly off your main, is make-up water. If you find it necessary to soften your boiler make-up water, this is softened make-up water. Finally, if you recover steam from your absorber and collect it in a receiver tank and use it for a portion of your cooling tower make-up water, you should consider this as part of your make-up and analyze it.

Your water treater must know the condition of the water he is being asked to treat. In addition, he must know if an external treatment—such as softening, filtering, disinfecting or pH control—is being used. Once you give him this information, he can design your program with these external treatment factors in mind and check regularly to confirm their presence in the make-up water. A typical analysis of a potable make-up water includes:

- Conductivity (micromhos)
- pH
- Calcium Hardness as $CaCO_3$, ppm
- Magnesium Hardness as $CaCO_3$, ppm
- Total Hardness as $CaCO_3$, ppm
- Phenolthalein Alkalinity as $CaCO_3$, ppm
- Methyl Orange Alkalinity as $CaCO_3$, ppm

- Chloride as C1, ppm
- Silica as SiO_2, ppm
- Iron as Ferrous, ppm

Special factors peculiar to your water supply (such as sulphur, turbidity, etc.) should be included in your analysis.

If you can secure this make-up water analysis from your local water supply company, this could save you money. If not, contact a local analytical laboratory and find out how large a sample they need for a water treatment profile that would include all the items listed above plus any items they know needs analysis from their experience in your area. They may provide you with a bottle, or you can use your own providing it has not been used already to hold something other than plain water from your plant. Run the water for several minutes and then rinse the bottle several times before you fill it to the top. Label the bottle and get it to the laboratory the same day you fill it. Don't let it lay around or heat up or you could get bad results due to growths. If your water is externally treated (i.e.; softening, filtering, chlorination or pH control), provide a before and after sample and label them clearly for the laboratory. If you use external treatment, check your analysis reports and see if you are accomplishing the softening, filtering, chlorination and pH control. If not, you will want to address this issue with your suppliers.

EVALUATING YOUR CURRENT PROGRAM & CONSIDERING AVAILABLE OPTIONS

Once you get your report back and your survey is complete you are almost ready to provide a water treatment vendor with sufficient information for him to recommend a responsive program. Before you contact any suppliers, you have a few more decisions to make.

On the assumption that you have a current water treatment program, now is the time to revaluate it to determine whether or not you wish to continue it. When you checked your records for the survey, you should have uncovered any untreated systems, water treatment related repairs, i.e.; boiler tubes, condenser vessel tubes, condensate receiver, condensate piping, pumps, and so on. Ask your current vendor about any problems you discovered, keeping several key issues in mind:

If your people are administering the water treatment program under the guidance of your vendor, improper chemical control levels are your responsibility. If the vendor has been advising you or your predecessor or your staff about any shortcomings in the application of chemicals, bleed-off, blow-down and systems leaks, etc. he has done his job.

On the other hand, you may have a full service program where the proper chemical control levels are set up in advance and then administered by your water treatment vendor. In this instance, the vendor must be held more responsible unless steam trap leaks, condensate losses, level control or external treatment problems compromise his program.

Water treatment can only put off the inevitable corrosive deterioration of the materials of construction in your heat exchanger system. Water treatment is an investment you make to inhibit the destructive tendencies of water, not to totally eliminate them. In addition, your water treatment program should permit you to run your equipment economically by reducing mineral scale, growths, and other deposits that insulate heat exchange surfaces and increase your fuel and electric costs.

Once you review your program and evaluate its performance, you must decide whether or not to continue with your current vendor, as is, or continue but make some changes. You may also wish to switch from a program where you administer the chemicals to a program where a vendor administers the chemicals. If you have the personnel available to administer the chemical, do it. Full-service water treatment vendors show up at your building and are motivated to get in and out quickly. They are usually unaware of what operational problems or changes took place between their visits. And although they can fill a chemical feeder and adjust the dose by test, they usually react to a symptom, rather than look for the reason.

Using Automated Equipment

There is so much uncomplicated automated equipment available today that there is no excuse to burden your staff with small manual shock dosing of chemicals or manual bleed-off and blow-downs. Your staff should spend most of their water treatment time performing chemical tests to adjust chemical feed and correcting make-up losses, not climbing up ladders to manually apply biocides or loading your by-pass feeders to add shock doses of chemicals on a daily basis. Some automated areas are:

- Contact head water meters installed on make-up water lines that signal resetting timers to which chemical feed pumps and or blow-down and bleed valves are connected. These devices automatically maintain chemical and conductivity levels because the make-up water is proportional to the load.
- Continuous or intermittent conductivity and pH sensors that activate bleed-off and blow-down valves and chemical feed pumps when present conductivity or pH levels are exceeded.
- Biocide controller timers that feed two biocides alternately each week. In addition, they shut down bleed off for a preset period of time after biocide feed to permit sufficient contact time for a proper growth kill.
- Reverse conductivity controls that feed chemical to closed loops based on a low conductivity reading that indicates a drop in chemical levels.
- Level controls with alarms to alert your staff that a chemical drum has a low level of liquid and needs refilling.
- No-flow controls to shut chemical feed systems down when the heat exchange system is inactive.
- Time-out devices that shut down electrical circuits if a pump is on for too long a period of time and there is a danger of overfeed.

All the equipment listed above as well as the pumps used to inject the chemicals are available from independent manufacturers who can always repair the equipment or supply you with spare parts. Many water treaters use the same manufacturers and put their "private label" brand name on the product. Insist on the manufacturers name, address, equipment, nomenclature part numbers, etc. This will allow you to continue your use of the equipment, regardless of who is your water treatment vendor. All equipment should be considered universal in the sense that it should work with anyone's program. The only change that might be required is wiring for additional pumps or an increase in a pump's output capacity because of chemical concentration changes.

In the future, automated equipment and sensing devices will be microprocessor integrated to a point that on-site monitoring and off-site computers will be able to control chemical programs on a daily basis.

Understanding Chemical Water Additives

Now that you are more familiar with your water treatment requirements and made some basic decisions about how you want to handle your program, you should be familiar with the current state-of-the-art chemicals used for water treatment. While this chapter is not meant to be a primer in theoretical chemistry, there are some water treatment terms and concepts you must understand.

Total Dissolves Solids (TDS) and Conductivity

All normal make-up water contains dissolved solids. These solids become dissolved in the water as it goes through its usual earth-to-the-atmosphere-and-back cycle. The amount of dissolved solids the water picks up varies between geographic areas. The water treatment chemist uses the conductivity or specific conductance of the water as an easy way to measure the amount of total dissolved solids. He does this in units called micromhos.

When you boil off water in a steam boiler or evaporate it in a cooling tower, all the dissolved solids remain behind in the boiler or condenser water system. This water is usually replaced automatically. Each time you replace or make-up a full system's volume of water, you bring in another concentration of dissolved solids. If this process were to continue without moderation, eventually you would overconcentrate the dissolved solids in the boiler or condenser water system, and the solids would precipitate and deposit themselves all over your equipment. Because of inverse solubility, the first place these deposits would form is on the hottest surfaces such as boiler or condenser tubes.

Your water treater can inhibit these deposits using chemicals; however, he cannot stop them unless he also limits the amount of dissolved solids in your operating system. Therefore, your water treater examines the actual minerals and other solids that constitute your dissolved solids and makes a judgment as to how much of them can be concentrated before you will have deposit problems. He relates this amount to conductivity readings in your make-water and sets a limit of micromho units to be allowed in your operating equipment. This limit is called the allowable cycles of concentration. The actual number of cycles of concentration is determined by this formula:

$$\frac{\text{Conductivity in micromhos in the operating system}}{\text{Conductivity in micromhos in the make-up water}} = \text{cycles of concentration in the operating systems}$$

Note: Chlorides can usually be used in place of conductivity to determine cycles when they are unaffected by zeolite softener malfunctions.

The water treater then specifies a bleed-off or blow-down schedule to relieve the operating equipment of dissolved solids and regulates it by measuring the conductivity in the operating system. If the cycles of concentration are higher than specified, more blow-down or bleed is required. If the cycles of concentration are too low, the blow-down or bleed-off schedule is reduced. Blow-down or bleed-off is best when it is continuous or done frequently in small amounts. Infrequent, large blow-downs or bleed-offs are not as effective because the overconcentration is not being controlled on a steady basis. Inconsistent bleed or blow-down is not only ineffective, it also makes it impossible to maintain chemical control levels.

Chemical Control Levels.

All chemical treatment programs must require that a specific level of chemical be maintained in the operating system. This level is frequently called a residual. The required residual of a specific chemical can vary from water treatment program to water treatment program, depending on the other components in the program, the make-up water being treated, and the type of system.

To be sure that the amount of chemical needed for protection is present in the operating system, the water treatment chemist tests for a specific amount of a chemical substance. Rather than use a term like ounces per gallon to measure the amount, the water treatment chemist uses parts per million (ppm) or milligrams/liter (mg/1). The water treatment chemist uses ppm because it is more appropriate to the use rate of his products. Five parts per million could refer to 5 gallons in a million gallons or 5 pounds in a million pounds. Trying to express such a use rate in ounces per gallon would not be practical.

The pH Factor

pH is a scale devised by the chemist to describe or characterize the acid or alkaline nature of a solution. The units of the pH scale start at 1 and end at 14. Concentrated acids like sulfuric acid have pH values of 1 or less, and concentrated alkalis like sodium hydroxide have pH values of 14. The pH of drinking water is commonly thought of as neutral with a pH value of 7, but most of the make-up water you encounter will probably have a pH between 6 and 8. It is important to understand that low pH make-water usually has corrosive tendencies and high pH water usually has scaling tendencies. It's your water treater's job to regulate the pH in your operating system so it is high enough not to encourage corrosion and low enough not to encourage scaling. This explanation is an oversimplification of a very complicated chemical problem that has been the subject of many books and research projects. Many complicated indexes such as the Ryznar and the Langelier have been developed to help characterize make-up waters tendencies towards scale or corrosion.

Environmental Considerations

There are a number of environmentally acceptable chemicals used alone and in conjunction with other chemicals to treat water in heating and cooling systems. Tables 15-4, 15-5, and 15-6 should familiarize you with some of these chemicals and how they are used:

Many of the chemicals listed in Tables 15-4 through 15-6 can be applied to the same make-up water. If possible, try to avoid the ploy used by some water treaters who try to get you to specify one chemical, only theirs. Most of the time their objective is to eliminate the competition. There are many corrosion inhibitors, biocides, dispersant, scale inhibitors, alkalinity builders, pH controllers, etc. Your objective should be to make sure that the ones you need are all present in the program. Don't be bashful about asking your water treatment vendor specifically what chemicals his products contain. In most states, there are laws requiring chemical disclosure due to regulations of E.P.A., OSHA and Right-To-Know legislation.

Chromate, a heavy metal that is an excellent corrosion inhibitor with a long and successful water treatment history, has not been included in Tables 15-4 through 15-6. Initially, the E.P.A. placed a 5 ppm limit on chromium discharge because of its toxicity. This virtually eliminated it for use by itself, but some suppliers persisted and others combined it at low levels with other inhibitors. Next, the E.P.A. banned its use in comfort cooling towers based on studies they conducted that they felt proved that the drift from cooling towers treated with chromium caused cancer. They have made it illegal for water treatment vendors to sell chromium for use in comfort cooling towers. By the way, comfort cooling towers that provide computer cooling also are still comfort cooling towers. If chromate is cariogenic and toxic and banned in cooling towers then why should any responsible engineer want to use it in boilers or closed loop systems.

Be suspicious of any vendor who tells you that his program is strictly proprietary and he cannot divulge any specific chemical information. Keep in mind that most good water treatment products contain more than one component. Also remember that a combination product with factory preset ratios of many components is usually an easier program to administer. A water treatment chemical user like a large industrial plant, co-generator or powerhouse often can justify the cost of a member of the plant staff just to administer individual chemical component products at the proper levels. This is not usually the case in most maintenance departments where water treatment is a part-time responsibility.

Chemical	Main Function(s)	Comments
Phosphates	Scale Inhibition, some corrosion inhibition	Should be used with a sludge conditioner and caustic.
Caustic	Scale and Inhibition Corrosion	Too much can cause corrosion in high pressure boilers. Required to drive many reactions in a calcium phosphate sludge forming programs and to help remove magnesium.
Chelants	Scale Inhibition, Iron Oxide Control	Not economical or safe unless softened boiler feed water is very well controlled. Requires special feed techniques. Should be used with a polymer dispersant.
Dispersant and Sludge Conditioners or Polymers	Fluidizes Sludge, Distorts Crystal Formation to Inhibit Scale	Usually are combinations of long and short chain or heavy and light molecular weight organic polymers. Older types were natural organics, i.e.; lignen and tanin.
Sulfites	Prevent Oxygen Pitting	Ties up free oxygen. Sometimes referred to as an oxygen scavenger. Should be fed to storage section of deairator or into condensate make-up receiver to protect feed water system. Catalyzed for use with low temperature feed water.
Hydrazine	Prevents oxygen pitting and steam line corrosion	A toxic product with a low flash point. Used mostly where low dissolved solids are required in very high pressure boilers. Can corrode copper if overfed. Used at 15% to 35% solutions it's less dangerous.
Volatile Neutralizing Amines	Neutralizes carbonic acid in the steam condensate steam systems.	Not to be confused with the older filming amines which can cause difficulties in older systems.

Table 15-4. Basic Chemicals Used in Steam Boilers

Chemical	Main Function(s)	Comments
Nitrite	Corrosion inhibitor for ferrous metals	Film forming inhibitor usually combined with borate for buffering. Can biologically foul chilled water systems.
Nitrate	Corrosion inhibitor for soft soldered joints	Very useful in systems composed of copper piping and copper coils where solder is used extensively.
Borate	Corrosion Inhibitor	Usually combined with other inhibitors primarily as a buffer to elevate pH into more conducive inhibitor range.
Silicates	Corrosion Inhibitor	Reacts with iron oxide to form a protective coating and elevates pH.
Molybdates	Corrosion Inhibitor	Environmentally acceptable heavy metal. Sometimes combined with silicates and other components. Forms passive films.
Sulfite	Prevents oxygen pitting	Good for tying up oxygen, but very little else and difficult to maintain reasonable levels.
Azoles	Copper corrosion inhibitor	Used as an adjunct with most other inhibitors because of its specific copper protection ability at a low use levels.

Note: Closed loop systems such as chillers and hot water boilers are not supposed to be taking in large amounts of make-up water. If they do, the cost of keeping up the chemical treatment control limits will become prohibitive. These products which are usually formulated for corrosion inhibition not scale prevention and will not work properly in high make-up situations. If this is the case, find the leaks and fix them.

Table 15-5. Basic Chemicals Used in Closed Loop Systems

Chemical	Main Function(s)	Comments
Phosphates	Corrosion Inhibitor and threshold scale prevention	Used in conjunction with many other components.
Phosphonates	Scale Inhibition Iron Oxide Inhibition	Sometimes referred to as an organic phosphate. Can be aggressive to copper and should be used with a copper inhibitor like an azole.
Polyacrylates, Polymethacrylates, Polymaleic-Anhydride	Dispersant for suspended solids and Scale Control	Used in combination with other products to improve their scale inhibiting ability.
Zinc	Corrosion Inhibitor	Very quick film forming corrosion inhibitor, but must be used with other corrosion inhibitors to stabilize performance. Can precipitate in high pH. Limited use due to toxicity, but permitted at low levels.
Molybdate	Corrosion Inhibitor	Environmentally acceptable heavy metal inhibitor generally combined with other components. Forms passivating films.
Azoles: Tolytriazole, Benzotriazole, Mercaptobenzotriazole	Corrosion Inhibitor for copper and copper alloys	Family of chemicals that require only small amounts to reduce copper corrosion.
Biocides, Slimicides, Algaecides	Inhibit Growths of Algae and other Microorganisms	Two (2) different formulations (*type* not *concentration*) should be used alternately to prevent immunities. E.P.A. registration approved dose rates make them all similar performers, at specific pH levels. The use of an oxidizing and a non-oxidizing one helps. Some states require applicators license to apply.*

*Check this out in your location.

Table 15-6. Basic Chemicals Used in Cooling Towers

MONITORING YOUR WATER TREATMENT PROGRAM

Now that you are familiar with your equipment's water treatment requirements, the terms used to define a water treatment program, your specific problems and some of the latest chemicals in the field, it's still too early to invite in the vendors or simply review your ideas and questions with your current vendor. You still have to define what services you want from your vendor as well as how your staff will participate in the water treatment program.

Deciding Between Full Vendor Service vs. Staff Service

You now have to make a decision about using a full service program where the vendor applies the chemicals, or a staff service program where your people apply the program or some combination of the two. To make your decision you should keep in mind that proper administration of a chemical water treatment program requires the following:

- Chemical Testing of the boiler water, condensate, closed loops and condenser water.
- Biological testing of condenser water and closed chilled loops.
- Adjustment of the chemical dose mixing ratios.
- Adjustment of the chemical equipment feed rates.
- Blow-down and bleed-off regulation.
- Chemical testing of External treatment, i.e.; softeners.
- Stock control and ordering of chemicals and test reagents.
- Chemical Testing of the make-up water.
- Record keeping in a test log indicating adjustments and results.
- An in-line monitoring program such as corrosion coupons.

The frequency of these tasks usually depends on your scheduling of them. Blow-down and bleed-off are daily tasks unless you automate them. When starting up a system, testing should be daily until you establish consistent chemical and bleed-off or blow-down rates that satisfy your control limits. In general, test as often as you need to in order to keep your chemicals and control limits in their proper preset range. Every water treatment program must have a set of control limits prescribed in advance by your vendor or by your water treatment consultant. In most cases, it would be wise to establish a set of control limits and make them known to all the water treatment vendors bidding your requirement. This will provide you with "apples to apples" bids and a set of standards you can apply once your water treatment program gets underway.

By now you should realize that a full-service vendor program administered once a month by a service person, who just stops by long enough to fill up the feeder, is not quality water treatment unless you have only a few small closed-loop systems. If you have opted to do it yourself to be sure it's done properly, insist that your water treatment vendor comes in frequently and tests your system himself, on-site. At that time, he should generate a full analysis report and discuss the changes necessary to keep the program on track. Specifically, he should offer practical, easily understood recommendations to get any control limit out of range, back into range. In addition, he should be observing your system first hand to help you spot make-up water problems, i.e., a change in the chemical characteristics of the make-up or unnecessary extra make-up due to leaks or poor control. Obviously, any chemical overdoses or under doses should be pointed out and a course of action suggested. It would be wise for you to let your vendor and staff know that you expect this type of reporting from the vendor.

Don't let anyone get the false impression that you want only "good" reports. Your vendor's worth will be greatly diminished if he gets the impression that he has to "guild the lily" to keep everyone happy. Keep in mind that like other contractors he wants to keep your business, but he also has to keep your program on track. Don't be the kind of supervisor who jumps all over everybody as soon as there's a slight infraction. Discuss the problem with the vendor and your staff and work it out. Mechanical repairs requiring more budget than you have will compromise your water treatment program. When this happens, encourage everyone to do as well as they can. The analysis report provided in Figure 15-1 is typical of what you should expect from your vendor.

Testing Your System

Between your water treatment vendors' service visit and reports, your staff will be administering the water treatment program. In order for you to administer them effectively, they must use test procedures and keep a test entry log. All test kits should come with a full set of instructions that are simple to perform and easy to understand. Each test must be specifically related to a chemical or procedure in your water treatment program. In many cases, the instruction of the test kit manufacturer will only tell you how to perform a test on a sample and tell you how to interpret the results of the test in chemical terms; i.e., "the number of drops it took to change the sample color from pink to white, times 10 equals the ppm of P-alkalinity as $CaCO_3$." This means nothing specific to your staff unless they know which sample to test and how many ppm of P-alkalinity they are trying to maintain in the boiler water. They must also know what chemical in your program should be increased or decreased to raise or lower the ppm of P-alkalinity. The same sound water treatment practice applies to all the other chemicals in all the water treated systems in your plant.

In addition to testing and adjusting chemicals, you must test and adjust pH, blow-down, bleed-off, softeners, etc. Even though these items are not chemicals, you

FORM WA-G 8/78

North Eastern Chemicals
240 MARY STREET • SUITE 1
HACKENSACK, NEW JERSEY 07601
(201) 342-8774

WATER ANALYSIS

Name: Mr. John Doe, Building Maintenance Mgr. Phone: (212) 000-0000
Area Code Number Ext.

Company: ABC CORPORATION

Street: 10 Main Street

City: New York State: NY Zip: 10001

cc: Bill Smith, Engineer

INVENTORY					
Product	**Gals./Lbs.**	**Product**	**Gals./Lbs.**	**Product**	**Gals./Lbs.**
NE600MZ	85 gal	Hydrax 2	55gal		
NE2030	25 gal	Hydrax 4	45gal		

TESTS	RESULTS Steam					CONTROL LIMITS Steam			
	Tower1	Chiller	Boiler	Condensate		Tower	Chiller	Boiler2	Condensate
Molybdate	8	40				10	35		
Conductivity (mm)	500		1743	5		600-800		1600-1800	Record
pH	8.3	9.3		7.9		8.0-8.5	9+		7.5-8.5
Phosphate			25					30-60	
P-Alkalinity		110	420				100+	300-600	
Sulfite			40					30-50	
Hardness			0					0	

RECOMMENDATIONS: Tower: Reduce bleed by 1/3. This will increase conductivity and bring up the level of molybdate.

Chiller: Chemical and pH levels are good. Test next week.

Boiler: Please increase Hydrax 2 dose by 25% to bring up phosphate levels.

Condensate: pH level is good. Continue current dose of Hydrax 4.

BY: Jerold I. Berkley DATE: January 4, 1990

Figure 15-1. Sample Water Analysis Report

cannot have a consistent water treatment program without keeping them within their establish control limits, i.e.; The amount of blow-down and bleed-off directly effects the amount of the make-up water entering your system. The make-up water effects pH, hardness, alkalinity, and dissolved oxygen—to mention just a few items. Therefore, the consistency of your chemical treatment to adjust pH, counteract hardness, increase or decrease alkalinity, and scavenge dissolved oxygen are all subject to your bleed-off and blow-down rates. If these rates are inconsistent, your chemical dose and protection levels will also be inconsistent.

The following is a typical chemical test log that you may use or adapt to keep track of your boiler and cooling tower tests, dose adjustments, and bleed-off/blow-down:

Date	Tester	Equip.	Test	Control Limits	Test Results	Action Taken
1/3/90	Jolin C	Boiler #1	Nitrite	800-1000	700	Add 3 gallons NE 2030 to Boiler #1
1/6/90	George R	50 Ton Tower	Molybdate	8-10	6	Increase NE 600MZ pump output 30% to setting #6
1/6/90	George R	50 Ton Tower	Conductivity	600-800	900	Increase bleed timer 10% to 1½ minutes

EVALUATING THE NEW PROGRAM'S EFFECTIVENESS

Now that you have your program in place, how do you know it's working? Unfortunately, water treatment insufficiencies are not readily apparent. You can head off problems and your vendor can make changes to help because there are many "sign posts" you can look for along the way, but you will have to go out of your way to do it. The old "if it's not broken don't fix it" adage is not a good approach in water treatment. Boiler replacement, retubing of condensers, or replacement of condensate systems represent enormous expenses. Don't wait until it's too late. Develop a standard inspection system as part of your regular maintenance procedures.

Seasonal Inspection Procedures

Keep a record of head pressures and temperature differentials. Abnormal conditions could indicate a build-up of deposits such as mineral scale or a biological fouling. Make sure your pressure and temperature gauges are working. Machines don't just shut down overnight. The temperature or pressure problems builds slowly a psi or degree at a time.

Inspect the water in your systems visually and with your nose. A badly biologically fouled chilled and condenser water stinks, but before it gets fouled, you can use a biocount test strip to see if your biocides are effective. These biocount test strips measure the quantity of biological growth colonies, usually in terms of colonies/milliliter. After adding the biocide, the count should decrease. Tower basins filled with slime are a pretty positive indication that there is slime throughout the system. Feel around the surfaces to see if the walls are clean or slimy. Climb up top and check your distribution troughs for algae, especially if they are uncovered and in direct sunlight. Consider covering them, as well as alternating biocides, increasing your dose rate or trying new biocides if you have growth problems.

Install several corrosion coupon racks in your cooling systems, i.e.; condenser and chilled water. These assemblies should be plastic (PVC or CPVC) if the pressure allows, because they will last longer and not corrode if they are shut down in between use. If the pressures are too high, use carbon steel or preferably stainless if possible. In either case, a minimum of two positions in the rack for coupons is needed. Usually a mild steel and copper coupon will offer test results that you can relate to the materials of construction in your system. The configuration is very similar in most racks, but be sure you install them so they remain flooded when your system shuts down, i.e.; overnight or just in light-load periods. You want the corrosion coupon to be similar in operating cycle to your actual system and you don't drain it everytime you shut down.

The object of the corrosion coupon study is to expose precisely and carefully preweighed metallic coupons to the treated water in your system. After a fixed period of time (e.g., 30, 60, 80, 120 days), these coupons are removed, specially stripped of the excess weight of corrosion products according to ASME standards, and weighed again. A specific mathematical formula is applied to the weight differential, and projected corrosion rate in mils/year is calculated. Although there are many arguments regarding how valuable these coupons really are, there certainly is an advantage in any program that can reduce these rates by several mils—providing the cost of the improvement is reasonable.

The most advantageous way to conduct these corrosion coupon studies is to do them frequently and for different lengths of time and look at the overall results. Always get the coupons back from the lab in a clean condition so you can evaluate the pitting and corrosion patterns for change. This brings me to what I believe is the single most important recommendation about coupon studies. Buy the rack and coupons from an independent lab and send them back to this lab for full analysis and report. That way you know the results are unprejudiced.

COMPLYING WITH REGULATIONS

There are many outside regulations that you must comply with when applying water treatment programs. It doesn't really make much difference whether your staff applies the program or if you have an outside vendor apply it. If there are chemicals on hand at your plant, you are responsible and liable. Make sure you check out all the various regulations that apply in your geographic area. The following will give you some ideas regarding where to look and what to do.

All cooling tower biocides must be federally approved and registered with the federal EPA and most likely with the state and possibly even the city in which you are located. If you are a New York State facility and buy biocides from a firm that manufactures them in Pennsylvania, there must be a federal EPA registration number on the label and the biocide must be registered in New York State. A Pennsylvania state registration is not sufficient.

In all cases, the biocides must be applied according to the label directions. In addition, the biocides containers should be triple rinsed and the rinsate discharge into the tower, not the sewer. Keep in mind that the overdose of biocide or misapplication is what an EPA inspector will be looking for to protect the environment. Remember to use the safety equipment (such as gloves and goggles) required on the label when handling the biocide.

The use of proper protective clothes and equipment, as defined by OSHA, applies to all chemicals, and it's your job to have the required safety equipment warnings and notices on hand at the chemical location and also to educate your staff in proper chemical handling. This was always part of your OSHA responsibility and now it is also part of your Right-to-Know law responsibility. A yearly refresher course and an initial introductory course for newly hired engineers is also your responsibility. In addition to conducting the course for water treatment chemicals, you should conduct it for all chemicals (e.g., caustic cleaners, solvent cleaners, antifreezes, battery acids, etc). I strongly suggest that you have a sign-in log at the course for your staff in case you ever have to prove that you conducted the course. A simple questionnaire (quiz) indicating that your staff understood will go a long way towards being sure that you are getting your safety message across. Engineers who do not use the safety equipment you make available should be disciplined for their own health and also as an enforcement method to reduce your liability. Keep in mind that your liability extends not only to your staff, but to the community and other workers and pedestrians in your facility.

Community "Right-to-Know" laws as dictated by SARA III and local regulation require that the ingredients of your chemicals be made known to all those in possible contact with them. A specific number of ingredients must be listed on your drum labels along with their CAS numbers. CAS numbers are a universal chemical classification system that specifically define the chemical formula of an ingredient. They prevent confusion and help with personal health treatment or pollution control in case of ingestion or spills. If you have to fill out forms, your chemical supplier should give you these numbers or supply you with ingredient labels that have them listed. Watch out for vendors who claim, "We are sorry, but our ingredients are proprietary." In case of spill, fire or accidental exposure or ingestion, "proprietary" won't help. You will be exposing yourself, your associates, your pedestrians, and your company to liability if you cannot take reasonable steps in an emergency. There is nothing in the water treatment that is such an innovation that absolute secrecy is needed. If your vendor won't help you comply, get another vendor.

This checklist will help you develop a water treatment file with most of the things you need to comply with regulations. For each product used you should have:

- area of application, average application rates, and the average amount of stock on hand.
- hazardous and non-hazardous ingredients and safety procedures for storage, handling, and spills.
- special licenses or permits required to apply or store the products.
- OSHA data sheets as well as the vendors' company literature.
- all logs recording dose, activity, etc. by your staff and service reports from the vendor.

As new health, safety, disclosure and right-to-know regulations arise, ask your water treatment vendor to keep you apprised of what changes you should make to stay current. It is his job to help you stay up to date. This applies not only to regulations, but also to water treatment chemical technology and equipment. Every year, new methods come into the water treatment field, and your vendor should evaluate them and bring them to your attention—just as we have tried to bring the various factors involved in a good water treatment program to your attention.

It was not our intent in this chapter to propose that you use a specific water treatment program. Your specific make-up water, physical plant, operation, staff, and equipment and those of your counterparts all over the country are too numerous and varied to make one program universally proper. Make sure that when you choose a water treatment program, it is designed for your specific set of conditions and is sold and serviced by a qualified water treatment chemist.

16

General Building Maintenance: Fire Protection

Michael J. Kerr

As maintenance manager you are confronted with a multitude of problems that are not mechanical, electrical or structural in nature. Yet these problems are of utmost importance and if not addressed can result in the most disastrous results. Complete destruction as a result of a fire have closed many plants and created many situations that could have been eliminated or reduced in scope. This chapter deals with these situations.

This chapter deals with portable fire extinguishers, sprinkler and standpipe systems, fire alarms, fire suppression systems, flame proofing, and chemical problems and their solutions as well as employee training, fire prevention, and housekeeping.

PORTABLE FIRE EXTINGUISHERS

Selecting Extinguishers

The proper selection of extinguishers for a fire should be determined by the type of fire or hazard involved. Extinguishers are classified for use on certain types of fires and are rated for their extinguishing effectiveness. The specific classes of hazards are listed below;

- CLASS-A Ordinary combustibles, wood, cloth, paper
- CLASS-B Flammable liquids, oils, greases, flammable gases, plastics, rubber.

- CLASS-C Energized electrical equipment.
- CLASS-D Combustible metals.

Different types of extinguishing agents are;

- Ammonium phosphate, which is rated for ABC fires. The color of the powder is yellow.
- Sodium bicarbonate base, which is rated for BC fires. The color of the powder is white.
- Potassium Bicarbonate base (known as Purple K) is used on B and C fires. The color of the powder is purple.
- Sodium Chloride base (known as Metl-X) is used on class D fires such as magnesium, Titanium, Zirconium, Sodium, Lithium and Potassium.

Fire hazards are classified into three categories:

- Light hazards exist in areas where small fires can be expected, ie., classrooms, offices and halls.
- Ordinary hazards exist in areas where a middle-range type of fire can be expected, ie., small manufacturing plants, parking garages and small warehouses.
- Extra hazards exist in areas where there is a potential for severe fires, ie., manufacturing/processing with items such as flammable liquid handling, painting and dipping & large warehouses with combustible stock in excess of 12′ 0″.

Placement and Distribution of Extinguishers

Your maintenance staff should periodically check the applicable current codes indicating maximum travel distances to fire extinguishers. This helps assure that the most effective system of fire extinguishers are installed to ensure that fires do not unnecessarily spread. Survey the individual property for actual protection requirements. Extinguishers are most effective when they are readily available in sufficient number and with adequate extinguishing capacity. Travel distance and time is a main factor because it is the distance the user of the extinguisher will need to travel from the fire, get the extinguisher, and return to the fire before beginning extinguishing operations. In general, select locations that will:

- provide uniform distribution,
- provide easy access,
- be free from blocking by storage or other obstacles,
- be near paths of travel, entrance and exit doors,
- be visible, and
- be installed on a floor to floor basis.

Provide extinguishers to protect both the building structure, if combustible, (Class A) and the occupancy hazard (Class A,B,C or D). Locate them in plain view along normal paths of travel including exits from areas. Make sure an extinguisher in the area of a hazard is accessible without any danger to the operator. Standard maximum distance to an extinguisher from a hazard should not exceed:

- 75 feet for Class A hazards,
- 50 feet for Class B hazards,
- 75 feet for class C hazards and
- 75 feet for Class D hazards.

If there are highly combustible materials in either enclosed areas or small rooms, hang an appropriate extinguisher outside the room.

Selecting the best extinguisher for a given situation depends on:

- the nature of the combustibles that might be ignited
- the size, intensity, and speed of travel of any resulting fire,
- effectiveness of the extinguisher on the hazard and,
- the ease of use of the extinguisher.

Therefore, follow these guidelines:

- Locate the extinguisher near fire hazards for which they are suitable.
- Use extinguishers suitable for more than one class of fire.
- Clearly mark the intended use.
- Train employees in the proper use of extinguishers.

Make sure fire extinguishers are visible and conspicuous. Paint a red stripe around the extinguisher. If hung on columns, the red stripe can be painted all around the column so that it can be seen from all angles. In addition, post large signs indicating the location of fire extinguishers. The fire extinguisher must be readily accessible. Make sure stored material, stock, machinery and other obstructions do not block access to fire extinguishers. Keep extinguishers clean and ready for use. Take care that they are not in the normal path of material handling equipment, flow of people and generally not in the way of the normal facilities operation. If necessary, use guard rails, bumpers and other protective items to protect fire extinguishers. Use wall indentations, if possible.

Place extinguishers that weigh up to 40 lb at a height so that their tops are no more than 5′0″ above the floor. Extinguishers weighing over 40 lbs. should not be more than 3½′ above the floor.

Various portable equipment is available to be utilized in special conditions. For example, carts or hand trucks carrying multiple extinguishers, hose carts, various trucks, portable heat sensors, portable fire pumps, water storage tanks, etc. are available for temporary conditions or special situations in (or out) of the plant. In some cases, fire blankets can be used to smother small fires. They can be used on burning clothing and flammable liquid fires in small open containers.

Inspecting, Maintaining, and Recharging Extinguishers

Inspection, maintenance and recharging of extinguishers are important factors for ensuring proper operation at the time the extinguisher is used. Inspect all extinguishers frequently. Procedures for inspection and maintenance of extinguishers vary considerably, and therefore, should be done by a trained person who can perform the maintenance. Recharge all rechargeable-type extinguishers after use or as indicated by an inspection. When performing recharging, follow the recommendations of the manufacturer and use only those agents specified on the nameplate to recharge the extinguisher. Inspect extinguishers when initially installed and thereafter at one month intervals or more frequently if your maintenance procedures require. When inspecting extinguishers, make sure:

- The extinguisher is in its proper place.
- There are no obstructions to access or visibility.
- The seals are not broken.
- The weight is proper.
- There is no corrosion or dents.
- There is no leakage or clogged nozzle.

Record the date of the inspection and the initials of the person performing the inspection. Use extinguisher tags as a convenient means to record maintenance checks. A typical tag includes various information. (See Figure 16-1.)

For pump tank extinguishers, periodically check the water level by stroking the pump several times. The liquid is to be returned to the tank. Check the condition of operating parts, tank and foot bracket for corrosion and oil the piston rod packing. After use, refill with water or antifreeze solution. Protect from freezing.

For stored pressure water fire extinguishers, check the pressure, hose, and nozzle. When charging, fill to water level mark to allow sufficient room for pressurized air. Most models have an overflow tube which will fill to proper level. Protect from freezing.

For dry chemical fire extinguishers with cartridges, periodically weigh the CO^2 cartridge (ideally at semi-annual intervals). If less then 1/2 oz. of the appropriate weight, replace the cartridge. Check hose, nozzle, syphon tube, cap gaskets, container and dry chemical agent for caking and proper level. Moisture will cause dry chemical to cake. Wash extinguisher or parts only if the container can be thoroughly dried.

For dry chemical extinguishers, stored pressure type with combination handles, check the pressure gauge often. Clean the valve assembly thoroughly so that the valve will seat tight. Check manufacturers instructions for recharging. However, in all cases, pressurizing must be dry to eliminate moisture.

Recharge carbon dioxide fire extinguishers by an approved CO^2 filling facility. Inspections of the units should include weighing each. Recharge if less than 10% of

DO NOT REMOVE

FOR CITY, STATE AND FIRE INSURANCE INSPECTION

FULL WT. ____

D.O.T. CERT. # A-192

FIRE COMMAND COMPANY, INC.

475 Long Beach Blvd.
LONG BEACH, N.Y. 11561

(516) 889-1111

SERVICED BY ____

● AFFF/LD. STRM	ABC DRY CHEM ●
● CARBON DIOXIDE	STD. DRY CHEM ●
● PRES. WATER	PK DRY CHEM ●
● HALON 1211	●
● CO2 SYSTEM	DRY CHEM SYS ●
● HALON SYSTEM	WET CHEM SYS ●

NATIONAL ASSOCIATION OF FIRE EQUIPMENT DISTRIBUTORS — NAFED

☐ 1990/1991 ☐

VOID 1 YR. FROM MO PUNCHED; SYSTEMS 6 MOS.

SERVICED	NEW	RECHARGED

DEC. | NOV. | OCT. | SEPT. | AUG. | JULY | JUNE | MAY | APR. | MAR. | FEB | JAN.

____ (Model No.)

____ (Mfr.)

OWNER'S I.D NO. (if used) ____

REMARKS ____

MONTHLY INSPECTION RECORD

DATE	BY	DATE	BY

PRINTED IN U.S.A.

Courtesy of Fire Command Co., Inc., Long Beach, N.Y.

Figure 16-1. Typical Fire Extinguisher Maintenance Tags

rated weight. Check for general conditions. Note that when used, the discharge horn becomes very cold. It should not be touched. Also, if it is used in restricted areas, it can create an oxygen deficient condition. Do not use in areas where temperatures are over 120° F.

Hydrostatic Testing

Hydrostatic testing is a pressurization of the extinguisher to test cylinder structure. Testing is required every 5 or 12 years depending on the type extinguisher, cylinder or shell. Requirements for hydrostatic testing differ for each authority having jurisdiction. In general, CO_2 (Carbon Dioxide) extinguishers require 5-year tests. Dry chemicals require 12-year tests, with the exception of stainless steel cylinders, which require a 5-year test. Do not retest brass shells. These units should be replaced. Pressurized/water extinguishers require a 5-year test.

Hydrostatic testing must be performed by competent personnel who have suitable testing equipment and facilities. All traces of water and moisture must be removed by using special drying equipment. Record these tests on the extinguisher. For compressed gas cylinders and cartridges passing a hydrostatic test, stamp the month and year into the shoulder, top head, neck or footing of the cylinder. These tests are generally performed by outside contractors.

Operation and Use of Portable Fire Extinguishers

The operator must be familiar with all steps necessary to operate all extinguishers in the plant. Continuous training is essential. "Portable" is the designation for manual equipment used for small fires in the interim between the observation of a fire and the operation of automatic fire extinguishing equipment or the arrival of professional fire fighters. Obviously, the sooner the fire extinguisher is applied, the better. To expedite the extinguishing of the fire:

- Know location of all fire extinguishers
- Remove nozzle & restraining or locking devices (ie., release safety pin and seal)
- Hold portable units upright
- Start discharge (depress or squeeze handle to release agent)
- Use sweeping motion and extend the charge at least 6″ on all sides of the fire
- Do not start too close to the fire. This could cause a splash and allow fire to spread
- Attack the source not the top of the flames
- Do not start and stop flow; keep it continuous until the fire has been extinguished
- Use common sense when fighting a fire. Do not take unnecessary chances.

CHOOSING & OPERATING SPRINKLER SYSTEMS*

All sprinkler systems require qualified inspection service to insure their reliability. You should make sure that the following important components are incorporated in your system:

*All information on sprinkler systems was provided by Robert S. Kirk, of Inter-Boro Sprinkler Corp.

- Tamper switches installed to O.S. & Y. valves that signal an alarm when someone operates the valve either to open or close it.
- Water flow switches installed into the sprinkler system piping that sends a signal when the flow of water passes the switch.
- Quick opening devices, either accelerator or an exhauster, used to cause a dry pipe valve to operate more quickly.

The following sections describe specific types of sprinkler systems.

Wet Sprinkler Systems

The wet sprinkler system is a sprinkler installation that has all piping filled up to the sprinkler heads with water under pressure. When a fusible link is activated, water is immediately sprayed over the area below. For areas where freezing is a problem, antifreeze or a dry system can be utilized.

When placing a wet system into service

- Check system to be sure it is ready to be filled with water. If the system has been shut down because a head has opened, be sure the head has been replaced with one of proper rating.
- Open "vent valves" located at high points.
- Place "alarm cock" in CLOSED position. This will prevent sounding of alarm while flowing water fills system.
- Place "drain valve" in nearly closed position. A trickle of water should flow from drain valve during filling.
- Open " indicator post valve" slowly. When the system has filled, there will be a quieting of the sound of rushing water. Open valve fully then back off 1/4 turn.
- Observe the water flow at the vent valves. When a steady flow of water occurs (no air), close vent valves.
- Check the water flow by quickly opening the "drain valve" and closing it. The water pressure should drop about 10 pounds when the valve is opened and immediately return to full pressure when valve is closed. Excessive pressure drop indicates insufficient water flow.
- Open "alarm cock"; system is now in SET condition.
- Test-Open "drain valve" several turns. Water should flow, sounding the mechanical gong alarm and the electrical alarm.
- Close "drain valve." If the alarms have functioned properly the system is operational.

Dry Sprinkler Systems

The dry sprinkler system is a system designed the same way as a wet system but the system piping does not contain water. Instead, The piping contains air or nitrogen under pressure. When released from the system as from an open sprinkler head, the water that was held back at the dry alarm valve by the air or nitrogen will flow into the systems piping and out thru the open sprinkler head. When placing a dry system into service:

- Close the valve controlling water flow to the system. This may be located in the riser or it may be an underground valve with an indicator post. If a fire has occurred and water is flowing from opened sprinklers, obtain approval of the person in authority before closing the valve.
- Open the "drain valve" and allow the water to drain from the sprinkler system piping.
- Open all "vent" and "drain" valves throughout the system. Vents will be located at the high points and drains at all trapped and low portions of the piping system.
- Manually push open the "velocity drip valve." Also open the "drain" valve for the "dry pipe valve body," if one is provided.
- Remove the cover plate from the "dry pipe valve" and carefully clean the rubber facings and seat surfaces of the internal air and water valves. **Do not** use rags or abrasive wiping materials. Wipe the seats clean with the bare fingers.
- Unlatch the "clapper" and carefully close the internal air and water valves.
- Replace the "dry pipe valve" cover and close the "drain" valve, if one is provided.
- Open the "priming cup valve" and "priming water valve" to admit priming seal water into the "dry pipe valve" to the level of the pipe connection. The priming water provides a more positive seal to prevent air from escaping past the air valve seat into the intermediate chamber.
- Drain excess water by opening "condensate drain valve." Close tightly when water no longer drains from valve.
- Open "air supply valve" and admit air to build up a few pounds of pressure in the system.
- Check all open vents and drains throughout the system to be sure all water has been forced from the low points. As soon as dry air exhausts at the various open points, the openings should be closed. Close "air supply valve."
- Replace any open sprinklers with new sprinkler heads of the proper rating.
- Open "air supply valve" and allow system air pressure to build up to the required pressure. The air pressure required to keep the internal valves closed varies directly with water supply pressure. Consult pressure setting tables.
- Open the system water supply valve slightly to obtain a small flow of water to the "dry pipe valve."
- When water is flowing clear at the drain valve, slowly close it allowing water pressure to gradually build up below the internal water valve as observed on the water pressure gauge.

- When water pressure has reached the maximum below the internal water valve, open the supply valve to full open position. Back off valve about a 1/4 turn from full open.
- Test the alarms. Open the "test valve," or if system has a three-position test cock, place cock in TEST position. Water should flow to the electrical alarm switch and also to the alarm water motor gong.
- If alarms have functioned properly, close the "test valve" or place the three-position test cock in ALARM position.

The alarm test tests the functioning of the alarm system only and does not indicate the condition of the "dry pipe valve." The "dry pipe valve" operation is tested by opening a vent valve to allow air in the piping system to escape, causing the "dry pipe valve" to trip. Is must then be reset, going through the procedure listed above.

Preaction Sprinkler Systems

A preaction system is a closed sprinkler system which also employs air in its system. This system has a supplemental fire detection system installed, such as heat sensors and or smoke detectors in the area covered by the sprinkler system. The actuation of the fire detection system will signal the preaction valve to open and allow water to flow in the system. If a fire is present, a sprinkler head will open to permit the water to extinguish the fire. The preaction system also prevents water damage to the area being protected in the event a sprinkler head becomes damaged. Air escapes and the alarm rings. The preaction valve does not open and permit water into the system piping until the fire detection system (smoke or heat sensors) signal is received at the panel.

Deluge Systems

A deluge system has open sprinklers and a deluge valve which will not permit water to flow through the system piping until the fire detection system is activated, which allows water to flow through all sprinkler heads, because all heads are the open type.

Combined Dry Pipe and Preaction Sprinkler Systems

A dry pipe and preaction system combines automatic sprinklers attached to a piping system containing air under pressure with a supplemental fire detection system installed in the same areas as the sprinklers. The fire detection system also serves as an automatic fire alarm system. The systems are constructed so that if the dry pipe system fails to operate, the fire detection system functions as an automatic fire alarm system. If the fire detection system fails to operate, the dry pipe system functions as a conventional automatic dry pipe system.

Combined Sprinkler and Standpipe Systems

Combination systems use the water piping that serves both outlets for fire department use and outlets for automatic sprinklers.

Maintaining Sprinkler Systems

Sprinkler inspections should be done on a regular basis. Table 16-1 shows what to do and how often you should perform inspections. Figure 16-2 provides a typical form for reporting the results of sprinkler inspections.

Sprinkler systems are generally very reliable & dependable. However, there are several guidelines that you should follow for all systems:

- Keep supply valve open.
- Do not shut supply valve off prematurely during a fire.
- Take precautions to avoid freezing of wet system sprinkler piping.
- Check valves frequently in accordance with insurance company recommendations.
- Flush out system on a regular basis and assure you have debris clear water at intake by using filter screens, if necessary.
- Check pitch of pipes and eliminate low spots.
- Do not store materials closer than 18″ from sprinkler heads. Keep at least 3′ 0″ away from bales, big cartons, and so on.
- Check for corrosion of sprinklers in all areas where chemicals are processed.
- Continually verify that there is an adequate water supply and that water supply conditions have not changed since the installation.

PARTS	ACTIVITY	FREQUENCY
Flushing piping	Test	Every 5 years
Control valves	Inspection	Monthly
Indicator posts	Test	Quarterly
Valves-roadway	Test	Quarterly
Main drain	Flow Test	Quarterly
Open sprinklers	Test	Annually
Pressure gauges	Calibration test	Every 5 years
Sprinklers	Replacement	When needed
Water flow alarms	Test	Quarterly
Preaction	Test	Semi-annually
Deluge system	Test	Semi-annually

Table 16-1. Sprinkler System Inspection Checklist

AUTOMATIC FIRE SPRINKLER SYSTEM

Semi-Annual Inspection Report

ADDRESS: ______________________ d/b/a: ______________________

______________________ ______________________

Owner's Name & Address: Inspection performed by:

______________________ Name: ______________________

______________________ Title: ______________________

______________________ Lic#: ______________________

Date of Inspection: ______________________

Type of system: wet ____, dry ____, combination ________

Date of last complete system hydrostatic test (200 p.s.i.-2hrs.) ____

		Yes	No
EXTERIOR:	1. Siamese (F.D.) connection in good condition	________	________
	2. Clapper valve free and clear of obstructions	________	________
	3. Swivels turn freely	________	________
	4. Proper caps . . . removable . . .	________	________
	5. Siamese is watertight (no visible leaks)	________	________
	6. Siamese is painted green	________	________
	7. Siamese location indicated with sign	________	________
	8. Siamese visible from street . . .	________	________

		Yes	No
INTERIOR:	1. O.S. & Y. valve open and labeled	________	________
	2. All valves in good working order and labeled	________	________
	3. Proper air and/or water pressure	________	________
	4. Sprinkler heads unobstructed .	________	________
	5. Proper heads in good condition	________	________
	6. Spare heads and wrench available	________	________

Courtesy of Fire Command Co., Inc., Long Beach, N.Y.

Figure 16-2. Typical Sprinkler Inspection Report

7. System is watertight (no visible leaks) .... ________ ________

If NO to *any* of the above, specify:

____

____

____

I hereby certify the above information to be true and correct.

Signature: ____________________
Title: ____________________
Date: ____________________

Figure 16-2. Typical Sprinkler Inspection Report (*continued*)

KEEPING STANDPIPE SYSTEMS COMPLIANT WITH REGULATIONS*

There are two different types of standpipe systems. All are subject to the approval of the local authority having jurisdiction.

- Wet standpipe—a system maintaining water pressure at all times.
- Dry standpipe—a system that has no water pressure in its lines until water is pumped into a siamese connection located on the exterior of the building being serviced by the standpipe.

Standpipe systems are hydraulically designed to provide water at the minimum residual pressure of 65 psi at the topmost outlet or sized by the pipe schedule indicated in Table 16.2 to produce the minimum of 65 psi residual pressure.

Acceptable water supplies include:

- Public water systems where water pressure is adequate.
- Pressure tanks.
- Gravity tanks.
- Automatic fire pumps.
- Manually controlled fire pumps with pressure tanks.
- Manually controlled fire pumps operated by remote control devices at each hose station.

*Submitted by Robert S. Kirk, Inter-Boro Sprinkler Corp.

Total accumulated flow-GPM	Total distance of piping from furthest outlet		
	50′	50-100′	100′
100	2″	2-1/2″	3″
101-500	4″	4″	6″
501-750	5″	5″	6″
751-1250	6″	6″	6″
1251 & over	8″	8″	8″

Table 16-2. Minimum Nominal Pipe Size in Inches for Standpipe and Supply

Regulations for Siamese Connections

Siamese connections (fire department connections) are connections used by the local fire department to pump water into the standpipe system. There may be one or more siamese connections on any system.

A siamese connection must comply with the following regulations:

- Be accessible to the Fire Department.
- Have an approved check valve at point of joining the siamese piping and the system piping.
- Have an automatic ball drip valve on the dry side between the standpipe system and the check valve (to protect siamese from freezing in cold weather.)
- Be approved type, equipped with caps to protect internal threaded swivel fitting from becoming damaged.
- Have its fire department connection located on street side of building. If building is served by two streets, then two siamese connections will be installed.
- Post appropriate signs over siamese connection indicating portion of building that is protected by siamese and what type of siamese it is (i.e., standpipe, sprinkler, or combination).
- Be hydrostatically tested at time of installation at 200 psi for a period of 2 hours with no leakage permitted. A flow test shall be conducted to insure that there is no blockage in system piping.
- Be regularly inspected to insure proper working condition
- Connections should be readily visible and accessible. They should not be obstructed by vegetation or equipment.
- Caps to be in place and swivels should enjoy freedom of movement. Hose threads to be clean and in good condition.
- Verify that check valve in fire department connection piping is not leaking.

Maintaining Hoses and Hydrants

Make sure at least 100′, preferably 150′, of 2-1/2″ woven jacket rubber-lined hose is always available so it can be attached to hydrant with minimal delay. Length of the hose depends on local conditions. Sufficient hose should be available if required. Do not forget to include roof areas, outlining buildings, outside storage areas, etc.

Inspect all fire hose monthly and schedule it for replacement at ten-year intervals. Test the pressure of hydrants and hoses yearly. The drip valve is designed to drain the entire barrel down to the level of the valve seat. Drain all hydrants every fall to prevent freezing. You can add antifreeze to hydrant barrel in fall, after draining or pumping to protect hydrant.

Installing Valves and Check Valves

Isolation valves may be installed on standpipe systems. This is done to allow work to be performed on a branch line without putting the entire standpipe system out of service. Check valves of an approved type may also be installed on the standpipe system. Check valves are usually placed at a source of water coming into the system such as water service to the system, or siamese connections. These valves prevent the water from escaping the system when the pressure drops or when the fire department connections are not in use.

All valves that are installed on the standpipe system must be supervised in the open position by one of the following methods;

- Central station or remote station signal system
- Locking the valve open with a chain and lock
- Providing tamper switches that are connected to a central alarm system

Installing Water Flow Switches

When required by the local authority having jurisdiction, an approved type water flow switch should be installed in the standpipe system on every riser.

Complying with Drain Regulations

The standpipe system should have a means of draining. The drain valves should be located at the lowest point possible downstream of the isolation valves. Use the following schedule for proper drain sizing:

STANDPIPE SIZE	SIZE OF DRAIN CONNECTION
Up to 2″	3/4″
2½″ up to 3½″	1-1½″
4″ & larger	2″

See Figure 16-3 for sample standpipe inspection report

STANDPIPE SYSTEM

Semi-Annual Inspection Report

ADDRESS: ______________ d/b/a: ______________

______________ ______________

Owner's Name & Address: Inspection performed by:

______________ Name: ______________

______________ Title: ______________

______________ Lic#: ______________

Date of last complete system hydrostatic test (200 p.s.i.-2hrs.) ____

Date of Inspection: ______________

		Yes	No
EXTERIOR:	1. Siamese (F.D.) connection in good condition	______	______
	2. Clapper valve free and clear of obstructions	______	______
	3. Swivels turn freely	______	______
	4. Proper caps . . . removable . . .	______	______
	5. Siamese is watertight (no visible leaks)	______	______
	6. Siamese is painted red	______	______
	7. Siamese location indicated with sign	______	______
	8. Siamese visible from street . . .	______	______

Courtesy of Fire Command Co., Inc., Long Beach, N.Y.

Figure 16-3. Typical Standpipe Inspection Report

		Yes	No
INTERIOR:	1. O.S. & Y. valve open and labeled	______	______
	2. All valves in good working order	______	______
	3. System is watertight (no visible leaks)	______	______
	4. Hoses are in good condition . .	______	______
	5. Hoses are properly racked	______	______
	6. Proper nozzles are connected .	______	______
	7. Hydrostatic pressure test - 50 p.s.i. at top story outlet	______	______

If NO to *any* of the above, specify:

____

____

____

I hereby certify the above information to be true and correct.

Signature: ______________

Title: ______________

Date: ______________

Figure 16-3. Typical Standpipe Inspection Report (*continued*)

FLAME RESISTANT AND FIRE RETARDANT TREATMENTS

It is not always possible to select materials that are considered noncombustible. In such cases, the use of various fire retardant or flame resistant treatments on combustible materials would provide some measure of fire protection, but terminology such as *fire retardant, fire resistant, flame retardant, flame resistant* and *flame proof* are generally not understood and misused.

Fire resistant signifies the ability of a structure, material, or assembly to resist the effects of a large-scale severe fire exposure. It refers to a specific time period from standard time-temperature curves. The term should not be used regarding *fire retardant* treatment of combustibles. Another abused term *flame proof,* should not be used.

Fire retardant is used in conjunction with chemicals, paints, or coatings applied to combustible building materials. It generally represents a lesser degree of protection than the term *fire resistant. Flame resistant* and *flame retardant* are generally applicable to decorative materials.

Using Fire Retardant Paint*

The use of class A/1 intumescent fire retardant paints or coatings in hallways, offices, breakrooms and stairwells can eliminate the spread of fires and the development of toxic smoke. Intumescent paints and coatings swell to 300 to 400 times their actual thickness in a fire, thereby stopping the flames' spread and decreasing the development of toxic fumes, such as those that may be developed by burning conventional paints. The toxic threat to occupants of structures with intumescent fire retardant paints are equal or less than those with conventional paints.

Intumescent fire retardant paints range from economical flat latexes to high-tech polyurethane varnish systems with a large range of colors.

Making Wood Fire Retardant

There is no way to make wood fire resistive, but it can be treated to retard both flame spread and the rate it contributes fuel to a fire. Both the application of fire retardant coatings and impregnation can be applied to wood. Consult the U.L. ratings and engineering data. Also, emphasize the importance of renewing exterior coatings, and schedule these as appropriate. Note, however, that some of the disadvantages of pressure impregnation are that the weight of the wood can be increased and the strength of the wood can be reduced. In addition, the chemicals can increase corrosion on bolts, fastenings, etc. and may affect the paints, stains, and other finishes that are applied to the wood. However, new fire retardants for wood are available which can eliminate this problem.

Christmas trees have created many fires and are extremely flammable. Particularly when cut and left indoors where heat and low humidity dry the wood, the danger accelerates. Few, if any, chemicals are effective for the flameproofing of Christmas trees. The best way to reduce the hazard is to keep the tree indoors for as short a time as possible, have the trunk cut off at least 1″ above the original cut and place it in water with a water level kept above the cut.

Choosing Fabric-Covered Wall Panels for Fire Retardant Characteristics**

Fabric-covered acoustical wall panels have become very popular because they reduce noise levels, provide a broad aesthetic appeal, and provide a tackable surface. You must not only consider the rating of the fabric and the acoustal substrate on which the fabric is laminated. You must also consider the composite wall panel, including the adhesive. Many acoustical wall panels being used do not comply with flame spread and smoke development ratings because they are not tested as a composite. Acoustical wall panels are considered interior finishes by building code stan-

*Courtesy of Samuel K. Harpwick Jr., Ocean Coatings
**Courtesy of Thomas Fritz, Armstrong World Industries

dards and therefore fall under the requirements for flame spread and smoke development for all building materials (ASTM E 84 Tunnel Test).

Another test has emerged on the building code scene in the last few years. The room fire test is required for all textile wall coverings, including fabric covered wall panels, in non-sprinklered areas. Because this test is much closer to the real-life situation than the tunnel test, it is necessary to use finished composite acoustical wall panels. Therefore, the test will be able to provide an accurate representation of full scale fire performance. Some manufacturers may not test the composite because some building codes do not call for the ratings of their composite panels. The three model codes on which local codes are based are the Basic Building Code (Northeast & Midwest), Standard Code (South) and Uniform Code (West). A fourth code that sometimes comes into play is the Life Safety Code (NFPA standard 101).

Standardization has been attempted. For example, a 450 maximum smoke development requirement applies to all three model codes. However, building codes generally specify different flame spread and smoke development for different building areas. An office may require materials that have a flame spread rating of 200 or less, while a corridor of 25 or less. Many specifiers tend to specify the same rating throughout the project, the lowest one mentioned in the code.

Because many suppliers do not realize that the adhesive used to bond fabrics and substrates in acoustical wall panels can significantly reduce the fire performance in the composite unit, you should ask them to test the composite wall panel in the form it will actually be used. Unfortunately, many local codes do not require this type of performance data. In those cases, it is even more important for specifiers to insist on it. Also, make sure you do not allow lower rated units to be installed when performing routine maintenance in your facility.

Making Fabrics Fire Retardant

Chemical treatment can inhibit the rapid spread of flame in fabrics or it can prevent flames and depress the afterglow. However, the application of chemicals cannot make fabrics non combustible. For high temperatures (above 500°F) only asbestos or chrome-tanned leather is required. However, for small sparks,small flames and temperatures less than 500°F, retardant canvas is generally used.

Two treatments are possible to make fabrics fire retardant: water-soluble and weather resistant or water-insoluble. Water-soluble treatments require re-application after each laundering. This is not required for fabrics that have had water-insoluble chemicals applied or satisfactory applications of weather resistant treatments.

Theatre scenery, drapes and curtains that are used in public assembly are generally required by law to be made flame retardant.Likewise fabrics used in nursing homes, hotels, hospitals, etc. should be treated with appropriate chemicals. The results of a treated material can be checked to confirm that the chemical application is satisfactory.

Making Plastics Fire Retardant

There is no current treatment for plastics that will add significantly to the fire resistance that will not affect its color, strength and form even under only moderate fire exposure. Plastics can be made partially flame retardant (i.e., reduce flame spread) by applying coatings or with a chemical change.

FIRE SUPPRESSION SYSTEMS*

Halon 1301 Systems

Computer rooms, electronic data processing areas, and other hazard areas may require fire protection and agents that will provide for a fast knock-down and suppression of a fire, but will not cause problems of additional damage due to the nature of the fire extinguishant. The result of water released by a fire sprinkler system which is successful in extinguishing a fire is often considerable damage to the floor of the buildings or area where the fire originated, and surrounding floors and adjacent work areas including subsequent water damage to sensitive electronic hardware, not to mention furniture, files, and other important business items.

As more and more vital business information is transmitted through systems deemed indispensable, you will be required to provide automatic fire suppression systems that operate 24 hours a day, 7 days a week. Typically, the hazard area is protected by a fixed system consisting of a central storage tank filled with a fire extinguishant called Halon 1301, or Bromotrifluromethane. The 1301 following the acronym Halon (Halongenated Hydro-Carbon) represents the number of atoms in the compound molecule. The make-up of the extinguishant Halon 1301 is 1-part carbon, 3-parts florine, 0-parts chlorine and 1-part bromide. The Halon 1301 storage tank is connected to a fixed pipe distribution network, which allows the released 1301 gas into the protected hazard area. Gas distribution throughout the hazard is accomplished by nozzles, which direct the flow of Halon to all internal areas of the hazard.

Fires are detected through a number of different types of automatic detection devices, which include ionization and photoelectric smoke detectors and rate compensated heat actuation devices. The fire suppression systems should also have a means of manually activating the suppression system, which is usually accomplished through a wall-mounted manual pull station.

Upon release of Halon 1301 during a fire condition, a number of important operations should occur automatically. First, an alarm sequence should occur, which will alert all area occupants that a fire condition has resulted, and an automatic release of Halon 1301 will be occurring, or has already occurred. Typically there are both audio and visual indicators which announce both the pre-discharge and dis-

*Courtesy of James Goerl

charge alarms separately. Systems have now been developed which provide a voice evacuation system count-down, announcing the time delay intervals leading to a system discharge.

As the system is discharging, an automatic shutdown of the air handling systems or HVAC units should also occur. This shutdown of the HVAC system will prevent the leakage of Halon 1301 from the protected area, and will provide for the most effective fire suppression system. At this time, all doors in the hazard area should close, and all dampers to the protected areas will also shut.

In many larger hazard areas, a graphic display option may be included with the system, which will illustrate all of the detector locations within the hazard. A graphic display will determine which detector(s) are in alarm, and can provide a source to begin a pre-discharge investigation of the source of the discharge.

Installing a Halon 1301 System

A Halon 1301 fire suppression system should always be installed, inspected and maintained by a factory-trained or certified distributor of equipment manufactured by an Underwriters Laboratories Listed 1301 system's equipment manufacturer. A manufacturer which sells a UL Listed or Factory Mutual approved suppression system requires all authorized distributors to complete factory certification training programs, as well as stock all necessary installation and servicing equipment for the systems the company installs. Proof of adequate insurance coverage is required by all manufacturers of fire suppression systems.

Due to the many design parameters and manufacturer's proprietary system requirements, as well as installation procedures which include fire protection contractor, as well as plumbing and electrical considerations, a complete comprehensive installation procedure is not possible. The following is a list of questions to ask a possible Halon 1301 systems distributor/potential contractor:

- What are the exact hazards being protected?
- What is the % of concentration and temperature of the hazard the system is being designed for?
- Is a connected reserve of Halon 1301 being supplied?
- What is the semi-annual inspection of the system going to cost?
- What previous installations have been completed?
- Does all hardware have UL and FM listing?
- Are mechanical & electrical drawings being supplied for system approvals?
- What subcontractors are being used on the job?
- Are signs and a sequence of operations being supplied?
- What is the system completion date?
- Is a full discharge test of the system included?
- Will employee training be provided?

- Does the installing contractor provide 24 hrs/day-7 days a week emergency service?
- Does the contractor have a UL listed Halon re-fill station?
- What is the cost of system recharge?
- Are all Halon hydraulic piping calculations and computer printouts being furnished?

You should also ask the following questions of the contractor on the system operation:

- Can the system be manually operated from a single location without electrical power?
- Are the Halon storage tanks located outside the protected area?
- Can the tanks be recharged without disturbing the equipment in the hazard?
- Does the system have an electrically supervised actuation system?
- Does the electrical actuation device have allowance for inspection, without a system discharge?
- What kind of detectors are included? Quantity?
- What type of alarms will there be? Quantity?
- Does the control panel provide complete supervision and 24-hour battery backup?
- Are automatic door closers and/or fire dampers included?
- Is the installation scheduled to be non-disrupting to normal operations?
- Is the insurance coverage of the contractor including completed operations?

Maintaining a Halon 1301 System

To assure that the Halon 1301 system will operate properly and effectively, the following outline of steps must be completed immediately after the system installation, and semi-annually thereafter. The following procedure is generic in nature, and may not truly reflect a specific manufacturer's own system guidelines.

- Check general appearance of all system components, checking all nameplates and instructions for legibility.
- Inspect nozzles for placement/blockage.
- Remove all tank actuators, and safely secure these devices.
- Make sure all pneumatic slave actuating tanks will fully extend to operate tanks, when system is in an operating mode.
- Activate all functions of the automatic detection and control system, and make sure all electrical actuators are energized. Check all alarms, damper closings, equipment shutdowns, and door closings to determine proper operation.

- Reset detection/control system.
- Disconnect all piping and fittings from the tanks, removing the tanks from the bracketing. Weigh all tanks and compare weight, with nameplate stamping.
- Check resultant weight, and compare to the appropriate manufacturer's temperature correction charts.
- Replace or recharge tanks shown to be below that of the required fill weight of the tank.
- Restore tanks to the appropriate piping network and re-bracket all tanks firmly in place.
- Reconnect all slave cylinder actuators.
- Check and reset all pressure switches.
- Reset all electrical actuators and reinstall on all tanks.
- Record the tank weight, pressure and date of maintenance on the tag attached to the Halon tank.

Once again, the above steps should be completed by a representative, factory trained by the manufacturer of the Halon 1301 system, and should not serve to replace actual manufacturer's procedures.

Wet Chemical Systems

Until the early 1980s, hood, duct and appliance protection for restaurant kitchen areas has consisted of an automatic dry chemical fire protection system. These systems are designed to operate in the event of a fire condition, and distribute the agent throughout the hood and duct-areas where a fire fighter using a hand portable fire extinguisher would have difficulty gaining accessibility. In the early 1980s, new agents developed by various fire protection manufacturers were introduced and these agents differed from those used at the time; the new agents were a liquid base. The new wet agents, potassium carbonate/water based, suppress fires by forming a combustion-resistant foam on the burning grease surface. This foam blanket is called saponification. Also, because the liquid agents are water based, the release of the agent on a fire, results in a cooling effect to help prevent fire reflashes, and allows for a fast clean-up.

A number of wet fire suppression systems are manufactured by a number of different companies, and the specific installation and maintenance procedures differ. The use of a contractor that has been factory trained or certified by the specific manufacturer of that system is always recommended, and in many areas, is a requirement. When contacting a company for installation of a wet chemical fire suppression system, follow these guidelines:

- Are the cylinders containing the extinguishant located in a workable location?

- Are all grease-producing appliances protected with overhead distribution nozzles, listed by the manufacturer for that specific type of appliance?
- Are there ample quantities of tanks to satisfy the total quantity of nozzles, or flow points required in the system?
- Are all tank valves properly connected to maintain a simultaneous discharge?
- Has the proper distribution piping network been used, according to the manufacturer's recommendations, and has the piping been properly secured within the hazard area?
- Are the manufacturer's limitations for all supply lines and branch lines being complied with?
- Has a maximum registering thermometer been used to determine the suitable fusible link detector ratings and have those detectors been appropriately placed?
- Are fusible links connected to the system control mechanism via wire rope and is the wire rope contained within pipe or EMT thin wall conduit?
- Are all wire rope directional changes accomplished by pulley elbows?
- Is there a means of manually actuating the system and is it located in a path of egress?
- Are the agent storage tanks filled?
- Do all nozzles have protective blow-off caps or nozzle orifice grease protection?
- Is there a means of fire shutdown for all appliances protected beneath the hood?
- Is protection being provided for the duct and plenum chamber?
- Is a full system discharge bag test being included?
- What is the cost of a system recharge?
- Will employee training be involved?
- Does the system have a UL listing?
- Are all plumbing connections (gas valve) and any electrical shutdown connections being completed by a qualified, licensed plumber or electrician?
- Does the installing contractor have complete liability insurance, including completed operations?

Maintenance of a Wet Chemical System

To provide for proper operation of a wet chemical system, perform the following outline of steps at six-month intervals and immediately following any hood/duct cleaning operation. The following procedure is generic in nature, and may not truly reflect a specific manufacturer's own system guidelines.

- Check the original installation, and make certain that no changes have been made to the components of the system or to the placement of the appliances.
- Make sure there are no signs of tampering, damage or corrosion on any of the system accessories.
- Disconnect system control heads/or cartridge if system is not of the stored pressure variety.
- Check cylinder pressure if system is stored pressure type, or weigh the propellent cartridge and determine the expected proper fill weight of the cartridge.
- Without interrupting any cooking operations, test the following devices to determine proper operation: fusible link detectors, remote pull station, automatic fuel shutdown, and automatic damper closing (if included).
- Replace all fusible links with the manufacturer's properly rated fusible links.
- Make sure there are no areas of excessive grease build up within the hazard area.
- Make sure all instruction signs are legible.
- Re-cock the system and add tension to the system.
- Reconnect all control hands or reassemble the control mechanism.
- Sign the date of maintenance on the inspection tag.

The above steps should be completed by a certified technician, factory trained by the system manufacturer, and should be completed in conjunction with NFPA-17A and NFPA-96.

Dry Chemical Systems

Since the hood and duct system first appeared in the late 1950s, there have been many advances made to this critical component of the restaurant kitchen environment. The first of this type of system was originally designed for the hood and duct areas only, and there was no appliance protection available until quite a number of years later. The fire extinguishant used in a dry chemical system is sodium bicarbonate, or common baking soda, which reacts to the hot grease to form a soap like blanket called saponification. This blanket prevents any further escape of the hot grease vapors, which results in the suppression of a fire.

There are a number of dry chemical system manufacturers, and with each system the specific installation and maintenance procedures differ, and each various system has its own proprietary installation and maintenance requirements. The use of a contractor that has been factory trained/certified by the specific manufacturer of that system is always recommended, and in many areas, is a requirement. Systems should always be installed and maintained following the guidelines set forth in NFPA-17 and NFPA-96. See the previous section on wet chemical systems for installation and maintenance guidelines.

Industrial Fire Suppression Systems

The topic of industrial fire suppression systems is a broad subject within the realm of fire protection, due in fact to the many types of industrial hazards faced, the various options and methods available for fire suppression, and the different fire extinguishants that could be used on similiar hazards, makes a summary of industrial fire protection difficult and extremely lengthy. Typical hazards that are faced in industry are flammable liquid storage rooms, stock rooms, printing machines, electrical motors, pump rooms, paint lockers, dip tanks, spray booths, wood finishing areas, and transformer rooms. The typical means of protecting these systems include total flooding, local application using a tankside type of discharge, or local application using an overhead type application. The fire extinguishants that can be used for industrial fire protection systems include multi-purpose dry chemical, so named for its tri-class fire fighting capabilities, Purple-K dry chemical, sodium bi-carbonate dry chemical, carbon dioxide, Halon 1211 and AFFF foam. Many of the typical hazards faced in an industrial environment are covered in the codes of the National Fire Protection Association. It is a requirement that systems be installed using the NFPA codes as a minimum criteria, and that additional local authority requirements also be adhered to.

Because many fire protection manufacturers have various types of fire suppression systems, each with its own specific installation and maintenance procedures, it is recommended to contact an installing/maintenance company that has been trained by the manufacturer, and that is using the manufacturer's required installation/ maintenance procedures.

The following brief outline represents guidelines of the types of questions that you should ask of any potential fire suppression systems installing company:

- Is the system UL listed or FM approved for this type of hazard?
- Has this system been approved by the insurance carrier, or local authority having jurisdiction?
- What is the exact hazard being protected?
- Is a connected reserve being supplied?
- What is the cost of the semi-annual inspection?
- What previous installations have been done?
- Are mechanical and electrical drawings being supplied?
- What subcontractors are being used on the job?
- Are signs and a sequence of operations being supplied?
- What is the system completion date?
- Is a full discharge test of the system included?
- Will employee training be provided?

- Does the installing contractor provide 24 hrs/day -7 days a week emergency service?
- What is the cost of system recharge?

Ask the following questions of the contractor on the system operation:

- Can the system be manually operated from a single location without electrical power?
- Are the storage tanks located outside the protected area?
- Can the tanks be recharged without disturbing the equipment in the hazard?
- Does the system have an electrically supervised actuation system?
- Does the electrical actuation device have allowance for inspection, without a system discharge?
- What kind of detectors are included? Quantity?
- What type of alarms will there be? Quantity?
- Does the control panel provide complete supervision and 24 hour battery backup?
- Are the automatic door closers and/or fire dampers included?
- Is the installation scheduled to be non-disrupting to normal operations?
- Is the insurance coverage of the contractor including completed operations?

Maintenance of an Industrial Fire Suppression System

To ensure that the industrial system will operate properly and effectively, complete the following steps immediately after installing the system and semi-annually thereafter. The following procedure is generic in nature, and may not truly reflect a specific manufacturer's own system guidelines.

- Check general appearance of all system components, checking all nameplates and instructions for legibility.
- Inspect nozzles for placement/blockage.
- Remove all tank actuators, and safely secure these devices.
- Make sure all pneumatic slave actuating tanks will fully extend to operate tanks, when system is in an operating mode.
- Activate *all* functions of the automatic detection and control system, and make sure all electrical actuators are energized. Check all alarms, damper closings, equipment shutdowns, and door closings to determine proper operation.
- Reset detection/control system.
- Disconnect all piping and fittings from the tanks, removing the tanks from the bracketing. Weigh all tanks and compare weight, with nameplate stamping.

- Check resultant weight, and compare it to the appropriate manufacturer's temperature correction charts.
- Replace or recharge tanks shown to be below that of the required fill weight of the tank.
- Restore tanks to the appropriate piping network, and re-bracket all tanks firmly in place.
- Re-connect all slave cylinder actuators.
- Check and reset all pressure switches.
- Reset all electrical actuators and reinstall on all tanks.
- Record the tank weight, pressure and date of maintenance on the tag attached to the tank.

Once again, the above steps should be completed by a representative factory trained by the manufacturer of the system, and should not serve to replace actual manufacturer's procedures.

CHOOSING AND MAINTAINING FIRE ALARM AND DETECTION SYSTEMS*

Proper selection of a fire detection device requires a detailed hazard analysis, since each type of detector is primarily designed for a specific type of fire. The key criteria to be considered include expected rate of fire growth, the expected products of combustion (flame, smoke or heat) and ambient conditions. The following sections describe the key types of detection devices and their primary intended uses.

Heat Detectors

Heat detectors are designed to respond to the convected thermal energy produced by fire. These devices are among the least expensive detectors available and provide the lowest false alarm rates. These devices are most suitable where rapid heat buildup is expected or life safety is not of prime concern. Also, heat detectors can perform in environments that may be unsuitable for other detectors. For example, dusty environments that may cause excessive false alarms in smoke detectors will not affect heat detectors.

Heat detectors are designed to respond to a pre-determined fixed temperature or at a specific rate of temperature rise. There are devices offered that provide both functions for added versatility.

---

*Information contributed by Mr. Lee Devito, Kidde-Fenwal

Thermostat Combination Detectors

The thermostats are spot detectors operating on a combination of the fixed temperature and rate-of-temperature-rise principles. The fixed temperature fusible element responds to slow developing, smoldering fires when the design temperature of 135°F or 197°F is reached. Rapidly developing fires cause the rate-of-temperature-rise feature to operate when the temperature rate of increase in the detector area exceeds 15°F per minute. The rate-of-rise device is designed to compensate for normal ambient temperature change.

Combination rate-of-rise/fixed temperature thermostats can be subject to false alarms if located where ambient temperature change exceeds 15°F per minute, such as near boilers and heating vents.

Rate-Compensated Heat Detectors

Rate-compensated heat detectors will respond when the design temperature rating is reached and when temperatures are rising rapidly. The device resets automatically and is impervious to momentary temperature changes. These devices have proven to be eminently reliable and are standard in the industrial field.

The rate compensation feature will not normally change the temperature at which the device responds, but will instead cause actuation at about the same temperature, regardless of the rate at which the temperature is rising. Without rate compensation, the device would experience "thermal lag" during a rapidly accelerating fire condition. Rate compensation, therefore, provides a faster response.

By offering a wide range of fixed temperature design ratings, rate-compensated detectors can provide response to fire in areas with high normal ambient temperatures. Explosion-proof construction makes these detectors suitable for hazardous indoor and outdoor environments.

Continuous Strip Detectors

Continuous detection systems are composed of wire-like temperature sensors providing complete coverage of the hazard. Monitored by a special control unit, every increment of the long-line loop (up to 300 feet per circuit) functions independently as a detector. In large hazard areas, strip detectors can provide high reliability at less installation cost than spot detector systems. The detection runs are sealed against weather.

Typical applications:

- Floating roof tanks - Strip detectors effectively protect the tank perimeter area that moves with the fill level of the tank.
- Conveyers - The entire length of conveyers are protected to prevent ignition of lubricants and material by sensing pre-fire heat build-up. Vehicle fire sup-

pression - When placed around components subject to high heat, strip detectors can prevent costly repairs and down time to heavy duty vehicles.

- Cryogenic storage - Continuous strip cold detectors can sense leaks of LNG, propane and other liquids of low boiling points.
- Process vessels - Strip detectors can monitor surface temperatures and warn of hot spots developing.
- Cable trays and raceways - When placed with cable runs, strip detectors can monitor the entire length.

Smoke Detectors

Smoke detectors tend to be more expensive than heat detectors, but provide faster response times to most fires. Photoelectric and ionization devices are offered. Each has advantages and limitations.

Both are suitable for early warning of slowly developing fires in an indoor environment when life safety is involved. Both types can be used in hazards that may produce different types of fires. In integrated fire suppression systems, these devices are often cross-zoned with adjacent detectors. If a detector on one circuit alarms, the control unit may activate audible alarms and auxiliary functions. However, a second detector must alarm before automatic release of suppression agent is initiated, reducing the possibility of accidental discharge.

Smoke detectors can also be cross-zoned with other types of devices. For example, ionization detectors may initiate general alarm while actuation of the suppression system is caused by less sensitive fixed temperature heat detectors.

Smoke detectors are suitable for a wide range of indoor hazards. Typical applications include:

- General office areas
- Computer and data processing rooms
- Retail stores
- Theaters
- Schools
- Hospitals

Photoelectric Smoke Detectors

Photoelectric smoke detectors are available with 135°F fixed temperature heat sensor. Designed for commercial industrial and institutional use, the detector will respond to smoke entry from any direction. An insect screen is provided to reduce nuisance alarms, and an LED pulses for visual supervision. The LED remains lit during alarm conditions.

Advantages and disadvantages of photoelectric smoke detectors as compared to ionization detectors are provided below.

Advantages:

- Fast response to smoldering fires.
- Fast response to cold, stale smoke that may have traveled a considerable distance.
- Unaffected by air velocity.
- Can be used where controlled fires producing low level products of combustion are present, such as kitchens and engine rooms.

Disadvantages of photoelectric smoke detectors:

- Slower response to fast flaming type fire.
- Slower response to products of combustion with diameters smaller than the wavelength of the light source (about 0.4 microns).
- Slower response to products of combustion that absorb light rather than scatter it, for example, black sooty smoke.
- Slower response to smoke traveling at low velocity (10-15 feet per minute).

Ionization Smoke Detectors

Designed for commercial, industrial and institutional use, the detector will respond to smoke entry from any direction. An insect screen is provided to reduce nuisance alarms, and an LED pulse for visual supervision. The LED remains lit during alarm conditions.

Advantages and disadvantages of ionization detectors as compared to photoelectric detectors are provided below.

Advantages:

- Fast response to fast flaming fires.
- Fast response to fires producing small aerosols and no flame.
- Not affected by color of smoke.
- Sensitive to wide range of aerosols produced by fire (0.001 to 2.0 microns).

Disadvantages:

- Slower response to heavy smoldering fires and cold, stale smoke.
- Too sensitive at air velocities above 1000 feet per minute.
- Sensitivity can be affected by sudden pressure, temperature and humidity changes.
- Sensitivity is affected by altitude and radiation sources that may be present in the ambient environment.

Flame Detectors

Flame detectors are "line of sight" devices usually designed to respond to light radiation in the ultraviolet and/or infrared ends of the spectrum. These devices have the fastest response times and the highest false alarm rate of any detector. Also, they are expensive relative to other detection systems. Flame detectors are often used in high risk areas where there is a danger of extremely rapid fire build up. The devices are effective outdoors where other detectors would not provide adequate detection reliability.

The sensitivity of ultraviolet and infrared detectors are subject to the inverse law of optics. If the distance between the detector and the fire is doubled, the area of a small fire must be quadrupled to have an equal response. Therefore, the distance between optical flame detectors and the hazard should be minimized if immediate response is required. Also, flame detectors are ineffective if their cone of vision is blocked.

Typical applications:

- Flammable liquid storage
- Explosive material storage
- Fuel loading platforms
- Hyperbaric chambers

Gas Detectors

Gas detectors provide warning of combustible gas before accumulations reach dangerous levels. Gas detection systems have a sensor located in the hazard area and a special control unit at a remote location. The control units can be provided in single or multiple zone configurations. Meters or digital displays indicate gas levels in the hazard area. Typically, alarms annunciate when the gas concentration reaches 25% and 50% of the lower flammable limit of the most detectable gas or vapor that could leak in the protected area. Place gas detectors and use system alarms after performing a careful engineering study.

Typical applications:

- Compression and pumping facilities
- Drilling and production platforms
- LNG/LPG processing and storage
- Gas turbines
- Solvent vapor hazards
- Fuel loading facilities
- Sewage treatment plants
- Oil refineries

Duct Detection

Air duct detection is intended to perform a primary function other than transmit an alarm. Its first function is to stop the spread of smoke within a building through the air handling system. (A secondary function may be to transmit an alarm. However, the air duct detector is not intended to supplant a fire alarm system.) This normally is accomplished by shutting off the circulating fans, closing smoke/fire dampers, etc.

Installing and Maintaining the Fire Command Station

This system is encountered increasingly today. It is commonly known as a voice communication system. This system consists of speakers located in specific areas throughout the building which are connected to and controlled from the main fire alarm console in a area designated as the fire command station.

A trained employee should assume responsibility of the command station until the fire department arrives, at which time, the chief officer will then take charge. This system is also used during fire operations to assist the chief in communicating with firemen on various floors. This system also aides in evacuation of the occupants or relocating to areas away from the fire floor which controls a possible panic situation.

Using Sprinkler Alarms

Sprinkler water flow and supervisory alarm devices are designed and installed so that they cannot be readily tampered with, opened or removed from the sprinkler system without initiating a signal. Water flow devices transmit a signal upon a flow in the sprinkler system to either the fire alarm system and/or central station.

Establishing Fire Department Connections

It is important that the fire alarm system located in the building transmits to a direct source. Upon a signal received, the fire department will then be notified to respond to the location. This can be accomplished by a central station, direct line to municipal fire alarm system or automatic dialer.

Inspection and Maintenance of Fire Alarm Systems

Fire alarm test and inspections should be done on a regular basis. The person assigned to this task should be qualified and trained in this field.

The following equipment should be checked when each test is performed:

- Anunciation panel/main panel
- Each zone or panel
- Smoke detectors
- Heat detectors

- Pull stations
- Audible alarms (bells, horns)
- Strobes and horns, if installed
- Automatic smoke door releases
- Backup power, battery/generator
- Signal received by central station or fire department
- Water flow switch activating

Maintenance should include cleaning the alarm panel and smoke detectors to remove dust and debris. Panel and detectors should be vacuumed or blown free of dust using moisture-free air available in pressurized containers.

Keep records in a log of what procedures were performed, date, name, title, company, license number, etc. A report should note any defects and what measures were taken to correct them. Establish a training program for maintenance of these systems.

FIRE PREVENTION AND HOUSEKEEPING*

It has been said that there are three causes of fire - men, women and children. If people could be completely separated from potential fire hazards, the problem of fires would probably be solved. However, people must live and work around fire hazards all the time, so it is obviously impractical to remove the people to solve the problem.

Much attention is paid to warning personnel that a fire has started, and to the extinguishment of that fire. Detection and alarm systems, portable extinguishers, sprinkler systems, special protection systems, building construction features and fire department response all work to minimize the effects of a fire that has already begun. However, none of these features act to *prevent* a fire from starting in the first place. The prevention of fire is obviously a very good way to completely eliminate the unwanted effects of fire—death and destruction.

Given that people must live and work around potential fire hazards, other approaches must be taken to achieve fire prevention. Fire begins when four basic elements come together - fuel, oxygen, an ignition source, and an uncontrolled chemical reaction. Elimination of oxygen can be used to extinguish fires which have already started, but use of oxygen elimination to prevent fires is impractical. An environment with enough oxygen eliminated to prevent fire would be unsuitable for human life. Further, elimination of the uncontrolled chemical reaction of fire is an excellent method of fire extinguishment, but one cannot eliminate a reaction which has not yet begun. If the chemical reaction of fire has already started, it is too late to worry about fire prevention. Therefore, we are left with elimination of fuel and ignition sources or separation of the two as a means to prevent fires.

*Courtesy of Mr. Stan Slonski.

Separating Fuels from Ignition Sources

Fuels for potential fires can be found anywhere we look. Some of these fuels must remain, because they are a part of our life and work. Separation of these fuels from ignition sources is necessary to prevent fires. Other fuels are not necessary to keep around, and disposal of this waste can eliminate fires from this source.

Fuels can be divided into two distinct classes. Class A fuels are solids which are carbonaceous in nature. Wood, paper, cloth, rubber, most plastics and many synthetic materials fall into this class. Since many of these are common, ordinary materials, this class of fuel is often referred to as "ordinary combustibles."

Class A fuels which cannot be eliminated from a hazard include building construction materials, such as wood beams and studs, wall coverings and carpeting. Examples of other fuels which cannot be eliminated are necessary furniture, papers and books, personal belongings such as coats, and office equipment. Class A fuels which can be eliminated from a hazard area include paper waste, unnecessary packing material, kitchen waste and other rubbish.

Class B fuels are liquids which are capable of burning. These are often known as "combustible liquids" or "flammable liquids." The difference is a matter of degree, and both kinds of liquids deserve attention when thinking about fire prevention. As with Class A fuels, Class B fuels may or may not be absolutely necessary to have around. Examples of Class B fuels including gasoline, kerosene, paints, lubricating and cutting oils, and things as mundane as rubber cement. All Class B fuels must be treated with respect. Proper storage is a must. Only the minimum quantities necessary should be kept. All excess quantities should be properly disposed of.

Types of Ignition Sources

Most ordinary combustibles, combustible liquids and flammable liquids require heating to a point far above room temperatures to ignite. Many sources of heat exist in the modern building. As with the fuels themselves, some of these heat sources are deliberate and cannot be eliminated. Separation of fuels from these heat sources is the only way to achieve fire prevention. Examples of ignition sources which cannot be eliminated include building heating plants, hot surfaces associated with plant processes, electrical and lighting equipment which becomes hot, open flames such as welding torches, and cooking equipment.

Many ignition sources can be strictly curtailed or eliminated. Examples include matches, lighters, and smoking, sparks or heat from malfunctioning electrical equipment, and electrical connections which become hot, such as extension cords and poorly made splices.

Housekeeping Fire Prevention Procedures

Housekeeping to prevent fires is a combination of diligence, patience and common sense. Housekeeping to prevent fires is not the exclusive purview of the building maintenance department. Rather, it begins with the habits of every employee. Maintenance should take the lead in defining and promoting housekeeping requirements. Maintenance personnel should also set an example by their own conduct that others may follow.

Rules should be set up for housekeeping. Some of these rules should be:

- Allow smoking only in designated areas.
- Fully extinguish all smoking materials upon completion of smoking.
- Never leave burning smoking materials unattended.
- Avoid accumulation of excess rubbish. Place all waste in approved containers.
- Store all flammable and combustible liquids in their original container or in other containers approved for use.
- Use only the minimum amount of flammable or combustible liquid at the workplace at one time. Store bulk quantities of such liquids in a specially prepared area.
- Notify management immediately of all hazards which could cause a fire to ignite, such as defective electrical equipment. Do not use such equipment, under any circumstances, until defects are corrected.

Housekeeping continues with regular inspections. These inspections, conducted at regular intervals, should look at all areas where rubbish can accumulate, and should also examine all equipment which is a potential ignition source. As the maintenance manager, you should know the building(s) you are responsible for well enough to identify and track the specific areas of greatest risk.

Conduct regular inspections with a checklist. Prepare a specific checklist for each building, floor or area. The checklist should cover items of fire prevention specific to that area, by room, area or machine number, or by description.

The following are suggestions for items for such a checklist. Inspect the following:

- Smoking
  - Make sure smoking is permitted in approved areas.
  - Post "No Smoking" signs in areas where smoking is prohibited.
  - Post signs indicating that smoking is permitted in a certain specific area.
  - Provide approved ashtrays in abundance in the designated smoking area.
  - Provide approved cigarette dowsing trays, filled with sand or similar material, in areas where smoking is prohibited, for the convenience of guests and others. Post signs indicating the purpose of the containers, and reiterating the "No Smoking" rule for the area.
  - Make sure smoking areas contain the minimum possible amount of ordinary combustibles as furnishings.
  - Make sure smoking areas do not contain accumulations of rubbish.

- Make sure smoking areas do not contain flammable or combustible liquids of any kind.

- Flammable liquids
  - Check for spills or accumulations of flammable liquids. Clean up such spills immediately.
  - Ensure that all flammable liquids are stored in original containers or other containers approved for use.
  - Check that excessive quantities of flammable liquids are not kept at work stations.
  - Make sure that no flammable liquids are stored near actual or potential sources of ignition.
  - Check to see that rags used with flammable liquids are kept only in approved containers.
- Rubbish and other unnecessary accumulation
  - Check for accumulation of rubbish, especially in office areas, around equipment, in storage areas, and near exits. All rubbish should be disposed of outside the building in containers provided for that purpose.
  - Check for excessive accumulation of paper, books, and files in aisles and in office areas. Proper storage, such as shelves and file cabinets, should be provided. Pay special attention to areas under stairways. These are natural areas for accumulation, and a fire here could seriously impede exit access.
- Necessary quantities of ordinary combustibles
  - Check that boxes and other packing material are stored only in areas designated for this purpose.
  - Make sure that rags are picked up and disposed of.
- Electrical equipment
  - Inspect all electric lights for clearance to combustibles.
  - Check all other electrical equipment for proper operation and clearance to combustibles.
  - Inspect all spark producing appliances for spark shields.
  - Check electric wiring for excessive heat. This may be a sign of deteriorated connections.
  - Check all appliances with heated surfaces for clearance to combustibles, and to ensure that flammable liquids are not located nearby.
- Heating equipment
  - Thoroughly inspect the building heating plant for proper operation at regular intervals.
  - Inspect chimneys and smokestacks for leaks.

- Thoroughly clean boiler rooms of all combustibles. Do not use boiler rooms as storage areas.
- Make sure boilers and other heaters do not have any accumulation of waste, other combustibles, or flammable liquids nearby.
- Check for spills or leaks of fuel near heating equipment. Clean up immediately.
- Make sure space heaters are used sparingly. Ensure proper clearance to combustibles is maintained. If flammable liquid is stored anywhere near a space heater of any type, the heater should be shut down immediately and the situation corrected. Wherever possible, space heaters should not be used without personnel in attendance.

- Other
  - Inspect all areas where lint and combustible dusts can accumulate. Have all such residue cleaned up immediately.
  - If welding or cutting is to be done, make sure proper precautions are taken. A "hot permit" is a good idea for all such work.
  - Inspect the area surrounding cooking equipment for excessive grease accumulation, rubbish accumulation, and combustible storage.
  - Regularly inspect and clean exhaust ductwork, especially that which serves cooking appliances and dust collectors.

Each building is different, as is each fire hazard. Proper identification of trouble spots, combined with personnel awareness, prevention rules, and regular inspections will result in a building with greatly increased protection from a fire breaking out.

TRAINING EMPLOYEES IN FIRE PREVENTION TECHNIQUES*

Every single employee, not just maintenance personnel, should be trained in the basics of fire prevention of fire safety. It is the responsibility of each person to see to it that he or she is safe, and is not putting his or her fellow employees at risk. These tasks can only be accomplished with proper training.

Do not underestimate the value of this training. The benefits in increased safety and reduction of workload for maintenance personnel will be very great. Training should be more than just perfunctory reading or lecturing. Actual demonstrations represent the single best way for concepts to be transmitted. There is an old saying in training - "tell 'em what you're gonna tell 'em, tell 'em, tell 'em what you told 'em." This may sound unkind - that people are only capable of learning through repetition. This is not the intent at all. Rather, reinforced learning experiences have been demonstrated over and over to be very valuable.

---

*Couresty of Mr. Stan Slonski.

Using Portable Extinguishers

All employees should be trained in the proper use of portable fire extinguishers. This training should take the form of viewing a film or videotape of extinguisher use, having a discussion and question-and-answer period with a qualified instructor, and then witnessing and being allowed to extinguish actual fires. This training need not be costly or difficult to arrange. Most distributors of fire equipment who deal with portable fire extinguishers are set up to perform such training. Planning can minimize the cost and maximize the impact of this program. Consider the following suggestions:

- Conduct the training during the mildest weather season possible. The actual fire training will be outdoors, and trainees should be as comfortable as possible.
- Plan the entire session - film or videotape, discussion, demonstration, and actual handling of extinguishers so that it takes no more than 90 minutes. This will minimize the interruption to the employees' work shift.
- To minimize costs, schedule back-to-back sessions at the end of one shift and the beginning of another.
- For maximum effectiveness, limit sessions to no more than 25 people.
- Make sure the company conducting the session demonstrates the use of all popular types of portable extinguishers on both Class A and Class B fires.
- Require each employee to fight a Class B flammable liquid fire with the type of extinguisher most commonly found in the building. Most buildings will be equipped with ABC dry chemical portable extinguishers.
- To save money, use those extinguishers which are due for hydrostatic test or for six-year maintenance teardown for the demonstration. These extinguishers need to be emptied, and in this way the emptying of the extinguishers accomplishes another purpose.
- Whenever an employee is fighting a fire, make sure an individual of the demonstrating company stands at the ready with a backup portable extinguisher, in case of problems.
- Check with your local fire department about their rules for outside burning. Even if such burning is prohibited, an exception can usually be made for training purposes. You and the fire equipment distributor should work together to seek all necessary approvals.
- Invite your local fire department to attend the session, and an invitation from you would probably be graciously accepted.
- Conduct the session well away from any structure, and conduct it on pavement. Second best is a cindered or graveled area. Suitable sites are remote corners of parking lots, or little-used access roads.

- If there is any possibility of traffic in the area, set up a sawhorse, cones or something similar to divert the traffic.
- Do not set up the demonstration in an area where significant quantities of smoke and clouds of the extinguishing agent can drift onto neighboring properties.
- If possible, set up the demonstration in an area where clouds of dry chemical will not drift over parking lots and settle onto their cars. Your employees will appreciate this.
- Check with your insurance company when planning the training. They may have films and brochures available. Check also on your liability in case of injury. Most insurance companies recognize, and cover, legitimate training needs.
- Keep complete and accurate records of the training which each employee received. Also, consider setting up a video camera to tape the session, so that new hires may see the tape during their orientation.

Exit Precautions

All employees should be trained in exiting the building. This may seem like showing someone the obvious, but it is surprising how many people make the wrong decision when faced with an emergency, and they, their colleagues, and their company may suffer for it.

- Make sure all employees are made aware of their primary and secondary means of egress from their work area(s) and break area(s).
- Designate one employee from each area or department with the responsibility for seeing to it that all employees exit the area in the event of an emergency.
- Make sure everyone gathers at a designated place outside the building in the event of an evacuation. One employee should have the responsibility to check that all employees from a designated area are accounted for, and to report to management if someone is not accounted for.
- Instruct employees not to return to their desks, workplaces, lockers, etc. to recover personal belongings. Time is of the essence when evacuating in an emergency, and lives have been lost when people stopped to pick up belongings before attempting to exit.
- Train designated employees and drill them on the proper procedures for shutting down special equipment, such as production machinery and computers. Make sure they know the exact time and signal to indicate when to carry out the procedures. Drilling these procedures is considered by many to be absolutely necessary to ensure that the procedures will be carried out properly, and only when necessary. Planning ahead for a drilled shutdown will be much cheaper than having a vital step overlooked in an emergency, or having an un-

planned, unwanted shutdown take place because people are not sure what to do in a minor crisis.

- Drill all employees to maximize the possibility that all will go smoothly in an emergency situation. Once an evacuation plan is developed, a drill will indicate better than anything else if the plan will work as designed. Employees will be more confident that they will be safe in an emergency. Unforeseen problems can be worked out before an emergency strikes. Drills such as this can be costly interruptions to work and to production. But they can pay off handsome dividends, and should be seriously considered.

Housekeeping Precautions

All employees should also be trained in proper housekeeping. Everyone should understand that housekeeping is a way of life for everyone, and is not just the responsibility of maintenance personnel and janitorial services. This training should follow portable extinguisher training, since what the trainees learn about classes of fire applies in no small measure to housekeeping. One of the best ways to conduct housekeeping training is to conduct "walking tours" of various areas of the building.

Emphasize to all employees that they have the right and the duty to note and point out to management any sort of housekeeping hazard, even those not in their work area or department. Sometimes, fresh eyes can spot hazards which are missed by individuals in the same area each day.

Training Maintenance Employees

Much of the responsibility for finding and correcting fire hazards will fall to maintenance people in the course of their work. Training in this endeavor will mostly come on the job, as more experienced maintenance employees work with less experienced personnel. All maintenance people should be trained that finding and correcting fire hazards is a positive part of their job. Do not blame, chastise, or embarrass employees working in areas with correctable fire hazards because of the hazards. This attitude will lead to the hiding of hazards, instead of open discovery and correction of those same hazards.

Maintenance employees should also attend manufacturers training schools on the equipment they will be required to maintain. Ask the manufacturer if fire safety will be covered. If not, insist the manufacturer add information on fire safety to the curriculum.

Ongoing Training

Training in fire safety should not stop with the items described above. Rather, recurrent training will consistently reinforce the idea that fire prevention and fire safety is everyone's business. Just because some employees have been through one portable extinguisher session one year does not mean that they should be excused the next. Additional experience with extinguishers will be valuable if that employee is faced with a serious fire situation.

Additionally, the employer's responsibility for training does not stop with the workplace. Consider holding a training session on fire safety in the home. Easy to use material is available from the National Fire Protection Association on home fire prevention, Operation EDITH (Exit Drills In The Home), and home smoke detector and fire extinguisher use. If employees can avoid fires and fire hazards in their own homes, they will be better employees in the future.

Index

F

I

K

L

N

O

P

R